Texts in Applied Mathematics 25

Springer
New York
Berlin
Heidelberg
Barcelona
Budapest
Hong Kong
London
Milan
Paris
Santa Clara
Singapore
Tokyo

Texts in Applied Mathematics

1. *Sirovich:* Introduction to Applied Mathematics.
2. *Wiggins:* Introduction to Applied Nonlinear Dynamical Systems and Chaos.
3. *Hale/Koçak:* Dynamics and Bifurcations.
4. *Chorin/Marsden:* A Mathematical Introduction to Fluid Mechanics, 3rd ed.
5. *Hubbard/West:* Differential Equations: A Dynamical Systems Approach: Ordinary Differential Equations.
6. *Sontag:* Mathematical Control Theory: Deterministic Finite Dimensional Systems.
7. *Perko:* Differential Equations and Dynamical Systems, 2nd ed.
8. *Seaborn:* Hypergeometric Functions and Their Applications.
9. *Pipkin:* A Course on Integral Equations.
10. *Hoppensteadt/Peskin:* Mathematics in Medicine and the Life Sciences.
11. *Braun:* Differential Equations and Their Applications, 4th ed.
12. *Stoer/Bulirsch:* Introduction to Numerical Analysis, 2nd ed.
13. *Renardy/Rogers:* A First Graduate Course in Partial Differential Equations.
14. *Banks:* Growth and Diffusion Phenomena: Mathematical Frameworks and Applications.
15. *Brenner/Scott:* The Mathematical Theory of Finite Element Methods.
16. *Van de Velde:* Concurrent Scientific Computing.
17. *Marsden/Ratiu:* Introduction to Mechanics and Symmetry.
18. *Hubbard/West:* Differential Equations: A Dynamical Systems Approach: Higher-Dimensional Systems.
19. *Kaplan/Glass*: Understanding Nonlinear Dynamics.
20. *Holmes:* Introduction to Perturbation Methods.
21. *Curtain/Zwart:* An Introduction to Infinite-Dimensional Linear Systems Theory.
22. *Thomas:* Numerical Partial Differential Equations: Finite Difference Methods.
23. *Taylor:* Partial Differential Equations: Basic Theory.
24. *Merkin:* Introduction to the Theory of Stability.
25. *Naber:* Topology, Geometry, and Gauge Fields: Foundations.

Gregory L. Naber

Topology, Geometry, and Gauge Fields

Foundations

With 55 Illustrations

Springer

Gregory L. Naber
Department of Mathematics
and Statistics
California State University, Chico
Chico, CA 95929-0525
USA

Series Editors

Mathematics Subject Classification (1991): 22E70, 58G05, 81T13, 53C80, 58B30, 81-99

Library of Congress Cataloging-in-Publication Data
Naber, Gregory L., 1948–
Topology, geometry, and gauge fields : foundations / Gregory L. Naber.
p. cm. – (Texts in applied mathematics; 25)
Includes bibliographical references and index.
ISBN 0-387-94946-1
1. Topology. 2. Geometry. 3. Gauge fields (Physics) 4. Mathematical physics. I. Title. II. Series.
QC20.7.T65N33 1997
516.3′62–dc21 96-49166

Printed on acid-free paper.

Production managed by Timothy Taylor; manufacturing supervised by Johanna Tschebull.
Camera-ready copy prepared from the author's LaTeX files.
Printed and bound by Maple-Vail Book Manufacturing Group, York, PA.
Printed in the United States of America.

9 8 7 6 5 4 3 2 1

ISBN 0-387-94946-1 Springer-Verlag New York Berlin Heidelberg SPIN 10557813

This book is dedicated, with love and gratitude, to my mom, Marguerite Naber, and to the memory of my dad, Bernard Naber.

Poor payment for so great a debt.

Series Preface

Mathematics is playing an ever more important role in the physical and biological sciences, provoking a blurring of boundaries between scientific disciplines and a resurgence of interest in the modern as well as the classical techniques of applied mathematics. This renewal of interest, both in research and teaching, has led to the establishment of the series: *Texts in Applied Mathematics* (*TAM*).

The development of new courses is a natural consequence of a high level of excitement on the research frontier as newer techniques, such as numerical and symbolic computer systems, dynamical systems, and chaos, mix with and reinforce the traditional methods of applied mathematics. Thus, the purpose of this textbook series is to meet the current and future needs of these advances and encourage the teaching of new courses.

TAM will publish textbooks suitable for use in advanced undergraduate and beginning graduate courses, and will complement the *Applied mathematical Sciences* (*AMS*) series, which will focus on advanced textbooks and research level monographs.

Preface

In Egypt, geometry was created to measure the land. Similar motivations, on a somewhat larger scale, led Gauss to the intrinsic differential geometry of surfaces in space. Newton created the calculus to study the motion of physical objects (apples, planets, etc.) and Poincaré was similarly impelled toward his deep and far-reaching topological view of dynamical systems. This symbiosis between mathematics and the study of the physical universe, which nourished both for thousands of years, began to weaken, however, in the early years of this century. Mathematics was increasingly taken with the power of abstraction and physicists had no time to pursue charming generalizations in the hope that the path might lead somewhere. And so, the two parted company. Nature, however, disapproved of the divorce and periodically arranged for the disaffected parties to be brought together once again. Differential geometry and Einstein's general theory of relativity are, by now, virtually inseparable and some of the offspring of this union have been spectacular (e.g., the singularity theorems of Stephen Hawking and Roger Penrose). Much of modern functional analysis has its roots in the quantum mechanics of Heisenberg and Schroedinger and the same can be said of the theory of group representations. Even so, the reconciliations have often been uneasy and ephemeral.

The past two decades, however, have witnessed an extraordinary and quite unexpected confluence of ideas. Two great streams of thought, one from physics and the other from mathematics, which flowed peacefully along, mindless of each other, for forty years are now seen to be but tributaries of the same remarkable sea of ideas. From Dirac's initial attempts, in the 1930's, to form a picture of the electromagnetic field consistent with quantum mechanics, through the quantum electrodynamics of Feynman, Dyson, Schwinger and Tomonaga, the Yang-Mills model of isotopic spin, Weinberg and Salam's electroweak theory and more recent excursions into quantum chromodynamics and quantum gravity, the problem of quantizing classical field theory has occupied center stage in theoretical physics. The star players in this drama have been objects that physicists call *gauge fields*. Concurrent with these activities in the physics community, mathematicians were engaged in deep investigations of the topology and geometry of differentiable manifolds. A long and rather arduous process of distillation eventually led to the appropriate objects of study: *fiber bundles, connections* on them, and the *curvature* of such connections. An extraordinary level of depth and refinement was achieved that culminated in the theory of what are called *characteristic classes.*

It was not until the early 1970's, however, that dawn broke and, in the clear light of day, it was recognized that a gauge field in the sense of the physicists is essentially nothing other than the curvature of a connection on some fiber bundle. Once made, however, this observation precipitated a furious storm of activity in both camps that produced mathematics of remarkable depth and beauty, profound insights into the nature of the physical world and, just like the good old days, an intense interaction between physics and mathematics that remains unabated today.

This is a book on topology and geometry and, like any book on subjects as vast as these, it has a point-of-view that guided the selection of topics. Our point-of-view is that the rekindled interest that mathematics and physics are showing in each other should be fostered and that this is best accomplished by allowing them to cohabit. The goal was to weave together rudimentary notions from the classical gauge theory of physicists with the topological and geometrical concepts that became the mathematical models of these notions. We ask the reader to come to us with some vague sense of what an electromagnetic field might be, a willingness to accept a few of the more elementary pronouncements of quantum mechanics, a solid background in real analysis (e.g., Chapters 1-3 of [**Sp1**]) and linear algebra (e.g., Chapters I-X of [**Lang**]) and some of the vocabulary of modern algebra (e.g., Chapters XIV and XV of [**Lang**]). To such a reader we offer an excursion that begins at sea level, traverses some lovely territory and terminates, if not on Everest, at least in the foothills.

Chapter 0 is intended to provide something along the lines of an initial aerial view of the terrain. Here we introduce, as a prototypical gauge theory problem, Dirac's famous magnetic monopole and the classical quantum mechanical description of the motion of a charged particle in the field of the monopole. This description is complicated by the fact that there is no globally defined vector potential for the monopole's field. Topological considerations enter through Dirac's ingenious notion of a "string" and his observation that one can indeed find vector potentials on the complement of any such string. In this way one can indeed find *two* vector potentials for the field whose domains exhaust all of space except the location of the monopole. On the intersections of these domains the potential functions do not agree, but differ only by a gradient term so that the corresponding wavefunctions for the particle traversing the field differ only by a phase factor $e^{\mathbf{i}\,\theta(x,y,z)}$. The significance of these phase factors is brought home by a discussion of the Aharonov-Bohm experiment and the Dirac Quantization Condition. We conclude that each potential function dictates the phase of the wavefunction and that keeping track of this phase factor as the particle traverses its path is crucial, particularly when there is more than one such particle and these traverse different paths, encountering different vector potentials and so acquiring different phases. The result of any subsequent interaction between two such particles is an interference pattern determined by the phase difference of their wavefunctions.

This problem of keeping track of a particle's phase as it traverses its path through an electromagnetic field has a lovely geometrical formulation. One imagines the space (or spacetime) in which the path lives and "above" each of its points a copy of the unit circle $S^1 = \{ e^{\mathbf{i}\theta} : \theta \in \mathbb{R} \} \subseteq \mathbb{C}$ acting as a sort of notebook in which to record the phase of any charge whose path happens to pass through the point "below". These circles, glued together topologically in some appropriate way, constitute what is called a "circle bundle", or "bundle of phases" over the space containing the path. Keeping track of the particle's phase as it traverses its path then amounts to "lifting" the path to a curve through the bundle of phases.

As it happens, such "bundles" and, indeed, such path lifting procedures (called "connections") arose quite independently of any physics in topology and geometry. As a warm-up for the more general, more abstract constructions to follow in the text we build in Chapter 0, essentially from scratch, the complex Hopf bundle $S^1 \to S^3 \to S^2$ and briefly describe how two locally defined vector potentials for the monopole of lowest strength determine a connection on it (called its "natural connection"). Replacing the complex numbers everywhere in this construction with the quaternions yields another Hopf bundle $S^3 \to S^7 \to S^4$ whose analogous "natural connection" will eventually be seen to represent a so-called BPST instanton (or pseudoparticle). These also arose first in physics as particular instances of what are called "Yang-Mills fields" and we briefly describe the physical motivation. We conclude this introductory chapter with a few remarks on the "moduli space" (i.e., set of gauge equivalence classes) of such instantons and the more general instanton moduli spaces studied by Simon Donaldson *en route* to his 1986 Fields Medal.

The exposition in Chapter 0 is somewhat informal since the goal is motivational. The mathematics begins in earnest in Chapter 1 with an introduction to topological spaces. The emphasis here is on a detailed understanding of those particular spaces that have a role to play in gauge theory (e.g., spheres, projective spaces, the classical groups, etc.) and not on the fine tuning of definitions. We regret that the reader in search of a connected, locally connected, almost regular, Urysohn space will have to look elsewhere, but, by way of compensation, we can offer at least five topologically equivalent models of the special unitary group $SU(2)$. Since locally trivial bundles and group actions permeate gauge theory we discuss these at some considerable length.

Homotopy theory is a vast, subtle and difficult subject, but one that has had a profound impact not only on topology, but on theoretical physics as well (we will find that the Dirac Quantization Condition is, in some sense, equivalent to the fact that principal circle bundles over the 2-sphere are in one-to-one correspondence with the elements in the fundamental group of the circle). Our Chapter 2 is a modest introduction to just those aspects of the subject that we will need to call upon. The central result is a Homotopy Lifting Theorem for locally trivial bundles (Theorem 2.4.1). The proof is rather intricate, but the dividends are substantial. Included among these are all of the homotopy groups $\pi_n(S^1)$ of the circle as well as the classification theorem for principal G-bundles over the n-sphere proved in the next chapter.

In Chapter 3 locally trivial bundles and group actions coalesce into the notion of a (continuous) principal bundle. Major results include the one-to-one correspondence between local cross-sections and local trivializations, the triviality of any principal bundle over a disc (which depends on the Homotopy Lifting Theorem in Chapter 2) and a Reconstruction Theorem (Theorem 3.3.4) which shows that the entire structure of the principal bundle is implicit in any complete set of transition functions. This last result is significant not only because it is the link between the physicist's local, coordinate description of gauge fields and the global, coordinate-free bundle description, but also because it is an essential ingredient in the proof of what might reasonably be regarded as the topological heart of the book. This is Theorem 3.4.3 in which we show that, for pathwise connected groups G, the set of equivalence classes of principal G-bundles over S^n, $n \geq 2$, is in one-to-one correspondence with the elements of the homotopy group $\pi_{n-1}(G)$.

The first six sections of Chapter 4 contain a rather detailed introduction to differentiable manifolds, vector fields and 1-forms. In Section 4.7 we begin with a few general results on Lie groups and their Lie algebras, but soon make the decision to restrict our attention to matrix Lie groups. This substantially simplifies much of the subsequent development and eliminates nothing of real interest in physics. The next order of business is to explicitly calculate the Lie algebras of those particular matrix Lie groups of interest to us and this we do in some considerable detail. The section concludes by introducing the pivotal notion of a fundamental vector field on a principal G-bundle associated with each element in the Lie algebra of G.

Section 4.8 introduces the general notion of a vector-valued 1-form on a manifold. The Cartan (or, canonical) 1-form $\boldsymbol{\Theta}$ on a Lie group G is a Lie algebra-valued 1-form on G and we provide explicit calculations of $\boldsymbol{\Theta}$ for all of the Lie groups G of interest to us. The calculation for $SU(2)$ contains a bit of a surprise. One of its components is essentially identical to the 1-form representing a Dirac magnetic monopole of lowest strength! The quaternionic analogue of $SU(2)$ is $Sp(2)$ and, remarkably enough, the Cartan 1-form for $Sp(2)$ also has a component that is essentially identical to an object that arose independently in the physics literature. This is the famous BPST instanton solution to the Yang-Mills equations. Section 4.8 concludes with some rather detailed calculations of various coordinate expressions and properties of these 1-forms.

Orientability, Riemannian metrics and 2-forms (real and vector-valued) are introduced in the last two sections of Chapter 4. We show that the n-sphere S^n is locally conformally equivalent to Euclidean n-space $\mathbb{R}^n$

(a result that will be crucial, in Chapter 5, to defining the notion of an anti-self-dual connection on the Hopf bundle $S^3 \to S^7 \to S^4$). Exterior derivatives and various types of wedge products for vector-valued 1-forms are described in Section 4.10 which concludes with a number of quite detailed calculations of concrete examples. These examples will be the focus of much of our attention in the final chapter.

Connections on principal bundles are introduced as Lie algebra-valued 1-forms on the bundle space in Section 5.1. Pullbacks of these by local cross-sections of the bundle are called gauge potentials and these are the objects most frequently encountered in the physics literature. We show that a connection is completely determined by a sufficiently large family of gauge potentials and also by what is called its distribution of horizontal spaces. Next we prove a theorem on pullbacks of connection forms by bundle maps and use it, and a natural left action of $SL(2, \mathbb{H})$ on S^7, to manufacture a large supply of connection forms on the Hopf bundle $S^3 \to S^7 \to S^4$. The corresponding gauge potentials are put into a simple standard form by appealing to a rather nontrivial algebraic result known as the Iwasawa Decomposition of $SL(2, \mathbb{H})$. Still in Section 5.1, we introduce a global version of the physicist's notion of gauge equivalence. The set of all gauge equivalence classes of connections on a given bundle is then called the moduli space of connections on that bundle. Finally in this section we show how a connection on a principal bundle determines the long sought after path lifting procedure from the base to the bundle space and thereby a notion of parallel translation from one fiber to another.

In Section 5.2 we define the curvature of a connection on a principal bundle to be its covariant exterior derivative. Such curvature 2-forms are called gauge fields in the physics literature. These are generally calculated from the Cartan Structure Equation which is our Theorem 5.2.1. Pullbacks of the curvature by local cross-sections are the local field strengths of physics and we derive their transformation properties and a number of computational formulas and concrete examples. The crucial difference between Abelian and non-Abelian gauge groups now becomes clear. Only in the Abelian case are the local field strengths gauge invariant so that they can be patched together to give a globally defined field strength 2-form. The section concludes with a brief discussion of flat connections.

Section 5.3 on the Yang-Mills functional is a bit unusual (and might even be thought of as Section 0.6). In the hope of providing some sense of the physical origins of not only the BPST potentials we have seen, but also the notion of (anti-) self-duality which is yet to come, we temporarily revert to the more informal style of Chapter 0. References are provided for those who wish to see all of this done rigorously. Section 5.4 is essentially algebraic. Here we introduce a special case of the Hodge star operator of linear algebra and use it to define self-dual and anti-self-dual 2-forms on any 4-dimensional, oriented, Riemannian manifold. The fact that these notions are invariant under orientation preserving conformal diffeomorphism allows us to define what is meant by an anti-self-dual connection on the Hopf bundle $S^3 \to S^7 \to S^4$ and to write down lots of examples (from the BPST potentials). The set of gauge equivalence classes of such connections is, in Section 5.5, called the moduli space of anti-self-dual connections on the Hopf bundle. Appealing to our earlier description of the BPST potentials and a deep result of Atiyah, Hitchin and Singer we identify this moduli space with the open half-space $(0, \infty) \times \mathbb{R}^4$ in $\mathbb{R}^5$. This is diffeomorphic to the open 5-dimensional ball B^5 in $\mathbb{R}^5$ and, using the Cartan decomposition of $SL(2, \mathbb{H})$, we find a parametrization of the moduli space which naturally identifies it with B^5. The closed disc D^5 in $\mathbb{R}^5$ is a compactification of the moduli space in which the base S^4 of the Hopf bundle appears as the boundary of the moduli space and corresponds intuitively to the set of "concentrated" connections. The section concludes with a sketch of how Donaldson generalized this simple, elegant picture to a vastly more general context and thereby proved deep and unexpected results on the topology of differentiable 4-manifolds.

Section 5.6 is another brief excursion into the murky waters of physical motivation. A gauge field is something akin to the old Newtonian concept of a "force" in that certain particles will "respond" to it by experiencing changes in their wavefunctions. In Section 5.6 we isolate the proper mathematical device for modeling the wavefunctions of these particles that are "coupled to" a gauge field and, in Section 5.7, we rigorously build them from the principal bundle on which the gauge field is defined. These are called matter fields and can be regarded either as cross-sections of an associated vector bundle or as equivariant maps on the original principal bundle space. Field equations describing the quantitative response of the matter field to the gauge field are formulated by physicists in terms of a covariant exterior derivative determined by the connection that represents the gauge field. This automatically "couples" the two fields and, moreover, is necessary to ensure the gauge invariance of the resulting theory. We introduce this derivative and derive a few computational formulas in Section 5.8. The actual business of writing down field equations and sorting out their predictions is best left to the physicists. Nevertheless, the geometry and topology of the interaction between gauge fields and matter fields probe the deepest levels of our current understanding of the fundamental processes at work in the world and we intend to take up this story in a subsequent volume, entitled **Topology, Geometry and Gauge Fields: Interactions**, currently in preparation and also to be published by Springer-Verlag.

The book concludes with a brief Appendix on the role of the special unitary group $SU(2)$ as the double cover of the rotation group $SO(3)$. This material is not required in the text, but it does go a long way toward explaining the privileged position that $SU(2)$ occupies in theoretical physics.

Perhaps alone in this era of enlightened pedagogy, the author remains a true believer in the value of routine calculation. Readers of a different persuasion may find more of this than they care to see in the text and more than they wish to do in the Exercises. There are, by the way, well over 400 Exercises. Each is an integral part of the development and each is located in the text at precisely the point at which it can be done with optimal benefit. We encourage the reader to pause *en route* and take the opportunity that these present to join in the fun.

Gregory L. Naber
1997

Acknowledgments

To the California State University, Chico, and, more particularly, its Department of Mathematics and Statistics, Graduate School, and CSU Research Program go my sincere thanks for the support, financial and otherwise, that they provided throughout the period during which this book was being written. For Debora, my wife and my partner in this work, who typed the manuscript, provided unwavering support and encouragement, and kept the world and its distractions at bay when that was necessary, sincere thanks seem inadequate, but I offer them anyway with love and admiration for all that she does so well.

Contents

0

Physical and Geometrical Motivation

0.1 Introduction

It sometimes transpires that mathematics and physics, pursuing quite different agendas, find that their intellectual wanderings have converged upon the same fundamental idea and that, once it is recognized that this has occurred, each breathes new life into the other. The classic example is the symbiosis between General Relativity and Differential Geometry. As the Singularity Theorems of Hawking and Penrose (see [**N2**]) amply attest, the results of such an interaction can be spectacular. The story we have to tell is of another such confluence of ideas, more recent and perhaps even more profound. Our purpose in this preliminary chapter is to trace the physical and geometrical origins of the notion of a "gauge field" (known to mathematicians as the "curvature" of a "connection on a principal bundle"). We will not be much concerned yet with rigorously defining the terms we use, nor will we bother to prove most of our assertions. Indeed, much of the remainder of the book is devoted to these very tasks. We hope only to offer something in the way of motivation.

0.2 Dirac's Magnetic Monopole

We ask the reader to recall (or accept blindly, on faith) that a point electric charge q residing at the origin of some inertial frame of reference determines an electric field $\vec{E}$ described in that frame by **Coulomb's Law**: $\vec{E} = (q/\rho^2)\hat{e}_\rho$, $\rho \neq 0$ (we employ standard spherical coordinates ρ, ϕ, θ as indicated in Figure 0.2.1 with unit coordinate vectors denoted $\hat{e}_\rho, \hat{e}_\phi$ and $\hat{e}_\theta$). The magnetic field $\vec{B}$ associated with q is identically zero in this frame: $\vec{B} = \vec{0}$, $\rho \neq 0$. On $\mathbb{R}^3 - 0$ (by which we mean $\mathbb{R}^3 - \{(0,0,0)\}$), $\vec{E}$ and $\vec{B}$ satisfy the so-called **static, source-free Maxwell equations**

$$\text{div } \vec{E} = 0 \qquad \text{div } \vec{B} = 0 \qquad \text{curl } \vec{E} = \vec{0} \qquad \text{curl } \vec{B} = \vec{0}. \tag{0.2.1}$$

Although the "magnetic analogue" of an electric charge has never been observed in nature, Paul Dirac ([**Dir1**], [**Dir2**]) felt (and so do we) that

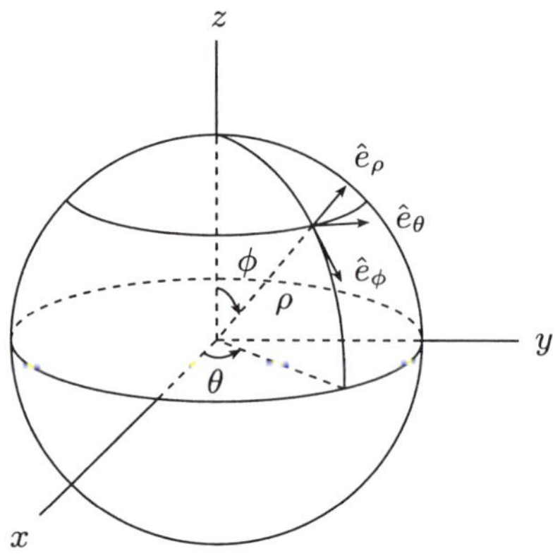

Figure 0.2.1

such an object is worth thinking about anyway. Thus, we consider a (hypothetical) point particle at rest at the origin of some inertial frame that determines an electromagnetic field described in that frame by

$$\vec{E} = \vec{0}, \quad \vec{B} = \frac{g}{\rho^2}\,\hat{e}_\rho, \quad \rho \neq 0, \tag{0.2.2}$$

where g is a constant (the strength of this so-called **magnetic monopole**). $\vec{E}$ and $\vec{B}$ clearly also satisfy the static, source-free Maxwell equations (0.2.1). In particular,

$$\text{div } \vec{B} = 0 \ \text{ on } \ \mathbb{R}^3 - 0, \tag{0.2.3}$$

$$\text{curl } \vec{B} = \vec{0} \ \text{ on } \ \mathbb{R}^3 - 0. \tag{0.2.4}$$

Since $\mathbb{R}^3 - 0$ is simply connected, (0.2.4) and standard results from vector analysis guarantee the existence of a scalar potential function for $\vec{B}$, i.e., there exists a smooth (i.e., C^∞) function $V : \mathbb{R}^3 - 0 \to \mathbb{R}$ whose gradient is $\vec{B} : \nabla V = \vec{B}$, on $\mathbb{R}^3 - 0$. Indeed, it is a simple matter to verify that $V(\rho, \phi, \theta) = -g/\rho$ will serve as such a scalar potential for $\vec{B}$. Now, for the Coulomb (electric) field, the existence of a scalar potential is a matter of considerable significance, but, for reasons that we hope to make clear shortly, in the case of a magnetic field it is the existence of a vector potential (a smooth vector field $\vec{A}$ with curl $\vec{A} = \vec{B}$) that is desirable. Now, (0.2.3) is surely a necessary condition for the existence of a vector potential (the divergence of a curl is zero), but it is *not sufficient even on a simply connected region.* Indeed, a bit of simple vector calculus shows that the field $\vec{B}$ given by (0.2.2) cannot have a vector potential on $\mathbb{R}^3 - 0$. To see this let us assume to the contrary that there does exist a vector field $\vec{A}$ that is smooth on $\mathbb{R}^3 - 0$ and satisfies curl $\vec{A} = \vec{B}$ there. Fix some sphere S of radius R about the origin, let C be its equator and S^+ and S^- its upper and lower hemispheres. Orient C counterclockwise and orient S, S^+ and S^- with the outward unit normal vector, i.e., $\hat{e}_\rho$. Now, a simple calculation gives

$$\begin{aligned}\iint_S \operatorname{curl}\vec{A}\cdot d\vec{S} &= \iint_S \vec{B}\cdot d\vec{S} = \iint_S \left(\frac{g}{\rho^2}\hat{e}_\rho\right)\cdot\hat{e}_\rho\, dS \\ &= \frac{g}{R^2}\iint_S dS = \frac{g}{R^2}(4\pi R^2) = 4\pi g.\end{aligned}$$

On the other hand, Stokes' Theorem (valid since $\vec{A}$ is assumed smooth on S) gives

$$\begin{aligned}\iint_S \operatorname{curl}\vec{A}\cdot d\vec{S} &= \iint_{S^+}\operatorname{curl}\vec{A}\cdot d\vec{S} + \iint_{S^-}\operatorname{curl}\vec{A}\cdot d\vec{S} \\ &= \oint_C \vec{A}\cdot d\vec{r} + \oint_{-C}\vec{A}\cdot d\vec{r} \\ &= \oint_C \vec{A}\cdot d\vec{r} - \oint_C \vec{A}\cdot d\vec{r} = 0,\end{aligned}$$

so we have a contradiction. No smooth vector potential for $\vec{B}$ exists on $\mathbb{R}^3 - 0$. Soon we will discuss at length how truly unfortunate this state of affairs is, but first we affix blame. What, fundamentally, is responsible for the failure of such a vector potential to exist and can anything be done to circumvent the difficulty?

We have already mentioned that (0.2.4) and the simple connectivity of $\mathbb{R}^3 - 0$ together imply the existence of a scalar potential for $\vec{B}$ on $\mathbb{R}^3 - 0$. Now, simple connectivity is a topological condition on the domain of $\vec{B}$. The precise definition (which we get around to in Section 2.2) is that the "fundamental group" $\pi_1(\mathbb{R}^3 - 0)$ of $\mathbb{R}^3 - 0$ is trivial: $\pi_1(\mathbb{R}^3 - 0) = 0$. Intuitively, this means that every loop (closed curve) in $\mathbb{R}^3 - 0$ is equivalent to a trivial (constant) loop in the sense that it can be continuously shrunk to a point without leaving $\mathbb{R}^3 - 0$. We have also seen that the vanishing of this fundamental group, together with the necessary condition (0.2.3) for the existence of a vector potential is, regrettably, not sufficient to yield a vector potential. If, however, one is prepared to take one further topological step it is possible to obtain a condition which, together with vanishing of the divergence, implies the existence of a vector potential. This step involves the use of another topological invariant (like the fundamental group) called the "second homotopy group" that we will define precisely in Section 2.5. Intuitively, the triviality of this group for some simply connected open set U in $\mathbb{R}^3$ amounts to the assertion that any 2-dimensional sphere in U

encloses only points of U (obviously false for $\mathbb{R}^3 - 0$). One can then prove the following:

> *Let U be an open, simply connected subset of $\mathbb{R}^3$ and $\vec{F}$ a smooth vector field on U. If $\operatorname{div}\vec{F} = 0$ on U and if the second homotopy group $\pi_2(U)$ is trivial ($\pi_2(U) = 0$), then there exists a smooth vector field $\vec{A}$ on U with $\operatorname{curl}\vec{A} = \vec{F}$.*

Thus, we have found the culprit. It is the topology of $\vec{B}$'s domain ($\mathbb{R}^3 - 0$) that prevents the monopole field from having a vector potential. So, what is to be done? If the failure of the vector potential to exist is, in fact, a matter of concern (and we will shortly attempt to convince you that it is), can one find some way around this topological difficulty? Indeed, one can and Dirac has shown us the way.

Let us imagine a continuous curve in $\mathbb{R}^3$ that begins at the origin, does not intersect itself and proceeds off to infinity in some direction. In the physics literature such curves are known as **Dirac strings**. The complement U of such a string is an open subset of $\mathbb{R}^3$ that is simply connected (a loop that surrounds the string can be continuously lifted around the origin and then shrunk to a point) and, moreover, has $\pi_2(U) = 0$ since no sphere *in* U can enclose points of the string. Thus, our monopole field $\vec{B}$, *if restricted to* $U = \mathbb{R}^3$-string, will have a vector potential on this set. Choosing two strings that have only the origin in common we can cover $\mathbb{R}^3 - 0$ with two open sets, on each of which $\vec{B}$ has a vector potential.

It would seem worthwhile at this point to write out a concrete example. The nonpositive z-axis $Z_- = \{(0,0,z) \in \mathbb{R}^3 : z \leq 0\}$ is a Dirac string so, on its complement $U_+ = \mathbb{R}^3 - Z_-$, the existence of a vector potential for $\vec{B}$ is assured. Indeed, with spherical coordinates (ρ, ϕ, θ) as in Figure 0.2.1, a simple calculation shows that the curl of

$$\vec{A}_+(\rho,\phi,\theta) = \frac{g}{\rho\sin\phi}(1-\cos\phi)\hat{e}_\theta \tag{0.2.5}$$

on U_+ is $\vec{B}$ (notice that this is smooth on U_+ despite the $\sin\phi$ in the denominator because $(1-\cos\phi)/\sin\phi$ is actually analytic at $\phi = 0$). Similarly, if $Z_+ = \{(0,0,z) \in \mathbb{R}^3 : z \geq 0\}$ and $U_- = \mathbb{R}^3 - Z_+$, then

$$\vec{A}_-(\rho,\phi,\theta) = \frac{-g}{\rho\sin\phi}(1+\cos\phi)\hat{e}_\theta \tag{0.2.6}$$

is a smooth vector potential for $\vec{B}$ on U_-. Taken together the domains of $\vec{A}_\pm$ fill up all of $\mathbb{R}^3 - 0 = U_+ \cup U_-$. Of course, on the overlap $U_+ \cap U_-$ ($\mathbb{R}^3$ minus the z-axis), $\vec{A}_+$ and $\vec{A}_-$ do not agree (if they did, they would define a vector potential on $\mathbb{R}^3 - 0$ and this, we know, does not exist). Indeed, on $U_+ \cap U_-$, (0.2.5) and (0.2.6) give $\vec{A}_+ - \vec{A}_- = (2g/\rho\sin\phi)\hat{e}_\theta$ which, as a simple calculation shows, is the gradient of $2g\theta$. Thus, on their common domain, $\vec{A}_+$ and $\vec{A}_-$ differ by a gradient:

$$\vec{A}_+ - \vec{A}_- = \nabla(2g\theta) \text{ on } U_+ \cap U_- . \tag{0.2.7}$$

But why make such a fuss over these vector potentials? Anyone with a bit of experience in undergraduate electromagnetic theory will know that such potential functions are, in that context, regarded as very convenient computational tools, but with no real physical significance of their own. The reason for this attitude is quite simple. Vector potentials, even when they exist, are highly nonunique. If $\vec{A}$ satisfies $\operatorname{curl}\vec{A} = \vec{B}$, then, since the curl of a gradient is identically zero, so does $\vec{A} + \nabla\Omega$ for *any* smooth real-valued function Ω. Nevertheless, this view changed dramatically with the advent of quantum mechanics and we need to understand the reason. To do so it will be necessary to consider a somewhat more complicated system than an isolated monopole.

We consider again a magnetic monopole situated at the origin of some inertial frame. Now we introduce into the vicinity of the monopole a moving electric charge q (which we regard as a "test charge" whose own electromagnetic field has negligible effect on the monopole). Classically, the motion of the charge is governed by Newton's Second Law and the so-called Lorentz Force Law (which describes how the charge responds to $\vec{B}$).

Remark: Although the details of this classical motion are not required for our purposes, they are quite interesting. The reader may find it entertaining to modify the usual procedure from calculus for solving the Kepler problem to show that, in general, the charge is constrained to move on a *cone* whose vertex is at the location of the monopole.

The current view of this system is rather different, however. The charge is not thought of as a "point" particle at all, but rather as a quantum mechanical object described by its wavefunction $\psi(x, y, z, t)$. This is a complex-valued function of space (x, y, z) and time (t) that is believed to contain all of the physically measurable information about the charge. For example, the probability of finding the charge in some region R of space at some instant t of time is computed by integrating $|\psi|^2 = \psi\bar{\psi}$ over R. The wavefunction ψ for q is found by solving the so-called Schroedinger equation for the monopole/charge system. Now, the Schroedinger equation for a given system is constructed by writing down the classical Hamiltonian for the system and employing what are called "correspondence rules" to replace each classical quantity in the Hamiltonian with an appropriate operator. The details need not concern us. The only feature relevant to our investigation is that the Hamiltonian for a charge in an electromagnetic field involves, in an essential way, the vector potential $\vec{A}$ for the electromagnetic field. Of course, this vector potential is not unique. One can show that replacing $\vec{A}$ by $\vec{A} + \nabla\Omega$ in the Schroedinger equation replaces the solution ψ by $e^{\mathbf{i}q\Omega}\psi$:

$$\vec{A} \longrightarrow \vec{A} + \nabla\Omega \Longrightarrow \psi \longrightarrow e^{\mathbf{i}q\Omega}\psi \tag{0.2.8}$$

(with apologies to our physicist readers we will, whenever possible, choose units so that as many physical constants as possible are 1). Now, Ω is real-valued so each $e^{\mathbf{i}q\Omega}$ is a complex number of modulus one. Thus, $\vec{A} \to \vec{A} + \nabla\Omega$ changes only the phase and not the modulus (amplitude) of the wavefunction ψ. For quite some time it was felt that such phase changes in the wavefunction were of no physical significance since all of the physically measurable quantities associated with the charge q depend only on the squared modulus $|\psi|^2$ and this is the same for ψ and $e^{\mathbf{i}q\Omega}\psi$. However, in 1959, Aharonov and Bohm [**AB**] suggested that, while the phase of a single charge may well be unmeasurable, the *relative* phase of two charged particles that interact should have observable consequences. They proposed an experiment that went roughly as follows: A beam of electrons is split into two partial beams that pass around opposite sides of a solenoid (this is a very tightly wound coil of wire through which a current passes, creating a magnetic field that is *confined inside the coil*). Beyond the solenoid the beams are recombined and detected at a screen. The result is a typical interference pattern that manifests itself experimentally as a variation from point to point on the screen of the probability of detecting a particle there. One observes this interference pattern when there is no current flowing through the coil, so that the magnetic field in the solenoid is zero, and then again when there is a current and hence a nonzero magnetic field inside the coil. Since the electrons pass *around* the coil and the magnetic field is confined *inside* the coil, any shift in the interference pattern in these two cases cannot be attributed to the magnetic field (which the electrons do not encounter). The vector potential, on the other hand, is generally nonzero outside the solenoid even though the magnetic field in this region is always zero. One could then only conclude that this vector potential induces different phase shifts on the two partial beams before they are recombined and that these relative phase changes account for the altered interference pattern. This experiment has, in fact, been performed (first by R. G. Chambers in 1960) with results that confirmed the expectations of Aharonov and Bohm.

We see then that, in quantum mechanics, it is no longer possible to regard the vector potential as a convenient, but expendible computational device. Thus, the failure of a global vector potential for the monopole to exist is more than just annoying. Nevertheless, it is a fact and our only option would seem to be to find some way around it. What we require is some other mathematical device for doing the vector potential's job, that is, keeping track of the phase changes experienced by a charge as it traverses its trajectory through the monopole's field.

Remark: Before hunting down this device we pause to point out another remarkable consequence of (0.2.8), first noticed by Dirac. Return to the two local vector potentials $\vec{A}_+$ and $\vec{A}_-$ for the monopole given by (0.2.5) and (0.2.6) on their respective domains U_+ and U_-. Denote by ψ_+ and ψ_- the wavefunctions for our charge determined (via the Schroedinger equation) by $\vec{A}_+$ and $\vec{A}_-$. On $U_+ \cap U_-$, (0.2.7) gives $\vec{A}_+ = \vec{A}_- + \nabla(2g\theta)$ so by (0.2.8), $\psi_+ = e^{\mathbf{i}(2qg\theta)}\psi_-$. But on $U_+ \cap U_-$ (which contains the circle $(\rho, \phi, \theta) = (1, \pi/2, \theta)$) both ψ_+ and ψ_- assign exactly one complex value to each point at each time. Thus, for each fixed t, the change $\theta \to \theta + 2\pi$ must leave both ψ_+ and ψ_- unchanged. However, this then implies that $\theta \to \theta + 2\pi$ must leave $e^{\mathbf{i}(2qg\theta)}$ unchanged, whereas,

$$e^{\mathbf{i}(2qg(\theta+2\pi))} = e^{\mathbf{i}(2qg\theta)}e^{\mathbf{i}(4qg\pi)} .$$

Consequently, we must have $e^{\mathbf{i}(4qg\pi)} = 1$. But $e^{\mathbf{i}(4qg\pi)} = \cos(4qg\pi) + \mathbf{i}\sin(4qg\pi)$ so this is possible only if $4qg\pi = 2n\pi$ for some integer n. We conclude that

$$qg = \tfrac{1}{2}n \quad \text{for some integer } n. \qquad (0.2.9)$$

This is the celebrated **Dirac quantization condition** and is interpreted as asserting that if even a single magnetic monopole (strength g) exists, then charge must be "quantized", i.e., come only in integer multiples of some basic quantity of charge ($q = n(1/2g)$). Since charge is, indeed, quantized in nature and since no other plausible explanation for this fact has ever been offered, the existence of magnetic monopoles becomes a rather tantalizing possibility. We will eventually see that the Dirac quantization condition is the physical manifestation of certain purely topological facts related to the classification of principal $U(1)$-bundles over S^2 by elements of the fundamental group $\pi_1(U(1))$ of the circle $U(1)$.

With this digression behind us we set off in search of the "mathematical device for doing the vector potential's job". This search will eventually lead us through some rather exotic topological and geometrical territory, the precise mapping of which is a principal objective of this book. Nevertheless, the basic idea is simple enough and we devote the remainder of this chapter to a quick aerial view.

Notice that, at each point on its trajectory, the phase of a charged particle is represented by an element $e^{\mathbf{i}\theta}$ of the unit circle S^1 in the complex plane $\mathbb{C}$. The trajectory itself lives in 3-space $\mathbb{R}^3$. Imagine a copy of S^1 setting "above" each point of $\mathbb{R}^3$, acting as something of a notebook in which to record the phase of a charge whose trajectory happens to pass through that point. Admittedly, this is not easy to visualize. You might try suppressing one spatial dimension (which is fine for charges moving in a plane) and thinking of each circle S^1 as the closed interval $[0, 2\pi]$ with its endpoints "identified" (glued together). In your picture, do not actually glue the endpoints together; just try to keep in mind that they are "really" the same point (e.g., give them the same name). What one sees then is a sort of "bundle" of these intervals/circles, one "fiber" atop each point in the plane (Figure 0.2.2).

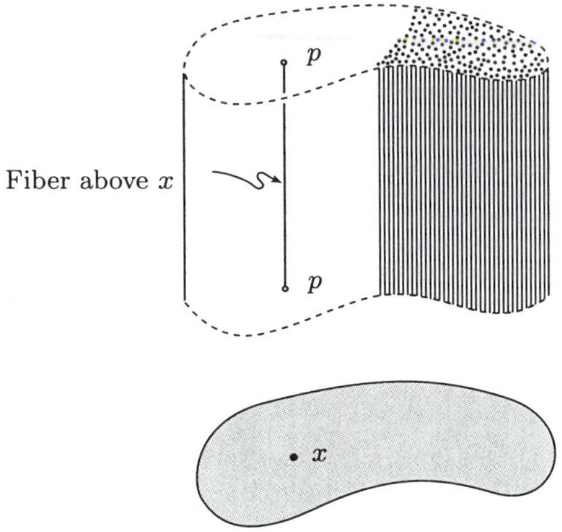

Figure 0.2.2

One thinks of this bundle as a "space of phases". A charge moving through $\mathbb{R}^3$ has, at each point x_0, a phase represented by a point in the fiber above x_0. A phase change at a point then amounts to a rotation of the fiber above that point (which can be accomplished by multiplying every element of the fiber S^1 by some fixed element of S^1, i.e., by an "action" of S^1 on itself). Disregarding the modulus of the wavefunction ψ for the moment, (0.2.8) suggests a phase change that varies from point to point along a trajectory. Moreover, keeping track of the charge's phase as it traverses its trajectory (the vector potential's job) can be viewed as a "lifting problem": Given the trajectory of the charge in space (a curve) and the value of the phase at some point, our problem is to specify a curve through the bundle space (space of phases) that sets "above" the trajectory, takes the known value of the phase at the given point and at all other points records the evolution of the phase dictated by the field through which the charge is moving (e.g., that of a monopole).

One would expect that, in some sense, the phase should vary continuously, or even smoothly, along the trajectory and it is here that the mathematical work really begins. The fibers S^1 must be "tied together" in some topological sense so that the notion of a continuous or differentiable curve is meaningful. One finds that, when the field through which the charge moves is defined on *all* of $\mathbb{R}^3$ (or any "contractible' subset of $\mathbb{R}^3$), there is essentially only one reasonable way to do this. When, as in the case of the monopole, this field is defined only on a ("noncontractible") subset of $\mathbb{R}^3$, then this uniqueness is lost and it is not clear how the tying together should be done (or even that it should be done the same way for monopoles of different strength g). A simple analogy may clarify matters here. There are at least two obvious ways to build a bundle of intervals above a circle. A simple stack gives a cylinder (Figure 0.2.3 (a)) and a stack with a 180° twist gives a Möbius strip (Figure 0.2.3 (b)) and these are genuinely different creatures.

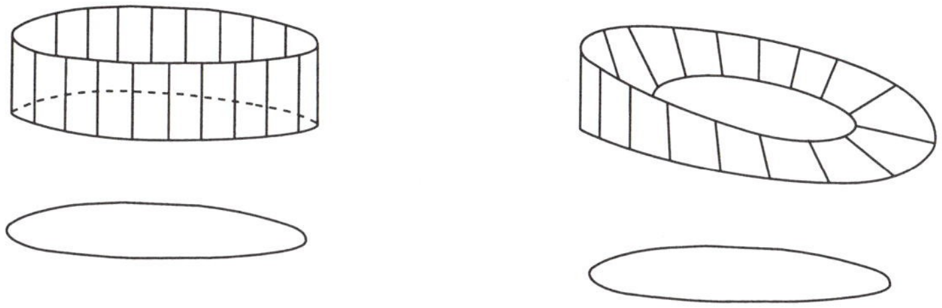

(a) (b)

Figure 0.2.3

To understand this situation properly we need to put aside the physics for awhile and do some mathematics. This we take up in the next section. First though it is essential that we rephrase the information we have accumulated thus far in a language that, unlike the simple vector calculus we have employed to this point, generalizes to higher dimensional situations. This is the language of differential forms, which we will discuss in detail in Chapter 4. For the time being it will be sufficient to think of differential forms in the intuitive terms in which they are introduced in calculus. We offer a brief synopsis of the formalism.

Let U be an open set in $\mathbb{R}^3$. A "0-form" on U is simply a real-valued function $f : U \to \mathbb{R}$ that is C^∞ (i.e., continuous with continuous partial derivatives of all orders). A "1-form" on U will eventually be defined precisely as a certain type of real-valued linear transformation on vectors in $\mathbb{R}^3$. For example, any 0-form f determines a 1-form df, called its exterior derivative, or differential, whose value at a vector v is the directional derivative of f in the direction v, i.e., $df(v) = \nabla f \cdot v$. Expressed in terms of standard coordinates on $\mathbb{R}^3$, $df = (\partial f / \partial x)dx + (\partial f / \partial y)dy + (\partial f / \partial z)dz$. Any 1-form

$\boldsymbol{\alpha}$ on U can be expressed in standard coordinates as $\boldsymbol{\alpha}= f_1dx+f_2dy+f_3dz$, where f_1, f_2 and f_3 are 0-forms. There is an obvious one-to-one correspondence between 1-forms and vector fields that carries $\boldsymbol{\alpha}$ to the vector field $\vec{\alpha}$ with component functions $< f_1, f_2, f_3 >$. Both the vector field and the 1-form can be expressed equally well in any other coordinate system on $\mathbb{R}^3$ (e.g., spherical).

A "2-form" on U is a certain type of real-valued bilinear map on pairs of vectors in $\mathbb{R}^3$. For example, if $\boldsymbol{\alpha}$ and $\boldsymbol{\beta}$ are 1-forms, their "wedge product" $\boldsymbol{\alpha} \wedge \boldsymbol{\beta}$ is a 2-form defined by $(\boldsymbol{\alpha} \wedge \boldsymbol{\beta})(v, w) = \boldsymbol{\alpha}(v)\boldsymbol{\beta}(w) - \boldsymbol{\alpha}(w)\boldsymbol{\beta}(v)$. Expressed in standard coordinates a 2-form $\boldsymbol{\Omega}$ looks like $h_1\, dy \wedge dz + h_2\, dz \wedge dx + h_3\, dx \wedge dy$. These are, of course, also in one-to-one correspondence with vector fields on U. Indeed, if $\boldsymbol{\alpha}$ and $\boldsymbol{\beta}$ are 1-forms corresponding to vector fields $\vec{\alpha}$ and $\vec{\beta}$, then the 2-form $\boldsymbol{\alpha}\wedge\boldsymbol{\beta}$ corresponds to the cross product $\vec{\alpha}\times\vec{\beta}$. There is a natural way to extend the exterior differentiation operation d to 1-forms $\boldsymbol{\alpha}$, the result being a 2-form $d\boldsymbol{\alpha}$. Moreover, if $\boldsymbol{\alpha}$ corresponds to the vector field $\vec{\alpha}$, then $d\boldsymbol{\alpha}$ corresponds to the curl of $\vec{\alpha}$.

"3-forms" on U result from yet one more extension of the wedge product and exterior differentiation operators. In standard coordinates they have the form $f(x, y, z)\, dx \wedge dy \wedge dz$ and so are in one-to-one correspondence with C^∞ functions $f(x, y, z)$ on U. In fact, if $\boldsymbol{\Omega}$ is a 2-form corresponding to the vector field $\vec{\Omega}$, then $d\boldsymbol{\Omega}$ corresponds to the divergence of $\vec{\Omega}$. One of the principal virtues of differential forms is this elegant consolidation of the basic operations of vector calculus (div, grad, curl).

Let us return now to the monopole field $\vec{B} = (g\,/\,\rho^2\,)\,\hat{e}_\rho = (g/\rho^2) < \sin\phi\cos\theta, \sin\phi\sin\theta, \cos\phi > = (g/\rho^3) < x, y, z >$ on $U = \mathbb{R}^3 - 0$. We introduce the corresponding 2-form $\boldsymbol{F} = (g/\rho^3)(x\, dy \wedge dz + y\, dz \wedge dx + z\, dx \wedge dy)$, where $\rho = (x^2+y^2+z^2)^{1/2}$. The role of a vector potential is now played by a 1-form $\boldsymbol{A}$ that satisfies $d\boldsymbol{A} = \boldsymbol{F}$. Of course, such a 1-form does not exist on all of $U = \mathbb{R}^3 - 0$. However, a simple calculation shows that, on their respective domains of U_+ and U_-, the 1-forms

$$\boldsymbol{A}_+ = \frac{g}{\rho}\,\frac{1}{z+\rho}\,(x\, dy - y\, dx)$$

and

$$\boldsymbol{A}_- = \frac{g}{\rho}\,\frac{1}{z-\rho}\,(x\, dy - y\, dx)$$

have the property that $d\boldsymbol{A}_+ = \boldsymbol{F}$ and $d\boldsymbol{A}_- = \boldsymbol{F}$. An interesting thing happens when these are converted to spherical coordinates:

$$\boldsymbol{A}_+ = g(1 - \cos\phi)d\theta$$

and

$$\boldsymbol{A}_- = -g(1 + \cos\phi)d\theta\,.$$

The interesting part is that they are *independent of* ρ and so both can be regarded as 1-forms on open sets in the $\phi\theta$-plane $\mathbb{R}^2$. But the spherical coordinate transformation $x = \rho \sin\phi \cos\theta$, $y = \rho \sin\phi \sin\theta$, $z = \rho \cos\phi$, with ρ held fixed at 1 identifies these open sets in the $\phi\theta$-plane with the unit sphere S^2 in $\mathbb{R}^3$, minus its south pole $(0,0,-1)$ for $\boldsymbol{A}_+$ and minus the north pole $(0,0,1)$ for $\boldsymbol{A}_-$. Since the task we ask these potentials to perform is that of keeping track of the phase of a charged particle as it moves through the field of the monopole, it would appear that we need only keep track of how this phase varies with ϕ and θ, i.e., on S^2. In light of this we adjust slightly the "lifting problem" proposed earlier as an approach to keeping track of our charge's phase. We henceforth seek a "bundle" of circles S^1 over S^2 (rather than $\mathbb{R}^3 - 0$) and a "path lifting procedure" that lifts a curve in S^2 (the radial projection of the charge's trajectory into the sphere) to a curve in the bundle space (space of phases). The mathematical machinery for accomplishing all of this ("principal fiber bundles" and "connections" on them) was, unbeknownst to Dirac, being developed almost simultaneously with his earliest ruminations on magnetic monopoles.

0.3 The Hopf Bundle

Dirac published his first paper on magnetic monopoles [**Dir1**] in 1931. In that same year Heinz Hopf [**Hopf**] announced some startling results on the higher homotopy groups of the spheres. Although it would not become clear for many years that the purely topological work of Hopf had any bearing on Dirac's ideas, the two are, in fact, intimately related and we need to understand the reason.

Hopf was studying continuous maps between spheres of various dimensions and so we begin with a few preliminaries on the 1−, 2−, and 3-dimensional cases. The 1-sphere, or circle, is the set $S^1 = \{e^{\mathbf{i}\xi} : \xi \in \mathbb{R}\}$ of complex numbers of modulus one. Since S^1 is closed under complex multiplication ($e^{\mathbf{i}\xi_1} e^{\mathbf{i}\xi_2} = e^{\mathbf{i}(\xi_1+\xi_2)}$) and inversion ($(e^{\mathbf{i}\xi})^{-1} = e^{-\mathbf{i}\xi}$), it forms an Abelian group. Moreover, these two operations are smooth in the sense that they are restrictions to S^1 of smooth mappings from $\mathbb{C} \times \mathbb{C}$ to $\mathbb{C}$ and from $\mathbb{C} - 0$ to $\mathbb{C} - 0$, respectively. S^1 therefore qualifies as a "Lie group" (defined precisely in Section 4.3) and will eventually emerge as the so-called "gauge group" of our bundle (although, in this context, it is more customary to denote the circle $U(1)$, rather than S^1).

Denoting the usual norm in $\mathbb{R}^3$ by $\| \ \|$ (so that, if $p = (p^1, p^2, p^3) \in \mathbb{R}^3$, $\| p \|^2 = (p^1)^2 + (p^2)^2 + (p^3)^2$), the 2-sphere S^2 is the subset $\{p \in \mathbb{R}^3 : \| p \| = 1\}$ of $\mathbb{R}^3$. We will have need of two standard stereographic projection maps on S^2. First let $U_S = S^2 - (0,0,1)$ be S^2 minus its north pole. Define a map $\varphi_S : U_S \to \mathbb{R}^2$ by $\varphi_S(p) = \varphi_S(p^1, p^2, p^3) = \left(\frac{p^1}{1-p^3}, \frac{p^2}{1-p^3} \right)$. Thus, φ_S carries p onto the intersection with the horizontal ($xy-$) plane of the straight line joining (0,0,1) and p (see Figure 0.3.1). φ_S is continuous,

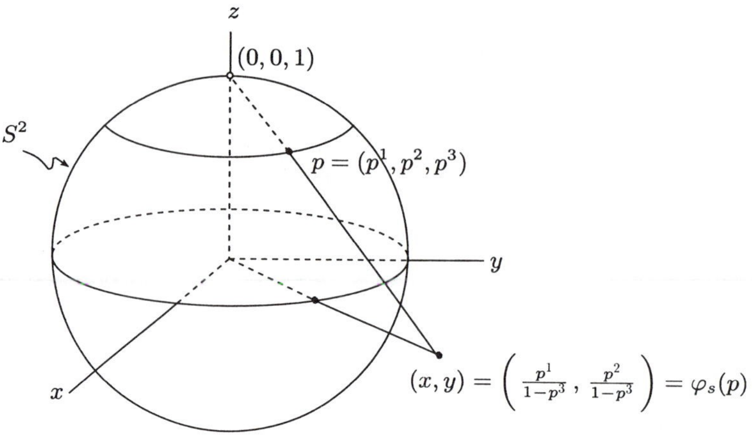

Figure 0.3.1

one-to-one and onto and has an inverse $\varphi_S^{-1} : \mathbb{R}^2 \to U_S$ given by

$$\begin{aligned} \varphi_S^{-1}(x,y) &= \left(\frac{2x}{x^2+y^2+1}, \frac{2y}{x^2+y^2+1}, \frac{x^2+y^2-1}{x^2+y^2+1}\right) \\ &= \left(\frac{z+\bar{z}}{z\bar{z}+1}, \frac{z-\bar{z}}{\mathbf{i}(z\bar{z}+1)}, \frac{z\bar{z}-1}{z\bar{z}+1}\right), \end{aligned} \tag{0.3.1}$$

where, for the second equality, we have identified (x, y) with the complex number $z = x + \mathbf{i}y$. φ_S^{-1} is also continuous so φ_S is a "homeomorphism" (continuous bijection with a continuous inverse). In fact, φ_S^{-1} is clearly C^∞ as a map from $\mathbb{R}^2$ into $\mathbb{R}^3$ and φ_S is the restriction to U_S of a C^∞ map from $\mathbb{R}^3$ minus the positive z-axis into $\mathbb{R}^2$. Thus, φ_S is actually a "diffeomorphism" (C^∞ bijection with a C^∞ inverse). Identifying the xy-plane with the complex plane $\mathbb{C}$ one can, in the usual way, adjoin a "point at infinity" to obtain the extended complex plane $\mathbb{C}^* = \mathbb{C} \cup \{\infty\}$ (the precise mechanism for doing this is called "1-point compactification" and will be described in Section 1.4). One can then extend φ_S in the obvious way to a homeomorphism $\varphi_S^* : S^2 \to \mathbb{C}^*$ (in this guise, S^2 is the Riemann sphere of complex analysis).

Similarly, one can define a stereographic projection from the south pole of S^2: Let $U_N = S^2 - (0, 0, -1)$ and define $\varphi_N : U_N \to \mathbb{R}^2$ by

$$\varphi_N(p) = \varphi_N(p^1, p^2, p^3) = \left(\frac{p^1}{1+p^3}, \frac{p^2}{1+p^3}\right).$$

We leave it to the reader to mimic our discussion of φ_S for φ_N and to show that, if $p \in U_S \cap U_N$ and $\varphi_S(p) = z$, then $\varphi_N(p) = 1/\bar{z}$.

One can define the 3-sphere S^3 in an entirely analogous manner as the set of $p = (p^1, p^2, p^3, p^4)$ in $\mathbb{R}^4$ with $\| p \|^2 = (p^1)^2 + (p^2)^2 + (p^3)^2 + (p^4)^2 = 1$. However, for our purposes it will be much more convenient to identify $\mathbb{R}^4$ with $\mathbb{C}^2$ via the correspondence $(x^1, y^1, x^2, y^2) \leftrightarrow (x^1 + \mathbf{i}y^1, x^2 + \mathbf{i}y^2)$ and take

$$S^3 = \{ (z^1, z^2) \in \mathbb{C}^2 : \ |z^1|^2 + |z^2|^2 = 1 \} ,$$

where $|z| = |x + \mathbf{i}y| = \sqrt{x^2 + y^2}$ is the modulus of z. A useful parametrization of the points of S^3 is obtained as follows: Let $z^1 = r_1 e^{\mathbf{i}\xi_1}$ and $z_2 = r_2 e^{\mathbf{i}\xi_2}$. Then, since ${r_1}^2 + {r_2}^2 = 1$ and r_1 and r_2 are both non-negative, there is some ϕ in $[0, \pi]$ such that $r_1 = \cos(\phi/2)$ and $r_2 = \sin(\phi/2)$. Thus,

$$S^3 = \left\{ (\cos\frac{\phi}{2} e^{\mathbf{i}\xi_1}, \sin\frac{\phi}{2} e^{\mathbf{i}\xi_2}) : \ 0 \le \frac{\phi}{2} \le \frac{\pi}{2}, \ \xi_1, \xi_2 \in \mathbb{R} \right\}.$$

We attempt a "visualization" of S^3 along the following lines: First note that, just as for S^2, we may regard S^3 as the 1-point compactification $(\mathbb{R}^3)^* = \mathbb{R}^3 \cup \{\infty\}$ of $\mathbb{R}^3$ via some stereographic projection (from, say, (0,0,0,1)). Now consider the subset T of S^3 defined by $T = \{ (z^1, z^2) \in S^3 : |z^1| = |z^2| \}$. Then $|z^1|^2 + |z^2|^2 = 1$ and $|z^1| = |z^2|$ imply $|z^1| = |z^2| = \sqrt{2}/2$ (so $\phi/2 = \pi/4$) and therefore

$$T = \left\{ (\frac{\sqrt{2}}{2} e^{\mathbf{i}\xi_1}, \frac{\sqrt{2}}{2} e^{\mathbf{i}\xi_2}) : \ \xi_1, \xi_2 \in \mathbb{R} \right\}$$

and this is clearly a copy of the torus (a torus is, after all, just a Cartesian product of two circles, one prescribing a lattitude, the other a longitude; see Section 1.3 for a more detailed discussion if you wish). Next let $K_1 = \{ (z^1, z^2) \in S^3 : \ |z^1| \le |z^2| \}$. Now, $|z^1| \le |z^2|$ implies $\cos(\phi/2) \le \sin(\phi/2)$ so $\pi/4 \le \phi/2 \le \pi/2$. $\phi/2 = \pi/4$ gives the torus $T \subseteq K_1$. $\phi/2 = \pi/2$ gives $z^1 = 0$ with z^2 on the unit circle so this is $\{0\} \times S^1$, a copy of the circle. Any fixed $\phi/2$ in $(\pi/4, \pi/2)$ gives another torus (just as for T above) so K_1 is a solid torus with boundary T. View this as layers of 2-dimensional tori beginning with T and collapsing onto a central circle as $\phi/2$ increases from $\pi/4$ to $\pi/2$ (Figure 0.3.2). Next let $K_2 = \{ (z^1, z^2) \in S^3 : \ |z^1| \ge |z^2| \}$. This is the subset of S^3 corresponding to $0 \le \phi/2 \le \pi/4$ which, just as for K_1, is a solid torus bounded by T with layers of 2-dimensional tori collapsing onto a central circle $S^1 \times \{0\}$. Thus, $S^3 = K_1 \cup K_2$ expresses S^3 as the union of two solid tori which intersect only along their common boundary T. To fit all of this into a single picture we begin with the central circle of K_1 ($\phi/2 = \pi/2$) and, as $\phi/2$ decreases from $\pi/2$ to $\pi/4$, expand through 2-dimensional tori out to T. As $\phi/2$ continues to decrease from $\pi/4$

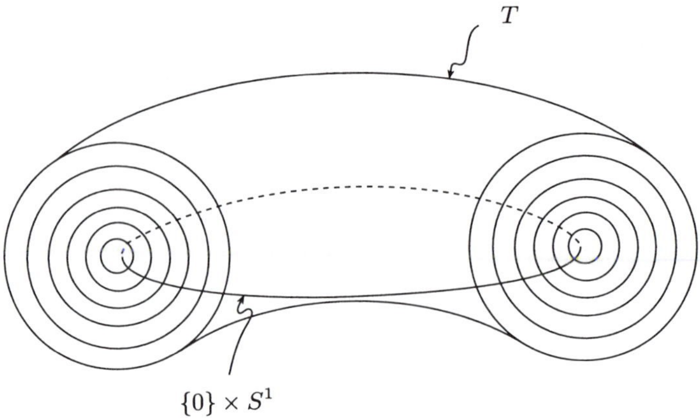

Figure 0.3.2

to 0 one obtains K_2, layered with 2-dimensional tori that expand to what appears to be a straight line in $\mathbb{R}^3$, but is actually a circle in S^3 through the point at infinity (see Figure 0.3.3).

Now let $p = (z^1, z^2)$ be in S^3 and g in $U(1)$ (recall that $U(1)$ is just S^1,

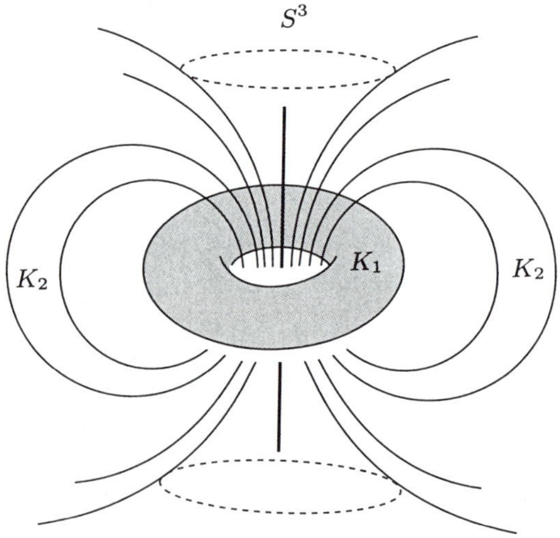

Figure 0.3.3

but it now begins to assume its role as the "gauge group" so we opt for the more traditional notation in this context). Observe that if we define

$$p \cdot g = (z^1, z^2) \cdot g = (z^1 g, z^2 g),$$

then $p \cdot g$ is also in S^3. Writing matters out in coordinates makes it clear that the map $(p, g) \to p \cdot g$ of $S^3 \times U(1)$ to S^3 is C^∞. Moreover, if $g_1, g_2 \in U(1)$ and if we denote by e the identity element $e^{\mathbf{i}0}$ in $U(1)$, it is clear that

$$(p \cdot g_1) \cdot g_2 = p \cdot (g_1 g_2) \quad \text{and} \quad p \cdot e = p$$

for all p in S^3. These few properties qualify the map $(p, g) \to p \cdot g$ as what we will later (Sections 1.6 and 4.3) call a "(C^∞) right action of a Lie group ($U(1)$) on a manifold (S^3)". For any fixed $p \in S^3$ we define the **orbit** of p under this action to be the subset $\{p \cdot g : g \in U(1)\}$ of S^3 obtained by letting everything in $U(1)$ act on p. The orbit of p surely contains p and is, in fact, just a copy of the circle S^1 inside S^3 and through p. Indeed, one can show that each orbit is a **great circle** on S^3 (i.e., the intersection of $S^3 \subseteq \mathbb{R}^4$ with a 2-dimensional plane in $\mathbb{R}^4$) which actually lies on one of the tori mentioned above that layer S^3 (see Figure 0.3.4).

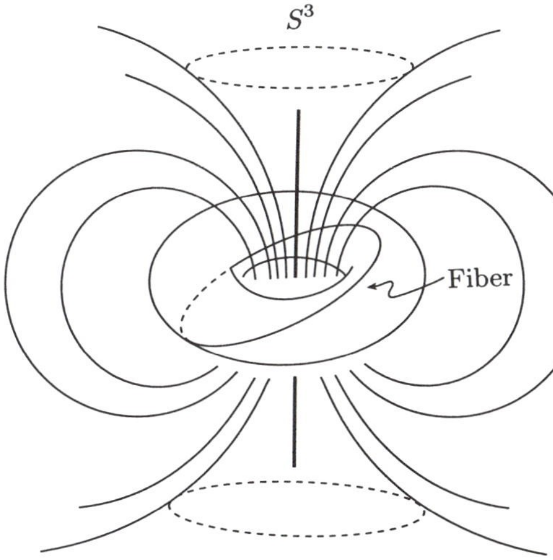

Figure 0.3.4

These orbits (for various $p \in S^3$) are easily seen to be either disjoint or identical and so, since they exhaust all of S^3, one can define an equivalence relation on S^3 whose equivalence classes are precisely the orbits.

A very fruitful attitude in mathematics is that objects that are "equivalent" (in some sense) should not be distinguished, i.e., should be thought of as "really" the same object. Isomorphic vector spaces spring immediately to mind, but it goes much deeper than this. We will encounter this phenomenon of "identifying" distinct objects in many contexts as we proceed (quotient groups, quotient topologies, etc.). For the present we wish only to indicate how one can model an entire orbit in S^3 by a single point in another (quotient) space and how the structure of S^3 is thereby elucidated. It's really quite simple. Note first that two points (z^1, z^2) and (w^1, w^2) of S^3 that lie in the same orbit have ratios z^1/z^2 and w^1/w^2 that are the same extended complex number (as usual, we take z^1/z^2 to be $\infty \in \mathbb{C}^*$ if $z^2 = 0$). The converse is a simple exercise in complex arithmetic: If $(z^1, z^2), (w^1, w^2) \in S^3$ and $z^1/z^2 = w^1/w^2$, then there exists a $g \in U(1)$ such that $(w^1, w^2) = (z^1 g, z^2 g) = (z^1, z^2) \cdot g$. Thus, the orbits are in one-to-one correspondence with the elements of $\mathbb{C}^*$ and these, via stereographic projection, are in one-to-one correspondence with the elements of S^2. Let us make this a bit more formal: Define a map $\mathcal{P} : S^3 \to S^2$ by

$$\mathcal{P}(z^1, z^2) = (\varphi_S^*)^{-1}\left(\frac{z^1}{z^2}\right).$$

A bit of arithmetic with (0.3.1) gives

$$\mathcal{P}(z^1, z^2) = \left(z^1\bar{z}^2 + \bar{z}^1 z^2,\ -\mathbf{i}\, z^1\bar{z}^2 + \mathbf{i}\,\bar{z}^1 z^2,\ |z^1|^2 - |z^2|^2 \right) \tag{0.3.2}$$

for all $(z^1, z^2) \in S^3$. For the record we write this out in terms of real coordinates given by $(z^1, z^2) = (x^1 + \mathbf{i}\, y^1,\ x^2 + \mathbf{i}\, y^2)$,

$$\begin{aligned}\mathcal{P}(x^1, y^1, x^2, y^2) = \big(& 2x^1x^2 + 2y^1y^2,\ 2x^2y^1 - 2x^1y^2, \\ & (x^1)^2 + (y^1)^2 - (x^2)^2 - (y^2)^2 \big),\end{aligned} \tag{0.3.3}$$

and in terms of the parameters ϕ, ξ_1 and ξ_2 given by $(z^1, z^2) = (\cos(\phi/2)\, e^{\mathbf{i}\xi_1}, \sin(\phi/2)\, e^{\mathbf{i}\xi_2})$,

$$\mathcal{P}(\phi, \xi_1, \xi_2) = (\sin\phi\cos(\xi_1 - \xi_2),\ \sin\phi\sin(\xi_1 - \xi_2),\ \cos\phi). \tag{0.3.4}$$

Notice that, letting $\xi_1 - \xi_2 = \theta$, $\mathcal{P}$ carries (ϕ, ξ_1, ξ_2) onto

$$\mathcal{P}(\phi, \xi_1, \xi_2) = (\sin\phi\cos\theta,\ \sin\phi\sin\theta,\ \cos\phi), \tag{0.3.5}$$

i.e., the point of S^2 with standard spherical coordinates $(\phi, \theta) = (\phi, \xi_1 - \xi_2)$.

The map from $\mathbb{R}^4$ to $\mathbb{R}^3$ defined by (0.3.3) is obviously C^∞ (quadratic coordinate functions) and so $\mathcal{P}$ (its restriction to S^3) is C^∞. $\mathcal{P}$ maps S^3 onto S^2 and, for any $x \in S^2$, the **fiber** $\mathcal{P}^{-1}(x)$ of $\mathcal{P}$ above x is the orbit $\{(z^1, z^2) \cdot g : g \in U(1)\}$ of any (z^1, z^2) with $\mathcal{P}(z^1, z^2) = x$. Thus, in a precise sense, $\mathcal{P} : S^3 \to S^2$ "identifies" each orbit in S^3 with a single point in S^2.

Remarks: Hopf's construction of $\mathcal{P} : S^3 \to S^2$ was motivated by his interest in what are called the "higher homotopy groups" of the spheres (see Section 2.5). Although this is not our major concern at the moment, we point out that $\mathcal{P}$ was the first example of a continuous map $S^m \to S^n$

with $m > n$ that is not "nullhomotopic" (Section 2.3). From this it follows that the homotopy group $\pi_3(S^2)$ is not trivial and this came as quite a surprize in the 1930's.

Our real interest in $\mathcal{P} : S^3 \to S^2$ for the present resides in certain additional structure that we now proceed to describe. Specifically, we show that $\mathcal{P}$ provides S^3 with the structure of a "principal $U(1)$-bundle over S^2". In order to give the discussion some focus we would like to record the general definition of a principal bundle, although some of the terms used will not be formally introduced for some time. For each such term we have included below a parenthetical reference to the object with which it should be identified in our present context.

Let X be a manifold (e.g., S^2) and G a Lie group (e.g., $U(1)$). A $\mathbf{C^\infty}$ **(smooth) principal bundle over** X with **structure group** G (or, simply, a **smooth** G**-bundle over** X) consists of a manifold P (e.g., S^3), a smooth map $\mathcal{P}$ of P onto X and a smooth right action $(p, g) \to p \cdot g$ of G on P (e.g., $((z^1, z^2), g) \to (z^1, z^2) \cdot g = (z^1 g, z^2 g))$, all of which satisfy the following conditions:

1. The action of G on P preserves the fibers of $\mathcal{P}$, i.e.,

$$\mathcal{P}(p \cdot g) = \mathcal{P}(p) \text{ for all } p \in P \text{ and all } g \in G, \tag{0.3.6}$$

2. **(Local Triviality)** For each x_0 in X there exists an open set V containing x_0 and a diffeomorphism $\Psi : \mathcal{P}^{-1}(V) \to V \times G$ of the form $\Psi(p) = (\mathcal{P}(p), \psi(p))$, where $\psi : \mathcal{P}^{-1}(V) \to G$ satisfies

$$\psi(p \cdot g) = \psi(p)g \text{ for all } p \in \mathcal{P}^{-1}(V) \text{ and all } g \in G. \tag{0.3.7}$$

P is called the **bundle space**, X the **base space** and $\mathcal{P}$ the **projection** of the bundle. For the example we have under consideration, condition (1) follows at once from the way in which we defined $\mathcal{P} : S^3 \to S^2$. Thus, we need only verify the local triviality condition (2). The motivation here is as follows: Notice the analogy between the Hopf map $\mathcal{P} : S^3 \to S^2$ and the standard projection $S^2 \times U(1) \to S^2$ of the Cartesian product $S^2 \times U(1)$ onto its first factor. Each is a smooth map onto S^2 and each has the property that its fibers slice the domain up into a disjoint union of circles, one "above" each point of S^2. However, the circles are "glued together" differently in S^3 and $S^2 \times U(1)$ (more precisely, these two are *not* homeomorphic, a fact we will prove in Section 2.4 by showing that they have different fundamental groups). The thrust of condition (2) is that, while S^3 and $S^2 \times U(1)$ are not *globally* the same, they are, in fact, *locally* diffeomorphic, and in a way that respects the group action.

The proof is actually quite simple. We consider again the open sets $U_S = S^2 - (0,0,1)$ and $U_N = S^2 - (0,0,-1)$ on S^2. These two together cover every point in S^2. Moreover, $\mathcal{P}^{-1}(U_S) = \{(z^1, z^2) \in S^3 : z^2 \neq 0\}$ and $\mathcal{P}^{-1}(U_N) = \{(z^1, z^2) \in S^3 : z^1 \neq 0\}$ (each is the complement in S^3 of one of the two limiting circles, i.e., degenerate tori, discussed earlier). Define maps $\Psi_S : \mathcal{P}^{-1}(U_S) \to U_S \times U(1)$ and $\Psi_N : \mathcal{P}^{-1}(U_N) \to U_N \times U(1)$ by $\Psi_S(z^1, z^2) = (\mathcal{P}(z^1, z^2), z^2/|z^2|)$, and $\Psi_N(z^1, z^2) = (\mathcal{P}(z^1, z^2), z^1/|z^1|)$. Written out in real coordinates these are easily seen to be C^∞ on their respective domains. The easiest way to see that they are diffeomorphisms is to simply write down their smooth inverses $\Psi_S^{-1} : U_S \times U(1) \to \mathcal{P}^{-1}(U_S)$ and $\Psi_N^{-1} : U_N \times U(1) \to \mathcal{P}^{-1}(U_N)$. These are easily seen to be given by

$$\Psi_S^{-1}(x, g) = (z^1, z^2) \cdot \left(g \frac{|z^2|}{z^2} \right) \quad \text{and} \quad \Psi_N^{-1}(x, g) = (z^1, z^2) \cdot \left(g \frac{|z^1|}{z^1} \right),$$

where, in each case, (z^1, z^2) is *any* element of $\mathcal{P}^{-1}(x)$. Observe that $\Psi_S(z^1, z^2) = (\mathcal{P}(z^1, z^2), \psi_S(z^1, z^2))$, where $\psi_S(z^1, z^2) = z^2/|z^2|$ and ψ_S satisfies $\psi_S((z^1, z^2) \cdot g) = \psi_S(z^1 g, z^2 g) = z^2 g / |z^2 g| = (z^2/|z^2|)g = \psi_S(z^1, z^2)g$, which is the property required in the local triviality condition (2). Similarly, $\Psi_N(z^1, z^2) = (\mathcal{P}(z^1, z^2), \psi_N(z^1, z^2))$, where $\psi_N(z^1, z^2) = z^1/|z^1|$ satisfies $\psi_N((z^1, z^2) \cdot g) = \psi_N(z^1, z^2)g$. Local triviality has therefore been established and we are free to refer to $\mathcal{P} : S^3 \to S^2$ as the **Hopf bundle**.

The Hopf bundle is a principal $U(1)$-bundle over S^2. We have actually already encountered another specimen of this same species. The standard projection of the product $S^2 \times U(1)$ onto S^2 with the obvious action of $U(1)$ on $S^2 \times U(1)$ (i.e., $((x, g_1), g_2) \to (x, g_1 g_2)$) trivially satisfies the required conditions (1) and (2) (for (2) one can take V to be all of S^2). This is called the **trivial $U(1)$-bundle over S^2**. Any principal $U(1)$-bundle over S^2 is *locally* the same as this trivial bundle. There are lots of examples and we will eventually (Section 3.4) find a complete classification of them in terms of the fundamental group $\pi_1(U(1))$ of the circle. The diffeomorphisms $\Psi : \mathcal{P}^{-1}(V) \to V \times U(1)$ are called **local trivializations** of the bundle and the V's are called **trivializing neighborhoods**. What distinguishes one such bundle from another is how these local trivializations overlap on the intersections of their domains. One can keep track of this by computing what are called the "transition functions" of the bundle. This we now do for the Hopf bundle.

Fix an $x \in U_S \cap U_N$ and consider the fiber $\mathcal{P}^{-1}(x)$ above x in S^3. Ψ_N identifies $\mathcal{P}^{-1}(x)$ with a copy of $U(1)$ via ψ_N and Ψ_S does the same via ψ_S. Let $\psi_{S,x}, \psi_{N,x} : \mathcal{P}^{-1}(x) \to U(1)$ be defined by $\psi_{S,x} = \psi_S | \mathcal{P}^{-1}(x)$ and $\psi_{N,x} = \psi_N | \mathcal{P}^{-1}(x)$. Then $\psi_{S,x} \circ \psi_{N,x}^{-1} : U(1) \to U(1)$ is a diffeomorphism that describes the relationship between the ways in which $U(1)$ is "glued onto" $\mathcal{P}^{-1}(x)$ by the two local trivializations Ψ_S and Ψ_N (see Figure 0.3.5). By selecting any $(z^1, z^2) \in \mathcal{P}^{-1}(x)$ and manipulating our definitions a bit

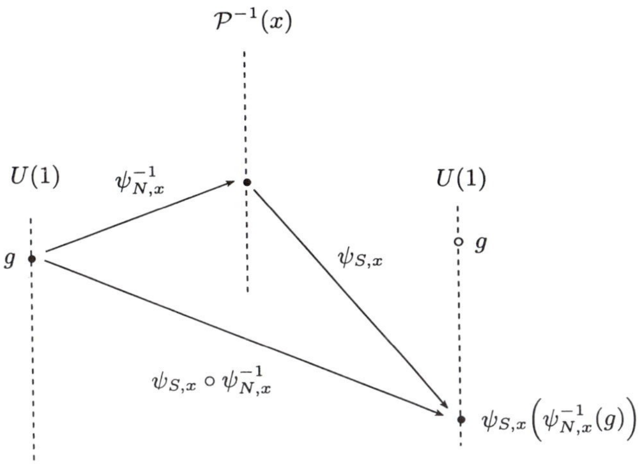

Figure 0.3.5

one soon finds that

$$\psi_{S,x} \circ \psi_{N,x}^{-1}(g) = \left(\frac{z^2/\,|z^2|}{z^1/\,|z^1|}\right) g$$

for all g in $U(1)$. Similarly,

$$\psi_{N,x} \circ \psi_{S,x}^{-1}(g) = \left(\frac{z^1/\,|z^1|}{z^2/\,|z^2|}\right) g\,.$$

Now, both $(z^2/\,|\,z^2\,|)/(z^1/\,|\,z^1\,|)$ and $(z^1/\,|\,z^1\,|)/(z^2/\,|\,z^2\,|)$ are elements of $U(1)$ (remember that $x \in U_S \cap U_N$) so we may define two maps g_{SN} : $U_N \cap U_S \to U(1)$ and $g_{NS} : U_S \cap U_N \to U(1)$ by

$$g_{SN}(x) = \frac{z^2/\,|z^2|}{z^1/\,|z^1|} \quad \text{and} \quad g_{NS}(x) = \frac{z^1/\,|z^1|}{z^2/\,|z^2|} = (\,g_{SN}(x)\,)^{-1}$$

for any $(z^1, z^2) \in \mathcal{P}^{-1}(x)$. Then we find that

$$\psi_{S,x} \circ \psi_{N,x}^{-1}(g) = g_{SN}(x)g \quad \text{and} \quad \psi_{N,x} \circ \psi_{S,x}^{-1}(g) = g_{NS}(x)g\,.$$

These maps $\{g_{SN}, g_{NS}\}$ are called the **transition functions** of the Hopf bundle and we will eventually show that they completely characterize it among the principal $U(1)$-bundles over S^2 (Section 3.4). These maps are particularly attractive when written in terms of the parameters ϕ, ξ_1 and

ξ_2 given by $z^1 = \cos(\phi/2)e^{\mathbf{i}\xi_1}$ and $z^2 = \sin(\phi/2)e^{\mathbf{i}\xi_2}$. Letting $\xi_1 - \xi_2 = \theta$ as in (0.3.5) we find that

$$g_{SN}(\sin\phi\cos\theta,\ \sin\phi\sin\theta,\ \cos\phi) = e^{-\mathbf{i}\theta} \tag{0.3.8}$$

$$g_{NS}(\sin\phi\cos\theta,\ \sin\phi\sin\theta,\ \cos\phi) = e^{\mathbf{i}\theta}\,. \tag{0.3.9}$$

0.4 Connections on Principal Bundles

Perhaps we should pause to recapitulate. Section 0.2 ended with some rather vague mutterings about an appropriate replacement for the classical vector potential of a monopole consisting of some sort of "bundle of circles above S^2" and a procedure for lifting paths in S^2 to that bundle space. In Section 0.3 we found that such bundles actually arise in nature (so to speak) and are of considerable importance in areas not (apparently) related to mathematical physics. However, we also saw that there are, in fact, many different ways to construct such circle bundles over the 2-sphere and it is not clear how one should make a selection from among these. But here's a coincidence for you. Monopole field strengths g are "quantized" (Dirac Quantization Condition). In effect, there is one monopole for each integer (assuming there are any monopoles at all, of course). On the other hand, we have also pointed out that the principal $U(1)$-bundles over S^2 are classified by the elements of the fundamental group $\pi_1(U(1))$ of the circle and, as we will prove in Section 2.4, this is just the group of integers. In effect, there is one principle $U(1)$-bundle over S^2 for each integer. This tantalizing one-to-one correspondence between monopoles and principal $U(1)$-bundles over S^2 suggests that the monopole strength may dictate the choice of the bundle with which to model it. Precisely how this choice is dictated is to be found in the details of the path lifting procedure to which we have repeatedly alluded. We will consider here only the simplest nontrivial case.

The Dirac Quantization Condition (0.2.9) asserts that, for any charge q and any monopole strength g, one must have $qg = (1/2)n$ for some integer n. For a charge of unit strength ($q = 1$) this becomes $g = (1/2)n$ so that the smallest positive value for g (in the units we have tacitly adopted) is

$$g = \frac{1}{2}\,. \tag{0.4.1}$$

For this case, the potential 1-forms for the monopole are

$$\boldsymbol{A}_N = \frac{1}{2}(1-\cos\phi)d\theta \quad \text{on} \quad U_N \subseteq S^2\,, \tag{0.4.2}$$

and

$$\boldsymbol{A}_S = -\frac{1}{2}(1+\cos\phi)d\theta \quad \text{on} \quad U_S \subseteq S^2\,. \tag{0.4.3}$$

($\boldsymbol{A}_N = \boldsymbol{A}_+ \,|U_N$ and $\boldsymbol{A}_S = \boldsymbol{A}_- \,|U_S$). Thus, on $U_S \cap U_N$, $\boldsymbol{A}_N - \boldsymbol{A}_S = d\theta$ so

$$\boldsymbol{A}_N = \boldsymbol{A}_S + d\theta \quad \text{on} \quad U_S \cap U_N \,. \tag{0.4.4}$$

At this point we must beg the reader's indulgence. We are about to do something which is (quite properly) considered to be in poor taste. We are going to introduce what will appear to be a totally unnecessary complication. For reasons that we will attempt to explain once the deed is done, we replace the real-valued 1-forms $\boldsymbol{A}_N$ and $\boldsymbol{A}_S$ by the pure imaginary 1-forms $\boldsymbol{\mathcal{A}}_N$ and $\boldsymbol{\mathcal{A}}_S$ defined by

$$\boldsymbol{\mathcal{A}}_N = -\mathbf{i}\,\boldsymbol{A}_N \quad \text{on} \quad U_N \quad \text{and} \quad \boldsymbol{\mathcal{A}}_S = -\mathbf{i}\,\boldsymbol{A}_S \quad \text{on} \quad U_S \,. \tag{0.4.5}$$

Now, (0.4.4) becomes $\boldsymbol{\mathcal{A}}_N = \boldsymbol{\mathcal{A}}_S - \mathbf{i}\,d\theta$ which, for no apparent reason at all, we prefer to write as

$$\boldsymbol{\mathcal{A}}_N = e^{\mathbf{i}\theta}\,\boldsymbol{\mathcal{A}}_S\,e^{-\mathbf{i}\theta} + e^{\mathbf{i}\theta}\,de^{-\mathbf{i}\theta} \,. \tag{0.4.6}$$

All of this is algebraically quite trivial, of course, but the motivation is no doubt obscure (although one cannot help but notice that the transition functions for the Hopf bundle (0.3.8) and (0.3.9) have put in an appearence). Keep in mind that our purpose in this preliminary chapter is to illustrate with the simplest case the general framework of gauge theory and that the process of generalization often requires that the instance being generalized undergo some cosmetic surgery first (witness the derivative of $f : \mathbb{R} \to \mathbb{R}$ at $a \in \mathbb{R}$ as a *number* $f'(a)$, versus the derivative of $f : \mathbb{R}^n \to \mathbb{R}^m$ at $a \in \mathbb{R}^n$ as a *linear transformation* $Df_a : \mathbb{R}^n \to \mathbb{R}^m$). The process which led to the appropriate generalization in our case was particularly long and arduous and did not reach fruition until the 1950's with the work of Ehresmann [**Ehr**]. Ehresmann was attempting to generalize to the context of bundles such classical notions from differential geometry as "connection", "parallel translation" and "curvature", all of which had been elegantly formulated by Elie Cartan in terms of the so-called "frame bundle". There are, in fact, three different ways of describing the generalization that eventually materialized, all of which will be discussed in detail in Chapter 5. One of these deals directly with path lifting procedures, another with "distributions" on the bundle space and the third with "Lie algebra-valued 1-forms". The details here are sufficiently technical that even a brief synopsis of the general situation would, we feel, only serve to muddy the waters. In the case of immediate concern to us, however, it is possible to gain some intuitive appreciation of what is going on.

Locally defined 1-forms on S^2 (such as $\boldsymbol{A}_N$ and $\boldsymbol{A}_S$) cannot, unless they *agree* on the intersection of their domains, be spliced together into a globally defined 1-form on all of S^2. The essence of Ehresmann's construction is that locally defined "Lie algebra-valued 1-forms on S^2" can, if they satisfy a certain *consistency condition* on the intersection of their domains, always be spliced together into a globally defined "Lie algebra-valued 1-form" on a principal $U(1)$-bundle over S^2. The consistency condition involves the transition functions of the bundle.

So, what exactly is a "Lie algebra-valued 1-form" and why should anyone care? We will eventually (Section 4.7) show that any Lie group G (e.g., $U(1)$) has associated with it an algebraic object called its Lie algebra $\mathcal{G}$. One can think of $\mathcal{G}$ simply as the tangent space at the identity element in G. For $U(1)$ this object is generally identified with the set of pure imaginary numbers $\mathrm{Im}\,\mathbb{C} = \{\mathbf{i}\theta : \theta \in \mathbb{R}\}$ with the vector space structure it inherits from $\mathbb{C}$ (the tangent space to the circle at 1 is, of course, just a copy of $\mathbb{R}$, but the isomorphic space $\mathrm{Im}\,\mathbb{C}$ is more convenient because its elements can be "exponentiated" to give the elements $e^{\mathbf{i}\theta}$ of $U(1)$). In our context, therefore, a Lie algebra-valued 1-form is simply a pure imaginary-valued 1-form and these differ trivially from ordinary 1-forms. Now, suppose that one is given a principal $U(1)$-bundle over S^2. We will eventually (Section 3.4) see that one can always take the trivializing neighborhoods to be U_N and U_S. Suppose also that one has two Lie algebra-valued 1-forms $\boldsymbol{\mathcal{A}}_1$ and $\boldsymbol{\mathcal{A}}_2$ defined on U_N and U_S, respectively, and that, on $U_S \cap U_N$, they are related by

$$\boldsymbol{\mathcal{A}}_2 = g_{12}^{-1}\boldsymbol{\mathcal{A}}_1 g_{12} + g_{12}^{-1}dg_{12}\,, \tag{0.4.7}$$

where $g_{12} : U_S \cap U_N \to U(1)$ is the corresponding transition function of the bundle (while it may seem silly not to cancel the g_{12}^{-1} and g_{12} in the first term, this would require commuting one of the products and this will not be possible for the non-Abelian generalizations we have in mind). Then Ehresmann's result asserts that $\boldsymbol{\mathcal{A}}_1$ and $\boldsymbol{\mathcal{A}}_2$ determine a unique Lie algebra-valued 1-form $\boldsymbol{\omega}$ on the entire bundle space. Although it would not be cost-effective at this point to spell out precisely how $\boldsymbol{\mathcal{A}}_1$ and $\boldsymbol{\mathcal{A}}_2$ "determine" $\boldsymbol{\omega}$, for the cogniscenti we point out that $\boldsymbol{\omega}$ "pulls back" to $\boldsymbol{\mathcal{A}}_1$ and $\boldsymbol{\mathcal{A}}_2$ via the natural local cross-sections on U_N and U_S. This said, we observe that (0.4.6) is just (0.4.7) with the transition function of the Hopf bundle. We conclude that the monopole potentials $\boldsymbol{\mathcal{A}}_S$ and $\boldsymbol{\mathcal{A}}_N$ uniquely determine a Lie algebra-valued 1-form $\boldsymbol{\omega}$ on S^3. Although we will not go into the details at the moment, Ehresmann's result also guarantees that $\boldsymbol{\omega}$ is related in various natural ways to the structure of the bundle (group action, projection, etc.).

But still, why should anyone care? What does this have to do with the path lifting procedure we are in search of for the monopole? By way of explanation we offer the following (admittedly rather terse) synopsis of Section 5.1 for the special case of the Hopf bundle. The Lie algebra $\mathrm{Im}\,\mathbb{C}$ of $U(1)$ is the tangent space to the circle at $1 \in U(1)$. The tangent space to any point in $U(1)$ can be identified with the Lie algebra by "rotating" it around the circle. Now, each point p in S^3 (the bundle space of the Hopf bundle) has through it a fiber of the Hopf map, diffeomorphic to $U(1)$. The

3-dimensional tangent space $T_p(S^3)$ to S^3 at p, consisting of all velocity vectors at p to smooth curves in S^3 through p, therefore has a subspace isomorphic to the Lie algebra of $U(1)$ (vectors tangent to the fiber). Now, just as an ordinary 1-form is defined precisely as a real-valued operator on tangent vectors (Section 0.2), so a Lie algebra-valued 1-form is an operator carrying tangent vectors to elements of the Lie algebra. Thus, a Lie algebra-valued 1-form $\boldsymbol{\omega}$ on S^3 assigns to every $p \in S^3$ a linear transformation $\boldsymbol{\omega}_p$ from $T_p(S^3)$ to the copy of the Lie algebra Im $\mathbb{C}$ inside $T_p(S^3)$ (think of $\boldsymbol{\omega}_p$ as a kind of "projection"). The kernel $\ker \boldsymbol{\omega}_p$ of this map is a 2-dimensional subspace of $T_p(S^3)$ (as p varies over S^3 the kernels $\ker \boldsymbol{\omega}_p$ collectively determine what is called a 2-dimensional "distribution" on S^3). One can show that $\mathcal{P} : S^3 \rightarrow S^2$ (or, more precisely, its derivative at p) carries $\ker \boldsymbol{\omega}_p$ isomorphically onto the tangent plane $T_{\mathcal{P}(p)}(S^2)$ to S^2 at $\mathcal{P}(p)$. Now, along a smooth curve in S^2 each velocity vector lifts by one of these isomorphisms to a unique vector setting "above" in S^3. Since everything in sight is smooth these lifted vectors can, given an initial condition, be "fitted" with a unique integral curve that lifts the original curve in S^2.

These Lie algebra-valued 1-forms (or the corresponding path lifting procedures, or the corresponding distributions on the bundle space) are called **connections** on the bundle (or, in the physics literature, **gauge potentials**). We conclude then that the Hopf bundle admits a connection $\boldsymbol{\omega}$ whose description in terms of local 1-forms on S^2 consists precisely of the (imaginary) potentials for the Dirac monopole field. The corresponding path lifting procedure to S^3 "does the job" of the classical vector potential for the monopole. The exterior derivative $\boldsymbol{\Omega} = d\boldsymbol{\omega}$ of this 1-form is called the **curvature** of the connection, or the **gauge field strength**, and corresponds ("pulls back") to the monopole field $-\mathbf{i}F$ on S^2 (in more general contexts, the curvature is gotten by computing the "covariant exterior derivative" of the connection form, but for $U(1)$-bundles this coincides with the usual exterior derivative). Monopoles of different strengths are modeled by connections on different $U(1)$-bundles over S^2. We have already observed that such bundles are in one-to-one correspondence with the elements of the fundamental group $\pi_1(U(1))$ of the circle, i.e., with the integers.

0.5 Non-Abelian Gauge Fields and Topology

You are sitting in a room with a friend and a ping-pong ball (perfectly spherical and perfectly white— the ping-pong ball, not the friend). The conversation gets around to Newtonian mechanics. You toss the ball to your friend. Both of you agree that, given the speed and direction of the toss, $\vec{F} = m\vec{A}$ and the formula for the gravitational attraction at the surface of the earth ($\vec{F} = -mg\vec{k}$, if the positive z-direction is up), you could calculate the motion of the ball, at least if air resistence is neglected. But then you ask your friend: "As the ball was traveling toward you, was it spinning?" "Not a fair question", he responds. After all, the ball is perfectly spherical and perfectly white. How is your friend supposed to know if it's spinning? And, besides, it doesn't matter anyway. The trajectory of the ball is determined entirely by the motion of its center of mass and we've already calculated that. Any internal spinning of the ball is irrelevant to its motion through space. Of course, this internal spinning might well be relevant in other contexts, e.g., if the ball interacts (collides) with another ping-pong ball traveling through the room. Moreover, if we believe in the conservation of angular momentum, any changes in the internal spin state of the ball would have to be accounted for by some force being exerted on it, such as its interaction with the atmosphere in the room, and we have, at least for the moment, neglected such interactions in our calculations. It would seem proper then to regard any intrinsic spinning of the ball about some axis as part of the "internal structure" of the ball, not relevant to its motion through space, but conceivably relevant in other situations.

The phase of a charged particle moving in an electromagnetic field (e.g., a monopole field) is quite like the internal spinning of our ping-pong ball. We have seen that a phase change alters the wavefunction of the charge only by a factor of modulus one and so does not effect the probability of finding the particle at any particular location, i.e., does not effect its motion through space. Nevertheless, when two charges interact (in, for example, the Aharonov-Bohm experiment), phase differences are of crucial significance to the outcome. The gauge potential (connection), which mediates phase changes in the charge along various paths through the electromagnetic field, is the analogue of the room's atmosphere, which is the agency ("force") responsible for any alteration in the ball's internal spinning.

The current dogma in particle physics is that elementary particles are distinguished, one from another, precisely by this sort of internal structure. A proton and a neutron, for example, are regarded as but two states of a single particle, differing only in the value of an "internal quantum number" called **isotopic spin**. In the absence of an electromagnetic (gauge) field with which to interact, they are indistinguishable. Each aspect of a particle's internal state is modeled, at each point in the particle's history, by some sort of mathematical object (a complex number of modulus one for the phase, a pair of complex numbers whose squared moduli sum to one for isotopic spin, etc.) and a group whose elements transform one state into another ($U(1)$ for the phase and, for isotopic spin, the group $SU(2)$ of complex 2×2 matrices that are unitary and have determinant one). A bundle is built in which to "keep track" of the particle's internal state (generally over a 4-dimensional manifold which can accomodate the particle's "history"). Finally, connections on the bundle are studied as models of those physical phenomena that can mediate changes in the internal state. Not all connections are of physical interest, of course, just as not all 1-forms represent realistic electromagnetic potentials. Those that are of interest satisfy a set of partial differential equations called the **Yang-Mills equations**, developed by Yang and Mills [**YM**] in 1954 as a nonlinear generalization of Maxwell's equations.

But if your interests are in mathematics and not particle physics, why should you care about any of this? There is, of course, the simple fact that the topology of the bundle, the geometry of the connections and the analysis of the partial differential equations are all deep and beautiful. There is more, however, especially for those who incline toward topology. Dimension four is quite special. For bundles over closed, oriented, Riemannian 4-manifolds (such as the 4-sphere S^4), one can isolate a class of connections, called **(anti-) self-dual**, that necessarily satisfy the Yang-Mills equations. The collection of all such, modulo a natural equivalence relation (gauge equivalence), is called the **moduli space** $\mathcal{M}$ of the bundle and its study (initiated by Simon Donaldson [**Don**]) has led to astonishing insights into the structure of smooth 4-manifolds. Although we are not so presumptuous as to view the following chapters as an introduction to Donaldson theory, we do feel that many of the salient features of that work are clearly visible in the particular example that we have chosen to examine in detail. We conclude our motivational chapter with a brief synopsis of how this example arises and what we intend to do with it.

The structure of the Hopf bundle is inextricably bound up with the properties of the complex numbers. The base S^2 is the extended complex plane, the fiber S^1 consists of the unit complex numbers and the total space S^3 can be thought of as those pairs of complex numbers in $\mathbb{C}^2 = \mathbb{R}^4$ whose squared moduli sum to 1. We will find that, like $\mathbb{R}^2$, Euclidean 4-space $\mathbb{R}^4$ admits a multiplicative structure, lacking only commutativity among the desirable properties of complex multiplication. This familiar quaternion structure on $\mathbb{R}^4$ permits the construction of an analogous Hopf bundle over S^4 (the 1-point compactification of $\mathbb{R}^4$) with fiber S^3 (homeomorphic to $SU(2)$) and total space $S^7 \subseteq \mathbb{R}^8$. Both Hopf bundles, complex and quaternionic, admit natural connections, the former being that associated with the Dirac monopole of lowest strength. Our primary interest, however, resides in the latter which, when written in terms of the natural trivializations of the bundle, gives rise to the famous instanton solutions to the Yang-Mills equations discovered by Belavin, Polyakov, Schwartz and Tyupkin in [**BPST**].

The BPST instantons were originally called *pseudoparticles* and were *not* viewed as the coordinate expressions for a globally defined connection on a bundle. Indeed, there is nary a bundle to be found in [**BPST**], where the perspective is the more traditional one of mathematical physics: Given a set of partial differential equations on $\mathbb{R}^4$ (those of Yang-Mills) for an object of interest (an $SU(2)$ gauge potential) one sets oneself the task of finding solutions that satisfy certain physically desirable asymptotic conditions. Only later was it shown (by Karen Uhlenbeck) that these asymptotic conditions suffice to guarantee the existence of a smooth extension of the solutions to the "point at infinity", i.e., to S^4. Somewhat more precisely, this remarkable *Removable Singularities Theorem* of Uhlenbeck [**Uhl**] asserts that, for any Yang-Mills potential on $\mathbb{R}^4$ with "finite action" (i.e., finite total field strength/curvature, computed as an integral over $\mathbb{R}^4$) there exists an $SU(2)$-principal bundle over S^4 and a connection on it which, when written in terms of some trivialization of the bundle is just the given potential. Moreover, it is the asymptotic behavior of the potential as $\| x \| \to \infty$ that determines the bundle on which this connection is defined so that these asymptotic boundary conditions are directly encoded in the topology. We will see that the behavior of the solutions found in [**BPST**] dictates the quaternionic Hopf bundle.

We arrive at the BPST instanton connections on the Hopf bundle by a different route (via the "Cartan canonical 1-form on $Sp(2)$"). Once these are in hand, however, it is a simple matter to use a basic property of the (anti-) self-dual equations they satisfy ("conformal invariance") to write down an entire 5-parameter family of such connections. A surprizing and very deep theorem of Atiyah, Hitchin and Singer [**AHS**], based on techniques from algebraic geometry and the Penrose "Twistor Program", asserts that every element of the moduli space $\mathcal{M}$ is uniquely represented by a connection in this 5-parameter family. From this one obtains a concrete realization of $\mathcal{M}$ as the open unit ball in $\mathbb{R}^5$. In particular, $\mathcal{M}$ is a 5-dimensional manifold with a natural compactification (the closed unit ball in $\mathbb{R}^5$) whose boundary is a copy of the base space S^4. Donaldson has shown that many features of this simple picture persist in a much more general context. Specifically, let X be a closed, oriented, simply connected, Riemannian 4-manifold with positive definite intersection form and $\mathcal{M}$ the analogous moduli space (here we are intentionally being a bit vague; see Section 5.5 for a more precise statement). Then there are finitely many points $p_1, \ldots, p_m$ in $\mathcal{M}$ such that $\mathcal{M} - \{p_1, \ldots, p_m\}$ is a smooth 5-dimensional manifold (m can be computed from the topology of X and the nature of $\mathcal{M}$ near each p_i, although not smooth, is known). Furthermore, $\mathcal{M}$ has a natural compactification $\bar{\mathcal{M}} = \mathcal{M} \cup X$ in which X appears as the boundary of $\mathcal{M}$. The topologies of $\mathcal{M}$ and X are therefore tightly intertwined and disentangling this relationship is the aim of gauge-theoretic techniques in topology.

1

Topological Spaces

1.1 Topologies and Continuous Maps

We begin by recording a few items from real analysis (our canonical reference for this material is [**Sp1**], Chapters 1-3, which should be consulted for details as the need arises). For any positive integer n, **Euclidean n-space** $\mathbb{R}^n = \{(x^1, \ldots, x^n) : x^i \in \mathbb{R},\ i = 1, \ldots, n\}$ is the set of all ordered n-tuples of real numbers with its usual vector space structure ($x + y = (x^1, \ldots, x^n) + (y^1, \ldots, y^n) = (x^1 + y^1, \ldots, x^n + y^n)$ and $ax = a(x^1, \ldots, x^n) = (ax^1, \ldots, ax^n)$) and norm ($\|x\| = ((x^1)^2 + \cdots + (x^n)^2)^{1/2}$). An **open rectangle in $\mathbb{R}^n$** is a subset of the form $(a_1, b_1) \times \cdots \times (a_n, b_n)$, where each (a_i, b_i), $i = 1, \ldots, n$, is an open interval in the real line $\mathbb{R}$. If r is a positive real number and $p \in \mathbb{R}^n$, then the **open ball of radius r about p** is $U_r(p) = \{x \in \mathbb{R}^n : \|x - p\| < r\}$. A subset U of $\mathbb{R}^n$ is **open in $\mathbb{R}^n$** if, for each $p \in U$, there exists an $r > 0$ such that $U_r(p) \subseteq U$ (equivalently, if, for each $p \in U$, there exists an open rectangle R in $\mathbb{R}^n$ with $p \in R \subseteq U$). The collection of all open subsets of $\mathbb{R}^n$ has the following properties: (a) The empty set $\emptyset$ and all of $\mathbb{R}^n$ are both open in $\mathbb{R}^n$. (b) If $\{U_\alpha : \alpha \in \mathcal{A}\}$ is any collection of open sets in $\mathbb{R}^n$ (indexed by some set $\mathcal{A}$), then the union $\bigcup_{\alpha \in \mathcal{A}} U_\alpha$ is also open in $\mathbb{R}^n$. (c) If $\{U_1, \ldots, U_k\}$ is any finite collection of open sets in $\mathbb{R}^n$, then the intersection $U_1 \cap \cdots \cap U_k$ is also open in $\mathbb{R}^n$. Moreover, one can prove that a map $f : \mathbb{R}^n \to \mathbb{R}^m$ from one Euclidean space to another is continuous if and only if (henceforth abbreviated "iff") $f^{-1}(U)$ is open in $\mathbb{R}^n$ for each open subset U of $\mathbb{R}^m$.

The notion of a topological space distills the essential features from this discussion of $\mathbb{R}^n$ and permits one to introduce a meaningful idea of continuity in a vastly more general context. Let X be an arbitrary nonempty set. A **topology** for X is a collection $\mathcal{T}$ of subsets of X that has the following properties:

(a) $\emptyset \in \mathcal{T}$ and $X \in \mathcal{T}$.

(b) If $U_\alpha \in \mathcal{T}$ for each $\alpha \in \mathcal{A}$, then $\bigcup_{\alpha \in \mathcal{A}} U_\alpha \in \mathcal{T}$.

(c) If $U_1, \ldots, U_k$ are in $\mathcal{T}$, then $U_1 \cap \cdots \cap U_k \in \mathcal{T}$.

The pair $(X, \mathcal{T})$ consisting of X and a topology $\mathcal{T}$ for X is called a **topological space** (although we will adhere to the custom of referring to X

itself as a topological space when it is clear from the context that only one topology $\mathcal{T}$ is involved). The elements of $\mathcal{T}$ are called the **open sets of** $(X, \mathcal{T})$, or simply **open in** X. If X and Y are both topological spaces, then a map $f : X \to Y$ of X into Y is said to be **continuous** if $f^{-1}(U)$ is open in X whenever U is open in Y.

Exercise 1.1.1 Let X, Y and Z be topological spaces and suppose $f : X \to Y$ and $g : Y \to Z$ are both continuous. Show that the composition $g \circ f : X \to Z$ is continuous.

Exercise 1.1.2 A subset C of a topological space X is said to be **closed in** X if its complement $X - C$ is open in X. Prove the following:

(a) $\emptyset$ and X are both closed in X.

(b) If C_α is closed in X for every $\alpha \in \mathcal{A}$, then $\bigcap_{\alpha \in \mathcal{A}} C_\alpha$ is closed in X.

(c) If $C_1, \ldots, C_k$ are all closed in X, then $C_1 \cup \cdots \cup C_k$ is closed in X.

Exercise 1.1.3 Let X and Y be topological spaces. Show that a map $f : X \to Y$ is continuous iff $f^{-1}(C)$ is closed in X whenever C is closed in Y.

The open sets in $\mathbb{R}^n$ defined at the beginning of this section constitute the **usual** (or **Euclidean**) **topology** for $\mathbb{R}^n$ and is the only topology on $\mathbb{R}^n$ of any interest to us. It has a rich, beautiful and very deep structure, some of which we will uncover as we proceed. At the other end of the spectrum are examples of topological spaces about which essentially nothing of interest can be said. Such spaces do have a tendency to arise now and then in meaningful discussions, however, so we shall not shun them. Thus, we consider an arbitrary nonempty set X. The collection of *all* subsets of X, usually denoted 2^X and called the **power set** of X, surely contains $\emptyset$ and X and is closed under arbitrary unions and finite intersections. Consequently, 2^X is a topology for X, called the **discrete topology** for X, in which every subset of X is open. It follows that any map from X to some other topological space Y is necessarily continuous. In particular, if Y happens also to have the discrete topology, then any map in either direction is continuous. Suppose further that X and Y, both with the discrete topology, have the same cardinality, i.e., that there exists a one-to-one map h of X onto Y. Then both $h : X \to Y$ and $h^{-1} : Y \to X$ are continuous. The existence of such a bijection that "preserves open sets" in both directions is reminiscent of the notion of an isomorphism from linear algebra (a bijection that preserves the linear structure in both directions) and leads us to formulate a definition.

Let X and Y be topological spaces. A continuous, one-to-one map h of X onto Y for which $h^{-1} : Y \to X$ is also continuous is called a **homeomorphism** and, if such a map exists, we say that X and Y are **homeomorphic**, or **topologically equivalent**, and write $X \cong Y$. Thus, we have seen that two discrete spaces X and Y are homeomorphic iff they have the same cardinality.

Remark: One might reasonably argue that this is an extraordinarily uninteresting result. On the other hand, there is a sense in which it is a topological theorem *par excellence.* An entire class of topological spaces (the discrete ones) is completely characterized up to topological equivalence by a single invariant (the cardinality) and this is the ideal to which all of topology aspires. The ideal is rarely achieved, however, and never again without a great deal of labor. One's experience in linear algebra, for example, might lead one to conjecture that two Euclidean spaces $\mathbb{R}^n$ and $\mathbb{R}^m$ are homeomorphic iff they have the same dimension ($n = m$). This is, indeed, the case, but it is far from being obvious and, in fact, is a very deep theorem of Brouwer (see Chapter $\overline{\text{XVII}}$ of [**Dug**]).

Exercise 1.1.4 Let X be a topological space and Homeo(X) the set of all homeomorphisms of X onto X. Show that, under the operation of composition $\circ$, Homeo(X) forms a group, called the **homeomorphism group of** X.

Many of our most important examples of topological spaces will arise naturally as subsets of some Euclidean space $\mathbb{R}^n$ with the topology they "inherit" from $\mathbb{R}^n$ in the sense of the following definition. Let $(X', \mathcal{T}')$ be a topological space and $X \subseteq X'$ a subset of X'. Define a collection $\mathcal{T}$ of subsets of X by

$$\mathcal{T} = \{X \cap U' : U' \in \mathcal{T}'\} .$$

Then, since $X \cap \emptyset = \emptyset$, $X \cap X' = X$, $\bigcup_{\alpha \in \mathcal{A}}(X \cap U'_\alpha) = X \cap (\bigcup_{\alpha \in \mathcal{A}} U'_\alpha)$ and $(X \cap U'_1) \cap \cdots \cap (X \cap U'_k) = X \cap (U'_1 \cap \cdots \cap U'_k)$, $\mathcal{T}$ is a topology for X. $\mathcal{T}$ is called the **relative topology** for X and with it X is a **(topological) subspace of** X'. Before proceeding with some of the examples of real interest to us here, we record a few elementary observations.

Exercise 1.1.5 Show that if X is a subspace of X' and X' is, in turn, a subspace of X'', then X is a subspace of X''.

Lemma 1.1.1 *Let X be a subspace of X' and $f : X' \to Y$ a continuous map. Then the restriction $f \mid X : X \to Y$ of f to X is continuous. In particular, the* **inclusion map** *$\iota : X \hookrightarrow X'$ defined by $\iota(x) = x$ for each $x \in X$ is continuous.*

Proof: Let U be open in Y. Since f is continuous, $f^{-1}(U)$ is open in X' and therefore $X \cap f^{-1}(U)$ is open in X. But $X \cap f^{-1}(U) = (f \mid X)^{-1}(U)$ so $(f \mid X)^{-1}(U)$ is open in X and $f \mid X$ is continuous. The inclusion map is the restriction to X of the identity map $id : X' \to X'$, which is clearly continuous. ■

Reversing the point of view in Lemma 1.1.1, one may be given a continuous map $g : X \to Y$ and ask whether or not there is a continuous map $f : X' \to Y$ with $f \mid X = g$. Should such an f exist it is called a **continuous extension** of g to X' and g is said to **extend continuously** to X'. The existence of continuous extensions is a central problem in topology and one that we will encounter repeatedly.

Lemma 1.1.2 *Let Y be a subspace of Y'. If $f : X \to Y'$ is a continuous map with $f(X) \subseteq Y$, then, regarded as a map into Y, $f : X \to Y$ is continuous. On the other hand, if $f : X \to Y$ is a continuous map, then, regarded as a map into Y', $f : X \to Y'$ is continuous.*

Exercise 1.1.6 Prove Lemma 1.1.2. ■

These few results are particularly useful when applied to subspaces of Euclidean spaces since they assure us that any map known, from real analysis, to be continuous on some subset X of $\mathbb{R}^n$ and taking values in some subset Y of $\mathbb{R}^m$ will, in fact, be a continuous map of the topological subspace X of $\mathbb{R}^n$ into the topological subspace Y of $\mathbb{R}^m$ (see Theorem 1-8 of [**Sp1**]). In general we will adopt the convention that a map defined on a subset A of some topological space X is said to be **continuous on** A if, when A is given the relative topology from X, it is continuous as a map on the topological space A. We now proceed to manufacture a long list of examples that will play a fundamental role in virtually all of the work we have to do. We begin with the circle (1-sphere) S^1. As a subspace of $\mathbb{R}^2$ it is given by $S^1 = \{ (x^1, x^2) \in \mathbb{R}^2 : (x^1)^2 + (x^2)^2 = 1 \}$ (see Figure 1.1.1). Let $N = (0, 1)$ be the "north pole" of S^1 and set $U_S = S^1 - \{N\}$. Being the intersection with S^1 of the open set $\mathbb{R}^2 - \{N\}$ in $\mathbb{R}^2$, U_S is open in S^1. Define a map $\varphi_S : U_S \to \mathbb{R}$ by

$$\varphi_S\left(x^1, x^2\right) = \frac{x^1}{1 - x^2} \,. \tag{1.1.1}$$

Geometrically, $\varphi_S(x^1, x^2)$ is the intersection with the x^1-axis ($x^2 = 0$) of the straight line in $\mathbb{R}^2$ joining N and (x^1, x^2) and φ_S is called the **stereographic projection from** N (see Figure 1.1.1). Since the rational function $(x^1, x^2) \to \frac{x^1}{1-x^2}$ is continuous on the open subspace of $\mathbb{R}^2$ with $x^2 \neq 1$, its restriction to U_S, i.e., φ_S, is continuous. It is, moreover, one-to-one and maps onto $\mathbb{R}$. In fact, it is a simple matter to write down its inverse

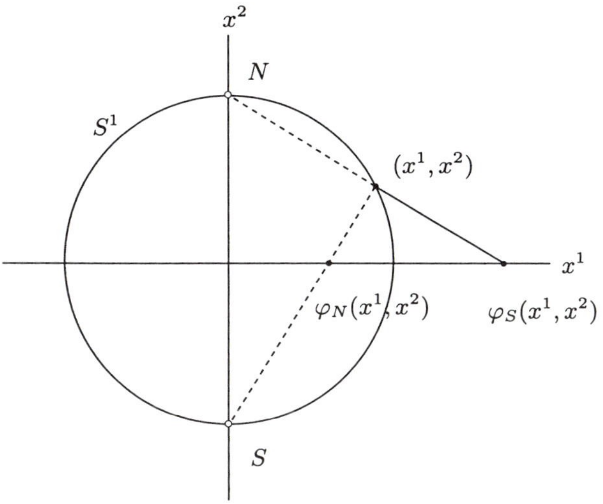

Figure 1.1.1

$\varphi_S^{-1} : \mathbb{R} \to U_S$:

$$\varphi_S^{-1}(y) = \left(\frac{2y}{y^2+1}, \frac{y^2-1}{y^2+1} \right) \tag{1.1.2}$$

(intersect the line joining N and $(y, 0)$ with S^1). Observe that φ_S^{-1} has continuous coordinate functions and so defines a continuous map of $\mathbb{R}$ into $\mathbb{R}^2$ whose image lies in U_S. Thus, $\varphi_S^{-1} : \mathbb{R} \to U_S$ is continuous. Consequently, φ_S is a homeomorphism of U_S onto $\mathbb{R}$.

Similarly, letting $S = (0, -1)$ be the "south pole" of S^1 and $U_N = S^1 - \{S\}$ one defines a homeomorphism $\varphi_N : U_N \to \mathbb{R}$ (**stereographic projection from** S) by

$$\varphi_N\left(x^1, x^2\right) = \frac{x^1}{1+x^2} \tag{1.1.3}$$

and calculates its inverse $\varphi_N^{-1} : \mathbb{R} \to U_N$ to find that

$$\varphi_N^{-1}(y) = \left(\frac{2y}{y^2+1}, \frac{1-y^2}{y^2+1} \right). \tag{1.1.4}$$

Note that $U_N \cap U_S = S^1 - \{N, S\}$ and $\varphi_N(U_N \cap U_S) = \varphi_S(U_N \cap U_S) = \mathbb{R} - \{0\}$. Thus, $\varphi_S \circ \varphi_N^{-1} : \mathbb{R} - \{0\} \to \mathbb{R} - \{0\}$ and $\varphi_N \circ \varphi_S^{-1} : \mathbb{R} - \{0\} \to \mathbb{R} - \{0\}$ and a simple calculation gives

$$\varphi_S \circ \varphi_N^{-1}(y) = y^{-1} = \varphi_N \circ \varphi_S^{-1}(y). \tag{1.1.5}$$

Exercise 1.1.7 Verify (1.1.5).

We will see shortly (Section 1.4) that S^1 itself is not homeomorphic to $\mathbb{R}$. However, we have just proved that it is "locally" homeomorphic to $\mathbb{R}$ in the sense that every point in S^1 is contained in some open subset of S^1 (either U_N or U_S) that is homeomorphic to $\mathbb{R}$. This sort of situation will arise so frequently in our work that it merits a few definitions.

Let X be an arbitrary topological space and n a positive integer. An **n-dimensional chart on** X is a pair (U, φ), where U is an open subset of X and φ is a homeomorphism of U onto some open subset of $\mathbb{R}^n$. X is said to be **locally Euclidean** if there exists a positive integer n such that, for each $x \in X$, there is an n-dimensional chart (U, φ) on X with $x \in U$; (U, φ) is then called a **chart at** $x \in X$. If (U_1, φ_1) and (U_2, φ_2) are two n-dimensional charts on X with $U_1 \cap U_2 \neq \emptyset$, then $\varphi_1 \circ \varphi_2^{-1} : \varphi_2(U_1 \cap U_2) \to \varphi_1(U_1 \cap U_2)$ and $\varphi_2 \circ \varphi_1^{-1} : \varphi_1(U_1 \cap U_2) \to \varphi_2(U_1 \cap U_2)$ are homeomorphisms between open subsets of $\mathbb{R}^n$ (see Figure 1.1.2) and are called the **overlap functions** for the two charts. Note that they are, in general, inverses of each other ($(\varphi_1 \circ \varphi_2^{-1})^{-1} = \varphi_2 \circ \varphi_1^{-1}$), but for the charts (U_N, φ_N) and (U_S, φ_S) on S^1 described above they happen to coincide.

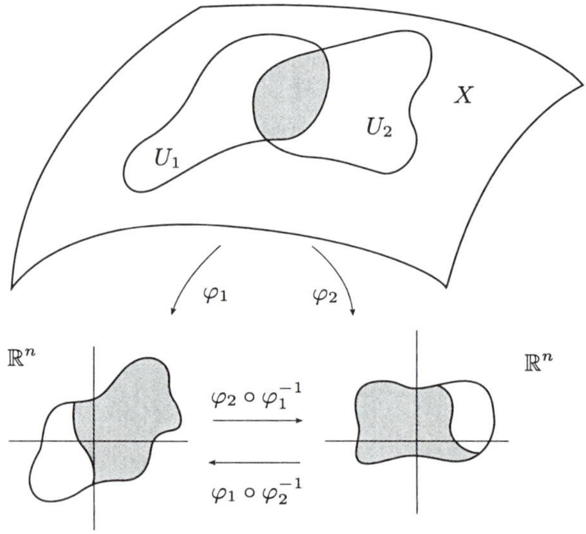

Figure 1.1.2

Before leaving S^1 we make one additional observation. S^1 is a subspace of $\mathbb{R}^2$ and $\mathbb{R}^2$ has more structure than we have thus far used. Specifically, one can define complex multiplication on $\mathbb{R}^2$, thereby converting it into the complex plane $\mathbb{C}$. S^1 is then identified with the set $S^1 = \{e^{\mathbf{i}\theta} : \theta \in \mathbb{R}\}$ of complex numbers of modulus 1 and this set is closed under complex multiplication ($e^{\mathbf{i}\theta_1} e^{\mathbf{i}\theta_2} = e^{\mathbf{i}(\theta_1+\theta_2)}$) and inversion ($(e^{\mathbf{i}\theta})^{-1} = e^{\mathbf{i}(-\theta)}$). Thus, under the operation of complex multiplication, S^1 is a group (in fact, an Abelian group). For future reference we write out complex multiplication and inversion in terms of real and imaginary parts (i.e., identifying a complex number $x + y\mathbf{i}$ with an ordered pair (x, y)):

$$(x^1, y^1)(x^2, y^2) = (x^1x^2 - y^1y^2, x^1y^2 + x^2y^1) \tag{1.1.6}$$

and

$$(x, y)^{-1} = \left(\frac{x}{x^2 + y^2}, \frac{-y}{x^2 + y^2} \right). \tag{1.1.7}$$

We will see somewhat later (Sections 1.6 and 4.3) that the continuity (respectively, smoothness) of these maps gives S^1 the structure of a "topological group" (respectively, "Lie group").

The next order of business is to generalize much of what we have just done for the circle S^1 to the case of the 2-dimensional sphere. The 2-sphere S^2 is, by definition, the topological subspace $S^2 = \{(x^1, x^2, x^3) \in \mathbb{R}^3 : (x^1)^2 + (x^2)^2 + (x^3)^2 = 1\}$ of $\mathbb{R}^3$.

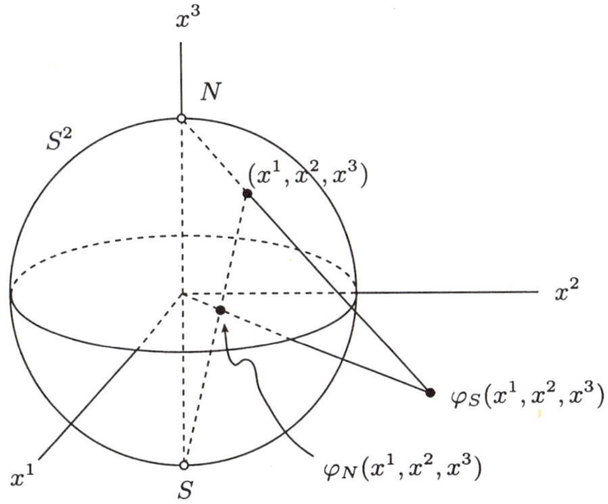

Figure 1.1.3

Again we let $N = (0, 0, 1)$ and $S = (0, 0, -1)$ be the north and south poles of S^2 and define $U_S = S^2 - \{N\}$ and $U_N = S^2 - \{S\}$. Introduce stereographic projection maps $\varphi_S : U_S \to \mathbb{R}^2$ and $\varphi_N : U_N \to \mathbb{R}^2$ defined by

$$\varphi_S(x^1, x^2, x^3) = \left(\tfrac{x^1}{1-x^3}, \tfrac{x^2}{1-x^3} \right) \quad \text{and} \quad \varphi_N(x^1, x^2, x^3) = \left(\tfrac{x^1}{1+x^3}, \tfrac{x^2}{1+x^3} \right)$$

($\varphi_S(x^1, x^2, x^3)$ is the intersection with $x^3 = 0$ of the straight line in $\mathbb{R}^3$ joining (x^1, x^2, x^3) and N, and similarly for $\varphi_N(x^1, x^2, x^3)$). These are continuous, one-to-one and onto $\mathbb{R}^2$ and their inverses $\varphi_S^{-1} : \mathbb{R}^2 \to U_S$ and $\varphi_N^{-1} : \mathbb{R}^2 \to U_N$ are easily seen to be given, for each $y = (y^1, y^2) \in \mathbb{R}^2$, by

$$\varphi_S^{-1}(y) = (1 + \|y\|^2)^{-1}(2y_1, 2y_2, \|y\|^2 - 1)$$

and

$$\varphi_N^{-1}(y) = (1 + \|y\|^2)^{-1}(2y_1, 2y_2, 1 - \|y\|^2).$$

Since these are also continuous, (U_S, φ_S) and (U_N, φ_N) are 2-dimensional charts on S^2. The overlap functions $\varphi_S \circ \varphi_N^{-1} : \mathbb{R}^2 - \{(0,0)\} \to \mathbb{R}^2 - \{(0,0)\}$ and $\varphi_N \circ \varphi_S^{-1} : \mathbb{R}^2 - \{(0,0)\} \to \mathbb{R}^2 - \{(0,0)\}$ are given, for each $y = (y^1, y^2) \neq (0,0)$, by $\varphi_S \circ \varphi_N^{-1}(y) = \frac{1}{\|y\|^2} y = \varphi_N \circ \varphi_S^{-1}(y)$. Notice that if we once again identify $\mathbb{R}^2$ with the complex plane $\mathbb{C}$ and y with a complex number $y^1 + y^2 \mathbf{i}$, then

$$\varphi_S \circ \varphi_N^{-1}(y) = \bar{y}^{-1} = \varphi_N \circ \varphi_S^{-1}(y) \tag{1.1.8}$$

and this is quite reminiscent of (1.1.5). In any case, S^2 is, like S^1, locally Euclidean, but, unlike S^1, there is no natural way to provide S^2 with a group structure.

Exercise 1.1.8 For any positive integer n define the **n-sphere** S^n to be the topological subspace of $\mathbb{R}^{n+1}$ given by $S^n = \{(x^1, \ldots, x^n, x^{n+1}) \in \mathbb{R}^{n+1} : (x^1)^2 + \cdots + (x^n)^2 + (x^{n+1})^2 = 1\}$. Let $N = (0, \ldots, 0, 1)$, $S = (0, \ldots, 0, -1)$, $U_S = S^n - \{N\}$ and $U_N = S^n - \{S\}$. Define $\varphi_S : U_S \to \mathbb{R}^n$ and $\varphi_N : U_N \to \mathbb{R}^n$ by

$$\varphi_S(x^1, \ldots, x^n, x^{n+1}) = \left(\frac{x^1}{1 - x^{n+1}}, \ldots, \frac{x^n}{1 - x^{n+1}} \right)$$

and

$$\varphi_N(x^1, \ldots, x^n, x^{n+1}) = \left(\frac{x^1}{1 + x^{n+1}}, \ldots, \frac{x^n}{1 + x^{n+1}} \right).$$

Show that (U_S, φ_S) and (U_N, φ_N) are n-dimensional charts on S^n and that $\varphi_S^{-1} : \mathbb{R}^n \to U_S$ and $\varphi_N^{-1} : \mathbb{R}^n \to U_N$ are given, for each $y = (y^1, \ldots, y^n) \in \mathbb{R}^n$, by

$$\varphi_S^{-1}(y) = (1 + \|y\|^2)^{-1}(2y^1, \ldots, 2y^n, \|y\|^2 - 1)$$

and

$$\varphi_N^{-1}(y) = (1 + \|y\|^2)^{-1}(2y^1, \ldots, 2y^n, 1 - \|y\|^2).$$

Show, furthermore, that the overlap functions are given, for each $y \in \mathbb{R}^n - \{(0, \ldots, 0)\}$, by

$$\varphi_S \circ \varphi_N^{-1}(y) = \frac{1}{\|y\|^2} y = \varphi_N \circ \varphi_S^{-1}(y).$$

The spheres S^3 and S^4 will play a particularly dominant role in our study and we must come to understand them rather well. This task is greatly facilitated by what might be regarded as something of an accident. S^3 is, of course, a subspace of $\mathbb{R}^4$ and S^4 is, by Exercise 1.1.8, locally homeomorphic to $\mathbb{R}^4$ and, as Euclidean spaces go, $\mathbb{R}^4$ is rather special. Like $\mathbb{R}^2$, and unlike virtually every other $\mathbb{R}^n$, it admits a natural multiplicative structure, the use of which makes clear many things that would otherwise be obscure. Pausing to introduce this multiplicative structure on $\mathbb{R}^4$ at this point will not only elucidate the topologies of S^3 and S^4, but will unify and clarify our analysis of the classical groups and projective spaces. Eventually, we will even use this material to describe the so-called BPST instantons on the 4-sphere. All in all, it is worth doing and we intend to do it.

This multiplicative structure on $\mathbb{R}^4$ to which we refer actually arises in a rather natural way, provided one is willing to look at $\mathbb{R}^4$ in a somewhat unnatural way. We begin by considering a remarkable set of 2×2 complex matrices:

$$\mathcal{R}^4 = \left\{ \begin{pmatrix} \alpha & \beta \\ -\bar{\beta} & \bar{\alpha} \end{pmatrix} : \alpha, \beta \in \mathbb{C} \right\}.$$

Observe that $\mathcal{R}^4$ is closed under matrix addition and multiplication by *real* scalars and so may be regarded as a real vector space. Moreover, with $\alpha = y^0 + y^1\mathbf{i}$ and $\beta = y^2 + y^3\mathbf{i}$ we have

$$\begin{pmatrix} \alpha & \beta \\ -\bar{\beta} & \bar{\alpha} \end{pmatrix} = \begin{pmatrix} y^0 + y^1\mathbf{i} & y^2 + y^3\mathbf{i} \\ -y^2 + y^3\mathbf{i} & y^0 - y^1\mathbf{i} \end{pmatrix}$$

$$= y^0 \begin{pmatrix} 1 & 0 \\ 0 & 1 \end{pmatrix} + y^1 \begin{pmatrix} \mathbf{i} & 0 \\ 0 & -\mathbf{i} \end{pmatrix} + y^2 \begin{pmatrix} 0 & 1 \\ -1 & 0 \end{pmatrix} + y^3 \begin{pmatrix} 0 & \mathbf{i} \\ \mathbf{i} & 0 \end{pmatrix}$$

so the four matrices indicated span $\mathcal{R}^4$. Since this last sum is clearly $\begin{pmatrix} 0 & 0 \\ 0 & 0 \end{pmatrix}$ iff $y^0 = y^1 = y^2 = y^3 = 0$, these matrices are linearly independent and so form a basis for $\mathcal{R}^4$. In particular, $\dim \mathcal{R}^4 = 4$. We introduce the following notation (momentarily asking the symbol "i" to do double duty):

$$\mathbf{1} = \begin{pmatrix} 1 & 0 \\ 0 & 1 \end{pmatrix}, \ \mathbf{i} = \begin{pmatrix} \mathbf{i} & 0 \\ 0 & -\mathbf{i} \end{pmatrix}, \ \mathbf{j} = \begin{pmatrix} 0 & 1 \\ -1 & 0 \end{pmatrix}, \ \mathbf{k} = \begin{pmatrix} 0 & \mathbf{i} \\ \mathbf{i} & 0 \end{pmatrix}.$$

The basis $\{\mathbf{1}, \mathbf{i}, \mathbf{j}, \mathbf{k}\}$ for $\mathcal{R}^4$ determines a **natural isomorphism** from $\mathcal{R}^4$ to $\mathbb{R}^4$ given by $y^0\mathbf{1} + y^1\mathbf{i} + y^2\mathbf{j} + y^3\mathbf{k} \leftrightarrow (y^0, y^1, y^2, y^3)$. If we define an inner product on $\mathcal{R}^4$ by declaring that the basis $\{\mathbf{1}, \mathbf{i}, \mathbf{j}, \mathbf{k}\}$ is orthonormal, then the norm of $y = y^0\mathbf{1} + y^1\mathbf{i} + y^2\mathbf{j} + y^3\mathbf{k}$ is given by $\| y \|^2 = (y^0)^2 + (y^1)^2 + (y^2)^2 + (y^3)^2$ so that the natural isomorphism is actually an isometry. Note also that $\| y \|^2$ is, in fact, the determinant of the matrix $\begin{pmatrix} \alpha & \beta \\ -\bar{\beta} & \bar{\alpha} \end{pmatrix}$. Except for cosmetic features, $\mathcal{R}^4$ is "really just" $\mathbb{R}^4$ and we shall round this picture out by assigning to $\mathcal{R}^4$ the topology that makes the natural isomorphism a homeomorphism, i.e., a subset U of $\mathcal{R}^4$ is open in $\mathcal{R}^4$ iff its image under the natural isomorphism is open in $\mathbb{R}^4$.

Now for the good part. Notice that, in addition to being closed under sums and real scalar multiples, $\mathcal{R}^4$ is actually closed under matrix multiplication since

$$\begin{pmatrix} \alpha & \beta \\ -\bar{\beta} & \bar{\alpha} \end{pmatrix} \begin{pmatrix} \gamma & \delta \\ -\bar{\delta} & \bar{\gamma} \end{pmatrix} = \begin{pmatrix} \alpha\gamma - \beta\bar{\delta} & \alpha\delta + \beta\bar{\gamma} \\ -\overline{(\alpha\delta + \beta\bar{\gamma})} & \overline{\alpha\gamma - \beta\bar{\delta}} \end{pmatrix}.$$

Moreover, $\mathcal{R}^4$ contains the inverse of each of its nonzero elements since

$$\begin{pmatrix} \alpha & \beta \\ -\bar{\beta} & \bar{\alpha} \end{pmatrix}^{-1} = \frac{1}{|\alpha|^2 + |\beta|^2} \begin{pmatrix} \bar{\alpha} & -\beta \\ \bar{\beta} & \alpha \end{pmatrix}, \qquad (1.1.9)$$

which is in $\mathcal{R}^4$. Thus, under matrix multiplication, the nonzero elements of $\mathcal{R}^4$ take on the structure of a (non-Abelian) group.

Exercise 1.1.9 Verify that, under matrix multiplication, the basis elements $\{\mathbf{1}, \mathbf{i}, \mathbf{j}, \mathbf{k}\}$ satisfy the following **commutation relations**

$$\begin{gathered} \mathbf{i}^2 = \mathbf{j}^2 = \mathbf{k}^2 = -\mathbf{1} \\ \mathbf{ij} = -\mathbf{ji} = \mathbf{k}, \ \mathbf{jk} = -\mathbf{kj} = \mathbf{i}, \ \mathbf{ki} = -\mathbf{ik} = \mathbf{j} \end{gathered} \qquad (1.1.10)$$

and that $\mathbf{1}$ is a multiplicative identity. Show, moreover, that these relations, together with the usual distributive and associative laws (see (1.1.11) below) completely determine the multiplication on $\mathcal{R}^4$.

It should by now be clear that what we have just constructed is a concrete model (representation) of the familiar **algebra of quaternions** $\mathbb{H}$, usually defined abstractly as a 4-dimensional real vector space on which is defined a multiplication $(x, y) \to xy$ which satisfies the following associative and distributive laws for all $x, y, z \in \mathbb{H}$ and all $a \in \mathbb{R}$

$$\begin{aligned} (xy)z &= x(yz) \\ x(y+z) &= xy + xz \\ (x+y)z &= xz + yz \\ a(xy) &= (ax)y = x(ay) \end{aligned} \qquad (1.1.11)$$

and in which there exists a distinguished basis $\{\mathbf{1}, \mathbf{i}, \mathbf{j}, \mathbf{k}\}$ which satisfies (1.1.10) and $\mathbf{1}x = x\mathbf{1} = x$ for all $x \in \mathbb{H}$. One can view $\mathbb{H}$ as given abstractly this way, or as the set $\mathcal{R}^4$ of matrices with familiar matrix operations, or as the vector space $\mathbb{R}^4$ with a multiplicative structure obtained by transferring that of $\mathcal{R}^4$ to $\mathbb{R}^4$ (see (1.1.12) below and apply the natural isomorphism).

We proceed to enumerate the basic algebraic properties of quaternions that we will require. First though, we observe that, since **1** is the multiplicative identity in $\mathbb{H}$, it does no real harm to omit it altogether and write a quaternion as $x = x^0 + x^1\mathbf{i} + x^2\mathbf{j} + x^3\mathbf{k}$. The **real part** of x is then $\mathrm{Re}\,(x) = x^0$, while its **imaginary part** is $\mathrm{Im}\,(x) = x^1\mathbf{i} + x^2\mathbf{j} + x^3\mathbf{k}$. Quaternions whose imaginary part is the zero vector are then called **real quaternions** and the set of all such is isomorphic to (and will be identified with) $\mathbb{R}$. $\mathrm{Im}\,\mathbb{H} = \{x^1\mathbf{i} + x^2\mathbf{j} + x^3\mathbf{k} : x^1, x^2, x^3 \in \mathbb{R}\}$ is the set of pure **imaginary quaternions**. The **conjugate** of x is the quaternion $\bar{x} = x^0 - x^1\mathbf{i} - x^2\mathbf{j} - x^3\mathbf{k}$ (we caution the reader that if x is thought of as a matrix in $\mathcal{R}^4$, then $\bar{x}$ is the *conjugate transpose* matrix). The product of two quaternions is best computed directly using the associative and distributive laws and the commutation relations (1.1.10), but it will be convenient to have the general result written out explicitly. Thus, if $x = x^0 + x^1\mathbf{i} + x^2\mathbf{j} + x^3\mathbf{k}$ and $y = y^0 + y^1\mathbf{i} + y^2\mathbf{j} + y^3\mathbf{k}$, one finds that

$$\begin{aligned} xy = {} & [x^0y^0 - x^1y^1 - x^2y^2 - x^3y^3] \; + \\ & [x^0y^1 + x^1y^0 + x^2y^3 - x^3y^2]\,\mathbf{i} \; + \\ & [x^0y^2 + x^2y^0 + x^3y^1 - x^1y^3]\,\mathbf{j} \; + \\ & [x^0y^3 + x^3y^0 + x^1y^2 - x^2y^1]\,\mathbf{k}\,. \end{aligned} \tag{1.1.12}$$

In particular, we note that $x\bar{x} = \bar{x}x = (x^0)^2 + (x^1)^2 + (x^2)^2 + (x^3)^2$. Defining the **modulus** of the quaternion x to be $|\,x\,| = ((x^0)^2 + (x^1)^2 + (x^2)^2 + (x^3)^2)^{1/2}$ we therefore have $x\bar{x} = \bar{x}x = |\,x\,|^2$.

Exercise 1.1.10 Let $x, y \in \mathbb{H}$ and $a, b \in \mathbb{R}$. Prove each of the following:

(a) $\overline{ax + by} = a\bar{x} + b\bar{y}$ (b) $\overline{(\bar{x})} = x$

(c) $|\,\bar{x}\,| = |\,x\,|$ (d) $|\,ax\,| = |\,a\,|\,|\,x\,|$

(e) $\overline{xy} = \bar{y}\bar{x}$ (f) $|\,xy\,| = |\,x\,|\,|\,y\,|$.

Exercise 1.1.11 Show that, for all $x, y \in \mathrm{Im}\,\mathbb{H}$,

(a) $\overline{xy} = yx$, and (b) $xy - yx = 2\,\mathrm{Im}\,(xy)$.

Next observe that if $y \in \mathbb{H}$ is not the zero vector, then $y\left(\frac{1}{|y|^2}\bar{y}\right) = \frac{1}{|y|^2}y\bar{y} = \frac{1}{|y|^2}\,|\,y\,|^2 = 1$ (the real quaternion 1) and, similarly, $\left(\frac{1}{|y|^2}\bar{y}\right)y = 1$. Thus, each nonzero element y of $\mathbb{H}$ has a multiplicative **inverse** defined by

$$y^{-1} = \frac{1}{|\,y\,|^2}\,\bar{y} \tag{1.1.13}$$

and so the nonzero quaternions form a group under quaternion multiplication. Notice also that the subset of $\mathbb{H}$ consisting of all elements of the form

$x^0 + x^1\mathbf{i}$ is a linear subspace of $\mathbb{H}$ that is also closed under multiplication and inversion (of nonzero elements), i.e., it is a subalgebra of $\mathbb{H}$. Moreover, since $\mathbf{i}^2 = -1$, this subalgebra is naturally isomorphic to the usual algebra $\mathbb{C}$ of complex numbers. Henceforth, we will identify this subalgebra with $\mathbb{C}$ (so that asking the symbol "$\mathbf{i}$" to do double duty a moment ago was not quite the notational *faux pas* it appeared).

An $x \in \mathbb{H}$ with $|x| = 1$ is called a **unit quaternion** and the set of all such is closed under multiplication (by Exercise 1.1.10 (f)) and inversion (because $|x| = 1$ implies $|x^{-1}| = \left|\frac{1}{|x|^2}\bar{x}\right| = |\bar{x}| = |x| = 1$) and is therefore a subgroup of the nonzero quaternions. We wish to determine what this **group of unit quaternions** looks like in the matrix model $\mathcal{R}^4$ of $\mathbb{H}$. Here the squared modulus of an element is its determinant so unit quaternions correspond to elements of $\mathcal{R}^4$ with determinant 1. According to (1.1.9) any $A \in \mathcal{R}^4$ with determinant 1 has an inverse A^{-1} that equals its conjugate transpose $\bar{A}^T$, i.e., is unitary. Now, the collection of all 2×2 complex matrices that are unitary and have determinant 1 is traditionally denoted $SU(2)$ and called the **special unitary group of order** 2. We show next that $SU(2)$ is entirely contained in $\mathcal{R}^4$.

Lemma 1.1.3 $SU(2) \subseteq \mathcal{R}^4$.

Proof: Let $A = \begin{pmatrix} \alpha & \beta \\ \gamma & \delta \end{pmatrix} \in SU(2)$. We must show that $\delta = \bar{\alpha}$ and $\gamma = -\bar{\beta}$. Now, $\det A = 1$ implies $\alpha\delta - \beta\gamma = 1$, while $A\bar{A}^T = id$ gives

$$\begin{pmatrix} \alpha\bar{\alpha} + \beta\bar{\beta} & \alpha\bar{\gamma} + \beta\bar{\delta} \\ \bar{\alpha}\gamma + \bar{\beta}\delta & \gamma\bar{\gamma} + \delta\bar{\delta} \end{pmatrix} = \begin{pmatrix} 1 & 0 \\ 0 & 1 \end{pmatrix}.$$

Thus, $\alpha\bar{\gamma} = -\beta\bar{\delta} \Rightarrow \alpha(\bar{\gamma}\gamma) = (-\beta\gamma)\bar{\delta} \Rightarrow \alpha(1 - \delta\bar{\delta}) = (1 - \alpha\delta)\bar{\delta} \Rightarrow \alpha - \alpha\delta\bar{\delta} = \bar{\delta} - \alpha\delta\bar{\delta} \Rightarrow \alpha = \bar{\delta}$.

Exercise 1.1.12 Show, in the same way, that $\gamma = -\bar{\beta}$. ∎

Now, we have just seen that the unit quaternions in $\mathcal{R}^4$ are all in $SU(2)$. Lemma 1.1.3 implies that any element of $SU(2)$, being in $\mathcal{R}^4$ and having determinant 1, is a unit quaternion. Thus, the group of unit quaternions in $\mathcal{R}^4$ is precisely $SU(2)$. With this we can reap our first topological dividend of the quaternion structure of $\mathcal{R}^4$.

Theorem 1.1.4 *The 3-sphere S^3 is homeomorphic to the subspace $SU(2)$ of $\mathcal{R}^4$.*

Proof: The natural homeomorphism of $\mathbb{R}^4$ onto $\mathcal{R}^4$ carries S^3 onto a subspace of $\mathcal{R}^4$ homeomorphic to S^3. Since this homeomorphism also preserves the norm of any vector in $\mathbb{R}^4$, the image consists precisely of the unit quaternions and we have just shown that the group of unit quaternions in $\mathcal{R}^4$ is $SU(2)$. ∎

The natural isomorphism of $\mathcal{R}^4$ onto $\mathbb{R}^4$ transfers the quaternion multiplication given by (1.1.12) to $\mathbb{R}^4$ and S^3 thereby acquires the group structure of $SU(2)$. Like S^1, but unlike S^2, the 3-sphere has a natural group structure and we will see later (Sections 1.6 and 4.3) that this provides us with perhaps our most important example of a "topological group" and, indeed, a "Lie group".

The quaternion structure of $\mathbb{R}^4$ also has something interesting to say about S^4. The 4-sphere S^4 is a subspace of $\mathbb{R}^5$ which, as the reader has shown in Exercise 1.1.8, is locally homeomorphic to $\mathbb{R}^4$. Specifically, there exist open subsets $U_S = S^4 - \{(0,0,0,0,1)\}$ and $U_N = S^4 - \{(0,0,0,0,-1)\}$ of S^4 and homeomorphisms $\varphi_S : U_S \to \mathbb{R}^4$ and $\varphi_N : U_N \to \mathbb{R}^4$ for which the overlap functions are given by $\varphi_S \circ \varphi_N^{-1}(y) = \frac{1}{\|y\|^2}\, y = \varphi_N \circ \varphi_S^{-1}(y)$ for all $y \in \mathbb{R}^4 - \{(0,0,0,0)\}$. Regarding such a y as a nonzero quaternion and noting that $\bar{y}^{-1} = \frac{1}{|\bar{y}|^2}\bar{\bar{y}} = \frac{1}{\|y\|^2}\, y$, the overlap functions can therefore be written

$$\varphi_S \circ \varphi_N^{-1}(y) = \bar{y}^{-1} = \varphi_N \circ \varphi_S^{-1}(y)\,, \tag{1.1.14}$$

which is entirely analogous to the cases of S^1 ((1.1.5)) and S^2 ((1.1.8)), provided, in (1.1.5), the "conjugate" of a real number is taken to be the same real number.

The examples that we have considered thus far (spheres) arise naturally as subspaces of some $\mathbb{R}^n$. While this is not the case for all topological spaces of interest, there is one further class of important examples that can be identified in a natural way with subspaces of Euclidean space. The idea is quite simple. An $m \times n$ real matrix

$$\begin{pmatrix} a^{11} & \cdots & a^{1n} \\ \vdots & & \vdots \\ a^{m1} & \cdots & a^{mn} \end{pmatrix} \tag{1.1.15}$$

is a rectangular array of mn real numbers. Now, arranging these numbers in a rectangle is convenient for some purposes (e.g., matrix products), but one might just as well arrange them as an mn-tuple by enumerating the rows in order:

$$\left(a^{11}, \ldots, a^{1n}, \ldots, a^{m1}, \ldots, a^{mn}\right)\,. \tag{1.1.16}$$

The map which assigns to an $m \times n$ matrix (1.1.15) the mn-tuple (1.1.16) is a linear isomorphism which identifies any particular set of $m \times n$ matrices with a subspace of $\mathbb{R}^{mn}$ and thereby provides it with a topology (that is, the one for which this map is a homeomorphism).

One can extend this idea to $m \times n$ complex matrices

$$\begin{pmatrix} z^{11} & \cdots & z^{1n} \\ \vdots & & \vdots \\ z^{m1} & \cdots & z^{mn} \end{pmatrix} = \begin{pmatrix} x^{11} + y^{11}\mathbf{i} & \cdots & x^{1n} + y^{1n}\mathbf{i} \\ \vdots & & \vdots \\ x^{m1} + y^{m1}\mathbf{i} & \cdots & x^{mn} + y^{mn}\mathbf{i} \end{pmatrix} \tag{1.1.17}$$

in the obvious way by stringing out the rows as above $\left((z^{11}, \ldots, z^{1n}, \ldots, z^{m1}, \ldots, z^{mn})\right)$ and then splitting each z^{ij} into real and imaginary parts:

$$\left(x^{11}, y^{11}, \ldots, x^{1n}, y^{1n}, \ldots, x^{m1}, y^{m1}, \ldots, x^{mn}, y^{mn}\right). \tag{1.1.18}$$

Thus, any set of complex $m \times n$ matrices acquires a topology as a subset of $\mathbb{R}^{2mn}$. One can even push this a bit further (and we will need to do so). An $m \times n$ matrix whose entries are quaternions

$$\left(q^{ij}\right)_{\substack{i=1,\ldots,m \\ j=1,\ldots,n}} = \left(x^{ij} + y^{ij}\mathbf{i} + u^{ij}\mathbf{j} + v^{ij}\mathbf{k}\right)_{\substack{i=1,\ldots,m \\ j=1,\ldots,n}} \tag{1.1.19}$$

is regarded as a $4mn$-tuple

$$\left(x^{11}, y^{11}, u^{11}, v^{11}, \ldots, x^{mn}, y^{mn}, u^{mn}, v^{mn}\right). \tag{1.1.20}$$

Consequently, any set of $m \times n$ quaternionic matrices has the topology of a subspace of $\mathbb{R}^{4mn}$. In particular, since the sets $\mathbb{C}^n$ and $\mathbb{H}^n$ of ordered n-tuples of complex numbers and quaternions, respectively, can be regarded as the sets of all $1 \times n$ (or $n \times 1$) complex and quaternionic matrices, they are thereby identified topologically with $\mathbb{R}^{2n}$ and $\mathbb{R}^{4n}$, respectively.

Now we isolate those particular collections of matrices that give rise, in the manner described above, to topological spaces of particular interest to us. To do so we will regard $\mathbb{R}^n$ as a real vector space in the usual way. Similarly, $\mathbb{C}^n$ is an n-dimensional complex vector space. The case of $\mathbb{H}^n$ requires some care because of the noncommutativity of quaternion (scalar) multiplication. In order to treat all three cases simultaneously we will exercise the same care for $\mathbb{R}$ and $\mathbb{C}$. Specifically, we let $\mathbb{F}$ denote one of $\mathbb{R}$, $\mathbb{C}$, or $\mathbb{H}$. Then, for each positive integer n, $\mathbb{F}^n = \{\xi = (\xi^1, \ldots, \xi^n) : \xi^i \in \mathbb{F},\ i = 1, \ldots, n\}$. We define an algebraic structure on $\mathbb{F}^n$ by adding coordinatewise $\left(\xi + \zeta = (\xi^1, \ldots, \xi^n) + (\zeta^1, \ldots, \zeta^n) = (\xi^1 + \zeta^1, \ldots, \xi^n + \zeta^n)\right)$ and scalar multiplying by any $a \in \mathbb{F}$ *on the right* $\left(\xi a = (\xi^1, \ldots, \xi^n)a = (\xi^1 a, \ldots, \xi^n a)\right)$.

Exercise 1.1.13 Show that $\mathbb{F}^n$ is an Abelian group under addition $+$ and satisfies $(\xi + \zeta)a = \xi a + \zeta a$, $\xi(a + b) = \xi a + \xi b$, $\xi(ab) = (\xi a)b$ and $\xi 1 = \xi$ for all $\xi, \zeta \in \mathbb{F}^n$ and $a, b \in \mathbb{F}$.

Were it not for the fact that $\mathbb{F} = \mathbb{H}$, being noncommutative, is only a division ring and not a field, Exercise 1.1.13 would show that $\mathbb{F}^n$ is a vector space over $\mathbb{F}$. Without this commutativity the proper terminology would be that $\mathbb{F}^n$ is a **right module over** $\mathbb{F}$ (although "right vector space over $\mathbb{F}$" is not unheard of and, in any case, the terminology used will not be important for us).

We define, in the usual way, the **standard basis** $\{e_1, \ldots, e_n\}$ for $\mathbb{F}^n$ to consist of the elements $e_1 = (1, 0, \ldots, 0, 0), \ldots, e_n = (0, 0, \ldots, 0, 1)$. Then any $\xi = (\xi^1, \ldots, \xi^n)$ can be written as $\xi = \sum_{i=1}^n e_i \xi^i$. We also define on $\mathbb{F}^n$ a bilinear form $<\,,\,>: \mathbb{F}^n \times \mathbb{F}^n \to \mathbb{F}$ by

$$< \xi, \zeta > = < (\xi^1, \ldots, \xi^n), (\zeta^1, \ldots, \zeta^n) > = \overline{\xi^1}\zeta^1 + \cdots + \overline{\xi^n}\zeta^n, \quad (1.1.21)$$

where $\overline{\xi^i}$ denotes the complex (quaternionic) conjugate if $\mathbb{F} = \mathbb{C}$ ($\mathbb{H}$) and $\overline{\xi^i} = \xi^i$ if $\mathbb{F} = \mathbb{R}$.

Exercise 1.1.14 Show that, if $\xi, \xi_1, \xi_2, \zeta, \zeta_1$ and ζ_2 are in $\mathbb{F}^n$ and a is in $\mathbb{F}$, then

$$\begin{aligned} < \xi_1 + \xi_2, \zeta > &= < \xi_1, \zeta > + < \xi_2, \zeta > \\ < \xi, \zeta_1 + \zeta_2 > &= < \xi, \zeta_1 > + < \xi, \zeta_2 > \\ < \xi, \zeta a > &= < \xi, \zeta > a \\ < \xi a, \zeta > &= \bar{a} < \xi, \zeta > \\ < \zeta, \xi > &= \overline{< \xi, \zeta >}. \end{aligned}$$

Show also that $< \xi, \zeta > = 0 \in \mathbb{F}$ iff $< \zeta, \xi > = 0$ and that $<\,,\,>$ is **nondegenerate** in the sense that $< \xi, \zeta > = 0$ for all $\xi \in \mathbb{F}^n$ iff $\zeta = (0, \ldots, 0) \in \mathbb{F}^n$.

A map $A : \mathbb{F}^n \to \mathbb{F}^n$ is said to be $\mathbb{F}$**-linear** if it satisfies $A(\xi + \zeta) = A(\xi) + A(\zeta)$ and $A(\xi a) = A(\xi)a$ for all $\xi, \zeta \in \mathbb{F}^n$ and all $a \in \mathbb{F}$. Such a map is completely determined by the $n \times n$ matrix $(A_{ij})_{i,j=1,\ldots,n}$ of A relative to $\{e_1, \ldots, e_n\}$, whose entries (in $\mathbb{F}$) are defined by $A(e_j) = \sum_{i=1}^n e_i A_{ij}$. Indeed, if $\xi = \sum_{j=1}^n e_j \xi^j$, then

$$A(\xi) = A\left(\sum_{j=1}^n e_j \xi^j\right) = \sum_{j=1}^n A(e_j \xi^j) = \sum_{j=1}^n A(e_j)\xi^j$$

$$= \sum_{j=1}^n \left(\sum_{i=1}^n e_i A_{ij}\right)\xi^j = \sum_{j=1}^n \left(\sum_{i=1}^n e_i (A_{ij}\xi^j)\right) = \sum_{i=1}^n e_i \left(\sum_{j=1}^n A_{ij}\xi^j\right).$$

Exercise 1.1.15 Show that if $B : \mathbb{F}^n \to \mathbb{F}^n$ is another $\mathbb{F}$-linear map, then so is $B \circ A : \mathbb{F}^n \to \mathbb{F}^n$ and that the matrix of $B \circ A$ relative to $\{e_1, \ldots, e_n\}$ is the product matrix

$$BA = \left(\sum_{i=1}^n B_{ki} A_{ij}\right)_{k,j=1,\ldots,n}.$$

From Exercise 1.1.15 it follows at once that an $\mathbb{F}$-linear map $A : \mathbb{F}^n \to \mathbb{F}^n$ is invertible (i.e., one-to-one and onto with an $\mathbb{F}$-linear inverse $A^{-1} : \mathbb{F}^n \to \mathbb{F}^n$) iff its matrix relative to $\{e_1, \ldots, e_n\}$ is invertible (i.e., has a matrix inverse). The collection of all invertible $n \times n$ matrices with entries in $\mathbb{F}$ is denoted $GL(n, \mathbb{F})$ and called the **general linear group of order n over $\mathbb{F}$**.

Exercise 1.1.16 Show that $GL(n, \mathbb{F})$ is, indeed, a group under matrix multiplication.

As a collection of $n \times n$ matrices, $GL(n, \mathbb{F})$ acquires a topology as a subspace of some Euclidean space. We claim that, in fact, $GL(n, \mathbb{R})$ is an open subset of $\mathbb{R}^{n^2}$, $GL(n, \mathbb{C})$ is an open subset of $\mathbb{R}^{2n^2}$ and $GL(n, \mathbb{H})$ is an open subset of $\mathbb{R}^{4n^2}$. This is particularly easy to see for $GL(n, \mathbb{R})$. Recall that an $n \times n$ real matrix is invertible iff its determinant is nonzero. Now, the determinant function det, defined on the set of all $n \times n$ real matrices (i.e., on $\mathbb{R}^{n^2}$), is a polynomial in the entries of the matrix (i.e., in the coordinates in $\mathbb{R}^{n^2}$) and so is continuous. The inverse image under this function of the open set $\mathbb{R} - \{0\}$ is therefore open and this is precisely $GL(n, \mathbb{R})$. Essentially the same argument works for $GL(n, \mathbb{C})$, but this time det is complex-valued.

Exercise 1.1.17 Write out the proof that $GL(n, \mathbb{C})$ is an open subspace of $\mathbb{R}^{2n^2}$. **Hint**: Recall that a map from $\mathbb{R}^{2n^2}$ into $\mathbb{R}^2$ is continuous if its coordinate functions are continuous.

$GL(n, \mathbb{H})$ presents something of a problem since the noncommutativity of $\mathbb{H}$ effectively blocks any meaningful notion of the "determinant" of a quaternionic matrix. We evade this difficulty as follows: As indicated earlier we identify $\mathbb{H}^n$ with $\mathbb{R}^{4n}$ via

$$\begin{aligned}(q^1, \ldots, q^n) = (x^1 + y^1\mathbf{i} + u^1\mathbf{j} + v^1\mathbf{k}, \ldots, x^n + y^n\mathbf{i} + u^n\mathbf{j} + v^n\mathbf{k}) \longrightarrow \\ (x^1, y^1, u^1, v^1, \ldots, x^n, y^n, u^n, v^n).\end{aligned}$$

Similarly, the set of $n \times n$ quaternionic matrices is identified with $\mathbb{R}^{4n^2}$ via

$$\begin{aligned}Q = \begin{pmatrix} q_{11} & \cdots & q_{1n} \\ \vdots & & \vdots \\ q_{n1} & \cdots & q_{nn} \end{pmatrix} &= (q_{ij}) \\ &= (x_{ij} + y_{ij}\mathbf{i} + u_{ij}\mathbf{j} + v_{ij}\mathbf{k}) \longrightarrow \\ &\quad (q_{11}, \ldots, q_{1n}, \ldots, q_{n1}, \ldots, q_{nn}) \longrightarrow \\ &\quad (x_{11}, y_{11}, u_{11}, v_{11}, \ldots, x_{1n}, y_{1n}, u_{1n}, v_{1n}, \ldots, \\ &\qquad x_{n1}, y_{n1}, u_{n1}, v_{n1}, \ldots, x_{nn}, y_{nn}, u_{nn}, v_{nn}).\end{aligned}$$

Now, identify Q with an $\mathbb{H}$-linear map $Q : \mathbb{H}^n \to \mathbb{H}^n$. Applying this linear transformation to the element $(a^1+b^1\mathbf{i}+c^1\mathbf{j}+d^1\mathbf{k},\ldots,a^n+b^n\mathbf{i}+c^n\mathbf{j}+d^n\mathbf{k})$ of $\mathbb{H}^n$ (i.e., multiplying the column vector with these entries by the matrix (q_{ij})) yields the element of $\mathbb{H}^n$ whose i^{th} coordinate is

$$\left[\sum_{j=1}^{n}(x_{ij}a^j - y_{ij}b^j - u_{ij}c^j - v_{ij}d^j)\right] + \left[\sum_{j=1}^{n}(x_{ij}b^j + y_{ij}a^j + u_{ij}d^j - v_{ij}c^j)\right]\mathbf{i} +$$

$$\left[\sum_{j=1}^{n}(x_{ij}c^j + u_{ij}a^j + v_{ij}b^j - y_{ij}d^j)\right]\mathbf{j} + \left[\sum_{j=1}^{n}(x_{ij}d^j + v_{ij}a^j + y_{ij}c^j - u_{ij}b^j)\right]\mathbf{k}.$$

Exercise 1.1.18 Perform these calculations.

Identifying this image point with an element of $\mathbb{R}^{4n}$ as indicated above (and writing it as a column vector) one finds that the result can be written as the following real matrix product:

$$\begin{pmatrix} x_{11} & -y_{11} & -u_{11} & -v_{11} & \cdots & x_{1n} & -y_{1n} & -u_{1n} & -v_{1n} \\ y_{11} & x_{11} & -v_{11} & u_{11} & \cdots & y_{1n} & x_{1n} & -v_{1n} & u_{1n} \\ u_{11} & v_{11} & x_{11} & -y_{11} & \cdots & u_{1n} & v_{1n} & x_{1n} & -y_{1n} \\ v_{11} & -u_{11} & y_{11} & x_{11} & \cdots & v_{1n} & -u_{1n} & y_{1n} & x_{1n} \\ \vdots & \vdots & \vdots & \vdots & & \vdots & \vdots & \vdots & \vdots \\ x_{n1} & -y_{n1} & -u_{n1} & -v_{n1} & \cdots & x_{nn} & -y_{nn} & -u_{nn} & -v_{nn} \\ y_{n1} & x_{n1} & -v_{n1} & u_{n1} & \cdots & y_{nn} & x_{nn} & -v_{nn} & u_{nn} \\ u_{n1} & v_{n1} & x_{n1} & -y_{n1} & \cdots & u_{nn} & v_{nn} & x_{nn} & -y_{nn} \\ v_{n1} & -u_{n1} & y_{n1} & x_{n1} & \cdots & v_{nn} & -u_{nn} & y_{nn} & x_{nn} \end{pmatrix} \begin{pmatrix} a^1 \\ b^1 \\ c^1 \\ d^1 \\ \vdots \\ a^n \\ b^n \\ c^n \\ d^n \end{pmatrix}$$

Now, (q_{ij}) is invertible (i.e., in $GL(n,\mathbb{H})$) iff $Q : \mathbb{H}^n \to \mathbb{H}^n$ has trivial kernel and this is the case iff the $4n \times 4n$ real matrix above has nonzero determinant. The map which sends $(x_{11}, y_{11}, u_{11}, v_{11}, \ldots, x_{nn}, y_{nn}, u_{nn}, v_{nn})$ in $\mathbb{R}^{4n^2}$ to this matrix is clearly continuous since it has continuous coordinate (entry) functions so its composition with the real determinant function is continuous. Thus, the set of $n \times n$ quaternionic matrices for which this determinant is nonzero is open and this is precisely $GL(n,\mathbb{H})$.

Thus, each general linear group $GL(n,\mathbb{F})$ is homeomorphic to an open set in some Euclidean space. These will eventually (Sections 1.6 and 4.7) supply more examples of "topological groups" and "Lie groups". The same is true of the remaining examples in this section.

We will be particularly interested in those $\mathbb{F}$-linear maps $A : \mathbb{F}^n \to \mathbb{F}^n$ that preserve the bilinear form $< , >$, i.e., that satisfy

$$< A(\xi), A(\zeta) > = < \xi, \zeta > \quad \text{for all } \xi, \zeta \in \mathbb{F}^n. \tag{1.1.22}$$

We determine a necessary and sufficient condition on the matrix (A_{ij}) of A to ensure that this is the case. Letting $\xi = \sum_{j=1}^n e_j \xi^j$ and $\zeta = \sum_{k=1}^n e_k \zeta^k$

we have $A(\xi) = \sum_{i=1}^n e_i(\sum_{j=1}^n A_{ij}\xi^j)$ and $A(\zeta) = \sum_{i=1}^n e_i(\sum_{k=1}^n A_{ik}\zeta^k)$ so

$$\begin{aligned}
<A(\xi), A(\zeta)> &= \left\langle (\sum_{j=1}^n A_{1j}\xi^j, \ldots, \sum_{j=1}^n A_{nj}\xi^j),\ (\sum_{k=1}^n A_{1k}\zeta^k, \ldots, \sum_{k=1}^n A_{nk}\zeta^k) \right\rangle \\
&= \overline{(\sum_{j=1}^n A_{1j}\xi^j)}(\sum_{k=1}^n A_{1k}\zeta^k) + \cdots + \overline{(\sum_{j=1}^n A_{nj}\xi^j)}(\sum_{k=1}^n A_{nk}\zeta^k) \\
&= (\sum_{j=1}^n \bar{\xi}^j \bar{A}_{1j})(\sum_{k=1}^n A_{1k}\zeta^k) + \cdots + (\sum_{j=1}^n \bar{\xi}^j \bar{A}_{nj})(\sum_{k=1}^n A_{nk}\zeta^k) \\
&= \sum_{j,k=1}^n (\bar{\xi}^j \bar{A}_{1j} A_{1k}\zeta^k) + \cdots + \sum_{j,k=1}^n (\bar{\xi}^j \bar{A}_{nj} A_{nk}\zeta^k) \\
&= \sum_{j,k=1}^n \bar{\xi}^j (\bar{A}_{1j} A_{1k} + \cdots + \bar{A}_{nj} A_{nk})\zeta^k .
\end{aligned}$$

But $< \xi, \zeta > = \sum_{j=1}^n \bar{\xi}^j \zeta^j = \sum_{j,k=1}^n \bar{\xi}^j \delta_{jk} \zeta^k$ (where δ_{jk} is the Kronecker delta, i.e., 1 if $j = k$, but 0 otherwise). Thus, $< A(\xi), A(\zeta) >$ can equal $< \xi, \zeta >$ *for all* ξ and ζ in $\mathbb{F}^n$ iff

$$\bar{A}_{1j} A_{1k} + \cdots + \bar{A}_{nj} A_{nk} = \delta_{jk}\,, \qquad j, k = 1, \ldots, n\,,$$

i.e.,

$$\sum_{i=1}^n \bar{A}_{ij} A_{ik} = \delta_{jk} \qquad j, k = 1, \ldots, n\,. \tag{1.1.23}$$

If we denote by A also the matrix (A_{ij}) of $A : \mathbb{F}^n \to \mathbb{F}^n$, then the left hand side of (1.1.23) is the jk-entry in the product $\bar{A}^T A$. The right-hand side of (1.1.23) is the jk-entry in the $n \times n$ identity matrix *id*. Thus, we find that (1.1.22) is equivalent to

$$\bar{A}^T A = id\,. \tag{1.1.24}$$

Exercise 1.1.19 Show that an $\mathbb{F}$-linear map $A : \mathbb{F}^n \to \mathbb{F}^n$ that satisfies (1.1.22) is necessarily invertible (i.e., has trivial kernel). Conclude that its matrix A is invertible and that

$$A^{-1} = \bar{A}^T\,. \tag{1.1.25}$$

For $\mathbb{F} = \mathbb{R}, \mathbb{C}$ or $\mathbb{H}$ and for any positive integer n we define the **$\mathbb{F}$-unitary group of order** n to be the set $U(n, \mathbb{F})$ of all $n \times n$ matrices A with entries in $\mathbb{F}$ that are invertible and satisfy $A^{-1} = \bar{A}^T$. When $\mathbb{F} = \mathbb{R}$,

$U(n, \mathbb{R})$ is generally called the **orthogonal group of order** n and denoted $O(n)$. When $\mathbb{F} = \mathbb{C}$, $U(n, \mathbb{C})$ is simply written $U(n)$ and called the **unitary group of order** n. Finally, when $\mathbb{F} = \mathbb{H}$, $U(n, \mathbb{H})$ is called the **symplectic group of order** n and denoted $Sp(n)$. As the terminology suggests, these are all groups under matrix multiplication. Indeed, this follows at once from

Exercise 1.1.20 Let $A, B : \mathbb{F}^n \to \mathbb{F}^n$ be two $\mathbb{F}$-linear maps that preserve $< , >$ (i.e., $< A(\xi), A(\zeta) > = < \xi, \zeta > = < B(\xi), B(\zeta) >$ for all $\xi, \zeta \in \mathbb{F}^n$). Show that $B \circ A$ and A^{-1} also preserve $< , >$.

A **basis** for $\mathbb{F}^n$ is a set of n elements $\{\xi_1, \ldots, \xi_n\}$ of $\mathbb{F}^n$ with the property that any $\xi \in \mathbb{F}^n$ can be uniquely written as $\xi = \xi_1 a_1 + \cdots + \xi_n a_n$, where $a_1, \ldots, a_n$ are in $\mathbb{F}$. The basis is said to be **orthonormal** if $< \xi_j, \xi_k > = \delta_{jk}$ for all $j, k = 1, \ldots, n$. Of course, the standard basis $\{e_1, \ldots, e_n\}$ is one such.

Exercise 1.1.21 Show that an $\mathbb{F}$-linear map $A : \mathbb{F}^n \to \mathbb{F}^n$ preserves $< , >$ iff it carries an orthonormal basis for $\mathbb{F}^n$ onto another orthonormal basis for $\mathbb{F}^n$. Conclude that the columns of any matrix in $U(n, \mathbb{F})$, when each is regarded as an element in $\mathbb{F}^n$, constitute an orthonormal basis for $\mathbb{F}^n$. Show that the same is true of the rows.

Exercise 1.1.22 Let ξ be any nonzero element in $\mathbb{F}^n$. Show that $< \xi, \xi >$ is a positive real number and that $\xi_1 = \xi(< \xi, \xi >)^{-1/2}$ satisfies $< \xi_1, \xi_1 > = 1$. Now mimic the usual Gram-Schmidt orthogonalization procedure (Chapter $\overline{\text{VI}}$ of [**Lang**]) to show that there exists an orthonormal basis $\{\xi_1, \ldots, \xi_n\}$ for $\mathbb{F}^n$ containing ξ_1.

We have already described a procedure for supplying each $U(n, \mathbb{F})$ with a topology ($O(n) \subseteq \mathbb{R}^{n^2}$, $U(n) \subseteq \mathbb{R}^{2n^2}$ and $Sp(n) \subseteq \mathbb{R}^{4n^2}$) and we will have much to say about these topological spaces as we proceed. We isolate further subspaces of these that will figure prominently in our work. Notice first that if A is in $O(n)$, then $AA^T = id$ so $\det(AA^T) = 1 \Rightarrow (\det A)(\det A^T) = 1 \Rightarrow (\det A)^2 = 1 \Rightarrow \det A = \pm 1$. The subset of $O(n)$ consisting of those A with determinant 1 is called the **special orthogonal group of order** n and is denoted

$$SO(n) = \{A \in O(n) : \det A = 1\}.$$

We provide $SO(n)$ with the relative topology it inherits from $O(n)$ (or, equivalently, from $\mathbb{R}^{n^2}$). Since $\det(AB) = (\det A)(\det B)$ and $\det(A^{-1}) = (\det A)^{-1}$, $SO(n)$ is a subgroup of $O(n)$.

Exercise 1.1.23 Show that, for any $A \in U(n)$, $\det A$ is a complex number of modulus 1.

The **special unitary group of order** n is defined by $SU(n) = \{A \in U(n) : \det A = 1\}$ and is a topological subspace as well as a subgroup of $U(n)$.

Exercise 1.1.24 Notice that we have now introduced $SU(2)$ twice; once as a subspace of $\mathcal{R}^4$ (Lemma 1.1.3) and just now as a subspace of $U(2) \subseteq \mathbb{R}^8$. Show that these are homeomorphic.

Exercise 1.1.25 Show that $U(1)$ is homeomorphic to S^1.

Exercise 1.1.26 Show that $Sp(1)$ is homeomorphic to $SU(2)$.

Exercise 1.1.27 Let $\mathbb{F}$ be either $\mathbb{R}$, $\mathbb{C}$, or $\mathbb{H}$ and denote by S the topological subspace of $\mathbb{F}^n$ given by $S = \{\xi \in \mathbb{F}^n : <\xi, \xi>= 1\}$. Show that S is homeomorphic to either S^{n-1}, S^{2n-1}, or S^{4n-1} depending on whether $\mathbb{F}$ is $\mathbb{R}$, $\mathbb{C}$, or $\mathbb{H}$, respectively.

Quaternionic matrices are rather difficult to compute with due to the noncommutativity of $\mathbb{H}$. We conclude this section by constructing alternate representations of the groups $GL(n, \mathbb{H})$ and $Sp(n)$ that are often more convenient. Note first that if $x = x^0 + x^1\mathbf{i} + x^2\mathbf{j} + x^3\mathbf{k}$ is any element of $\mathbb{H}$ and if we define $z^1 = x^0 + x^1\mathbf{i}$ and $z^2 = x^2 + x^3\mathbf{i}$, then $z^1 + z^2\mathbf{j} = x^0 + x^1\mathbf{i} + x^2\mathbf{j} + x^3\mathbf{ij} = x^0 + x^1\mathbf{i} + x^2\mathbf{j} + x^3\mathbf{k} = x$. Thus, we may identify $\mathbb{H}$ with $\mathbb{C}^2$ via the map that carries $x = x^0 + x^1\mathbf{i} + x^2\mathbf{j} + x^3\mathbf{k}$ to $(z^1, z^2) = (x^0 + x^1\mathbf{i}, x^2 + x^3\mathbf{i})$.

Now, suppose P is an $n \times n$ quaternionic matrix. By writing each entry in the form $z^1 + z^2\mathbf{j}$ we may write P itself in the form $P = A + B\mathbf{j}$, where A and B are $n \times n$ complex matrices. If $Q = C + D\mathbf{j}$ is another such $n \times n$ quaternionic matrix, then, since $\mathbf{j}C = \bar{C}\mathbf{j}$ and $D\mathbf{j} = \mathbf{j}\bar{D}$,

$$\begin{aligned} PQ &= (A + B\mathbf{j})(C + D\mathbf{j}) = AC + AD\mathbf{j} + (B\mathbf{j})C + (B\mathbf{j})(D\mathbf{j}) \\ &= AC + AD\mathbf{j} + B\bar{C}\mathbf{j} + (B\mathbf{j})(\mathbf{j}\bar{D}) \\ &= (AC - B\bar{D}) + (AD + B\bar{C})\mathbf{j}\,. \end{aligned}$$

Now, define a mapping ϕ from the algebra of $n \times n$ quaternionic matrices to the algebra of $2n \times 2n$ complex matrices as follows: For $P = A + B\mathbf{j}$, let

$$\phi(P) = \begin{pmatrix} A & B \\ -\bar{B} & \bar{A} \end{pmatrix} \tag{1.1.26}$$

(compare with the definition of $\mathcal{R}^4$).

Exercise 1.1.28 Show that ϕ is an isomorphism (of algebras) that preserves the conjugate transpose, i.e., $\phi(\bar{P}^T) = (\overline{\phi(P)})^T$. Conclude that $P \in Sp(n)$ iff $\phi(P) \in U(2n)$.

Thus, we may identify $Sp(n)$ with the set of all elements of $U(2n)$ of the form (1.1.26).

Exercise 1.1.29 Show that a $2n \times 2n$ complex matrix M has the form $\begin{pmatrix} A & B \\ -\bar{B} & \bar{A} \end{pmatrix}$ iff it satisfies $JMJ^{-1} = \bar{M}$, where

$$J = \begin{pmatrix} 0 & id \\ -id & 0 \end{pmatrix}$$

(here 0 is the $n \times n$ zero matrix and id is the $n \times n$ identity matrix). Show also that, if M is unitary, then the condition $JMJ^{-1} = \bar{M}$ is equivalent to $M^T JM = J$.

Thus, we may identify $Sp(n)$ algebraically with the subgroup of $U(2n)$ consisting of those elements M that satisfy $M^T JM = J$. $GL(n, \mathbb{H})$ is identified with the set of invertible $2n \times 2n$ matrices M that satisfy $JMJ^{-1} = \bar{M}$. We will also have occasion to consider the collection of all $n \times n$ quaternionic matrices P for which $\det \phi(P) = 1$. This is called the **quaternionic special linear group**, denoted $SL(n, \mathbb{H})$ and can be identified with the set of $2n \times 2n$ complex matrices M that satisfy $JMJ^{-1} = \bar{M}$ and $\det M = 1$.

Exercise 1.1.30 Show that $SL(n, \mathbb{H})$ is a subgroup of $GL(n, \mathbb{H})$.

We conclude by observing that our two views of quaternionic matrices are topologically consistent. We have supplied the set of $n \times n$ quaternionic matrices with a topology by identifying it with $\mathbb{R}^{4n^2}$ in the following way:

$$(q_{ij}) = (x_{ij} + y_{ij}\mathbf{i} + u_{ij}\mathbf{j} + v_{ij}\mathbf{k}) \longrightarrow$$
$$(x_{11}, y_{11}, u_{11}, v_{11}, \ldots, x_{nn}, y_{nn}, u_{nn}, v_{nn}) .$$

On the other hand, the corresponding set of $2n \times 2n$ complex matrices acquires its topology as a subset of $\mathbb{R}^{8n^2}$ as follows: With $A = (x_{ij} + y_{ij}\mathbf{i})$ and $B = (u_{ij} + v_{ij}\mathbf{i})$,

$$\begin{pmatrix} A & B \\ -\bar{B} & \bar{A} \end{pmatrix} \longrightarrow (x_{11}, y_{11}, \ldots, x_{1n}, y_{1n}, u_{11}, v_{11}, \ldots, u_{1n}, v_{1n}, \ldots,$$
$$-u_{n1}, v_{n1}, \ldots, -u_{nn}, v_{nn}, x_{n1}, -y_{n1}, \ldots, x_{nn}, -y_{nn}) .$$

The projection of this subset of $\mathbb{R}^{8n^2}$ onto the first $4n^2$ coordinates is clearly one-to-one and maps onto the first $\mathbb{R}^{4n^2}$ factor in $\mathbb{R}^{8n^2} = \mathbb{R}^{4n^2} \times \mathbb{R}^{4n^2}$. This projection is also linear and therefore a homeomorphism. Composing with the homeomorphism $(x_{11}, y_{11}, \ldots, x_{nn}, y_{nn}, u_{11}, v_{11}, \ldots, u_{nn}, v_{nn}) \to (x_{11}, y_{11}, u_{11}, v_{11}, \ldots, x_{nn}, y_{nn}, u_{nn}, v_{nn})$ gives the desired result.

1.2 Quotient Topologies and Projective Spaces

Many of the topological spaces of interest to us do not arise naturally as simple subspaces of some Euclidean space and so we must now begin to enlarge our collection of procedures for producing examples. Suppose first that one has a topological space X, a set Y and a mapping $\mathcal{Q} : X \to Y$ of X onto Y. Consider the collection of all subsets U of Y with the property that $\mathcal{Q}^{-1}(U)$ is open in X. Clearly, $\mathcal{Q}^{-1}(\emptyset) = \emptyset$ and, since $\mathcal{Q}$ is surjective, $\mathcal{Q}^{-1}(Y) = X$ so $\emptyset$ and Y are both in this set. Moreover, $\mathcal{Q}^{-1}(\bigcup_{\alpha \in \mathcal{A}} U_\alpha) = \bigcup_{\alpha \in \mathcal{A}} \mathcal{Q}^{-1}(U_\alpha)$ and $\mathcal{Q}^{-1}(U_1 \cap \cdots \cap U_k) = \mathcal{Q}^{-1}(U_1) \cap \cdots \cap \mathcal{Q}^{-1}(U_k)$ imply that this collection is closed under the formation of arbitrary unions and finite intersections. In other words,

$$\mathcal{T}_{\mathcal{Q}} = \left\{ U \subseteq Y : \mathcal{Q}^{-1}(U) \text{ is open in } X \right\}$$

is a topology on Y which we will call the **quotient topology on Y determined by the (surjective) map $\mathcal{Q} : X \to Y$**. Notice that, since $\mathcal{Q}^{-1}(Y - U) = X - \mathcal{Q}^{-1}(U)$, a subset of Y is closed in this topology iff its inverse image under $\mathcal{Q}$ is closed in X. Moreover, the map $\mathcal{Q} : X \to Y$, called the **quotient map**, is obviously continuous if Y has the topology $\mathcal{T}_{\mathcal{Q}}$. More is true, however.

Lemma 1.2.1 *Let X be a topological space, $\mathcal{Q} : X \to Y$ a surjection and suppose Y has the quotient topology determined by $\mathcal{Q}$. Then, for any topological space Z, a map $g : Y \to Z$ is continuous iff $g \circ \mathcal{Q} : X \to Z$ is continuous.*

Proof: If g is continuous, then so is $g \circ \mathcal{Q}$ by Exercise 1.1.1. Conversely, suppose $g \circ \mathcal{Q}$ is continuous. We show that g is continuous. Let V be an arbitrary open set in Z. Then $(g \circ \mathcal{Q})^{-1}(V) = \mathcal{Q}^{-1}(g^{-1}(V))$ is open in X. But then, by definition of $\mathcal{T}_{\mathcal{Q}}$, $g^{-1}(V)$ is open in Y so g is continuous. ∎

If Y has the quotient topology determined by some surjection $\mathcal{Q} : X \to Y$, then Y is called a **quotient space of X (by $\mathcal{Q}$)**. Thus, a map out of a quotient space is continuous iff its composition with the quotient map is continuous.

If $\mathcal{Q} : X \to Y$ is a quotient map, then, for any $y \in Y$, the subset $\mathcal{Q}^{-1}(y) = \{x \in X : \mathcal{Q}(x) = y\}$ is called the **fiber of $\mathcal{Q}$ over y**. We show now that any continuous map out of X that is constant on each fiber of $\mathcal{Q}$ "descends" to a continuous map on Y.

Lemma 1.2.2 *Let $\mathcal{Q} : X \to Y$ be a quotient map, Z a topological space and $f : X \to Z$ a continuous map with the properly that $f \mid \mathcal{Q}^{-1}(y)$ is a constant map for each $y \in Y$. Then there exists a unique continuous map $\bar{f} : Y \to Z$ such that $\bar{f} \circ \mathcal{Q} = f$.*

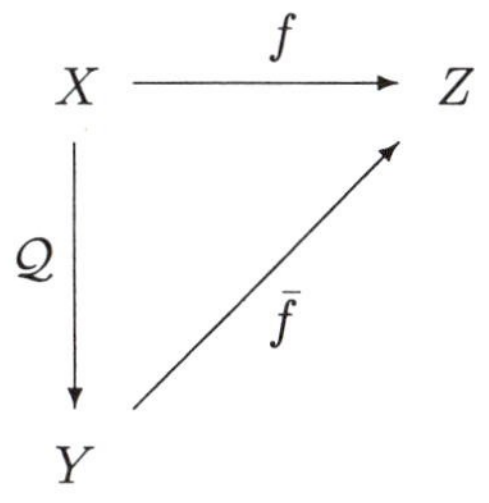

Proof: For each $y \in Y$ we define $\bar{f}(y)$ by $\bar{f}(y) = f(x)$ for any $x \in \mathcal{Q}^{-1}(y)$. $\bar{f}$ is well-defined because f is constant on the fibers of $\mathcal{Q}$. Moreover, for every x in X, $(\bar{f} \circ \mathcal{Q})(x) = \bar{f}(\mathcal{Q}(x)) = f(x)$ so $\bar{f} \circ \mathcal{Q} = f$. To show that $\bar{f}$ is continuous we consider an open set V in Z and $\bar{f}^{-1}(V)$ in Y. By definition of the quotient topology on Y we need only show that $\mathcal{Q}^{-1}(\bar{f}^{-1}(V))$ is open in X. But $\mathcal{Q}^{-1}(\bar{f}^{-1}(V)) = (\bar{f} \circ \mathcal{Q})^{-1}(V) = f^{-1}(V)$ and this is open in X because f is continuous. Finally, to prove uniqueness, suppose $\bar{f}' : Y \to Z$ also satisfies $\bar{f}' \circ \mathcal{Q} = f$. Then, for every $x \in X$, $\bar{f}'(\mathcal{Q}(x)) = \bar{f}(\mathcal{Q}(x))$. But $\mathcal{Q}$ is surjective so every $y \in Y$ is $\mathcal{Q}(x)$ for some $x \in X$ and $\bar{f}'(y) = \bar{f}(y)$ for every $y \in Y$. ∎

Quotient spaces arise most frequently in the following way: Let X be a topological space on which is defined some equivalence relation $\sim$. For each $x \in X$, the equivalence class containing x is denoted $[x]$ and the set of all such equivalence classes is written $X/\sim$. The canonical projection map $\mathcal{Q} : X \to X/\sim$ assigns to each $x \in X$ the equivalence class containing $x : \mathcal{Q}(x) = [x]$. Assigning to $X/\sim$ the quotient topology determined by $\mathcal{Q}$ gives a quotient space in which each equivalence class is represented by a single point ($\mathcal{Q}$ "identifies" the equivalence classes of $\sim$ to points). A collection of equivalence classes in $X/\sim$ is then open if the union of all of these equivalence classes (thought of as subsets of X) is open in X.

Very shortly we will consider in detail some important examples of the construction just described. First, however, we must point out that a quotient of a very nice topological space X can be a quite unimpressive specimen.

Exercise 1.2.1 Let X be the subspace $[0, 1]$ of $\mathbb{R}$. Define a relation $\sim$ on $[0, 1]$ as follows: $x \sim y$ iff $|x - y|$ is rational. Verify that $\sim$ is an equivalence relation on $[0, 1]$ and describe its equivalence classes. Provide $X/\sim$ with the quotient topology and show that its only open sets are $\emptyset$ and $X/\sim$ (the topology on a set that consists only of $\emptyset$ and the set itself is rightly called the **indiscrete topology** on that set and is utterly devoid of any redeeming social value).

The most serious indiscretion of the topology just described is that it does not satisfy the following, very desirable, condition (unless it does so vacuously). A topological space X is said to be **Hausdorff** if whenever x and y are distinct points of X there exist open sets U_x and U_y in X with $x \in U_x$, $y \in U_y$ and $U_x \cap U_y = \emptyset$ (distinct points can be "separated" by disjoint open sets). Surely, any subspace X of a Euclidean space is Hausdorff (intersect with X the open balls $U_{d/2}(x)$ and $U_{d/2}(y)$, where $d = \| x - y \|$). Although non -Hausdorff spaces do come up now and then, they are rather pathological and we shall avoid them whenever possible. In particular, when we construct examples we will take care to verify the Hausdorff condition explicitly. Notice that "Hausdorff" is a **topological property**, i.e., if X has this property and Y is homeomorphic to X, then Y must also have the same property.

Now we will construct the so-called **projective spaces**. These come in three varieties (real, complex and quaternionic), but it is possible to carry out all of the constructions at once. Let $\mathbb{F}$ denote one of $\mathbb{R}$, $\mathbb{C}$, or $\mathbb{H}$ and let $n \geq 2$ be an integer. We consider $\mathbb{F}^n$ with the structure described in the previous section and denote by 0 the zero element $(0, \ldots, 0)$ in $\mathbb{F}^n$. On the topological subspace $\mathbb{F}^n - \{0\}$ of $\mathbb{F}^n$ we define a relation $\sim$ as follows: $\zeta \sim \xi$ iff there exists a nonzero $a \in \mathbb{F}$ such that $\zeta = \xi a$.

Exercise 1.2.2 Show that $\sim$ is an equivalence relation on $\mathbb{F}^n - \{0\}$.

The equivalence class of ξ in $\mathbb{F}^n - \{0\}$ is

$$\begin{aligned}[\xi] = [\xi^1, \ldots, \xi^n] &= \{\xi a : a \in \mathbb{F} - \{0\}\} \\ &= \{ (\xi^1 a, \ldots, \xi^n a) : a \in \mathbb{F} - \{0\} \} .\end{aligned}$$

Note that if $\mathbb{F} = \mathbb{R}$ these are just straight lines through the origin in $\mathbb{R}^n$ with the origin then deleted. If $\mathbb{F} = \mathbb{C}$ or $\mathbb{H}$ they are called **complex** or **quaternionic lines through the origin in** $\mathbb{C}^n$ or $\mathbb{H}^n$, respectively (minus the origin). We denote by $\mathbb{FP}^{n-1}$ the quotient space $(\mathbb{F}^n - \{0\})/\sim$, i.e.,

$$\mathbb{FP}^{n-1} = \{ [\xi] : \xi \in \mathbb{F}^n - \{0\} \}$$

with the quotient topology determined by the projection $\mathcal{Q} : \mathbb{F}^n - \{0\} \to \mathbb{FP}^{n-1}$, $\mathcal{Q}(\xi) = [\xi]$. $\mathbb{RP}^{n-1}$, $\mathbb{CP}^{n-1}$ and $\mathbb{HP}^{n-1}$ are called, respectively, the **real**, **complex** and **quaternionic projective space** of **dimension** $n-1$.

There is another way of viewing these projective spaces that will be quite important. As in Exercise 1.1.27 we consider the subset $S = \{\xi \in \mathbb{F}^n : < \xi, \xi >= 1\}$ of $\mathbb{F}^n - \{0\}$ (it is homeomorphic to a sphere of some dimension). Let $\mathcal{P} = \mathcal{Q} \mid S$ be the restriction of $\mathcal{Q}$ to S. Then $\mathcal{P} : S \to \mathbb{FP}^{n-1}$ is continuous. We claim that it is also surjective. To see this let $\xi = (\xi^1, \ldots, \xi^n)$ be in $\mathbb{F}^n - \{0\}$. Then $< \xi, \xi >= \bar{\xi}^1 \xi^1 + \cdots + \bar{\xi}^n \xi^n = |\xi^1|^2 + \cdots + |\xi^n|^2$ is a positive real number. Now "normalize" ξ by dividing out $\sqrt{<\xi,\xi>}$, i.e., define $\zeta = \xi(<\xi,\xi>)^{-1/2}$. Then $<\zeta,\zeta>=1$ so ζ is an element of S. Moreover, $\zeta \sim \xi$ so $\mathcal{Q}(\zeta) = \mathcal{Q}(\xi)$, i.e., $\mathcal{P}(\zeta) = \mathcal{Q}(\xi)$. Now consider the following diagram of continuous maps

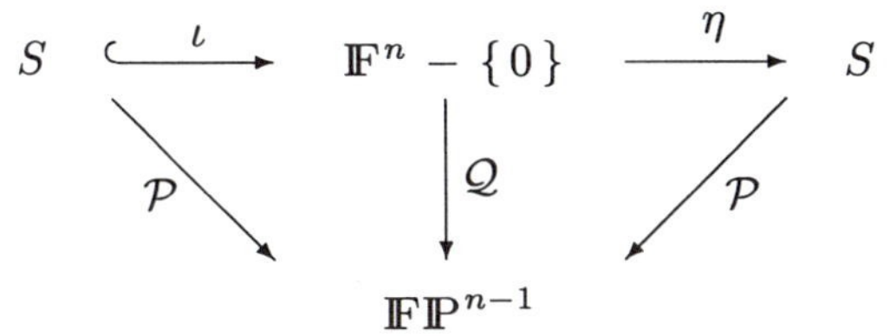

where $\mathbb{F}^n - \{0\} \xrightarrow{\eta} S$ normalizes elements of $\mathbb{F}^n - \{0\}$, i.e., sends ξ to $\xi(<\xi,\xi>)^{-1/2}$. This is easily seen to be continuous by writing out the map in terms of Euclidean coordinates and appealing to Lemma 1.1.2. We have just seen that this diagram commutes, i.e., $\mathcal{Q} \circ \iota = \mathcal{P}$ and $\mathcal{P} \circ \eta = \mathcal{Q}$. It follows that a subset U of $\mathbb{FP}^{n-1}$ is open iff $\mathcal{P}^{-1}(U)$ is open in S, i.e., that $\mathbb{FP}^{n-1}$ *also has the quotient topology determined by* $\mathcal{P} : S \to \mathbb{FP}^{n-1}$.

Exercise 1.2.3 Prove this.

This second description of $\mathbb{FP}^{n-1}$, as a quotient of the sphere S, will generally be the most convenient for our purposes. The fiber $\mathcal{P}^{-1}([\xi])$ above $[\xi] \in \mathbb{FP}^{n-1}$ is then the intersection with S of $\{\xi a : a \in \mathbb{F} - \{0\}\}$. Now, if $\xi \in S$, then $<\xi a, \xi a> = \bar{a} <\xi,\xi> a = \bar{a}a = |a|^2$ so ξa will be in S iff $|a|^2 = 1$. Thus, for every $\xi \in S$,

$$\mathcal{P}^{-1}([\xi]) = \{\xi a : a \in \mathbb{F} \text{ and } |a| = 1\}.$$

Thus, one obtains the fiber containing any $\xi \in S$ by multiplying it by all the unit elements of $\mathbb{F}$. For example, if $\mathbb{F} = \mathbb{R}$, then $S \cong S^{n-1}$ (Exercise 1.1.27) and an $a \in \mathbb{R}$ satisfies $|a| = 1$ iff $a = \pm 1$. Thus, for any $x = (x^1, \ldots, x^n) \in S^{n-1}$, $\mathcal{P}^{-1}([x]) = \{x, -x\}$ is a pair of antipodal points on S^{n-1}. $\mathbb{RP}^{n-1}$ can therefore be thought of as the $(n-1)$-sphere S^{n-1} with "antipodal points identified".

Next suppose $\mathbb{F} = \mathbb{C}$. An $a \in \mathbb{C}$ satisfies $|a| = 1$ iff a is on the unit circle $S^1 \subseteq \mathbb{C}$. Thus, for any $\xi_0 = (z_0^1, \ldots, z_0^n) \in S \subseteq \mathbb{C}^n$, $\mathcal{P}^{-1}([\xi_0]) = \{\xi_0 a : a \in S^1\} = \{(z_0^1 a, \ldots, z_0^n a) : a \in S^1\}$. If ξ_0 is fixed, this subspace of S is homeomorphic to S^1. We prove this as follows. Since $\xi_0 \in S$, some z_0^j is nonzero. Consider the map from $\mathbb{C}^n$ to $\mathbb{C}$ that carries $(z^1, \ldots, z^n)$ to $(z_0^j)^{-1} z^j$. Identifying $\mathbb{C}^n$ with $\mathbb{R}^{2n}$ and $\mathbb{C}$ with $\mathbb{R}^2$ as in Section 1.1 we show that this map is continuous. Indeed, writing $z^1 = x^1 + y^1\mathbf{i}, \ldots, z^n = x^n + y^n\mathbf{i}$ and $z_0^j = \alpha + \beta\mathbf{i}$, our map is the following composition of continuous maps:

$$(x^1, y^1, x^2, y^2, \ldots, x^n, y^n) \longrightarrow (x^j, y^j) \longrightarrow \left(\frac{\alpha x^j + \beta y^j}{\alpha^2 + \beta^2}, \frac{\alpha y^j - \beta x^j}{\alpha^2 + \beta^2}\right).$$

Thus, the restriction of this map to $\mathcal{P}^{-1}([\xi_0])$, which carries $(z_0^1 a, \ldots, z_0^n a)$ to a, is also continuous. This restriction is obviously also one-to-one and onto S^1. Its inverse $a \to (z_0^1 a, \ldots, z_0^n a)$ is also continuous since, writing $a = s + t\mathbf{i}$, it is just

$$(s,t) \longrightarrow \left(x_0^1 s - y_0^1 t, x_0^1 t + y_0^1 s, \ldots, x_0^n s - y_0^n t, x_0^n t + y_0^n s\right)$$

and this defines a continuous map from $\mathbb{R}^2$ to $\mathbb{R}^{2n}$. Thus, $(z_0^1 a, \ldots, z_0^n a) \to a$ is a homeomorphism of $\mathcal{P}^{-1}([\xi_0])$ onto S^1. Since $S \cong S^{2n-1}$ we may regard $\mathbb{C}\mathbb{P}^{n-1}$ as the result of decomposing S^{2n-1} into a disjoint union of circles S^1 and collapsing each S^1 to a point. We will prove shortly that $\mathbb{C}\mathbb{P}^1 \cong S^2$ so that the $n = 2$ case of this construction is precisely the Hopf bundle described in Section 0.3.

Finally, suppose $\mathbb{F} = \mathbb{H}$. An $a \in \mathbb{H}$ with $|a| = 1$ is just a unit quaternion and we have already seen (Theorem 1.1.4) that the set of these in $\mathbb{H}$ is homeomorphic to S^3.

Exercise 1.2.4 Show, just as for $\mathbb{C}\mathbb{P}^{n-1}$, that the fibers of $\mathcal{P} : S^{4n-1} \to \mathbb{H}\mathbb{P}^{n-1}$ are all homeomorphic to S^3.

The $n = 2$ case is again of particular interest. We will prove shortly that $\mathbb{H}\mathbb{P}^1 \cong S^4$ so we have $\mathcal{P} : S^7 \to S^4$ with fibers S^3. Eventually we will see that this map is related to the so-called "BPST instantons" in much the same way that the Hopf bundle of Chapter 0 is related to the Dirac magnetic monopole.

Next we show that the projective spaces $\mathbb{F}\mathbb{P}^{n-1}$ are all Hausdorff. For this we first fix a $\xi_0 \in S$ and define a map $\rho : \mathbb{F}\mathbb{P}^{n-1} \to \mathbb{R}$ by $\rho([\zeta]) = 1 - |<\zeta, \xi_0>|^2$. Here we are regarding $\mathbb{F}\mathbb{P}^{n-1}$ as a quotient of S so $\zeta \in S$. Note that the map is well-defined since $[\zeta'] = [\zeta]$ implies $\zeta' = \zeta a$ for some $a \in \mathbb{F}$ with $|a| = 1$ so $|<\zeta', \xi_0>|^2 = |<\zeta a, \xi_0>|^2 = |\bar{a}<\zeta, \xi_0>|^2 = |\bar{a}|^2 \, |<\zeta, \xi_0>|^2 = |<\zeta, \xi_0>|^2$. Also note that $\rho([\xi_0]) = 0$. We claim that if $[\zeta] \neq [\xi_0]$, then $\rho([\zeta]) \neq 0$. To see this, suppose $\rho([\zeta]) = 0$. Then $|<\zeta, \xi_0>|^2 = 1$ so $|<\zeta, \xi_0>| = 1$.

Exercise 1.2.5 Show that $\langle \xi_0 - \zeta<\zeta, \xi_0>, \xi_0 - \zeta<\zeta, \xi_0> \rangle = 0$.

Thus, $|\xi_0 - \zeta<\zeta, \xi_0>|^2 = 0$ so $\xi_0 = \zeta<\zeta, \xi_0>$ and therefore $\xi_0 \sim \zeta$ so $[\xi_0] = [\zeta]$. Consequently, $[\xi_0] \neq [\zeta]$ implies $\rho([\xi_0]) \neq \rho([\zeta])$. Finally, notice that ρ is continuous. Indeed, the map from S (a sphere) to $\mathbb{R}$ that carries $\zeta = (\zeta^1, \ldots, \zeta^n)$ to $1 - |<\zeta, \xi_0>|^2$ is seen to be continuous by writing it in Euclidean coordinates and is just the composition $\rho \circ \mathcal{P}$ so we may appeal to Lemma 1.2.1. With this it is easy to show that $\mathbb{F}\mathbb{P}^{n-1}$ is Hausdorff. Let $[\xi_0]$ and $[\zeta]$ be any two distinct points in $\mathbb{F}\mathbb{P}^{n-1}$. Use ξ_0 to define ρ as above. Then $\rho([\xi_0])$ and $\rho([\zeta])$ are distinct real numbers so we can find disjoint open intervals I_{ξ_0} and I_ζ in $\mathbb{R}$ containing $\rho([\xi_0])$

and $\rho([\zeta])$, respectively. By continuity, $U_{\xi_0} = \rho^{-1}(I_{\xi_0})$ and $U_\zeta = \rho^{-1}(I_\zeta)$ are open sets in $\mathbb{F}\mathbb{P}^{n-1}$, obviously disjoint and containing $[\xi_0]$ and $[\zeta]$ respectively.

We show next that each $\mathbb{F}\mathbb{P}^{n-1}$ is locally Euclidean. Note that if $\xi = (\xi^1, \ldots, \xi^n) \in S$, then $\xi^k \neq 0$ iff $\xi^k a \neq 0$ for every $a \in \mathbb{F}$ with $|a| = 1$. Thus, it makes sense to say that $[\xi] \in \mathbb{F}\mathbb{P}^{n-1}$ has $\xi^k \neq 0$. For each $k = 1, \ldots, n$, let

$$U_k = \left\{ [\xi] \in \mathbb{F}\mathbb{P}^{n-1} : \xi^k \neq 0 \right\} . \tag{1.2.1}$$

Then $\mathcal{P}^{-1}(U_k) = \{\xi \in S : \xi^k \neq 0\}$ and this is open in S. By definition of the quotient topology, U_k is therefore open in $\mathbb{F}\mathbb{P}^{n-1}$. We define a map $\varphi_k : U_k \to \mathbb{F}^{n-1}$ by

$$\begin{aligned} \varphi_k([\xi]) &= \varphi_k\left([\xi^1, \ldots, \xi^k, \ldots, \xi^n]\right) \\ &= \left(\xi^1(\xi^k)^{-1}, \ldots, \overset{\wedge}{1}, \ldots, \xi^n(\xi^k)^{-1}\right), \end{aligned} \tag{1.2.2}$$

where the $\wedge$ indicates that we delete the 1 in the kth slot. Observe that the map is well-defined since $[\zeta] = [\xi]$ with $\xi^k \neq 0$ implies $\zeta^k \neq 0$ and $\zeta = \xi a$ implies $\zeta^k = \xi^k a$ so $(\zeta^k)^{-1} = a^{-1}(\xi^k)^{-1}$ and therefore $\zeta^i(\zeta^k)^{-1} = (\xi^i a)(a^{-1}(\xi^k)^{-1}) = \xi^i(\xi^k)^{-1}$ for each $i = 1, \ldots, n$. We claim that φ_k is a homeomorphism of U_k onto $\mathbb{F}^{n-1}$ and leave surjectivity for the reader.

Exercise 1.2.6 Fix a $k = 1, \ldots, n$. Show that any $(y^1, \ldots, y^{n-1}) \in \mathbb{F}^{n-1}$ is $\varphi_k([\xi])$ for some $[\xi] \in U_k$.

To show that φ_k is one-to-one, suppose $[\xi], [\zeta] \in U_k$ with $\varphi_k([\xi]) = \varphi_k([\zeta])$. Then

$$\left(\xi^1(\xi^k)^{-1}, \ldots, \overset{\wedge}{1}, \ldots, \xi^n(\xi^k)^{-1}\right) = \left(\zeta^1(\zeta^k)^{-1}, \ldots, \overset{\wedge}{1}, \ldots, \zeta^n(\zeta^k)^{-1}\right)$$

so, for $i \neq k$, $\xi^i(\xi^k)^{-1} = \zeta^i(\zeta^k)^{-1}$, i.e., $\zeta^i = \xi^i((\xi^k)^{-1}\zeta^k)$. But $\zeta^k = \xi^k((\xi^k)^{-1}\zeta^k)$ is trivial so $\zeta = \xi((\xi^k)^{-1}\zeta^k)$. It follows that $\zeta \sim \xi$ so $[\zeta] = [\xi]$. To prove that φ_k is continuous we appeal to Lemma 1.2.1 and show that $\varphi_k \circ \mathcal{P} : \mathcal{P}^{-1}(U_k) \to \mathbb{F}^{n-1}$ is continuous (the subspace topology on U_k coincides with the quotient topology determined by $\mathcal{P} : \mathcal{P}^{-1}(U_k) \to U_k$). But $(\varphi_k \circ \mathcal{P})(\xi^1, \ldots, \xi^k, \ldots, \xi^n) = (\xi^1(\xi^k)^{-1}, \ldots, \overset{\wedge}{1}, \ldots, \xi^n(\xi^k)^{-1})$ and writing this out as a map from one Euclidean space to another makes it clear that $\varphi_k \circ \mathcal{P}$ is continuous. The inverse $\varphi_k^{-1} : \mathbb{F}^{n-1} \to U_k$ is given by

$$\varphi_k^{-1}\left(y^1, \ldots, y^{n-1}\right) = \left[y^1, \ldots, 1, \ldots, y^{n-1}\right] , \tag{1.2.3}$$

where the 1 is in the kth slot. Note that this is the composition

$$\left(y^1, \ldots, y^{n-1}\right) \longrightarrow \left(y^1, \ldots, 1, \ldots, y^{n-1}\right) \longrightarrow \left[y^1, \ldots, 1, \ldots, y^{n-1}\right] .$$

The first map of $\mathbb{F}^{n-1}$ into $\mathbb{F}^n - \{0\}$ is obviously continuous and the second is $\mathcal{Q}$, which is also continuous. Thus, φ_k^{-1} is continuous. We conclude that $\varphi_k : U_k \to \mathbb{F}^{n-1}$ is a homeomorphism and so, since $\mathbb{F}^{n-1}$ is homeomorphic to a Euclidean space and every point of $\mathbb{FP}^{n-1}$ is in some U_k, $\mathbb{FP}^{n-1}$ is locally Euclidean. The overlap functions for the charts (U_k, φ_k), $k = 1, \ldots, n$, are obtained as follows: Fix k and j in the range $1, \ldots, n$. Then

$$\varphi_k \circ \varphi_j^{-1} : \varphi_j(U_k \cap U_j) \longrightarrow \varphi_k(U_k \cap U_j)\,.$$

These are rather cumbersome to write down in full generality, but the pattern should be clear from

$$\begin{aligned} \varphi_2 \circ \varphi_1^{-1}(y^2, y^3, \ldots, y^n) &= \varphi_2([1, y^2, y^3, \ldots, y^n]) \\ &= ((y^2)^{-1}, y^3(y^2)^{-1}, \ldots, y^n(y^2)^{-1})\,. \end{aligned} \tag{1.2.4}$$

These overlap functions will be of particular interest to us when $n = 2$. In this case one has just two charts (U_1, φ_1) and (U_2, φ_2) on $\mathbb{FP}^1$. $\varphi_2 \circ \varphi_1^{-1} : \varphi_1(U_2 \cap U_1) \to \varphi_2(U_2 \cap U_1)$ is then given by $\varphi_2 \circ \varphi_1^{-1}(y) = \varphi_2([1, y]) = y^{-1}$ and similarly for $\varphi_1 \circ \varphi_2^{-1}$. Thus, $\varphi_1(U_2 \cap U_1) = \varphi_2(U_1 \cap U_2) = \mathbb{F} - \{0\}$, and

$$\varphi_2 \circ \varphi_1^{-1}(y) = y^{-1} = \varphi_1 \circ \varphi_2^{-1}(y), \quad y \in \mathbb{F} - \{0\}\,. \tag{1.2.5}$$

These overlap functions are certainly reminiscent of those for the spheres S^1, S^2 and S^4 recorded in (1.1.5), (1.1.8) and (1.1.14), except for the presence of the conjugate in the latter two. It will be convenient to remove this discrepency by a minor adjustment of one chart on $\mathbb{FP}^1$. Define $\bar{\varphi}_1 : U_1 \to \mathbb{F}$ by $\bar{\varphi}_1([\xi]) = \overline{\varphi_1([\xi])}$ (in $\mathbb{F} = \mathbb{R}$, $\bar{y} = y$). Then $(U_1, \bar{\varphi}_1)$ is clearly also a chart on $\mathbb{FP}^1$ and now we have

$$\varphi_2 \circ \bar{\varphi}_1^{-1}(y) = \bar{y}^{-1} = \bar{\varphi}_1 \circ \varphi_2^{-1}(y), \quad y \in \mathbb{F} - \{0\}\,. \tag{1.2.6}$$

With this and one more very important tool we can show that the similarity between the overlap functions for S^1, S^2 and S^4 and those for $\mathbb{RP}^1$, $\mathbb{CP}^1$ and $\mathbb{HP}^1$ is no coincidence.

Lemma 1.2.3 (The Glueing Lemma) *Let X and Y be topological spaces and assume that $X = A_1 \cup A_2$, where A_1 and A_2 are open (or closed) sets in X. Suppose $f_1 : A_1 \to Y$ and $f_2 : A_2 \to Y$ are continuous and that $f_1 | A_1 \cap A_2 = f_2 | A_1 \cap A_2$. Then the map $f : X \to Y$ defined by*

$$f(x) = \begin{cases} f_1(x), & x \in A_1 \\ f_2(x), & x \in A_2 \end{cases}$$

is continuous.

Proof: The result is trivial if $A_1 \cap A_2$ is empty so suppose $A_1 \cap A_2 \neq \emptyset$. Note that f is well-defined since $x \in A_1 \cap A_2$ implies $f_1(x) = f_2(x)$. Now suppose A_1 and A_2 are open. Let V be an arbitrary open set in Y. Then $f^{-1}(V) = f^{-1}(V) \cap X = f^{-1}(V) \cap (A_1 \cup A_2) = [f^{-1}(V) \cap A_1] \cup [f^{-1}(V) \cap A_2] = {f_1}^{-1}(V) \cup {f_2}^{-1}(V)$. f_1 is continuous so ${f_1}^{-1}(V)$ is open in A_1 and therefore also in X since A_1 is open in X. Similarly, ${f_2}^{-1}(V)$ is open in X so $f^{-1}(V) = {f_1}^{-1}(V) \cup {f_2}^{-1}(V)$ is open in X as required. If A_1 and A_2 are closed in X, the result follows in the same way by using Exercise 1.1.3.■

Now we prove that if $\mathbb{F} = \mathbb{R}$, $\mathbb{C}$, or $\mathbb{H}$, respectively, then $\mathbb{FP}^1$ is homeomorphic to S^1, S^2, or S^4, respectively, i.e.,

$$\mathbb{RP}^1 \cong S^1 \tag{1.2.7}$$

$$\mathbb{CP}^1 \cong S^2 \tag{1.2.8}$$

$$\mathbb{HP}^1 \cong S^4 . \tag{1.2.9}$$

We prove all three at once. Let (U_S, φ_S) and (U_N, φ_N) be the stereographic projection charts on S^1, S^2, or S^4 (with φ_S and φ_N regarded as maps to $\mathbb{R}$, $\mathbb{C}$, or $\mathbb{H}$, respectively). Then their overlap functions can be written $\varphi_S \circ {\varphi_N}^{-1}(y) = \bar{y}^{\,-1} = \varphi_N \circ {\varphi_S}^{-1}(y)$ for $y \in \mathbb{F} - \{0\}$ (again, $\bar{y} = y$ if $\mathbb{F} = \mathbb{R}$). Now let $(U_1, \bar{\varphi}_1)$ and (U_2, φ_2) be the charts on $\mathbb{FP}^1$ described above. The overlap functions are given by (1.2.6). Next consider the homeomorphisms ${\varphi_S}^{-1} \circ \varphi_2 : U_2 \to U_S$ and ${\varphi_N}^{-1} \circ \bar{\varphi}_1 : U_1 \to U_N$ and observe that, on $U_1 \cap U_2$, they agree. Indeed, $[\xi] \in U_1 \cap U_2$ implies $\varphi_2([\xi]) \in \varphi_2(U_1 \cap U_2) = \mathbb{F} - \{0\}$. But, on $\mathbb{F} - \{0\}$, $\bar{\varphi}_1 \circ {\varphi_2}^{-1} = \varphi_N \circ {\varphi_S}^{-1}$ so

$$\begin{aligned} \left(\bar{\varphi}_1 \circ {\varphi_2}^{-1}\right)(\varphi_2([\xi])) &= \left(\varphi_N \circ {\varphi_S}^{-1}\right)(\varphi_2([\xi])) \\ \bar{\varphi}_1([\xi]) &= \varphi_N \circ \left({\varphi_S}^{-1} \circ \varphi_2\right)([\xi]) \\ {\varphi_N}^{-1} \circ \bar{\varphi}_1([\xi]) &= {\varphi_S}^{-1} \circ \varphi_2([\xi]) . \end{aligned}$$

Now, $U_1 \cup U_2 = \mathbb{FP}^1$ and $U_N \cup U_S$ is the entire sphere. According to Lemma 1.2.3, the homeomorphisms ${\varphi_S}^{-1} \circ \varphi_2$ and ${\varphi_N}^{-1} \circ \bar{\varphi}_1$ determine a continuous map of $\mathbb{FP}^1$ to the sphere that is clearly one-to-one and onto. The inverse is determined in the same way by $({\varphi_S}^{-1} \circ \varphi_2)^{-1} = {\varphi_2}^{-1} \circ \varphi_S : U_S \to U_2$ and $({\varphi_N}^{-1} \circ \bar{\varphi}_1)^{-1} = {\bar{\varphi}_1}^{-1} \circ \varphi_N : U_N \to U_1$ and so it too is continuous and the result follows.

We observe that the homeomorphism $\mathbb{RP}^1 \cong S^1$ could easily have been anticipated on intuitive grounds. Indeed, $\mathbb{RP}^1$ can be viewed as the result of identifying antipodal points on S^1 and this could be accomplished in two stages, as indicated in Figure 1.2.1. Notice that, after the first stage, having identified points on the lower semicircle with their antipodes on the upper semicircle (and leaving the equator alone), we have a (space homeomorphic to a) closed interval. The second stage identifies the endpoints of this interval to get a circle again. We ask the reader to generalize.

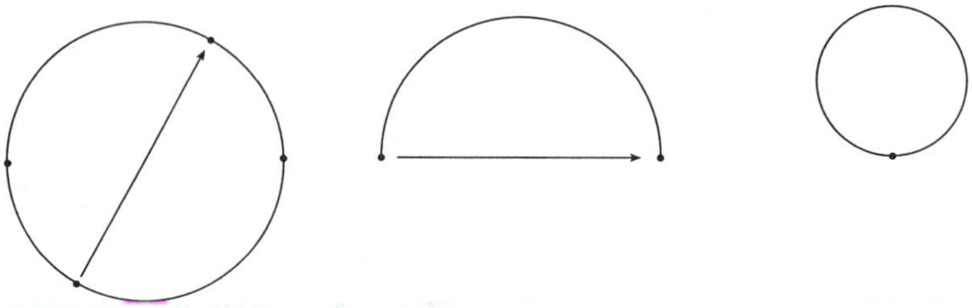

Figure 1.2.1

Exercise 1.2.7 For each positive integer n let D^n be the subspace $D^n = \{x \in \mathbb{R}^n : \| x \| \leq 1\}$ of $\mathbb{R}^n$. D^n is called the **n-dimensional disc** (or **ball**) in $\mathbb{R}^n$. The **boundary** ∂D^n **of** D^n is defined by $\partial D^n = \{x \in D^n : \| x \| = 1\}$ and is just the $(n-1)$-sphere S^{n-1} (when $n = 1$, $D^1 = [-1, 1]$ so we must stretch our terminology a bit and refer to $S^0 = \{-1, 1\}$ as the **0-sphere** in $\mathbb{R}$). Define an equivalence relation on D^n that identifies antipodal points on the boundary S^{n-1} and show that the quotient space of D^n by this relation is homeomorphic to $\mathbb{R}\mathbb{P}^n$.

We denote by I^n the n-dimensional cube $I^n = [0,1] \times \cdots \times [0,1] = \{(x^1, \ldots, x^n) \in \mathbb{R}^n : 0 \leq x^i \leq 1, \ i = 1, \ldots, n\}$ and by ∂I^n the subset consisting of those $(x^1, \ldots, x^n)$ for which some x^i is either 0 or 1. We wish to exhibit a useful homeomorphism of I^n onto D^n (Exercise 1.2.7) that carries ∂I^n onto $\partial D^n = S^{n-1}$. First observe that $(x^1, \ldots, x^n) \rightarrow (2x^1 - 1, \ldots, 2x^n - 1)$ carries I^n homeomorphically onto $\tilde{I}^n = \{(x^1, \ldots, x^n) \in \mathbb{R}^n : -1 \leq x^i \leq 1, \ i = 1, \ldots, n\}$ and, moreover, takes ∂I^n to $\partial \tilde{I}^n = \{(x^1, \ldots, x^n) \in \tilde{I}^n : x^i = -1 \text{ or } x^i = 1 \text{ for some } i = 1, \ldots, n\}$. Now we define two maps $h_1 : \tilde{I}^n \rightarrow D^n$ and $k_1 : D^n \rightarrow \tilde{I}^n$ by

$$\begin{aligned} h_1(x) &= h_1(x^1, \ldots, x^n) \\ &= \begin{cases} \dfrac{\max\{\,|x^i|\,\}}{\|\, x \,\|}(x^1, \ldots, x^n), & (x^1, \ldots, x^n) \neq (0, \ldots, 0) \\ (0, \ldots, 0), & (x^1, \ldots, x^n) = (0, \ldots, 0) \end{cases} \end{aligned}$$

and

$$\begin{aligned} k_1(y) &= k_1(y^1, \ldots, y^n) \\ &= \begin{cases} \dfrac{\|\, y \,\|}{\max\{\,|y^i|\,\}}(y^1, \ldots, y^n), & (y^1, \ldots, y^n) \neq (0, \ldots, 0) \\ (0, \ldots, 0), & (y^1, \ldots, y^n) = (0, \ldots, 0) \end{cases} . \end{aligned}$$

Observe that h_1 contracts $\tilde{I}^n$ radially onto D^n and that $h_1 \circ k_1 = id_{D^n}$

and $k_1 \circ h_1 = id_{\tilde{I}^n}$. The continuity of h_1 at $(0, \ldots, 0)$ follows at once from $\max\{\,|x^i|\,\} \leq \|x\|$. Similarly, the continuity of k_1 at $(0, \ldots, 0)$ is a consequence of $\|y\| \leq \sqrt{n}\max\{\,|y^i|\,\}$. Thus, h_1 and k_1 are inverse homeomorphisms. Furthermore, if $x \in \tilde{I}^n$ and some $|x^i| = 1$, then $h_1(x) = (1/\,\|x\|\,)x$ so $\|h_1(x)\| = 1$, i.e., $h_1(x) \in S^{n-1}$.

Exercise 1.2.8 Show that the composition

$$(x^1, \ldots, x^n) \longrightarrow (2x^1 - 1, \ldots, 2x^n - 1) \longrightarrow h_1(2x^1 - 1, \ldots, 2x^n - 1)$$

is a homeomorphism $\varphi : I^n \to D^n$ that carries ∂I^n onto S^{n-1}. Show also that φ carries the left/right half of I^n onto the left/right half of D^n (i.e., if $\varphi(x^1, \ldots, x^n) = (y^1, \ldots, y^n)$, then $0 \leq x^1 \leq 1/2$ implies $-1 \leq y^1 \leq 0$ and $1/2 \leq x^1 \leq 1$ implies $0 \leq y^1 \leq 1$).

1.3 Products and Local Products

If X is a subspace of $\mathbb{R}^n$ and Y is a subspace of $\mathbb{R}^m$, then the Cartesian product $X \times Y$ can be identified in a natural way with a subset of $\mathbb{R}^{n+m}$ and thereby acquires a subspace topology. For example, S^1 is a subspace of $\mathbb{R}^2$ and (-1,1) is a subspace of $\mathbb{R}$ so $S^1 \times (-1, 1)$ can be viewed as a subspace of $\mathbb{R}^3$, usually called a **cylinder** (see Figure 1.3.1).

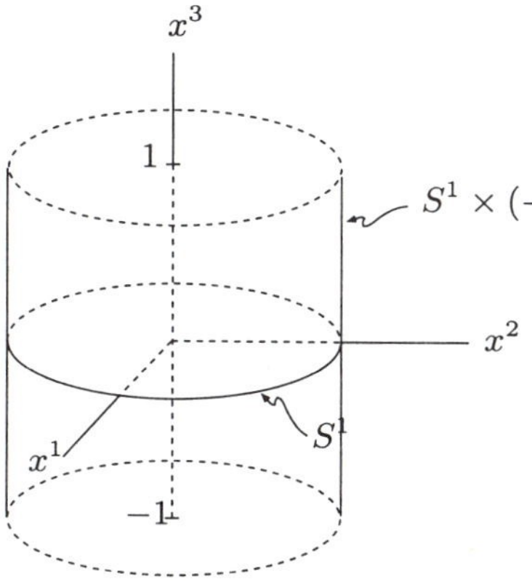

Figure 1.3.1

This process of forming product spaces is quite useful even when X and Y are not subspaces of Euclidean spaces, but in this case we must define the appropriate topology for $X \times Y$ rather than get it for free. To do this we will require a few preliminaries.

Let $(X, \mathcal{T})$ be an arbitrary topological space. A subcollection $\mathcal{B}$ of $\mathcal{T}$ is called a **basis for the topology** $\mathcal{T}$ if every open set in X (i.e., every element of $\mathcal{T}$) can be written as a union of members of $\mathcal{B}$. For example, the collection of all open balls $U_r(p)$, for $p \in \mathbb{R}^n$ and $r > 0$, is a basis for the usual topology of $\mathbb{R}^n$. Another basis for the topology of $\mathbb{R}^n$ is the collection of all open rectangles in $\mathbb{R}^n$. If X is a subspace of $\mathbb{R}^n$, then the collection of all intersections with X of open balls (or rectangles) in $\mathbb{R}^n$ is a basis for the topology of X. The collection of all points $\{y\}$ in a discrete space Y is a basis for the topology of Y. A **countable basis** is a basis $\mathcal{B} = \{U_k : k = 1, 2, \ldots\}$ that is in one-to-one correspondence with the positive integers.

Theorem 1.3.1 *Any subspace X of any Euclidean space $\mathbb{R}^n$ has a countable basis.*

Proof: It will suffice to prove that $\mathbb{R}^n$ itself has a countable basis for one can then intersect each of its elements with X to obtain a countable basis for X. Consider the collection $\mathcal{B}$ of all open balls $U_r(p)$ in $\mathbb{R}^n$, where $r > 0$ is rational and $p = (p^1, \ldots, p^n)$ has each coordinate p^i, $i = 1, \ldots, n$, rational. Since the set of rational numbers is countable, $\mathcal{B}$ is a countable collection of open sets in $\mathbb{R}^n$ so we need only show that it is a basis. For this it will be enough to show that for any open set U in $\mathbb{R}^n$ and any $x \in U$ there exists an element $U_r(p)$ of $\mathcal{B}$ with $x \in U_r(p) \subseteq U$ (for then U will be the union of all these elements of $\mathcal{B}$ as x varies over U). But $x \in U$ and U open in $\mathbb{R}^n$ implies that there exists an open ball $U_\epsilon(x)$ contained in U. The open ball $U_{\epsilon/4}(x)$ contains a point $p = (p^1, \ldots, p^n)$ with each p^i rational ($U_{\epsilon/4}(x)$ contains an open rectangle $(a_1, b_1) \times \cdots \times (a_n, b_n)$ and each (a_i, b_i) must contain a rational number p^i). Now choose a rational number r with $\epsilon/4 < r < \epsilon/2$. Then $\| x - p \| < \epsilon/4 < r$ implies $x \in U_r(p)$. Moreover, for any $y \in U_r(p)$, $\| y - x \| = \| y - p + p - x \| \leq \| y - p \| + \| p - x \| < r + \frac{\epsilon}{4} < \frac{\epsilon}{2} + \frac{\epsilon}{4} < \epsilon$ so $y \in U_\epsilon(x) \subseteq U$, i.e., $x \in U_r(p) \subseteq U$ as required. ∎

A topological space X is said to be **second countable** if there exists a countable basis for its topology. We have just seen that every subspace of a Euclidean space is second countable. A countable discrete space is second countable, but an uncountable discrete space is not.

Exercise 1.3.1 Show that second countability is a topological property, i.e., if X is second countable and Y is homeomorphic to X, then Y is also second countable.

Exercise 1.3.2 Show that every projective space $\mathbb{F}\mathbb{P}^{n-1}$ is second countable. **Hint:** There are n charts $(U_1, \varphi_1), \ldots, (U_n, \varphi_n)$ on $\mathbb{F}\mathbb{P}^{n-1}$ with the property that every point in $\mathbb{F}\mathbb{P}^{n-1}$ is contained in one of the open sets U_i, $i = 1, \ldots, n$, and each U_i is homeomorphic to an open set in a Euclidean space.

A **topological manifold** is a space X that is Hausdorff, locally Euclidean and second countable. Spheres and projective spaces are all topological manifolds, as will be virtually all of the examples of real interest to us.

To define product spaces in general we must reverse our point of view and ask when a collection $\mathcal{B}$ of subsets of some set X can be taken as a basis for *some* topology on X. The answer is simple enough.

Theorem 1.3.2 *Let X be a set and $\mathcal{B}$ a collection of subsets of X that satisfies the following condition:*

> *Whenever V and W are in $\mathcal{B}$ and $x \in V \cap W$, there exists a U in $\mathcal{B}$ with $x \in U \subseteq V \cap W$.*

Then the collection $\mathcal{T}_{\mathcal{B}}$ of subsets of X consisting of $\emptyset$, X and all unions of members of $\mathcal{B}$ is a topology for X.

Proof: Since $\emptyset$ and X are in $\mathcal{T}_{\mathcal{B}}$ by definition we need only show that $\mathcal{T}_{\mathcal{B}}$ is closed under the formation of arbitrary unions and finite intersections. The first is easy so we leave it to the reader.

Exercise 1.3.3 Show that the union of any collection of sets in $\mathcal{T}_{\mathcal{B}}$ is also in $\mathcal{T}_{\mathcal{B}}$.

Thus, let $U_1, \ldots, U_k$ be elements of $\mathcal{T}_{\mathcal{B}}$. We must show that $U_1 \cap \cdots \cap U_k$ can be written as a union of elements of $\mathcal{B}$ and for this it is enough to show that, for every $x \in U_1 \cap \cdots \cap U_k$, there exists a $U \in \mathcal{B}$ such that $x \in U \subseteq U_1 \cap \cdots \cap U_k$. We proceed by induction on k. Suppose first that $k = 2$. Thus, $x \in U_1 \cap U_2$, where U_1 and U_2 are in $\mathcal{T}_{\mathcal{B}}$. Write U_1 and U_2 as unions of elements of $\mathcal{B}$: $U_1 = \bigcup_\alpha V_\alpha$ and $U_2 = \bigcup_\beta W_\beta$. Then $U_1 \cap U_2 = (\bigcup_\alpha V_\alpha) \cap (\bigcup_\beta W_\beta)$. Thus, x is in some V_α and also in some W_β so $x \in V_\alpha \cap W_\beta$. By our hypothesis on $\mathcal{B}$, there exists a U in $\mathcal{B}$ with $x \in U \subseteq V_\alpha \cap W_\beta \subseteq (\bigcup_\alpha V_\alpha) \cap (\bigcup_\beta W_\beta) = U_1 \cap U_2$ as required.

Exercise 1.3.4 Finish the induction by assuming the result for intersections of $k-1$ elements of $\mathcal{T}_{\mathcal{B}}$ and proving it for $U_1 \cap \cdots \cap U_{k-1} \cap U_k$. ∎

Now we put Theorem 1.3.2 to use by defining a natural topology on the Cartesian product $X_1 \times \cdots \times X_n$ of a finite number of topological spaces. Although we will have no need to do so, it is possible to extend these ideas to define a topology on the product of an arbitrary (infinite) collection of spaces, but we caution the reader that this extension is not the "obvious" one (consult Chapter $\underline{\text{IV}}$ of [**Dug**] for details). Thus we let $(X_1, \mathcal{T}_1), \ldots, (X_n, \mathcal{T}_n)$ be a finite family of topological spaces and consider the set $X_1 \times \cdots \times X_n = \{(x^1, \ldots, x^n) : \ x^i \in X_i$ for $i =$

$1, \ldots, n\}$. Consider also the collection $\mathcal{B}$ of subsets of $X_1 \times \cdots \times X_n$ of the form $U_1 \times \cdots \times U_n$, where $U_i \in \mathcal{T}_i$ for each $i = 1, \ldots, n$. Note that if $V_1 \times \cdots \times V_n$ and $W_1 \times \cdots \times W_n$ are two elements of $\mathcal{B}$ and $(x^1, \ldots, x^n) \in (V_1 \times \cdots \times V_n) \cap (W_1 \times \cdots \times W_n)$, then $x^i \in V_i \cap W_i$ for each i. Moreover, each $V_i \cap W_i$ is open in X_i so $(V_1 \cap W_1) \times \cdots \times (V_n \cap W_n)$ is in $\mathcal{B}$ and $(x^1, \ldots, x^n) \in (V_1 \cap W_1) \times \cdots \times (V_n \cap W_n) \subseteq (V_1 \times \cdots \times V_n) \cap (W_1 \times \cdots \times W_n)$. $\mathcal{B}$ therefore satisfies the condition specified in Theorem 1.3.2. Since $\mathcal{B}$ also contains $\emptyset = \emptyset \times \cdots \times \emptyset$ and $X_1 \times \cdots \times X_n$, it is, in fact, a basis for the topology $\mathcal{T}_{\mathcal{B}}$ consisting of all subsets of $X_1 \times \cdots \times X_n$ that are unions of sets of the form $U_1 \times \cdots \times U_n$, with $U_i \in \mathcal{T}_i$ for $i = 1, \ldots, n$. $\mathcal{T}_{\mathcal{B}}$ is called the **product topology** on $X_1 \times \cdots \times X_n$. The $U_1 \times \cdots \times U_n$ are **basic open sets** in $X_1 \times \cdots \times X_n$.

Exercise 1.3.5 Show that if C_i is closed in X_i for each $i = 1, \ldots, n$, then $C_1 \times \cdots \times C_n$ is closed in the product topology on $X_1 \times \cdots \times X_n$.

Exercise 1.3.6 Show that the product topology on $\mathbb{R} \times \cdots \times \mathbb{R}$ (n factors) is the same as the usual Euclidean topology on $\mathbb{R}^n$ and conclude that $\mathbb{R}^n \cong \mathbb{R} \times \cdots \times \mathbb{R}$.

Theorem 1.3.3 *Let X_i be a subspace of Y_i for each $i = 1, \ldots, n$. Then the product topology on $X_1 \times \cdots \times X_n$ coincides with the relative topology that the subset $X_1 \times \cdots \times X_n$ inherits from the product topology on $Y_1 \times \cdots \times Y_n$.*

Proof: An open set in the product topology on $X_1 \times \cdots \times X_n$ is a union of sets of the form $U_1 \times \cdots \times U_n$, where each U_i is open in X_i, so, to show that it is also open in the relative topology, it will suffice to prove this for $U_1 \times \cdots \times U_n$. But X_i is a subspace of Y_i so each U_i is $X_i \cap U_i'$ for some open set U_i' in Y_i. Thus, $U_1 \times \cdots \times U_n = (X_1 \cap U_1') \times \cdots \times (X_n \cap U_n') = (X_1 \times \cdots \times X_n) \cap (U_1' \times \cdots \times U_n')$, which is the intersection with $X_1 \times \cdots \times X_n$ of an open set in $Y_1 \times \cdots \times Y_n$. Thus, $U_1 \times \cdots \times U_n$ is open in the relative topology.

Exercise 1.3.7 Show, similarly, that any subset of $X_1 \times \cdots \times X_n$ that is open in the relative topology is also open in the product topology. ∎

It follows, in particular, from Theorem 1.3.3 that if X_i is a subspace of some Euclidean space $\mathbb{R}^{n_i}$ for $i = 1, \ldots, k$, then the product topology on $X_1 \times \cdots \times X_k$ coincides with the topology that $X_1 \times \cdots \times X_k$ acquires as a subspace of $\mathbb{R}^{n_i} \times \cdots \times \mathbb{R}^{n_k} \cong \mathbb{R}^{n_1 + \cdots + n_k}$. Thus, for example, the cylinder $S^1 \times (-1, 1)$ may be treated either as a product space or as a subspace of $\mathbb{R}^3$. The same is true of the product $S^n \times S^m$ of any two spheres. In particular, the **torus** $S^1 \times S^1$ is a product of two circles, but also a subspace of $\mathbb{R}^4$. In this case there is yet another picture.

Exercise 1.3.8 Consider a circle in the xz-plane (in $\mathbb{R}^3$) of radius $r > 0$ about a point $(R, 0, 0)$, where $R > r$ (see Figure 1.3.2).

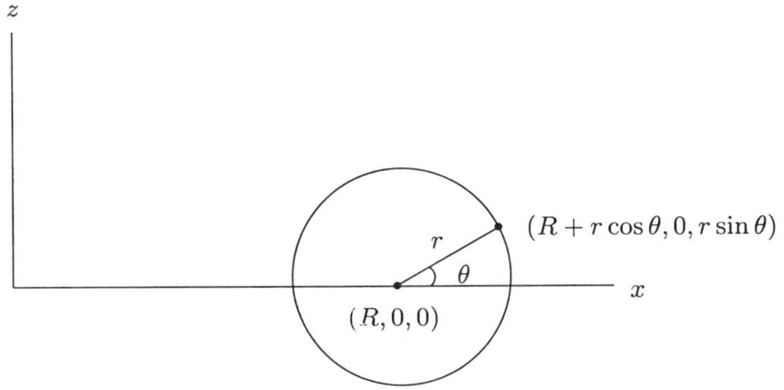

Figure 1.3.2

Any point on this circle has coordinates $(R + r\cos\theta, 0, r\sin\theta)$ for some θ in $[0, 2\pi]$. Now, revolve this circle about the z-axis to obtain a surface T in $\mathbb{R}^3$ and provide T with the subspace topology it inherits from $\mathbb{R}^3$ (see Figure 1.3.3).

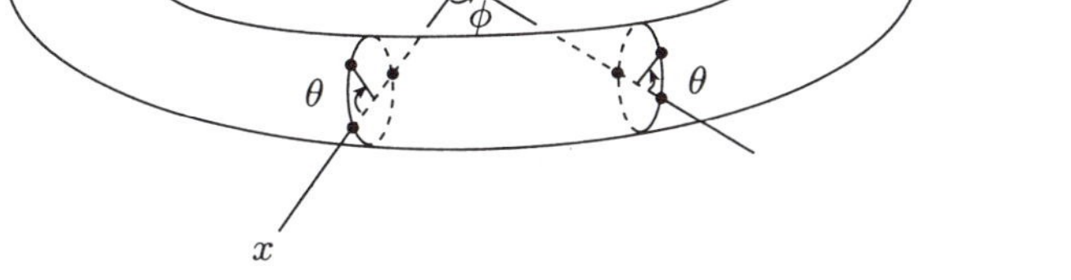

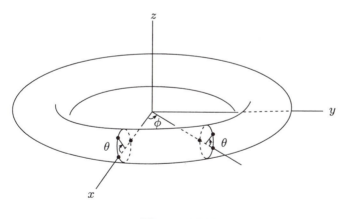

Figure 1.3.3

Notice that, for each point on the original circle, $x^2 + y^2$ and z remain constant during the rotation. Thus, if (x, y, z) is any point on T and ϕ denotes the angle through which the circle was rotated to arrive at this point, then $x = (R + r\cos\theta)\cos\phi$, $y = (R + r\cos\theta)\sin\phi$ and $z = r\sin\theta$, where $(R + r\cos\theta, 0, r\sin\theta)$ is the point on the original circle that arrives at (x, y, z) after the rotation through ϕ. T is thus the set of all points in $\mathbb{R}^3$ of the form $((R+r\cos\theta)\cos\phi, (R+r\cos\theta)\sin\phi, r\sin\theta)$ for $0 \leq \theta \leq 2\pi$ and $0 \leq \phi \leq 2\pi$. Show that T is homeomorphic to $S^1 \times S^1$.

A product $S^1 \times \cdots \times S^1$ of n circles is called an **n-dimensional torus.**

Let X be any space and consider the product $X \times [-1, 1]$. Define an equivalence relation $\sim$ on $X \times [-1, 1]$ by $(x_1, 1) \sim (x_2, 1)$ and $(x_1, -1) \sim (x_2, -1)$ for all $x_1, x_2 \in X$. The quotient space of $X \times [-1, 1]$ by this relation is denoted SX and called the **suspension of** X (see Figure 1.3.4). If X is Hausdorff, then so is SX. Denote by $\mathcal{Q} : X \times [-1, 1] \to SX$ the quotient map and by $<x, t>$ the point $\mathcal{Q}(x, t)$ in SX. We will show in Section 1.4 that $SS^{n-1} \cong S^n$.

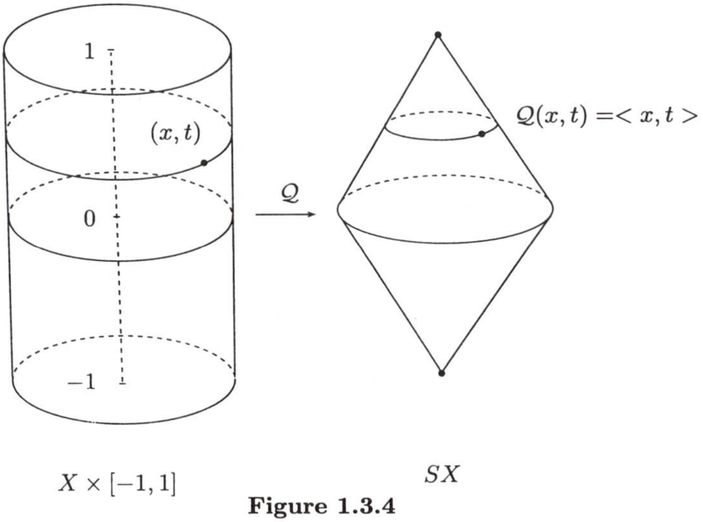

Figure 1.3.4

Exercise 1.3.9 Let $f : X \to Y$ be a continuous map. Define $Sf : SX \to SY$ (the **suspension of** f) by $Sf(<x, t>) = <f(x), t>$. Show that Sf is continuous. **Hint:** Lemma 1.2.2.

If $X_1 \times \cdots \times X_n$ is a product space, we define, for each $i = 1, \ldots, n$, the **projection onto the ith factor** X_i to be the map $\mathcal{P}^i : X_1 \times \cdots \times X_n \to X_i$ whose value at $(x^1, \ldots, x^n)$ is the ith coordinate $\mathcal{P}^i(x^1, \ldots, x^n) = x^i$. For any open set $U_i \subseteq X_i$, $(\mathcal{P}^i)^{-1}(U_i) = X_1 \times \cdots \times U_i \times \cdots \times X_n$ and this is open in $X_1 \times \cdots \times X_n$ so $\mathcal{P}^i$ is continuous. It is much more, however. For any spaces X and Y we say that a mapping $f : X \to Y$ is an **open mapping** if, whenever U is open in X, $f(U)$ is open in Y.

Exercise 1.3.10 Show that if $f : X \to Y$ is a continuous, open mapping of X onto Y and X is second countable, then Y is also second countable.

Exercise 1.3.11 Show that if $f : X \to Y$ is a continuous, open mapping of X onto Y, then Y has the quotient topology determined by f.

Lemma 1.3.4 *If $\{X_1, \ldots, X_n\}$ is a finite collection of topological spaces, then, for each $i = 1, \ldots, n$, the projection $\mathcal{P}^i : X_1 \times \cdots \times X_n \to X_i$ is a continuous, open surjection.*

Proof: We have already shown that $\mathcal{P}^i$ is continuous. It obviously maps onto X_i so all that remains is to show that it is an open mapping. For this, let U be open in $X_1 \times \cdots \times X_n$. If $U = \emptyset$, then $\mathcal{P}^i(U) = \emptyset$ is open so assume $U \neq \emptyset$. Then $\mathcal{P}^i(U) \neq \emptyset$. Let $p^i \in \mathcal{P}^i(U)$ be arbitrary. There exists a $(p^1, \ldots, p^i, \ldots, p^n) \in U$ with i^{th} coordinate p^i. By definition of the product topology, there is a basic open set $U_1 \times \cdots \times U_i \times \cdots \times U_n$ with $(p^1, \ldots, p^i, \ldots, p^n) \in U_1 \times \cdots \times U_i \times \cdots \times U_n \subseteq U$. Thus, U_i is open in X_i and $p^i \in U_i \subseteq \mathcal{P}^i(U)$ so $\mathcal{P}^i(U)$ is open in X_i. ∎

Exercise 1.3.12 A continuous map $f : X \to Y$ is a **closed mapping** if, whenever C is closed in X, $f(C)$ is closed in Y. Show that the projections $\mathcal{P}^i$ are, in general, *not* closed mappings. **Hint:** Look at $\mathbb{R}^2$.

Exercise 1.3.13 Show that if $f : X \to Y$ is a closed mapping of X onto Y, then Y has the quotient topology determined by f.

Lemma 1.3.5 *Let $\{X_1, \ldots, X_n\}$ be a finite collection of topological spaces and $f : X \to X_1 \times \cdots \times X_n$ a map from the topological space X into the product space $X_1 \times \cdots \times X_n$. Then f is continuous iff $\mathcal{P}^i \circ f : X \to X_i$ is continuous for each $i = 1, \ldots, n$ ($\mathcal{P}^i \circ f$ is called the* **i^{th} coordinate function of** *f and written $f^i = \mathcal{P}^i \circ f$).*

Proof: If f is continuous, then $f^i = \mathcal{P}^i \circ f$ is the composition of two continuous maps and so is continuous. Conversely, suppose $\mathcal{P}^i \circ f$ is continuous for each $i = 1, \ldots, n$. Notice that, for any basic open set $U_1 \times \cdots \times U_n$ in $X_1 \times \cdots \times X_n$, $U_1 \times \cdots \times U_n = (U_1 \times X_2 \times \cdots \times X_n) \cap (X_1 \times U_2 \times X_3 \times \cdots \times X_n) \cap \cdots \cap (X_1 \times \cdots \times X_{n-1} \times U_n) = (\mathcal{P}^1)^{-1}(U_1) \cap (\mathcal{P}^2)^{-1}(U_2) \cap \cdots \cap (\mathcal{P}^n)^{-1}(U_n)$ so $f^{-1}(U_1 \times \cdots \times U_n) = f^{-1}((\mathcal{P}^1)^{-1}(U_1)) \cap f^{-1}((\mathcal{P}^2)^{-1}(U_2)) \cap \cdots \cap f^{-1}((\mathcal{P}^n)^{-1}(U_n)) = (\mathcal{P}^1 \circ f)^{-1}(U_1) \cap (\mathcal{P}^2 \circ f)^{-1}(U_2) \cap \cdots \cap (\mathcal{P}^n \circ f)^{-1}(U_n)$, which is a finite intersection of open sets and is therefore open in X. Since any open set in $X_1 \times \cdots \times X_n$ is a union of basic open sets and since $f^{-1}(\bigcup_\alpha A_\alpha) = \bigcup_\alpha f^{-1}(A_\alpha)$, the result follows. ∎

Exercise 1.3.14 Suppose $f_1 : X_1 \to Y_1$ and $f_2 : X_2 \to Y_2$ are continuous maps. Define the **product map** $f_1 \times f_2 : X_1 \times X_2 \to Y_1 \times Y_2$ by $(f_1 \times f_2)(x_1, x_2) = (f_1(x_1), f_2(x_2))$.

(a) Show that $f_1 \times f_2$ is continuous.

(b) Show that, if f_1 and f_2 are open (closed) maps, then $f_1 \times f_2$ is an open (closed) map.

(c) Generalize the definition as well as (a) and (b) to larger (finite) products $f_1 \times \cdots \times f_n$.

Exercise 1.3.15 Let X, Y and Z be topological spaces. Show that $Y \times X \cong X \times Y$ and $(X \times Y) \times Z \cong X \times (Y \times Z)$. Generalize this commutativity and associativity to larger finite products.

Exercise 1.3.16 Let $\{X_1, \ldots, X_n\}$ be a finite collection of topological spaces. Show that $X_1 \times \cdots \times X_n$ is Hausdorff iff each X_i, $i = 1, \ldots, n$, is Hausdorff.

A fairly trivial, but useful, example of a product space is obtained as follows: Let X be an arbitrary space and Y a discrete space of cardinality $|Y|$. Then the product space $X \times Y$ is called the **disjoint union of** $|Y|$ **copies** of X. For example, if $X = S^1$ and $Y = \mathbb{Z}$ is the (discrete) subspace $\{\ldots, -2, -1, 0, 1, 2, \ldots\}$ of $\mathbb{R}$ consisting of the integers, then $S^1 \times \mathbb{Z}$ is the subspace of $\mathbb{R}^3$ consisting of a stack of countably many circles, one at each integer "height" (see Figure 1.3.5).

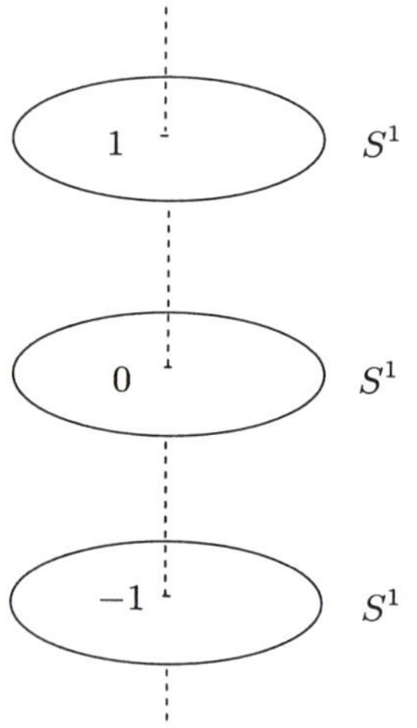

Figure 1.3.5

Notice that any product space $X \times Y$ contains a copy of X at each "height" y_0 in Y (namely, $X \times \{y_0\}$) and that copies at different heights are disjoint as sets. However, unless Y is discrete, these copies of X will generally not be open in $X \times Y$ so the union of these copies cannot reasonably be regarded as "disjoint" in the topological sense (contrast $S^1 \times \mathbb{Z}$ with the cylinder

$S^1 \times (-1,1)$ in Figure 1.3.1).

Most of the examples of real interest to us, while not simple product spaces, are "locally" product spaces in a sense that we now make precise. First, an example. Consider the map $\mathcal{P} : \mathbb{R} \to S^1$ defined by $\mathcal{P}(s) = e^{2\pi s \mathbf{i}} = (\cos 2\pi s, \sin 2\pi s)$. Then $\mathcal{P}$ is clearly continuous and surjective. It is also an open map (this can easily be verified directly, but will also follow from Lemma 1.3.6 below). To this extent, at least, it is analogous to the projection of a product space onto one of its factors. To visualize the map we view $\mathbb{R}$ as a helix above S^1 and $\mathcal{P}$ as a downward projection (see Figure 1.3.6).

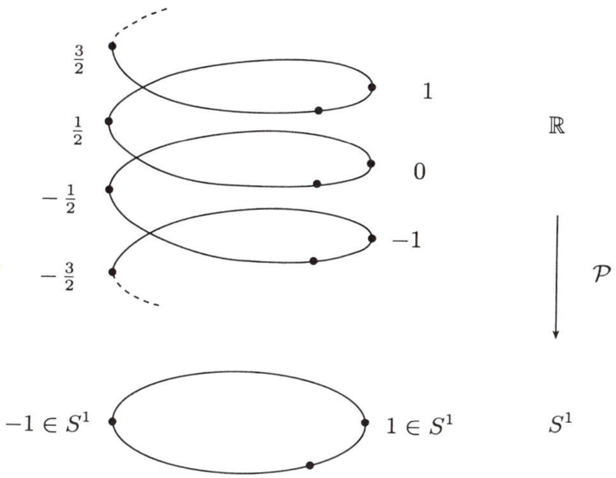

Figure 1.3.6

The fiber $\mathcal{P}^{-1}(x)$ above any $x \in S^1$ is then the set of integer translates of some real number s_0, e.g., $\mathcal{P}^{-1}(-1) = \{\frac{1}{2} + n : n = 0, \pm 1, \ldots\}$. Thus, any such fiber is a subspace of $\mathbb{R}$ homeomorphic to the space $\mathbb{Z}$ of integers. More interesting, however, is the following observation. Letting $V_1 = S^1 - \{1\}$ and $V_2 = S^1 - \{-1\}$ we note that $\mathcal{P}^{-1}(V_1) = \bigcup_{n=-\infty}^{\infty}(n, n+1)$ and $\mathcal{P}^{-1}(V_2) = \bigcup_{n=-\infty}^{\infty}(n - \frac{1}{2}, n + \frac{1}{2})$ and that each of these is a disjoint union of copies of V_1 and V_2, respectively. More precisely, we have the following result.

Exercise 1.3.17 Show that there exists a homeomorphism $\Phi_1 : V_1 \times \mathbb{Z} \to \mathcal{P}^{-1}(V_1)$ that "preserves fibers" in the sense that $\mathcal{P} \circ \Phi_1(x,y) = x$ for all $(x,y) \in V_1 \times \mathbb{Z}$. Similarly, there exists a homeomorphism $\Phi_2 : V_2 \times \mathbb{Z} \to \mathcal{P}^{-1}(V_2)$ such that $\mathcal{P} \circ \Phi_2(x,y) = x$ for all $(x,y) \in V_2 \times \mathbb{Z}$.

Exercise 1.3.18 Show that $\mathbb{R}$ is not homeomorphic to $S^1 \times \mathbb{Z}$. **Hint:**

In Section 1.5 we will prove that a nonempty subset of $\mathbb{R}$ that is both open and closed must be all of $\mathbb{R}$.

Thus, although $\mathbb{R}$ is not globally the product of S^1 and $\mathbb{Z}$ (Exercise 1.13.18), it is the union of two open sets ($\mathcal{P}^{-1}(V_1)$ and $\mathcal{P}^{-1}(V_2)$), each of which is homeomorphic to the product of an open set in S^1 and $\mathbb{Z}$. Moreover, if $\mathcal{P}^{-1}(V_i)$ is identified with $V_i \times \mathbb{Z}$ by Φ_i for $i = 1, 2$, then $\mathcal{P}$ is just the projection onto the first factor (Exercise 1.3.17). It is this phenomenon that we wish to capture with a definition.

A **locally trivial bundle** $(P, X, \mathcal{P}, Y)$ consists of a space P (the **bundle space**, or **total space**), a Hausdorff space X (the **base space**), a continuous map $\mathcal{P}$ of P onto X (the **projection**) and a Hausdorff space Y (the **fiber**) such that for each $x_0 \in X$ there exists an open set V in X containing x_0 and a homeomorphism $\Phi : V \times Y \to \mathcal{P}^{-1}(V)$ such that $\mathcal{P} \circ \Phi(x, y) = x$ for all $(x, y) \in V \times Y$ (we will see shortly that P is necessarily Hausdorff). V is called a **local trivializing neighborhood** in X and the pair (V, Φ) is a **local trivialization** of the bundle. One often simplifies the terminology by referring to $\mathcal{P} : P \to X$ as a locally trivial bundle (provided the fiber Y is clear from the context) or by expressing the entire object diagramatically as $Y \to P \xrightarrow{\mathcal{P}} X$, or just $Y \to P \to X$.

The map $\mathcal{P} : \mathbb{R} \to S^1$ defined by $\mathcal{P}(s) = e^{2\pi s \mathbf{i}}$ is a locally trivial bundle with fiber $\mathbb{Z}$. For any Hausdorff spaces X and Y one can take $P = X \times Y$ and let $\mathcal{P} : X \times Y \to X$ be the projection onto the first factor to obtain a locally trivial bundle with base X and fiber Y. Here one can take V to be all of X and Φ to be the identity map. This is called the **trivial bundle**, or **product bundle**.

Next we consider the real projective space $\mathbb{RP}^{n-1}$ as the quotient of the $(n-1)$-sphere S^{n-1} obtained by identifying antipodal points (Section 1.2). Specifically, we let $\mathcal{P} : S^{n-1} \to \mathbb{RP}^{n-1}$ be the quotient map, $\mathcal{P}(x) = [x]$ for every $x \in S^{n-1}$, and show that it is a locally trivial bundle whose fiber is the 2-point discrete space $\mathbb{Z}_2 = \{-1, 1\} \subseteq \mathbb{R}$ (any 2-point discrete space would do, of course, but this particular choice will be convenient somewhat later). Consider the various open hemispheres on S^{n-1}, i.e., for each $k = 1, \ldots, n$, let

$$U_k^+ = \left\{x = (x^1, \ldots, x^n) \in S^{n-1} : x^k > 0\right\}$$

and

$$U_k^- = \left\{x = (x^1, \ldots, x^n) \in S^{n-1} : x^k < 0\right\}.$$

Each of these is open in S^{n-1} and $U_k^+ \cap U_k^- = \emptyset$ for each k. Moreover, every point in S^{n-1} is in such a set. The restriction of $\mathcal{P}$ to any one of these is continuous and one-to-one ($x \sim y$ in S^{n-1} iff $y = \pm x$). We claim that each $\mathcal{P} \,|\, U_k^\pm$ is an open map. Indeed, if U is open in some $U_k^\pm$, then $\mathcal{P}^{-1}((\mathcal{P}|U_k^\pm)(U)) = \mathcal{P}^{-1}(\mathcal{P}(U)) = U \cup (-U)$, where $-U = \{-x : x \in U\}$.

Since $-U$ is open in S^{n-1}, so is $U \cup (-U)$ and therefore, by definition of the quotient topology, $(\mathcal{P} \mid U_k^{\pm})(U)$ is open in $\mathbb{RP}^{n-1}$. Thus, each $\mathcal{P} \mid U_k^{\pm}$ is a homeomorphism of $U_k^{\pm}$ onto $V_k = \mathcal{P}(U_k^+) = \mathcal{P}(U_k^-)$. Moreover, $\mathcal{P}^{-1}(V_k) = U_k^+ \cup U_k^-$ is a disjoint union of two open subsets of S^{n-1}, each of which is mapped homeomorphically onto V_k by $\mathcal{P}$. Now we define maps

$$\Phi_k : V_k \times \mathbb{Z}_2 \longrightarrow \mathcal{P}^{-1}(V_k) = U_k^+ \cup U_k^-$$

by

$$\Phi_k([x], 1) = (\mathcal{P}|U_k^+)^{-1}([x]) \quad \text{and} \quad \Phi_k([x], -1) = (\mathcal{P}|U_k^-)^{-1}([x])$$

for all $[x] \in V_k$. Now, $V_k \times \{1\}$ and $V_k \times \{-1\}$ are disjoint open sets in $V_k \times \mathbb{Z}_2$ whose union is all of $V_k \times \mathbb{Z}_2$. Moreover, the maps $([x], 1) \to [x]$ and $([x], -1) \to [x]$ are homeomorphisms of $V_k \times \{1\}$ and $V_k \times \{-1\}$, respectively, onto V_k. Thus, the compositions $([x], 1) \to [x] \to (\mathcal{P}|U_k^+)^{-1}([x])$ and $([x], -1) \to [x] \to (\mathcal{P}|U_k^-)^{-1}([x])$ are homeomorphisms and these are just $\Phi_k | V_k \times \{1\}$ and $\Phi_k | V_k \times \{-1\}$, respectively.

Exercise 1.3.19 Use this information to show that $\Phi_k : V_k \times \mathbb{Z}_2 \to \mathcal{P}^{-1}(V_k)$ is a homeomorphism and show also that $\mathcal{P} \circ \Phi_k([x], y) = [x]$ for all $([x], y)$ in $V_k \times \mathbb{Z}_2$.

Thus, $(S^{n-1}, \mathbb{RP}^{n-1}, \mathcal{P}, \mathbb{Z}_2)$ is a locally trivial bundle. We remark in passing that the argument we have just given, together with one additional property of $\mathbb{RP}^{n-1}$ to be introduced in Section 1.5 actually shows that $\mathcal{P} : S^{n-1} \to \mathbb{RP}^{n-1}$ is what is known as a "covering space" (Section 1.5). This is *not* true of the complex and quaternionic analogues to which we now turn.

Next we consider $\mathbb{CP}^{n-1}$ as the quotient of S^{2n-1} obtained by identifying to points the S^1-fibers of $\mathcal{P} : S^{2n-1} \to \mathbb{CP}^{n-1}$ (Section 1.2). We regard S^{2n-1} as the set of $\xi = (z^1, \ldots, z^n) \in \mathbb{C}^n$ with $<\xi, \xi> = \bar{z}^1 z^1 + \cdots + \bar{z}^n z^n = 1$ (Exercise 1.1.27). Our objective is to show that $(S^{2n-1}, \mathbb{CP}^{n-1}, \mathcal{P}, S^1)$ is a locally trivial bundle. For each $k = 1, \ldots, n$ we let $V_k = \{[\xi] \in \mathbb{CP}^{n-1} : z^k \neq 0\}$ (see (1.2.1)). Every element of $\mathbb{CP}^{n-1}$ is in such a V_k so it will suffice to define homeomorphisms $\Phi_k : V_k \times S^1 \to \mathcal{P}^{-1}(V_k)$ with $\mathcal{P} \circ \Phi_k([\xi], y) = [\xi]$ all $([\xi], y) \in V_k \times S^1$. Fix a $[\xi] = [z^1, \ldots, z^n] \in V_k$ and a $y \in S^1$. Since $z^k \neq 0$ we may consider

$$\left(z^1 (z^k)^{-1} |z^k| \, y, \ldots, z^n (z^k)^{-1} |z^k| \, y\right) \in \mathbb{C}^n.$$

Exercise 1.3.20 Show that if $[\xi] = [z^1, \ldots, z^n] = [w^1, \ldots, w^n]$ and $y \in$

S^1, then

$$\left(z^1(z^k)^{-1}\,|z^k|\,y\,,\,\ldots\,,\,z^n(z^k)^{-1}\,|z^k\,|\,y\right) = \left(w^1(w^k)^{-1}\,|w^k|\,y\,,\,\ldots\,,\,w^n(w^k)^{-1}\,|w^k|\,y\right).$$

Thus, we may define

$$\Phi_k\left([\,\xi\,],y\right) = \left(z^1(z^k)^{-1}\,|z^k|\,y\,,\,\ldots\,,\,z^n(z^k)^{-1}\,|z^k|\,y\right) \qquad (1.3.1)$$

where $(z^1,\ldots,z^n)$ is *any* point in $\mathcal{P}^{-1}([\,\xi\,])$. Observe that

$$\mathcal{P}\circ\Phi_k([\,\xi\,],y) = [z^1(z^k)^{-1}\,|z^k|\,y\,,\,\ldots\,,\,z^n(z^k)^{-1}\,|z^k|\,y] = [\,\xi\,]. \quad (1.3.2)$$

To prove that Φ_k is a bijection we show that the map $\Psi_k:\ \mathcal{P}^{-1}(V_k)\rightarrow V_k\times S^1$ defined by

$$\Psi_k(\xi) = \Psi_k(z^1,\ldots,z^n) = ([\,\xi\,],|z^k|^{-1}\,z^k) = (\mathcal{P}(\xi),|z^k|^{-1}\,z^k) \quad (1.3.3)$$

is its inverse:

$$\begin{aligned}\Phi_k\circ\Psi_k(\,\xi\,) &= \Phi_k\left([\,\xi\,],|z^k|^{-1}\,z^k\right)\\ &= \left(z^1(z^k)^{-1}\,|z^k|\,|z^k|^{-1}\,z^k\,,\,\ldots\,,\,z^n(z^k)^{-1}\,|z^k|\,|z^k|^{-1}\,z^k\right)\\ &= (z^1,\ldots,z^n)\\ &= \xi\,.\end{aligned}$$

Similarly,

$$\begin{aligned}\Psi_k\circ\Phi_k\left([\,\xi\,],y\right) &= \Psi_k\left(z^1(z^k)^{-1}\,|z^k|\,y\,,\,\ldots\,,\,z^n(z^k)^{-1}\,|z^k|\,y\right)\\ &= \Big(\mathcal{P}(z^1(z^k)^{-1}\,|z^k|\,y\,,\,\ldots\,,\,z^n(z^k)^{-1}\,|z^k|\,y),\\ &\qquad \left|z^k(z^k)^{-1}\,|z^k|\,y\right|^{-1}\,z^k(z^k)^{-1}\,|z^k|\,y\,\Big)\\ &= \left(\mathcal{P}(\,\xi\,),|z^k|^{-1}\,|z^k|\,y\right)\\ &= ([\,\xi\,],y)\,.\end{aligned}$$

Moreover, Ψ_k is continuous since its coordinate functions $\xi\rightarrow\mathcal{P}(\,\xi\,)$ and $\xi\rightarrow|\,z^k\,|^{-1}\,z^k$ are clearly continuous. Regarding Φ_k as a map into $\mathbb{C}^n$ whose image lies in $\mathcal{P}^{-1}(V_k)\subseteq S^{2n-1}$, its ith coordinate function is

$$([\,\xi\,],y)\longrightarrow z^i(z^k)^{-1}\,|z^k|\,y\,, \qquad (1.3.4)$$

where $(z^1,\ldots,z^n)$ is any point in S^{2n-1} with $\mathcal{P}(z^1,\ldots,z^n)=[\,\xi\,]$. Now, the mapping from $\{(z^1,\ldots,z^n)\in\mathbb{C}^n\,:\,z^k\neq0\}\times S^1$ to $\mathbb{C}$ given by $(z^1,\ldots,z^n,y)\rightarrow z^i(z^k)^{-1}\,|z^k|\,y$ is clearly continuous and takes the same value at all points $(w^1, \ldots, w^n, y)$ with $[w^1, \ldots, w^n] = [z^1, \ldots, z^n]$.

Exercise 1.3.21 Use this information and Lemma 1.2.2 to show that the map (1.3.4) is continuous for each $i = 1, \ldots, n$ and conclude that Φ_k is continuous.

Thus, $\Phi_k : V_k \times S^1 \to \mathcal{P}^{-1}(V_k)$ is a homeomorphism with $\mathcal{P} \circ \Phi_k([\xi], y) = [\xi]$ for all $([\xi], y) \in V_k \times S^1$ as required.

Exercise 1.3.22 Make whatever modifications might be required in these arguments to show that $(S^{4n-1}, \mathbb{HP}^{n-1}, \mathcal{P}, S^3)$ is a locally trivial bundle.

The locally trivial bundles $S^1 \to S^{2n-1} \to \mathbb{CP}^{n-1}$ and $S^3 \to S^{4n-1} \to \mathbb{HP}^{n-1}$ are called **Hopf bundles**. Of particular interest is the $n = 2$ case which gives (by (1.2.8) and (1.2.9)) $S^1 \to S^3 \to S^2$ and $S^3 \to S^7 \to S^4$. The first of these is the Hopf bundle used to model the Dirac magnetic monopole in Chapter 0 and the second will play a similar role in our discussion of BPST instantons in Chapter 4.

Bundles are of central importance in topology, geometry and mathematical physics and will be the major focus of our work. We conclude this brief introduction with the useful fact that, like the projection of a product space onto one of its factors, a bundle projection is always an open map.

Lemma 1.3.6 *Let $(P, X, \mathcal{P}, Y)$ be a locally trivial bundle. Then $\mathcal{P} : P \to X$ is a continuous, open surjection.*

Proof: Continuity and surjectivity are part of the definition so we need only show that $\mathcal{P}$ is an open map. Let U be open in P and consider $\mathcal{P}(U) \subseteq X$. Fix an $x_0 \in \mathcal{P}(U)$. It will suffice to find an open set V_{x_0} in X with $x_0 \in V_{x_0} \subseteq \mathcal{P}(U)$, for then $\mathcal{P}(U)$ will be the union of all such V_{x_0} as x_0 varies over $\mathcal{P}(U)$. By definition of a locally trivial bundle there exists an open set V in X containing x_0 and a homeomorphism $\Phi : V \times Y \to \mathcal{P}^{-1}(V)$ such that $\mathcal{P} \circ \Phi(x, y) = x$ for all $(x, y) \in V \times Y$. Now, $\mathcal{P}^{-1}(V) \cap U$ is an open set in P so $\Phi^{-1}(\mathcal{P}^{-1}(V) \cap U)$ is open in $V \times Y$. Letting $\mathcal{P}_V : V \times Y \to V$ be the projection onto the first factor of the product, Lemma 1.3.4 implies that $\mathcal{P}_V(\Phi^{-1}(\mathcal{P}^{-1}(V) \cap U))$ is open in V and therefore in X. Moreover, $\mathcal{P} \circ \Phi = \mathcal{P}_V$ implies that $\mathcal{P}_V(\Phi^{-1}(\mathcal{P}^{-1}(V) \cap U)) \subseteq \mathcal{P}(U)$ and $x_0 \in \mathcal{P}_V(\Phi^{-1}(\mathcal{P}^{-1}(V) \cap U))$ so the proof is complete. ∎

Exercise 1.3.23 Show that if the base X and fiber Y of a locally trivial bundle are Hausdorff, then the bundle space P is also Hausdorff.

Exercise 1.3.24 Let $(P, X, \mathcal{P}, Y)$ be a locally trivial bundle. Define an equivalence relation $\sim$ on P that identifies points in the same fiber of $\mathcal{P}$, i.e., $p_1 \sim p_2$ iff $\mathcal{P}(p_1) = \mathcal{P}(p_2)$. Show that X is homeomorphic to $P/\sim$.

1.4 Compactness Conditions

A collection $\{U_\alpha : \alpha \in \mathcal{A}\}$ of subsets of a space X is said to **cover** X if each $x \in X$ is an element of some U_α, i.e., if $X = \bigcup_{\alpha \in \mathcal{A}} U_\alpha$. If each U_α is an open set in X, then $\{U_\alpha : \alpha \in \mathcal{A}\}$ is an **open cover** of X. For example, the family $\{U_1(x) : x \in \mathbb{R}^n\}$ of all open balls of radius 1 in $\mathbb{R}^n$ is an open cover of $\mathbb{R}^n$. For any subspace X of $\mathbb{R}^n$, $\{U_1(x) \cap X : x \in X\}$ is an open cover of X. $\{U_S, U_N\}$ is an open cover of S^n, where $U_S = S^n - \{(0, \ldots, 0, 1)\}$ and $U_N = S^n - \{(0, \ldots, 0, -1)\}$. Another useful open cover of S^n consists of the open hemispheres $\{U_k^{\pm} : k = 1, \ldots, n+1\}$, where $U_k^+ = \{(x^1, \ldots, x^{n+1}) \in S^n : x^k > 0\}$ and $U_k^- = \{(x^1, \ldots, x^{n+1}) \in S^n : x^k < 0\}$. The projective space $\mathbb{F}\mathbb{P}^{n-1}$, $\mathbb{F} = \mathbb{R}$, $\mathbb{C}$, or $\mathbb{H}$, is covered by the family of all $U_k = \{[\xi^1, \ldots, \xi^n] \in \mathbb{F}\mathbb{P}^{n-1} : \xi^k \neq 0\}$, $k = 1, \ldots, n$, and these are all open in $\mathbb{F}\mathbb{P}^{n-1}$.

An open cover $\{U_\alpha : \alpha \in \mathcal{A}\}$ of X is said to be **finite** (respectively, **countable**) if the index set $\mathcal{A}$ is finite (respectively, countable), i.e., if it contains only finitely many (respectively, countably many) open sets. For most of the spaces of interest to us any open cover will be "reducible" to one that is either finite or countably infinite. More precisely, if $\{U_\alpha : \alpha \in \mathcal{A}\}$ is an open cover of X, then a subcollection $\{U_\alpha : \alpha \in \mathcal{A}'\}$, $\mathcal{A}' \subseteq \mathcal{A}$, is called a **subcover of** $\{U_\alpha : \alpha \in \mathcal{A}\}$ if it also covers X, i.e., if $\bigcup_{\alpha \in \mathcal{A}'} U_\alpha = X$.

Theorem 1.4.1 *Any open cover of a second countable space has a countable subcover.*

Proof: Suppose X is second countable. Let $\mathcal{B} = \{B_k : k = 1, 2, \ldots\}$ be a countable basis for the topology of X. Now, let $\mathcal{U} = \{U_\alpha : \alpha \in \mathcal{A}\}$ be an arbitrary open cover of X. Since each U_α is a union of elements of $\mathcal{B}$ we can select a (necessarily countable) subcollection $\{B_{k_1}, B_{k_2}, \ldots\}$ of $\mathcal{B}$ with the property that each B_{k_j} is contained in some element of $\mathcal{U}$ and $\bigcup_{j=1}^\infty B_{k_j} = X$. For each k_j, $j = 1, 2, \ldots$, select one U_{α_j} with $B_{k_j} \subseteq U_{\alpha_j}$. Then $\{U_{\alpha_1}, U_{\alpha_2}, \ldots\}$ is a countable subcollection of $\mathcal{U}$ that covers X since $\bigcup_{j=1}^\infty U_{\alpha_j} \supseteq \bigcup_{j=1}^\infty B_{k_j} = X$. ∎

Theorem 1.4.1 applies, in particular, to any subspace of any Euclidean space (Theorem 1.3.1) and to the projective spaces (Exercise 1.3.2).

For some topological spaces it is possible to go one step further and extract a finite subcover from any given open cover. A Hausdorff space X is said to be **compact** if every open cover of X has a finite subcover. The compact subspaces of $\mathbb{R}^n$ are well-known from analysis (see Corollary 1-7 of [**Sp 1**] for the proof of the following result).

Theorem 1.4.2 (Heine-Borel) *A subspace X of $\mathbb{R}^n$ is compact iff it is closed and bounded (i.e., closed and contained in some ball $U_r(0)$ about* *the origin in* $\mathbb{R}^n$).

Thus, for example, spheres are compact, but Euclidean spaces $\mathbb{R}^n$ themselves are not. Since compactness is surely a topological property (we prove more in Theorem 1.4.3) we find, in particular, that S^1 is not homeomorphic to $\mathbb{R}$ (a result promised in Section 1.1).

Exercise 1.4.1 Show that $O(n)$, $U(n)$, $Sp(n)$, $SO(n)$ and $SU(n)$ are all compact. **Hint:** Use Exercise 1.1.21 to show that any $U(n, \mathbb{F})$ is closed and bounded in the Euclidean space containing it.

Theorem 1.4.3 *Let* X *be a compact space and* $f : X \to Y$ *a continuous map of* X *onto a Hausdorff space* Y. *Then* Y *is compact.*

Proof: Let $\{U_\alpha : \alpha \in \mathcal{A}\}$ be an arbitrary open cover of Y. Then each $f^{-1}(U_\alpha)$ is open in X by continuity and $\{f^{-1}(U_\alpha) : \alpha \in \mathcal{A}\}$ covers X since f maps onto Y. Since X is compact we may select a finite subcover $\{f^{-1}(U_{\alpha_1}), \ldots, f^{-1}(U_{\alpha_k})\}$. Then $\{U_{\alpha_1}, \ldots, U_{\alpha_k}\}$ covers Y because f is surjective. Thus, we have produced a finite subcover of $\{U_\alpha : \alpha \in \mathcal{A}\}$. ∎

From this we conclude that all of the projective spaces $\mathbb{F}\mathbb{P}^{n-1}$, being Hausdorff and continuous images (quotients) of spheres, are compact.

Exercise 1.4.2 Show that a closed subspace of a compact space is compact. (More precisely, show that if X is compact and A is a closed subset of X, then, with the relative topology A inherits from X, A is also compact.)

Exercise 1.4.3 Show that a compact subspace A of a Hausdorff space Y is closed in Y.

Now suppose X is compact and $f : X \to Y$ is continuous, one-to-one and maps onto the Hausdorff space Y. If $U \subseteq X$ is open, then $X - U$ is closed and therefore compact by Exercise 1.4.2. Theorem 1.4.3 then implies that $f(X - U)$ is compact and so is closed in Y by Exercise 1.4.3. Since f is a bijection, $f(X - U) = f(X) - f(U) = Y - f(U)$ so $f(U)$ is open in Y. Thus, f is an open mapping and therefore $f^{-1} : Y \to X$ is continuous and we have proved the following very useful result.

Theorem 1.4.4 *A continuous bijection from a compact space onto a Hausdorff space is a homeomorphism.*

Notice that if a product space $X_1 \times \cdots \times X_n$ is compact, then each of the factor spaces X_i must be compact since the projection $\mathcal{P}^i : X_1 \times \cdots \times X_n \to X_i$ is a continuous surjection and X_i is Hausdorff by Exercise 1.3.16. Much more important is the fact that the converse is also true.

Theorem 1.4.5 *Let $X_1, \ldots, X_n$ be Hausdorff topological spaces and $X = X_1 \times \cdots \times X_n$ their product. Then X is compact iff each X_i, $i = 1, \ldots, n$, is compact.*

Proof: All that remains is to show that if $X_1, \ldots, X_n$ are compact, then so is $X_1 \times \cdots \times X_n$ and this will clearly follow by induction if we can prove the result when $n = 2$. To simplify the notation we suppose X and Y are compact and show that $X \times Y$ is compact. Let $\mathcal{U}$ be an arbitrary open cover of $X \times Y$. First we fix an $x \in X$ and consider $\{x\} \times Y \subseteq X \times Y$. We claim that there exists an open set U_x in X containing x such that $U_x \times Y$ is covered by finitely many of the open sets in $\mathcal{U}$. To see this we proceed as follows: For each (x, y) in $\{x\} \times Y$ select some basic open set $U_{(x,y)} \times V_{(x,y)}$ in $X \times Y$ containing (x, y) and contained in some element of $\mathcal{U}$. Then $\{V_{(x,y)} : y \in Y\}$ is an open cover of Y. Since Y is compact we can find a finite subcover $\{V_{(x,y_1)}, \ldots, V_{(x,y_k)}\}$. Let $U_x = U_{(x,y_1)} \cap \cdots \cap U_{(x,y_k)}$. Then U_x is open in X, $x \in U_x$ and

$$U_x \times Y \subseteq \left(U_{(x,y_1)} \times V_{(x,y_1)}\right) \cup \cdots \cup \left(U_{(x,y_k)} \times V_{(x,y_k)}\right) .$$

Each $U_{(x,y_i)} \times V_{(x,y_i)}$ is contained in some element U_i of $\mathcal{U}$ so $U_x \times Y \subseteq U_1 \cup \cdots \cup U_k$ as required.

Thus, we may select, for each $x \in X$, an open set U_x in X containing x and a finite subcollection $\mathcal{U}_x$ of $\mathcal{U}$ that covers $U_x \times Y$. Now, $\{U_x : x \in X\}$ is an open cover of X and X is compact so there is a finite subcover $\{U_{x_1}, \ldots, U_{x_\ell}\}$. $\mathcal{U}_{x_i}$ covers $U_{x_i} \times Y$ for each $i = 1, \ldots, \ell$ so $\mathcal{U}_{x_1} \cup \cdots \cup \mathcal{U}_{x_\ell}$ covers $X \times Y$ and this is a finite subcover of $\mathcal{U}$. ∎

As an application of Theorems 1.4.4 and 1.4.5 we prove that the suspension of any sphere is the sphere of dimension one greater, i.e.,

$$SS^{n-1} \cong S^n . \tag{1.4.1}$$

First we define a map f of $S^{n-1} \times [-1, 1]$ onto S^n as follows (f will carry $S^{n-1} \times [0, 1]$ onto the upper hemisphere and $S^{n-1} \times [-1, 0]$ onto the lower hemisphere): For $(x, t) \in S^{n-1} \times [0, 1]$, $(1 - t)x$ is in D^n so φ_N^{-1} (Exercise 1.1.8) carries this onto the upper hemisphere in S^n. Thus, we define $f_1 : S^{n-1} \times [0, 1] \to S^n$ by $f_1(x, t) = \varphi_N^{-1}((1-t)x)$. Similarly, define $f_2 : S^{n-1} \times [-1, 0] \to S^n$ by $f_2(x, t) = \varphi_S^{-1}((1 + t)x)$. Note that $f_1(x, 0) = \varphi_N^{-1}(x)$ and $f_2(x, 0) = \varphi_S^{-1}(x)$. Since $\| x \| = 1$, these are the same so, by the Glueing Lemma 1.2.3, f_1 and f_2 determine a continuous map $f : S^{n-1} \times [-1, 1] \to S^n$. Observe that f carries $S^{n-1} \times (-1, 1)$ homeomorphically onto $S^n - \{N, S\}$, $f(S^{n-1} \times \{-1\}) = \{S\}$ and $f(S^{n-1} \times \{1\}) = \{N\}$. By Lemma 1.2.2 there exists a unique continuous map $\bar{f} : SS^{n-1} \to S^n$ for which the diagram

$$\begin{array}{ccc} S^{n-1} \times [-1,1] & \xrightarrow{f} & S^n \\ \mathcal{Q} \downarrow & \nearrow \bar{f} & \\ SS^{n-1} & & \end{array}$$

commutes. Now, $S^{n-1} \times [-1,1]$ is compact by Theorem 1.4.5 so SS^{n-1} is compact by Theorem 1.4.3. Since $\bar{f}$ is bijective, Theorem 1.4.4 implies that $\bar{f}$ is a homeomorphism and this completes the proof of (1.4.1).

The compact subspaces of $\mathbb{R}^n$ have a particularly important property that we now wish to establish. First we define, for any subspace X of $\mathbb{R}^n$, any $x \in X$ and any $r > 0$, the **open ball** $U_r(x, X)$ **of radius** r **about** x **in** X by $U_r(x, X) = \{y \in X : \| y - x \| < r\} = U_r(x) \cap X$. For any $A \subseteq \mathbb{R}^n$, the **diameter of** A, written $\operatorname{diam}(A)$, is defined by $\operatorname{diam}(A) = \sup\{ \|y - x\| : x, y \in A\}$ if $A \neq \emptyset$ and $\operatorname{diam}(\emptyset) = 0$.

Exercise 1.4.4 Show that if X is a subspace of $\mathbb{R}^n$ and $A \subseteq X$ has $\operatorname{diam}(A) < r$, then $A \subseteq U_r(x, X)$ for any $x \in A$.

Theorem 1.4.6 *Let X be a compact subspace of $\mathbb{R}^n$. Then for each open cover $\mathcal{U}$ of X there exists a positive number $\lambda = \lambda(\mathcal{U})$, called a* **Lebesgue number for** $\mathcal{U}$, *with the property that any $A \subseteq X$ with* $\operatorname{diam}(A) < \lambda$ *is entirely contained in some element of $\mathcal{U}$.*

Proof: For each $x \in X$ choose $r(x) > 0$ so that $U_{r(x)}(x, X)$ is contained in some element of $\mathcal{U}$. Then $\{U_{\frac{1}{2}r(x)}(x, X) : x \in X\}$ is an open cover of X and so has a finite subcover $\{U_{\frac{1}{2}r(x_1)}(x_1, X), \ldots, U_{\frac{1}{2}r(x_k)}(x_k, X)\}$. Let $\lambda = \lambda(\mathcal{U}) = \min\{\frac{1}{2}r(x_1), \ldots, \frac{1}{2}r(x_k)\}$. We claim that λ is a Lebesgue number for $\mathcal{U}$. By Exercise 1.4.4 it will suffice to show that every $U_\lambda(x, X)$, $x \in X$, is contained in some element of $\mathcal{U}$. But any $x \in X$ is in some $U_{\frac{1}{2}r(x_i)}(x_i, X)$ so, for any $y \in U_\lambda(x, X)$, $\|y - x_i\| \leq \|y - x\| + \|x - x_i\| < \lambda + \frac{1}{2}r(x_i) \leq r(x_i)$ so $y \in U_{r(x_i)}(x_i, X)$ and therefore $U_\lambda(x, X) \subseteq U_{r(x_i)}(x_i, X)$. But $U_{r(x_i)}(x_i, X)$ is contained in some element of $\mathcal{U}$ so the result follows. ∎

Exercise 1.4.5 Use Theorem 1.4.6 to show that any continuous map from a compact subspace of $\mathbb{R}^n$ into some $\mathbb{R}^m$ is **uniformly continuous**. More precisely, let X be a compact subspace of $\mathbb{R}^n$, Y an arbitrary subspace of $\mathbb{R}^m$ and $f : X \to Y$ a continuous map. Show that, for every $\varepsilon > 0$, there exists a $\delta > 0$, depending only on ε, such that $f(U_\delta(x, X)) \subseteq U_\varepsilon(f(x), Y)$ for every $x \in X$. **Hint:** Let δ be a Lebesgue number for the open cover $\{f^{-1}(U_{\varepsilon/2}(y, Y)) : y \in Y\}$ of X.

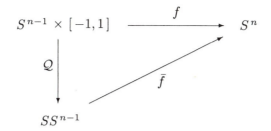

Euclidean space itself and many topological manifolds fail to be compact, but have a local version of this property that is useful. To discuss this property we must generalize a few familiar notions from analysis. We let X denote an arbitrary topological space. If $x \in X$, then a **neighborhood (nbd) of** x **in** X is a subset of X that contains an open set containing x. If $A \subseteq X$, then an **accumulation point of** A is an $x \in X$ with the property that every nbd of x in X contains some point of A other than x. Thus, for example, 0 is an accumulation point of $\{1, \frac{1}{2}, \frac{1}{3}, \dots\}$ in $\mathbb{R}$, but 1 is not. The set of all accumulation points of A in X is denoted A' and called the **derived set** of A. The **closure** of A in X, denoted $\bar{A}$, is the union of A and its set of accumulation points, i.e., $\bar{A} = A \cup A'$.

Lemma 1.4.7 *Let X be a topological space and A and B subsets of X. Then*

(a) $\bar{\emptyset} = \emptyset$ *and* $\bar{X} = X$.

(b) $A \subseteq B$ *implies* $\bar{A} \subseteq \bar{B}$.

(c) *A is closed in X iff* $\bar{A} = A$.

(d) $\bar{\bar{A}} = A$.

(e) $\overline{A \cup B} = \bar{A} \cup \bar{B}$.

(f) $\overline{A \cap B} \subseteq \bar{A} \cap \bar{B}$.

(g) *$\bar{A}$ is the intersection of all the closed subsets of X containing A.*

(h) *If X is a subspace of some $\mathbb{R}^n$ and $A \subseteq X$, then A' is the set of all points in X that are the limit of some sequence of points in X.*

Exercise 1.4.6 Prove Lemma 1.4.7 and show also that, in general, $\overline{A \cap B}$ need not equal $\bar{A} \cap \bar{B}$.

Exercise 1.4.7 A subset A of a space X is said to be **dense** in X if $\bar{A} = X$. X is said to be **separable** if there is a countable subset A of X that is dense in X. Show that any second countable space is separable.

Exercise 1.4.8 Show that if X is compact, $V \subseteq X$ is open and $x \in V$, then there exists an open set U in X with $x \in U \subseteq \bar{U} \subseteq V$ (in the jargon of point-set topology this shows that a compact space is **regular**).

Exercise 1.4.9 Let X and Y be topological spaces. Show that a map $f : X \to Y$ is continuous iff $f(\bar{A}) \subseteq \overline{f(A)}$ for every $A \subseteq X$.

Before introducing the promised local version of compactness we pause momentarily to use these last few ideas to generalize our Theorem 1.4.5 on the compactness of products to local products. We show that a bundle with compact base and fiber necessarily has a compact total space.

Theorem 1.4.8 *Let $(P, X, \mathcal{P}, Y)$ be a locally trivial bundle with compact base space X and compact fiber Y. Then the bundle space P is also compact.*

Proof: For each $x \in X$ we select a locally trivializing nbd V_x of x and a homeomorphism $\Phi_x : V_x \times Y \to \mathcal{P}^{-1}(V_x)$ such that $\mathcal{P} \circ \Phi_x$ is the projection of $V_x \times Y$ onto V_x. By Exercise 1.4.8 we can also select, for each $x \in X$, an open set U_x with $x \in U_x \subseteq \bar{U}_x \subseteq V_x$. The sets $\{U_x : x \in X\}$ cover X so we may select a finite subcover $\{U_{x_1}, \ldots, U_{x_n}\}$. Since $\bar{U}_{x_i}$ is closed in X and X is compact, $\bar{U}_{x_i}$ is also compact (Exercise 1.4.2). Thus, $\bar{U}_{x_i} \times Y$ is compact (Theorem 1.4.5) for each $i = 1, \ldots, n$. Since each Φ_{x_i} is a homeomorphism, $\Phi_{x_i}(\bar{U}_{x_i} \times Y)$ is compact for each $i = 1, \ldots, n$. But $\Phi_{x_i}(\bar{U}_{x_i} \times Y) = \mathcal{P}^{-1}(\bar{U}_{x_i})$ so each $\mathcal{P}^{-1}(\bar{U}_{x_i})$ is compact. Now observe that, since $X = \bigcup_{i=1}^n \bar{U}_{x_i}$, we have $P = \bigcup_{i=1}^n \mathcal{P}^{-1}(\bar{U}_{x_i})$.

Exercise 1.4.10 Complete the proof by showing that if a space can be written as a finite union of compact subspaces, then it also must be compact. ∎

Now, a subset U of a space X is said to be **relatively compact** if its closure $\bar{U}$ is compact. A Hausdorff space X is **locally compact** if each point in X has a relatively compact nbd. A compact space is certainly locally compact, but the converse is false. Indeed, $\mathbb{R}^n$ is not compact, but any $\overline{U_r(x)}$, being closed and bounded, is compact so $\mathbb{R}^n$ is locally compact. It follows that any locally Euclidean space (e.g., any topological manifold) is locally compact.

Lemma 1.4.9 *A locally compact space X has a basis consisting of relatively compact open sets. If X is also second countable, then it has a countable such basis.*

Proof: Let $\mathcal{B}$ be a basis for X (countable if X is second countable). Let $\mathcal{B}'$ be the subcollection of $\mathcal{B}$ consisting of all those basic open sets with compact closure. We show that $\mathcal{B}'$ is nonempty and, in fact, is actually a basis for X and this will prove the lemma. Let U be an arbitrary open set in X. It will suffice to find, for each $x \in U$, a $W \in \mathcal{B}'$ with $x \in W \subseteq U$. Select a nbd V of x with $\bar{V}$ compact. Then $U \cap V$ is an open set containing x so there is a W in $\mathcal{B}$ with $x \in W \subseteq U \cap V$. Then $\bar{W} \subseteq \overline{U \cap V} \subseteq \bar{V}$. Since $\bar{V}$ is compact and $\bar{W}$ is closed in $\bar{V}$, $\bar{W}$ is also compact. Thus, $W \in \mathcal{B}'$ and $x \in W \subseteq U \cap V \subseteq U$ as required. ∎

In analysis it is common to use stereographic projection to identify the 2-sphere S^2 with the "extended complex plane" (see Section 0.3). We show now that, like the plane, any locally compact space can be "compactified" by the addition of a single point. Thus, we let X be an arbitrary locally compact space and select some object ∞ that is not an element of X (the standard set-theoretic gimmick for doing this is to take ∞ to be the set X itself which, by the rules of the game in set theory, cannot be an element of X). On the set $X^* = X \cup \{\infty\}$ we define a topology by taking as open sets in X^* all of the open sets in X together with the complements in X^* of all the compact subsets of X.

Exercise 1.4.11 Verify that this collection of subsets of X^* is, indeed, a topology for X^* and that the relative topology that X inherits from X^* is precisely its original topology, i.e., that the subspace $X \subseteq X^*$ is homeomorphic to X.

To see that X^* is Hausdorff we let x and y be distinct points in X^*. If x and y are both in X, then there exist open sets U_x and U_y in X with $x \in U_x$, $y \in U_y$ and $U_x \cap U_y = \emptyset$. But U_x and U_y are also open in X^* so, in this case, the proof is complete. Thus, we need only show that $x \in X$ and $y = \infty$ can be separated by open sets in X^*. To see this let V be a relatively compact nbd of x in X. Then V and $X^* - \bar{V}$ are open sets in X^* with $x \in V$, $\infty \in X^* - \bar{V}$ and $V \cap (X - \bar{V}) = \emptyset$ as required. Finally, we show that X^* is compact. Let $\mathcal{U} = \{U_\alpha : \alpha \in \mathcal{A}\}$ be an open cover of X^*. Select some $U_{\alpha_0} \in \mathcal{U}$ with $\infty \in U_{\alpha_0}$. Since U_{α_0} is a nbd of ∞ in X^*, it is the complement of some compact set C in X. Select finitely many elements $\{U_{\alpha_1}, \ldots, U_{\alpha_k}\}$ of $\mathcal{U}$ that cover C. Then $\{U_{\alpha_0}, U_{\alpha_1}, \ldots, U_{\alpha_k}\}$ is a finite subcover of $\mathcal{U}$ so X^* is compact.

For any locally compact space X the compact space $X^* = X \cup \{\infty\}$ just constructed is called the **one-point compactification** of X.

Exercise 1.4.12 Show that X is an open, dense subspace of its one-point compactification X^*.

Exercise 1.4.13 Use a stereographic projection map (Exercise 1.1.8) to show that the one-point compactification of $\mathbb{R}^n$ is homeomorphic to S^n.

1.5 Connectivity and Covering Spaces

If a topological space X is the disjoint union of two nonempty open sets H and K, then the Glueing Lemma 1.2.3 implies that *any* continuous map on H and *any* continuous map on K can be glued together to give a continuous map on X. Since the maps on H and K need not bear any relationship to each other whatsoever, one regards X as consisting of two disconnected and topologically independent pieces.

We shall say that a topological space X is **disconnected** if X can be written as $X = H \cup K$, where H and K are disjoint, nonempty open sets in X. The pair $\{H, K\}$ is then called a **disconnection** of X. Notice that, since $H = X - K$ and $K = X - H$, the sets in a disconnection are closed as well as open.

Exercise 1.5.1 Show that a space X is disconnected iff it contains some nonempty, proper subset H that is both open and closed.

The subspace $\mathbb{Q}$ of $\mathbb{R}$ consisting of the rational numbers is disconnected since $\mathbb{Q} = \{x \in \mathbb{Q} : x < \sqrt{2}\} \cup \{x \in \mathbb{Q} : x > \sqrt{2}\}$ expresses $\mathbb{Q}$ as the disjoint union of two nonempty open subsets. A somewhat less trivial example is the orthogonal group $O(n)$. We have already seen (Section 1.1) that any $A \in O(n)$ has $\det A = \pm 1$. Now, $\det A$ is a polynomial in the entries of A and so, since $O(n)$ is a subspace of $\mathbb{R}^{n^2}$, det is a continuous real-valued function on $O(n)$. Thus, $\det^{-1}(-\infty, 0)$ and $\det^{-1}(0, \infty)$ are nonempty, disjoint open sets in $O(n)$ whose union is all of $O(n)$ and so $O(n)$ is disconnected. Note that $\det^{-1}(-\infty, 0) = \det^{-1}(-1)$ and $\det^{-1}(0, \infty) = \det^{-1}(1)$.

If a space X has no disconnection, i.e., cannot be written as the disjoint union of two nonempty open sets, then we will say that X is **connected**. We set about finding some important examples.

Lemma 1.5.1 *A subspace X of $\mathbb{R}$ is connected iff it is an interval (open, closed, or half-open).*

Proof: Suppose first that X is connected. We may assume that X contains more than one point since, if $X = \emptyset$, then $X = (x_0, x_0)$ for any $x_0 \in \mathbb{R}$ and if $X = \{x_0\}$, then $X = [x_0, x_0]$. Thus, if X were not an interval there would exist real numbers x, y and z with x and z in X, $x < y < z$, but $y \notin X$. But this implies that $X = [X \cap (-\infty, y)] \cup [X \cap (y, \infty)]$ so $\{X \cap (-\infty, y),\, X \cap (y, \infty)\}$ is a disconnection of X and this is a contradiction.

Now, let $X \subseteq \mathbb{R}$ be an interval. Again, we may assume that X contains more than one point. Suppose X were disconnected, i.e., $X = H \cup K$, where H and K are nonempty, disjoint open (and therefore closed) subsets of X. Choose $x \in H$ and $z \in K$. Then $x \neq z$ and, by relabeling H and K if necessary, we may assume $x < z$. Since X is an interval, $[x, z] \subseteq X$ so each point in $[x, z]$ is in either H or K. Let $y = \sup\{t \in [x, z] : t \in H\}$. Then $x \le y \le z$ so $y \in X$. Since H is the intersection with X of a closed set in $\mathbb{R}$, $t \in H$. Thus, $y < z$. But, by definition of y, $y + \varepsilon \in K$ for all sufficiently small $\varepsilon > 0$ (those for which $y + \varepsilon \le z$). Since K is the intersection with X of a closed set in $\mathbb{R}$, $y \in K$. Thus, $y \in H \cap K$ and this is a contradiction. Consequently, X cannot be disconnected, i.e., X is connected. ∎

Theorem 1.5.2 *The continuous image of a connected space is connected.*

Proof: Suppose X is connected and $f : X \to Y$ is a continuous map of X onto Y. If $\{H, K\}$ were a disconnection of Y, then $\{f^{-1}(H), f^{-1}(K)\}$ would be a disconnection of X and this is impossible so Y must be connected. ∎

Exercise 1.5.2 Let X be a topological space and Y a subspace of X that is connected (in its relative topology). Show that if Z is any subspace of X with $Y \subseteq Z \subseteq \bar{Y}$, then Z is connected. **Hint:** If Z were not connected there would exist closed subsets H and K *of* X whose union contains Z and whose intersections with Z are nonempty and disjoint.

A **path** in a space X is a continuous map $\alpha : [0,1] \to X$. The points $x_0 = \alpha(0)$ and $x_1 = \alpha(1)$ in X are called, respectively, the **initial** and **terminal points** of α and we say that α is a **path in** X **from** x_0 **to** x_1. According to Theorem 1.5.2, the image $\alpha([0,1])$ of α is a connected subspace of X. (Be careful to distinguish a path in X, which is a continuous map, from its image, which is a set of points.) A topological space X is **pathwise connected** if, for any two points x_0 and x_1 in X, there exists a path $\alpha : [0,1] \to X$ from $x_0 = \alpha(0)$ to $x_1 = \alpha(1)$.

Lemma 1.5.3 *A pathwise connected space is connected.*

Proof: Suppose X is pathwise connected. If X were not connected we could write $X = H \cup K$, where H and K are nonempty disjoint open subsets of X. Choose $x_0 \in H$ and $x_1 \in K$. By assumption, there exists a path $\alpha : [0,1] \to X$ from $x_0 = \alpha(0)$ to $x_1 = \alpha(1)$. But then $\{\alpha([0,1]) \cap H, \alpha([0,1]) \cap K\}$ is a disconnection of the image $\alpha([0,1])$ of α and this is impossible since $\alpha([0,1])$ is connected. ∎

From this it follows, for example, that $\mathbb{R}^n$ is connected since it is clearly pathwise connected (for any $x_0, x_1 \in \mathbb{R}^n$, $\alpha(s) = (1-s)x_0 + sx_1$, $0 \leq s \leq 1$, is a path from x_0 to x_1). More generally, let us say that a subset X of $\mathbb{R}^n$ is **convex** if it contains the line segment joining any two of its points, i.e., if $x_0, x_1 \in X$ implies $(1-s)x_0 + sx_1 \in X$ for all $0 \leq s \leq 1$. Then a convex subspace X of $\mathbb{R}^n$ is pathwise connected and therefore connected. In particular, any open or closed balls in $\mathbb{R}^n$ are connected. For any $n \geq 2$, $\mathbb{R}^n - \{p\}$ is pathwise connected for any $p \in \mathbb{R}^n$ (use two line segments, if necessary) so these "punctured" Euclidean spaces are connected. As another application of Lemma 1.5.3 we ask the reader to show that spheres are connected.

Exercise 1.5.3 Show that, for any $n \geq 1$, the n-sphere S^n is pathwise connected and therefore connected. **Hint:** Stereographic projections.

Exercise 1.5.4 Show that the continuous image of a pathwise connected space is pathwise connected.

From these last two results it follows that any projective space $\mathbb{F}\mathbb{P}^{n-1}$ is pathwise connected. Also note that $SU(2)$, being homeomorphic to S^3, is pathwise connected. By Exercise 1.1.26, the same is true of $Sp(1)$. Moreover, $U(1)$ is homeomorphic to S^1 (Exercise 1.1.25) so it too is pathwise connected. We will prove in Section 1.6 that $SU(n)$, $SO(n)$, $U(n)$ and $Sp(n)$ are connected for any n (we already know that $O(n)$ is not).

Exercise 1.5.5 Show that $\mathbb{R}^n$ is not homeomorphic to $\mathbb{R}$ for $n > 1$ and that S^1 is not homeomorphic to $[a, b]$ for any $a < b$ in $\mathbb{R}$. **Hint:** Suppose $h : \mathbb{R}^n \to \mathbb{R}$ is a homeomorphism and then delete a point p from $\mathbb{R}^n$.

It is often convenient to rephrase the definition of pathwise connectivity in terms of paths with some fixed initial point. The proof of the following result involves some ideas that will play a major role in Chapter 2.

Lemma 1.5.4 *Let X be a topological space and x_0 some fixed point in X. Then X is pathwise connected iff, for each x_1 in X, there exists a path in X from x_0 to x_1.*

Proof: Since the necessity is trivial we prove only the sufficiency. Thus, we assume that there is a path from x_0 to any point in X. Let $x_1, x_2 \in X$ be arbitrary. We must produce a path in X from x_1 to x_2. Let $\alpha : [0,1] \to X$ be a path from x_0 to x_1 and $\beta : [0,1] \to X$ a path in X from x_0 to x_2. Define a map $\alpha^{\leftarrow} : [0,1] \to X$ by $\alpha^{\leftarrow}(s) = \alpha(1-s)$. Then $\alpha^{\leftarrow}$ is a path in X from x_1 to x_0 ("α backwards"). Next define a map $\alpha^{\leftarrow}\beta : [0,1] \to X$ by

$$\alpha^{\leftarrow}\beta(s) = \begin{cases} \alpha^{\leftarrow}(2s), & 0 \le s \le \frac{1}{2} \\ \beta(2s-1), & \frac{1}{2} \le s \le 1 \end{cases} .$$

Then $\alpha^{\leftarrow}\beta$ is continuous by the Glueing Lemma 1.2.3 and satisfies $\alpha^{\leftarrow}\beta(0) = \alpha^{\leftarrow}(0) = \alpha(1) = x_1$ and $\alpha^{\leftarrow}\beta(1) = \beta(1) = x_2$ as required. ∎

Notice that the argument given in the proof of Lemma 1.5.4 shows that, even in a space that is not pathwise connected, any two points that can be joined by paths to a third point can, in fact, be joined to each other by a path.

The converse of Lemma 1.5.3 is not true and there is a standard example, known as the Topologist's Sine Curve, of a connected space that is not pathwise connected. We will not reproduce the example here (see Theorem 5.3, Chapter $\underline{\text{V}}$, of [**Dug**]), but will instead show that, for most of the spaces of interest to us (e.g., topological manifolds), this sort of thing cannot occur. Let us say that a topological space X is **locally connected** (respectively, **locally pathwise connected**) if, whenever x is in X and V is an open set containing x, there exists an open set U with $x \in U \subseteq V$ such that U, with its relative topology, is connected (respectively, pathwise connected). Notice that any locally Euclidean space obviously has both of these properties.

Theorem 1.5.5 *A connected, locally pathwise connected space is pathwise connected.*

Proof: Let X be connected and locally pathwise connected and fix some $x_0 \in X$ (we intend to apply Lemma 1.5.4). Denote by H the set of all points $x_1 \in X$ for which there exists some path in X from x_0 to x_1. We show that H is all of X and for this it will suffice to show that H is nonempty, open and closed (for then, if H were not all of X, $\{H, X - H\}$ would be a disconnection of X). $H \neq \emptyset$ is clear since $x_0 \in H$. To see that H is open, let $x_1 \in H$ be arbitrary. Since X is locally pathwise connected there exists an open set U in X containing x_1 which, in its relative topology, is pathwise connected. We claim that $U \subseteq H$. To see this let $x_2 \in U$ be arbitrary. Then there is a path in U from x_1 to x_2. But a continuous map into the subspace U of X is also continuous when thought of as a map into X so this gives a path in X from x_1 to x_2. Since $x_1 \in H$ there is a path in X from x_0 to x_1. Consequently, by the remark following the proof of Lemma 1.5.4, there is a path in X from x_0 to x_2, i.e., $x_2 \in H$. Thus, $x_1 \in U \subseteq H$ so H is open. Finally, we show that H is closed by proving $\bar{H} = H$. $\bar{H} \supseteq H$ is clear so we prove $\bar{H} \subseteq H$. Let $x_2 \in \bar{H}$ be arbitrary. As above, we let U be a pathwise connected nbd of x_2 in X. Then $U \cap H \neq \emptyset$. Choose some x_1 in $U \cap H$. Since x_1 is in H there is a path in X from x_0 to x_1. Since $x_1 \in U$ there is a path in U (and therefore in X) from x_1 to x_2. Thus, there is a path in X from x_0 to x_2 so $x_2 \in H$, i.e., $\bar{H} \subseteq H$ as required. ∎

Corollary 1.5.6 *A topological manifold is connected iff it is pathwise connected.*

To enlarge our collection of examples a bit more we consider the behavior of connectedness under the formation of products. First, a lemma.

Lemma 1.5.7 *Let X be a topological space and $\{X_\alpha : \alpha \in \mathcal{A}\}$ a family of connected subspaces of X with $X = \bigcup_{\alpha \in \mathcal{A}} X_\alpha$ and $\bigcap_{\alpha \in \mathcal{A}} X_\alpha \neq \emptyset$. Then X is connected.*

Proof: Suppose $X = H \cup K$, where H and K are disjoint open sets in X. Since X_α is connected and contained in $H \cup K$ for every α, each X_α must be contained entirely in either H or K. Since $\bigcap_{\alpha \in \mathcal{A}} X_\alpha \neq \emptyset$ and $H \cap K = \emptyset$, all of the X_α are contained in one of these sets. Without loss of generality, suppose $X_\alpha \subseteq H$ for every $\alpha \in \mathcal{A}$. Then $\bigcup_{\alpha \in \mathcal{A}} X_\alpha \subseteq H$ so $X \subseteq H$ and therefore $X = H$ so $K = \emptyset$. Consequently, X has no disconnection and must therefore be connected. ∎

Exercise 1.5.6 Show that if the X_α in Lemma 1.5.7 are all pathwise connected, then so is X.

Theorem 1.5.8 *Let $X_1, \ldots, X_k$ be topological spaces and $X = X_1 \times \cdots \times X_k$ the product space. Then X is connected iff each X_i, $i = 1, \ldots, k$, is connected.*

Proof: If X is connected, then, since X_i is the image of X under the projection $\mathcal{P}^i : X \to X_i$, Theorem 1.5.2 implies that X_i is also connected. For the converse, it will clearly suffice to prove the result for $k = 2$ and this we do by contradiction. Thus, we assume X_1 and X_2 are connected, but that $X_1 \times X_2 = H \cup K$, where H and K are nonempty, disjoint, open subsets of $X_1 \times X_2$. Choose $(a_1, b_1) \in H$ and $(a_2, b_2) \in K$. The subspaces $\{a_1\} \times X_2$ and $X_1 \times \{b_2\}$ are homeomorphic to X_2 and X_1, respectively, and are therefore connected. Moreover, $(\{a_1\} \times X_2) \cap (X_1 \times \{b_2\})$ is nonempty since it contains (a_1, b_2). By Lemma 1.5.7, the subspace $(\{a_1\} \times X_2) \cup (X_1 \times \{b_2\})$ of $X_1 \times X_2$ is connected. This, however, is impossible since this subspace intersects both H and K and so has a disconnection. Thus, $X_1 \times X_2$ is connected. ∎

Exercise 1.5.7 Show that $X_1 \times \cdots \times X_k$ is pathwise connected iff each X_i, $i = 1, \ldots, k$, is pathwise connected.

Thus, for example, the torus $S^1 \times S^1$, its higher dimensional analogues $S^1 \times \cdots \times S^1$ and, more generally, any product of spheres S^n, $n > 0$, is (pathwise) connected. One can go a step further and show that if the base and fiber of a locally trivial bundle are connected, then so is its total space.

Theorem 1.5.9 *Let $(P, X, \mathcal{P}, Y)$ be a locally trivial bundle with connected base X and connected fiber Y. Then the bundle space P is also connected.*

Proof: Suppose P were not connected so that $P = H \cup K$, where H and K are disjoint, nonempty, open subsets of P. Since each fiber $\mathcal{P}^{-1}(x)$, $x \in X$, is, as a subspace of P, homeomorphic to Y, it is connected and therefore must be entirely contained in one of H or K. Thus, $\mathcal{P}(H) \cap \mathcal{P}(K) = \emptyset$. But $\mathcal{P}$ is a continuous, open surjection (Lemma 1.3.6) so $\mathcal{P}(H)$ and $\mathcal{P}(K)$ are open and $X = \mathcal{P}(H) \cup \mathcal{P}(K)$. Since H and K are nonempty, so are $\mathcal{P}(H)$ and $\mathcal{P}(K)$. Thus, $\{\mathcal{P}(H), \mathcal{P}(K)\}$ is a disconnection of X and this contradicts the connectivity of X. ∎

Proving the analogue of Theorem 1.5.9 for pathwise connectedness raises some issues that will be of profound significance for us and that we have already encountered in the context of physics (Sections 0.2 and 0.4). Consider how one might go about proving that, if $(P, X, \mathcal{P}, Y)$ is a locally trivial bundle with X and Y pathwise connected, then P is also pathwise connected. Take two points p_0 and p_1 in P. If $\mathcal{P}(p_0) = \mathcal{P}(p_1)$, then p_0 and p_1 lie in a single fiber of $\mathcal{P}$ and this, being homeomorphic to Y, is pathwise connected so that one can find a path in that fiber (and therefore in P) from p_0 to p_1. Suppose then that $\mathcal{P}(p_0) \neq \mathcal{P}(p_1)$. Since X is pathwise connected there is a path $\alpha : [0,1] \to X$ from $\alpha(0) = \mathcal{P}(p_0)$ to $\alpha(1) = \mathcal{P}(p_1)$. The problem then is to "lift" α to a path in P that starts at p_0, i.e., to find a path $\tilde{\alpha} : [0,1] \to P$ such that $\tilde{\alpha}(0) = p_0$ and $\mathcal{P} \circ \tilde{\alpha}(s) = \alpha(s)$ for each s in $[0,1]$. Then $\mathcal{P} \circ \tilde{\alpha}(1) = \alpha(1) = \mathcal{P}(p_1)$ so $\tilde{\alpha}$ will end in the fiber containing p_1 and we can follow it by some path in this fiber to p_1 itself. Notice that such a lift would be easy to find in the trivial bundle $X \times Y$ ($\tilde{\alpha}(s) = (\alpha(s), p_0^2)$, where $p_0 = (p_0^1, p_0^2) \in X \times Y$) so the fact that P is locally like $X \times Y$ encourages us that this plan might succeed.

To carry out the program we have just outlined we introduce the general notion of a "lift" for maps into the base of a locally trivial bundle. Thus, we let $(P, X, \mathcal{P}, Y)$ be an arbitrary locally trivial bundle and $f : Z \to X$ a continuous map of some space Z into the base space X. A **lift of f to P** is a continuous map $\tilde{f} : Z \to P$ such that $\mathcal{P} \circ \tilde{f} = f$, i.e., such that the following diagram commutes:

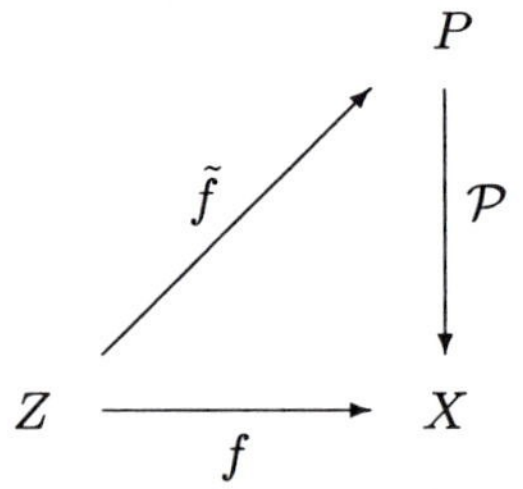

We hasten to point out that, in general, one cannot expect lifts to exist (although we prove shortly that *paths* always lift). Consider, for example, the locally trivial bundle $(\mathbb{R}, S^1, \mathcal{P}, \mathbb{Z})$, where $\mathcal{P}(s) = e^{2\pi s \mathbf{i}}$ and the identity map $id : S^1 \to S^1$:

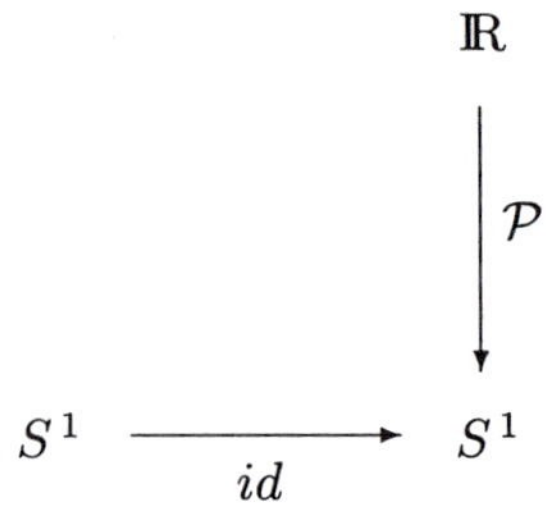

We claim that there is no continuous map $\tilde{f} : S^1 \to \mathbb{R}$ for which $\mathcal{P} \circ \tilde{f} = id$. Suppose there were such an $\tilde{f}$. Then the image $\tilde{f}(S^1)$ is compact and connected and so is a closed, bounded subinterval $[a, b]$ of $\mathbb{R}$. Now, $\tilde{f}$ cannot be one-to-one for then it would be a homeomorphism (Theorem

1.4.4), whereas S^1 is not homeomorphic to $[a,b]$ (Exercise 1.5.5). Thus, there exist points $x_0, x_1 \in S^1$ with $x_0 \neq x_1$, but $\tilde{f}(x_0) = \tilde{f}(x_1)$. But then $\mathcal{P} \circ \tilde{f}(x_0) = \mathcal{P} \circ \tilde{f}(x_1)$, i.e., $id\,(x_0) = id\,(x_1)$, so $x_0 = x_1$ and this is a contradiction.

We will be much concerned with the existence of lifts for various maps into the base of a locally trivial bundle. A particularly important case is that of the identity map on the base. A **(global) cross-section** of a locally trivial bundle $(P, X, \mathcal{P}, Y)$ is a lift of the identity map $id : X \to X$ to P, i.e., it is a continuous map $s : X \to P$ such that $\mathcal{P} \circ s = id$. Intuitively, a cross-section is a continuous selection of an element from each fiber $\mathcal{P}^{-1}(x)$, $x \in X$. Not every bundle has a cross-section, as we showed above for $(\mathbb{R}, S^1, \mathcal{P}, \mathbb{Z})$.

Now we return to the issue that motivated all of this. We again consider a locally trivial bundle $(P, X, \mathcal{P}, Y)$ and now a path $\alpha : [0,1] \to X$. We propose to show that α always lifts to a path $\tilde{\alpha}$ in P and that, moreover, one can start $\tilde{\alpha}$ at any point in the fiber above $\alpha(0)$. More precisely, we have the following result.

Theorem 1.5.10 **(Path Lifting Theorem)** *Let $(P, X, \mathcal{P}, Y)$ be a locally trivial bundle and $\alpha : [0,1] \to X$ a path in the base space X. Then, for any p in the fiber $\mathcal{P}^{-1}(\alpha(0))$ above $\alpha(0)$, there exists a lift $\tilde{\alpha} : [0,1] \to P$ of α to P with $\tilde{\alpha}(0) = p$.*

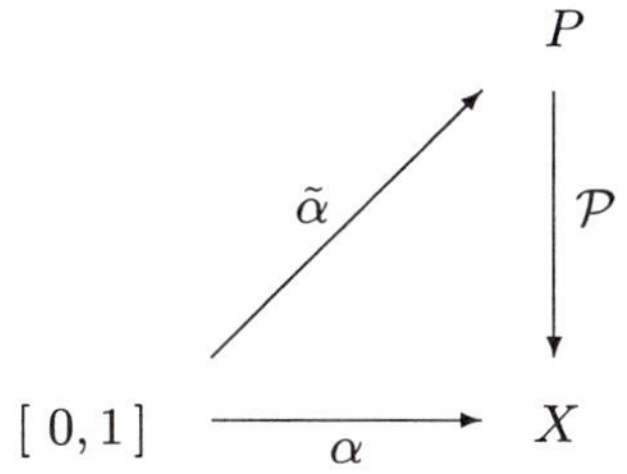

Proof: We wish to subdivide $[0,1]$ into subintervals with endpoints $0 = s_0 < s_1 < \cdots < s_{n-1} < s_n = 1$ in such a way that each $\alpha([s_{i-1}, s_i])$, $i = 1, \ldots, n$, is contained in some locally trivializing nbd V_i in X. This is done as follows: Cover X by locally trivializing nbds V and consider the open cover of $[0,1]$ consisting of all the corresponding $\alpha^{-1}(V)$. Select a Lebesgue number λ for this open cover (Theorem 1.4.6) and take n large enough that $\frac{1}{n} < \lambda$. Then we can let $s_i = \frac{i}{n}$ for $i = 0, 1, \ldots, n$. Now, we show by induction that for each $i = 0, 1, \ldots, n$ there exists a continuous map $\alpha_i : [0, s_i] \to P$ such that $\alpha_i(0) = p$ and $\mathcal{P} \circ \alpha_i = \alpha \mid [0, s_i]$. Then α_n will be the required lift $\tilde{\alpha}$. For $i = 0$ this is trivial; just define $\alpha_0(0) = p$. Now suppose $0 \leq k < n$ and that we have defined $\alpha_k : [0, s_k] \to X$ such that $\alpha_k(0) = p$ and $\mathcal{P} \circ \alpha_k = \alpha \mid [0, s_k]$. Then $\alpha([s_k, s_{k+1}])$ is contained in some locally trivializing nbd V_k of X. Let $\Phi_k : V_k \times Y \to \mathcal{P}^{-1}(V_k)$ be a homeomorphism satisfying $\mathcal{P} \circ \Phi_k(x,y) = x$ for $(x,y) \in V_k \times Y$. Now, $\mathcal{P} \circ \alpha_k(s_k) = \alpha(s_k)$ so $\alpha_k(s_k) \in \mathcal{P}^{-1}(\alpha(s_k)) \subseteq \mathcal{P}^{-1}(V_k)$. Thus, $\Phi_k^{-1}(\alpha_k(s_k)) \in V_k \times Y$ and so $\Phi_k^{-1}(\alpha_k(s_k)) = (\alpha(s_k), y_0)$ for some $y_0 \in Y$. Define $\alpha_{k+1} : [0, s_{k+1}] \to P$ as follows:

$$\alpha_{k+1}(s) = \begin{cases} \alpha_k(s), & s \in [0, s_k] \\ \Phi_k(\alpha(s), y_0), & s \in [s_k, s_{k+1}] \end{cases}.$$

Then, since $\Phi_k(\alpha(s_k), y_0) = \alpha_k(s_k)$, the Glueing Lemma 1.2.3 implies that α_{k+1} is continuous. Moreover, $\alpha_{k+1}(0) = \alpha(0) = p$ and $\mathcal{P} \circ \alpha_{k+1} = \alpha \mid [0, s_{k+1}]$ so the induction, and therefore the proof, is complete. ∎

Corollary 1.5.11 *Let $(P, X, \mathcal{P}, Y)$ be a locally trivial bundle with X and Y pathwise connected. Then P is also pathwise connected.*

Exercise 1.5.8 Prove Corollary 1.5.11. ∎

Lifts, even when they exist, are generally not unique. The path liftings described in Theorem 1.5.10, for example, can begin anywhere in the fiber above the initial point $\alpha(0)$ of the path being lifted. Even if this initial point is specified in advance, however, one can easily distort the lift "vertically", i.e., within the fiber at each point, and not alter its projection into X (re-examine the proof of Theorem 1.5.10 and devise various ways of doing this). Determining conditions and structures that specify unique lifts will be a matter of great interest to us and will eventually lead us to the notion of a "connection" ("gauge potential") on a principal bundle. For the present we limit ourselves to a rather obvious, but very important topological impediment to wandering around in the fibers — we make them discrete. This is accomplished with the notion of a "covering space", which we will show is a special type of locally trivial bundle.

Let X be a Hausdorff space. A **covering space for** X consists of a connected, Hausdorff space $\tilde{X}$ and a continuous map $\mathcal{P} : \tilde{X} \to X$ of $\tilde{X}$ onto X such that each $x \in X$ has a nbd V in X for which $\mathcal{P}^{-1}(V)$ is the disjoint union of a family of open sets S_α in $\tilde{X}$, each of which is mapped homeomorphically onto V by $\mathcal{P}$, i.e., each $\mathcal{P} | S_\alpha : S_\alpha \to V$ is a homeomorphism. Any such V is said to be **evenly covered** by $\mathcal{P}$ and the open sets S_α in $\tilde{X}$ are called **sheets over** V. We have already seen several examples of covering spaces in Section 1.3. The map $\mathcal{P} : \mathbb{R} \to S^1$ given by $\mathcal{P}(s) = e^{2\pi s \mathbf{i}}$ is one such ($V_1 = S^1 - \{1\}$ and $V_2 = S^1 - \{-1\}$ are evenly covered). We also showed in Section 1.3 that the quotient map $\mathcal{P} : S^{n-1} \to \mathbb{RP}^{n-1}$ is a covering space.

Exercise 1.5.9 Show that the map $\mathcal{P} : \mathbb{R}^n \to S^1 \times \cdots \times S^1$ (n factors) of $\mathbb{R}^n$ onto the n-torus given by $\mathcal{P}(x^1, \ldots, x^n) = (e^{2\pi x^1 \mathbf{i}}, \ldots, e^{2\pi x^n \mathbf{i}})$ is a covering space.

Exercise 1.5.10 For each $n = 1, 2, \ldots$, define $\mathcal{P}_n : S^1 \to S^1$ by $\mathcal{P}_n(z) = z^n$ for each $z \in S^1 \subseteq \mathbb{C}$. Show that these are all covering spaces.

We point out a few immediate consequences of the definition. First note that, for each $x \in X$, $\mathcal{P}^{-1}(x)$ is a discrete subspace of $\tilde{X}$. Indeed, if V is an evenly covered nbd of x in X, then each sheet S_α over V contains precisely one element of $\mathcal{P}^{-1}(x)$ ($\mathcal{P} \mid S_\alpha : S_\alpha \to V$ is a homeomorphism). Thus, $\mathcal{P}^{-1}(x) \cap S_\alpha$ is a single point, which, since S_α is open in $\tilde{X}$, must therefore be open in the subspace $\mathcal{P}^{-1}(x)$. Since each of its points is open, $\mathcal{P}^{-1}(x)$ is discrete. Thus, $\mathcal{P} : S^{2n-1} \to \mathbb{CP}^{n-1}$ and $\mathcal{P} : S^{4n-1} \to \mathbb{HP}^{n-1}$ are *not* covering spaces.

Next we show that all of the discrete subspaces $\mathcal{P}^{-1}(x)$, for $x \in X$, have the same cardinality (and are therefore homeomorphic). First observe that, since $\tilde{X}$ is connected and $\mathcal{P}$ is onto, X is also connected. Now, for every cardinal number Ω less than or equal to the cardinality of P let H_Ω be the set of all $x \in X$ for which $\mathcal{P}^{-1}(x)$ has cardinality Ω. We claim first that each H_Ω is open in X. To see this, let $x_0 \in H_\Omega$ and let V be an evenly covered nbd of x_0 in X. Every sheet over V contains precisely one element of $\mathcal{P}^{-1}(x_0)$ so the number of such sheets is also Ω. Thus, for every $x \in V$, $\mathcal{P}^{-1}(x)$ has the same cardinality as the number of sheets over V, i.e., Ω, so $x \in H_\Omega$. Thus, $x_0 \in V \subseteq H_\Omega$ so H_Ω is open in X.

Exercise 1.5.11 Show that each H_Ω is also closed in X.

Since X is connected, it follows that X must equal some one H_Ω and the result follows. Now, if we let D denote the discrete space whose cardinality is that of any $\mathcal{P}^{-1}(x)$ we can show that $(\tilde{X}, X, \mathcal{P}, D)$ is a locally trivial bundle with fiber D. Indeed, let V be an evenly covered nbd in X. Then $\mathcal{P}^{-1}(V) = \bigcup_{\alpha \in \mathcal{A}} S_\alpha$, where the S_α are disjoint open sets in $\tilde{X}$ and $\mathcal{P} | S_\alpha : S_\alpha \to V$ is a homeomorphism for each $\alpha \in \mathcal{A}$. Moreover, the cardinality of $\mathcal{A}$ is the same as that of D so we may select a bijection $\phi : \mathcal{A} \to D$. Now define a map $\Psi : \mathcal{P}^{-1}(V) \to V \times D$ by $\Psi(p) = (\mathcal{P}(p), \phi(\alpha))$, where α is the unique element of $\mathcal{A}$ with $p \in S_\alpha$.

Exercise 1.5.12 Show that $\Psi : \mathcal{P}^{-1}(V) \to V \times D$ is a homeomorphism.

Consequently, the inverse of this map, which we denote by $\Phi : V \times D \to \mathcal{P}^{-1}(V)$, is a homeomorphism. Since $\mathcal{P} \circ \Phi(x, y) = x$ for each $(x, y) \in V \times D$, (V, Φ) is a local trivialization and so $(\tilde{X}, X, \mathcal{P}, D)$ is a locally trivial bundle.

Thus, covering spaces are particular instances of locally trivial bundles (those with discrete fibers). In particular, any path in X lifts to $\tilde{X}$ by Theorem 1.5.10. We claim that in this case, however, the lift is unique once its initial point is specified. In fact, we prove much more.

Theorem 1.5.12 **(Unique Lifting Theorem)** *Let $\mathcal{P} : \tilde{X} \to X$ be a covering space, x_0 a point in X and $\tilde{x}_0$ a point in $\mathcal{P}^{-1}(x_0)$. Suppose Z is a connected space and $f : Z \to X$ is a continuous map with $f(z_0) = x_0$. If there is a lift $\tilde{f} : Z \to \tilde{X}$ of f to $\tilde{X}$ with $\tilde{f}(z_0) = \tilde{x}_0$, then this lift is unique.*

Proof: Suppose there are two continuous maps $\tilde{f}_1, \tilde{f}_2 : Z \to \tilde{X}$ that satisfy $\tilde{f}_1(z_0) = \tilde{f}_2(z_0) = \tilde{x}_0$ and $\mathcal{P} \circ \tilde{f}_1 = \mathcal{P} \circ \tilde{f}_2 = f$. Let $H = \{z \in Z : \tilde{f}_1(z) = \tilde{f}_2(z)\}$ and $K = Z - H = \{z \in Z : \tilde{f}_1(z) \neq \tilde{f}_2(z)\}$. We show that H and K are both open in Z so that, by connectedness of Z and the fact that $z_0 \in H$, $K = \emptyset$. Let $z_1 \in Z$ and let V be an evenly covered nbd of $f(z_1)$ in X. We consider two cases. Suppose first that $z_1 \in H$. Then $\tilde{f}_1(z_1) = \tilde{f}_2(z_1)$ lies in some sheet S over V. Then $U = \tilde{f}_1^{-1}(S) \cap \tilde{f}_2^{-1}(S)$ is an open nbd of z_1 in Z. Moreover, $\tilde{f}_1$ and $\tilde{f}_2$ both map U into S and $\mathcal{P}$ is a homeomorphism on S so $\mathcal{P} \circ \tilde{f}_1(z) = \mathcal{P} \circ \tilde{f}_2(z)$ $(= f(z))$ for every z in U implies $\tilde{f}_1(z) = \tilde{f}_2(z)$ for every z in U. Thus, $z_1 \in U \subseteq H$ so H is open. Next suppose $z_1 \in K$. Then $\tilde{f}_1(z_1) \neq \tilde{f}_2(z_1)$. But $\mathcal{P} \circ \tilde{f}_1(z_1) = \mathcal{P} \circ \tilde{f}_2(z_1)$ $(= f(z_1))$ so $\tilde{f}_1(z_1)$ and $\tilde{f}_2(z_1)$ must lie in different sheets S_1 and S_2 over V. Then $W = \tilde{f}_1^{-1}(S_1) \cap \tilde{f}_2^{-1}(S_2)$ is an open nbd of z_1 that $\tilde{f}_1$ and $\tilde{f}_2$ carry to different sheets over V. Since $S_1 \cap S_2 = \emptyset$, $\tilde{f}_1$ and $\tilde{f}_2$ disagree everywhere on W, i.e., $z_1 \in W \subseteq K$ so K is open and the proof is complete. ■

Corollary 1.5.13 *Let $\mathcal{P} : \tilde{X} \to X$ be a covering space, x_0 a point in X and $\tilde{x}_0$ a point in $\mathcal{P}^{-1}(x_0)$. Suppose $\alpha : [0,1] \to X$ is a path in X with $\alpha(0) = x_0$. Then there exists a unique lift $\tilde{\alpha} : [0,1] \to \tilde{X}$ of α to $\tilde{X}$ that satisfies $\alpha(0) = \tilde{x}_0$.*

Exercise 1.5.13 Let $\mathcal{P} : \tilde{X} \to X$ be a covering space. Suppose $\phi_1, \phi_2 : \tilde{X} \to \tilde{X}$ are continuous maps for which $\mathcal{P} \circ \phi_1 = \mathcal{P}$ and $\mathcal{P} \circ \phi_2 = \mathcal{P}$. Show that, if there exists a $p \in \tilde{X}$ for which $\phi_1(p) = \phi_2(p)$, then $\phi_1 = \phi_2$. **Hint:** Theorem 1.5.12.

We close this section with the observation that, while not every space of interest to us is (pathwise) connected, it is always possible to split a topological space up into maximal (pathwise) connected pieces. First consider an arbitrary space X and a fixed point $x_0 \in X$. Define the **component** $C(x_0)$ **of** x_0 **in** X to be the union of all the connected subspaces of X containing x_0. Since all of these contain $\{x_0\}$, Lemma 1.5.7 implies that $C(x_0)$ is a connected subspace of X. Moreover, if x_0 and x_1 are distinct points of X, then either $C(x_0) = C(x_1)$ or $C(x_0) \cap C(x_1) = \emptyset$, for otherwise $C(x_0) \cup C(x_1)$ would be a connected subspace of X containing x_0 and x_1 and larger than $C(x_0)$ or $C(x_1)$ and this is impossible. Thus, $\{C(x) : x \in X\}$ partitions X into disjoint, maximal connected subspaces. Since Exercise 1.5.2 implies that $\overline{C(x)}$ is also connected, we must have $\overline{C(x)} = C(x)$ so that each component of X is closed in X. They need not be open, however, as the following example shows.

Exercise 1.5.14 Show that any subspace of the space $\mathbb{Q}$ of rational numbers containing more than one point is disconnected so that the components in $\mathbb{Q}$ are its points (a space X with the property that $C(x_0) = \{x_0\}$ for each x_0 in X is said to be **totally disconnected**).

Exercise 1.5.15 Show that, in a locally connected space (e.g., a topological manifold), the components are open as well as closed.

To analogously carve an arbitrary space X into maximal pathwise connected pieces we proceed as follows: Define a relation $\sim$ on X by $x_0 \sim x_1$ iff there is a path α in X from x_0 to x_1.

Exercise 1.5.16 Show that $\sim$ is an equivalence relation on X. **Hint:** Re-examine the proof of Lemma 1.5.4.

The equivalence classes in X of this equivalence relation are called the **path components of** X. Each path component is pathwise connected and therefore connected so it is contained in some component of X.

1.6 Topological Groups and Group Actions

We have, on several occasions, pointed out that a particular topological space under consideration (e.g., S^1, S^3, $GL(n, \mathbb{F})$, $U(n, \mathbb{F})$, etc.) also happened to admit a natural group structure. In each case it is easy to check that the topology and the group structure are compatible in the sense that the group operations defined continuous maps. In this section we formalize and study this phenomenon.

A **topological group** is a Hausdorff topological space G that is also a group in which the operations of multiplication

$$(x, y) \longrightarrow xy : G \times G \longrightarrow G$$

and inversion

$$x \longrightarrow x^{-1} : G \longrightarrow G$$

are continuous. We generally denote the identity element in G by e.

Exercise 1.6.1 Show that if G is a Hausdorff topological space that is also a group, then G is a topological group iff the map $(x, y) \rightarrow x^{-1}y : G \times G \rightarrow G$ is continuous.

We have already seen a great many examples: The set $\mathbb{R}$ of real numbers with its usual additive group structure. The sets of nonzero real numbers, complex numbers and quaternions with their respective multiplicative group structures. Since any subgroup of a topological group is clearly also a topological group, one obtains such examples as the following: The discrete subgroup $\mathbb{Z}$ of integers in the additive group of real numbers. The discrete subgroup $\mathbb{Z}_2 = \{-1, 1\}$ of the multiplicative group of nonzero real numbers. S^1 as a subgroup of the nonzero complex numbers under complex multiplication. S^3 as the multiplicative subgroup of unit quaternions. The general linear groups $GL(n, \mathbb{R})$ and $GL(n, \mathbb{C})$ are easily seen to be topological groups by simply writing out the entries (coordinates) of the matrix product and inverse (and noting that the determinant that appears in the denominators of the latter are nonzero). Consequently, $O(n)$, $U(n)$, $SO(n)$ and $SU(n)$ are all topological groups. Multiplication in $GL(n, \mathbb{H})$ is clearly continuous. There is no analogous formula for the inverse in $GL(n, \mathbb{H})$, but we have seen that $GL(n, \mathbb{H})$ can be identified, algebraically and topologically, with a subgroup of $GL(2n, \mathbb{C})$ so it too is a topological group, as are its subgroups $Sp(n)$ and $SL(n, \mathbb{H})$.

Exercise 1.6.2 Show that if G_1 and G_2 are topological groups, then $G_1 \times G_2$, with the product topology and the direct product (i.e., coordinatewise) group structure, is also a topological group. Extend this by induction to arbitrary finite products $G_1 \times \cdots \times G_n$.

Thus, any torus $S^1 \times \cdots \times S^1$ is a topological group as are such things as $SU(2) \times U(1)$ (this group plays a fundamental role in the so-called "electroweak theory"). Notice that many of the examples we have described are, in addition to being topological groups, also locally Euclidean topological spaces. Such **locally Euclidean topological groups** are of profound significance in gauge theory.

For the general study of topological groups and their "actions" on other spaces we must begin by assembling a certain amount of simple, but important machinery. First we describe certain canonical homeomorphisms of any topological group G onto itself. Fix a $g \in G$ and define two maps $L_g : G \to G$ and $R_g : G \to G$, called **left** and **right multiplication by** g, respectively, by $L_g(x) = gx$ and $R_g(x) = xg$ for all x in G. Since L_g is the composition of $x \to (g, x)$ and the multiplication map on G, it is continuous. Since $L_{g^{-1}}$ is also continuous and is clearly the inverse of L_g, L_g is a homeomorphism of G onto G. Similarly, R_g is a homeomorphism of G onto G.

Exercise 1.6.3 Show that the inversion map $x \to x^{-1}$ is a homeomorphism of G onto G.

If A and B are any subsets of G we will write AB for the set of all products ab, where $a \in A$ and $b \in B$. If one of these sets consists of a single element g in G, then we write gB and Ag rather than $\{g\}B$ and $A\{g\}$, respectively. In particular, if H is a subgroup of G and $g \in G$, then gH and Hg are, respectively, the left and right cosets of H containing g. We denote by A^{-1} the set of all a^{-1} for $a \in A$, i.e., the image of A under the inversion map.

Exercise 1.6.4 Prove each of the following:

(a) A open (closed) $\Longrightarrow$ Ag and gA open (closed) for every $g \in G$.

(b) A open $\Longrightarrow$ AB and BA open for any $B \subseteq G$.

(c) A closed and B finite $\Longrightarrow$ AB and BA closed.

(d) A open (closed) $\Longrightarrow$ A^{-1} open (closed).

Now, if H is a subgroup of G (*not* necessarily a normal subgroup), then the set of left cosets gH of G with respect to H is denoted, as usual,

$$G/H = \{gH : g \in G\}$$

and the map that assigns to every $g \in G$ its coset gH is written

$$\mathcal{Q} : G \longrightarrow G/H .$$

With the quotient topology determined by $\mathcal{Q}$, G/H is called the **left coset space of G with respect to H**. Notice that $\mathcal{Q}$ is necessarily an open map since, if $U \subseteq G$ is open, then $\mathcal{Q}^{-1}(\mathcal{Q}(U)) = UH$, which is open by Exercise 1.6.4 (b). Thus, by definition of the quotient topology, $\mathcal{Q}(U)$ is open in G/H.

Lemma 1.6.1 *Let G be a topological group and H a subgroup of G. Then G/H is Hausdorff iff H is closed in G.*

Proof: First suppose that H is closed. Let x_1 and x_2 be distinct points in G/H. Choose $g_1 \in \mathcal{Q}^{-1}(x_1)$ and $g_2 \in \mathcal{Q}^{-1}(x_2)$. Then $g_1^{-1}g_2 \notin H$, for otherwise g_2 would be in g_1H so $\mathcal{Q}(g_2) = \mathcal{Q}(g_1)$, i.e., $x_2 = x_1$. Since H is closed we can select an open set W in G containing $g_1^{-1}g_2$ with $W \cap H = \emptyset$. By Exercise 1.6.1, there exist open sets U and V in G with $g_1 \in U$, $g_2 \in V$ and $U^{-1}V \subseteq W$. Since $\mathcal{Q}$ is an open map, $\mathcal{Q}(U)$ and $\mathcal{Q}(V)$ are open sets in G/H containing x_1 and x_2, respectively. We claim that they are disjoint. Indeed, suppose there is an $x_3 \in \mathcal{Q}(U) \cap \mathcal{Q}(V)$. Select $g_3 \in \mathcal{Q}^{-1}(x_3)$. Since $x_3 \in \mathcal{Q}(U)$, g_3H intersects U, i.e., there exists an $h \in H$ with $g_3h \in U$. Similarly, there exists a $k \in H$ with $g_3k \in V$. Thus, $(g_3h)^{-1}(g_3k) \in W$. But $(g_3h)^{-1}(g_3k) = h^{-1}(g_3^{-1}g_3)k = h^{-1}k \in H$ and this contradicts $W \cap H = \emptyset$. Thus, $\mathcal{Q}(U)$ and $\mathcal{Q}(V)$ are disjoint open sets in G/H containing x_1 and x_2 as required.

Exercise 1.6.5 Prove, conversely, that if G/H is Hausdorff, then H is closed in G. ∎

Proposition 1.6.2 *Let G be a topological group and H a closed normal subgroup of G. Then, with its canonical group structure ($(g_1H)(g_2H) = (g_1g_2)H$, $(gH)^{-1} = g^{-1}H$ and identity $eH = H$), the left coset space G/H is a topological group and the quotient map $\mathcal{Q}: G \to G/H$ is a continuous, open homomorphism with kernel H.*

Proof: Since H is normal, G/H is a group and, since H is closed, G/H is Hausdorff. We must show that $(g_1H, g_2H) \to (g_1g_2)H$ is a continuous map of $G/H \times G/H$ to G/H and $gH \to g^{-1}H$ is a continuous map of G/H to G/H. Consider the diagram

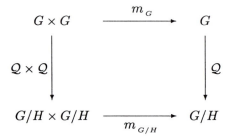

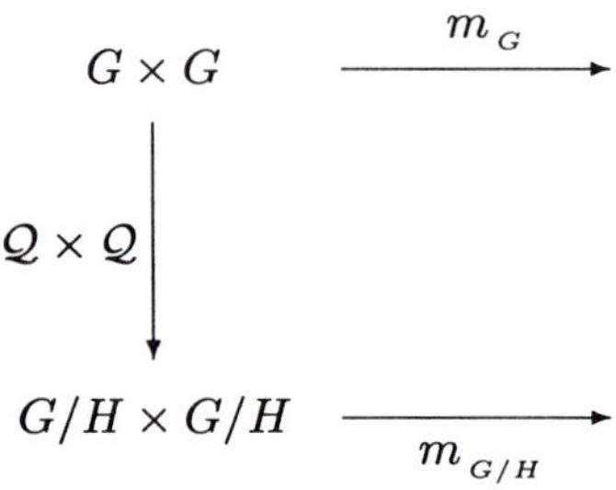

where $m_G(g_1, g_2) = g_1g_2$ and $m_{G/H}(g_1H, g_2H) = (g_1g_2)H$ are the group multiplications in G and G/H, respectively. The diagram commutes ($\mathcal{Q} \circ m_G = m_{G/H} \circ (\mathcal{Q} \times \mathcal{Q})$) by definition of $m_{G/H}$. Moreover, m_G is continuous, $\mathcal{Q}$ is continuous and open and $\mathcal{Q} \times \mathcal{Q}$ is open (Exercise 1.3.14 (b)). Thus, if $U \subseteq G/H$ is open, $m_{G/H}{}^{-1}(U) = (\mathcal{Q} \times \mathcal{Q})\left(m_G{}^{-1}(\mathcal{Q}^{-1}(U)\right)$ is open in $G/H \times G/H$ so $m_{G/H}$ is continuous.

Exercise 1.6.6 Show in the same way that inversion on G/H is continuous.

Since $\mathcal{Q}$ has already been shown to be a continuous open map and since it is a homomorphism with kernel H by definition of the group structure in G/H, the proof is complete. ∎

Remark: Right coset spaces are defined in an entirely analogous manner. Note, however, that inversion $g \to g^{-1}$ on G is a homeomorphism that, for any subgroup H, interchanges left and right cosets ($gH \to Hg^{-1}$) and therefore determines a homeomorphism of the left coset space onto the right coset space. Thus, anything topological we prove about G/H is equally true of the right coset space.

Suppose that G has a subgroup H that is open in G. By Exercise 1.6.4 (a) all of the left cosets gH are also open in G. But these cosets are pairwise disjoint and their union is G so each of them is also closed. In particular, H is closed. Moreover, since $\mathcal{Q} : G \to G/H$ is an open map, G/H must then be discrete.

Exercise 1.6.7 Show, conversely, that if G/H is discrete, then H is open (and therefore closed) in G.

In particular, if G contains a proper, open subgroup, then it cannot be connected. Notice also that any subgroup H of G that contains an open nbd U of the identity e must be open (since hU is a nbd of h contained in H for every $h \in H$). A simple, but useful consequence is that a connected topological group is generated by any nbd of the identity. Indeed, we prove the following.

Proposition 1.6.3 *Let G be a connected topological group and U an open subset of G containing the identity e. Then $G = \bigcup_{n=1}^{\infty} U^n$, where $U^1 = U$, $U^2 = UU, \ldots, U^n = U^{n-1}U$.*

Proof: U^{-1} is also an open set containing e and therefore so is $V = U \cap U^{-1}$. Moreover, V satisfies $V^{-1} = V$.

Exercise 1.6.8 Verify this.

Since $\bigcup_{n=1}^{\infty} V^n \subseteq \bigcup_{n=1}^{\infty} U^n \subseteq G$ it will suffice to prove that $\bigcup_{n=1}^{\infty} V^n = G$. But V is open, so each V^n is open (Exercise 1.6.4 (b)) and therefore $\bigcup_{n=1}^{\infty} V^n$ is open. Moreover, $\bigcup_{n=1}^{\infty} V^n$ is closed under the formation of products and (since $V^{-1} = V$) inverses and so it is a subgroup of G. Since G is connected we must have $\bigcup_{n=1}^{\infty} V^n = G$. ■

Theorem 1.6.4 *Let G be a topological group and H the component of G containing the identity e. Then H is a closed, connected, normal subgroup of G. If G is also locally Euclidean, then H is open.*

Proof: H is closed because components are always closed and it is connected by definition. To show that H is a subgroup of G we show that it is closed under the formation of inverses and products. First let $h \in H$. Then $L_{h^{-1}}(H)$ is a component of G ($L_{h^{-1}}$ is a homeomorphism) and it contains $h^{-1}h = e$ so, in fact, $L_{h^{-1}}(H) = H$. Thus, $h^{-1}e = h^{-1} \in H$ so H is closed under inversion. Next, let $h, k \in H$. Then $L_h(H)$ is a component of G. We have shown already that $h^{-1} \in H$ so $hh^{-1} = e \in L_h(H)$ and therefore $L_h(H) = H$. But then $k \in H$ implies $hk \in H$ as required. Thus, H is a subgroup of G. It is a normal subgroup since $g \in G$ implies $R_{g^{-1}}(L_g(H)) = gHg^{-1}$ is a component and $e \in H$ implies $geg^{-1} = e \in$

gHg^{-1} so $gHg^{-1} = H$. If G is locally Euclidean, it is locally connected and so its components are open (Exercise 1.5.15). ■

Exercise 1.6.9 Show that if G is a topological group and H is the component of G containing the identity e, then the components of G are precisely the left cosets of H in G.

If G is connected and H is an arbitrary closed subgroup of G, then, of course, G/H is also connected (being a continuous image of G). A rather surprising and quite useful fact is that if H and G/H are connected, then so is G.

Proposition 1.6.5 *Let G be a topological group and H a closed subgroup of G. If H and G/H are connected, then G is also connected.*

Proof: We suppose that H and G/H are connected, but that $G = A \cup B$, where A and B are nonempty, disjoint open sets in G. Without loss of generality we assume $e \in A$. Since H is connected, so are its left cosets $gH = L_g(H)$. Thus, since each coset meets either A or B, each must be contained entirely in one or the other. Consequently, each of A and B is a union of left cosets of H. If $\mathcal{Q} : G \to G/H$ is the quotient map, it follows that $\mathcal{Q}(A)$ and $\mathcal{Q}(B)$ are nonempty and disjoint. But $\mathcal{Q}$ is an open map so $\{\mathcal{Q}(A), \mathcal{Q}(B)\}$ is a disconnection of G/H and this is a contradiction. ■

Shortly we will use this proposition to show, for example, that $SO(n)$ is connected and, from this, that $O(n)$ has precisely two components.

In our discussion of the Hopf bundle in Section 0.3 we found that there was a natural, and physically significant, "action" of $U(1)$ on S^3. Such group actions are quite important to our study and we are now prepared to look into them in some detail. Thus, we let G be a topological group and Y some topological space. A **right action of G on Y** is a continuous map $\sigma : Y \times G \to Y$ which satisfies

1. $\sigma(y, e) = y$ for all $y \in Y$ (e is the identity in G), and
2. $\sigma(y, g_1 g_2) = \sigma(\sigma(y, g_1), g_2)$ for all $y \in Y$ and all $g_1, g_2 \in G$.

The following notation is generally more convenient. Writing $\sigma(y, g) = y \cdot g$, (1) and (2) become

$$y \cdot e = y \quad \text{for all} \quad y \in Y, \quad \text{and} \tag{1.6.1}$$

$$y \cdot (g_1 g_2) = (y \cdot g_1) \cdot g_2 \quad \text{for all} \quad y \in Y \quad \text{and all} \quad g_1, g_2 \in G. \tag{1.6.2}$$

Notice that if one defines, for each fixed $g \in G$, a map $\sigma_g : Y \to Y$ by $\sigma_g(y) = \sigma(y, g) = y \cdot g$, then σ_g is continuous (being the composition $y \to (y, g) \to \sigma(y, g)$), one-to-one ($y_1 \cdot g = y_2 \cdot g$ implies $(y_1 \cdot g) \cdot g^{-1} = (y_2 \cdot g) \cdot g^{-1}$

so $y_1 \cdot (gg^{-1}) = y_2 \cdot (gg^{-1})$ and therefore $y_1 \cdot e = y_2 \cdot e$ i.e., $y_1 = y_2$), onto Y ($y = (y \cdot g^{-1}) \cdot g$) and has a continuous inverse (namely, $\sigma_{g^{-1}}$). Thus, $\sigma_g : Y \to Y$ is a homeomorphism. In terms of these homeomorphisms σ_g, (1.6.1) and (1.6.2) become

$$\sigma_e = id_Y\,, \quad \text{and} \tag{1.6.3}$$

$$\sigma_{g_1 g_2} = \sigma_{g_2} \circ \sigma_{g_1} \tag{1.6.4}$$

(note the reversal of the g's in (1.6.4)).

Remark: One defines a **left action of** G **on** Y to be a continuous map $\rho : G \times Y \to Y$, $\rho(g, y) = g \cdot y$, that satisfies $e \cdot y = y$ and $(g_1 g_2) \cdot y = g_1 \cdot (g_2 \cdot y)$. Then $\rho_g : Y \to Y$, defined by $\rho_g(y) = g \cdot y$, is a homeomorphism and one has

$$\rho_e = id_Y\,, \quad \text{and} \tag{1.6.5}$$

$$\rho_{g_1 g_2} = \rho_{g_1} \circ \rho_{g_2} \tag{1.6.6}$$

(*no* reversal of the g's in (1.6.6)). In this case the assignment $g \to \rho_g$ is a homomorphism of G into the homeomorphism group Homeo (Y) (Exercise 1.1.4) and is often called a **representation of** G **in** Homeo (Y). All of the terms we define for right actions have obvious analogues for left actions which we leave it to the reader to formulate.

A right action σ of G on Y is said to be **effective** if $y \cdot g = y$ for *all* $y \in Y$ implies $g = e$, i.e., if $\sigma_g = id_Y$ iff $g = e$. The action is said to be **free** if $y \cdot g = y$ for *some* $y \in Y$ implies $g = e$. Obviously, a free action is effective, but we shall see that the converse is false. A right action of G on Y is said to be **transitive** if, given any two points $y_1, y_2 \in Y$, there exists a $g \in G$ such that $y_2 = y_1 \cdot g$. If there is no ambiguity as to which particular action is under consideration one often says that G **acts transitively** (**freely**, **effectively**) on Y. Given any $y \in Y$ we define the **orbit** of y under the action σ to be the subset $\{y \cdot g : g \in G\}$ of Y. Thus, the action is transitive if there is just one orbit, namely, all of Y. The **isotropy subgroup of** y under the action σ is the subset $\{g \in G : y \cdot g = y\}$ of G.

Exercise 1.6.10 Show that this is, indeed, a subgroup of G and is closed.

Thus, the action is free iff every isotropy subgroup is trivial and effective iff the intersection of all the isotropy subgroups is trivial.

We illustrate these definitions with a number of important examples. First observe that any topological group G acts on itself by right multiplication. That is, defining $\sigma : G \times G \to G$ by $\sigma(y, g) = yg$ for all $y, g \in G$, gives a right action of G on G. Such an action is obviously free ($yg = y$ implies $g = e$) and transitive ($y_2 = y_1(y_1^{-1} y_2)$). The example constructed in Section 0.3 is an action of $U(1)$ on S^3. One regards S^3 as the subset of $\mathbb{C}^2$ consisting of all (z^1, z^2) with $|z^1|^2 + |z^2|^2 = 1$ and identifies $U(1)$ with S^1 (Exercise 1.1.25). Then $((z^1, z^2), g) \to (z^1 g, z^2 g)$ is a right action. In an entirely analogous manner one can think of S^7 as the subset of $\mathbb{H}^2$ consisting of all (q^1, q^2) with $|q^1|^2 + |q^2|^2 = 1$ and identify $Sp(1) = SU(2)$ with the unit quaternions (Exercise 1.1.26 and Theorem 1.1.4). Then $((q^1, q^2), g) \to (q^1 g, q^2 g)$ is a right action of $Sp(1)$ on S^7. These last two examples are easy to generalize.

Exercise 1.6.11 Regard S^{2n-1} as the subspace of $\mathbb{C}^n$ consisting of all $(z^1, \ldots, z^n)$ with $|z^1|^2 + \cdots + |z^n|^2 = 1$ and define a map of $S^{2n-1} \times U(1)$ to S^{2n-1} by $((z^1, \ldots, z^n), g) \to (z^1, \ldots, z^n) \cdot g = (z^1 g, \ldots, z^n g)$. Show that this is a right action of $U(1)$ on S^{2n-1} that is free, but not transitive if $n \geq 2$.

Exercise 1.6.12 Regard S^{4n-1} as the subspace of $\mathbb{H}^n$ consisting of all $(q^1, \ldots, q^n)$ with $|q^1|^2 + \cdots + |q^n|^2 = 1$ and define a map of $S^{4n-1} \times Sp(1)$ to S^{4n-1} by $((q^1, \ldots, q^n), g) \to (q^1, \ldots, q^n) \cdot g = (q^1 g, \ldots, q^n g)$. Show that this is a right action of $Sp(1)$ on S^{4n-1} that is free, but not transitive if $n \geq 2$.

Notice that S^{2n-1} and S^{4n-1} are the total spaces of the complex and quaternionic Hopf bundles (Section 1.3) and that $U(1) \cong S^1$ and $Sp(1) \cong S^3$ are, respectively, the fibers of these bundles. What has occurred in these last examples is that the natural actions of $U(1)$ on $U(1)$ and $Sp(1)$ on $Sp(1)$ have been used to define actions on the bundle spaces "fiberwise". This is a recurrent theme and will eventually culminate (in Chapter 3) in the notion of a principal bundle.

Next we show that $O(n)$ acts transitively on S^{n-1} on the left. Each $A \in O(n)$ is an $n \times n$ orthogonal matrix which we identify with the matrix, relative to the standard basis $\{e_1, \ldots, e_n\}$ for $\mathbb{R}^n$, of an orthogonal linear transformation (also denoted A) on $\mathbb{R}^n$. We define $\rho : O(n) \times S^{n-1} \to S^{n-1}$ by $\rho(A, x) = A \cdot x = A(x)$. Then ρ is clearly continuous, $id \cdot x = x$ for all $x \in S^{n-1}$ and $(AB) \cdot x = A(B(x)) = A \cdot (B \cdot x)$, so ρ is a left action of $O(n)$ on S^{n-1}. We show that ρ is transitive. To see this first let $x_1 \in S^{n-1}$ be arbitrary. Let $\{x_1, x_2, \ldots, x_n\}$ be an orthonormal basis for $\mathbb{R}^n$ containing x_1 as its first element. For each $i = 1, \ldots, n$ write

$$x_i = \sum_{j=1}^{n} A_{ji} e_j$$

where the A_{ji} are constants. Then $A = (A_{ji})_{j,i=1,\ldots,n}$ is in $O(n)$ since it is the matrix of a linear transformation that carries one orthonormal basis onto another. Moreover,

$$A \cdot e_1 = A(e_1) = \sum_{j=1}^{n} A_{j1} e_j = x_1 \, .$$

We conclude that, given any $x_1 \in S^{n-1}$ there exists an $A \in O(n)$ such that $A \cdot e_1 = x_1$. Next suppose x_1 and y_1 are two arbitrary elements of S^{n-1}. Select $A \in O(n)$ with $A \cdot e_1 = x_1$ and $B \in O(n)$ with $B \cdot e_1 = y_1$. Then $B^{-1} \in O(n)$ and $B^{-1} \cdot y_1 = e_1$. Moreover, $AB^{-1} \in O(n)$ and $(AB^{-1}) \cdot y_1 = A \cdot (B^{-1} \cdot y_1) = A \cdot e_1 = x_1$ as required so this action of $O(n)$ on S^{n-1} is transitive. We will also need to calculate an isotropy subgroup of this action. We do this for the north pole $e_n \in S^{n-1}$. Consider the subset of $O(n)$ consisting of all elements of the form

$$\begin{pmatrix} & & & & 0 \\ & & & & 0 \\ & & \tilde{A} & & \\ & & & & \vdots \\ & & & & 0 \\ 0 & 0 & \cdots & 0 & 1 \end{pmatrix},$$

where $\tilde{A}$ is an $(n-1) \times (n-1)$ real matrix satisfying $\tilde{A}\tilde{A}^T = \tilde{A}^T\tilde{A} = id$. This is clearly a subgroup of $O(n)$ isomorphic to $O(n-1)$ and we will identify it with $O(n-1)$. We claim that $O(n-1)$ is the isotropy subgroup of $e_n \in S^{n-1}$ under the action ρ. To prove this, first observe that $A \in O(n-1)$ implies $A \cdot e_n = e_n$. Next suppose A is some element of $O(n)$ that satisfies $A \cdot e_n = e_n$. Then

$$e_n = A \cdot e_n = A(e_n) = \sum_{j=1}^{n} A_{jn} e_j = A_{1n} e_1 + \cdots + A_{nn} e_n .$$

Linear independence of $\{e_1, \ldots, e_n\}$ then implies that $A_{1n} = \cdots = A_{n-1\ n} = 0$ and $A_{nn} = 1$. Moreover, $AA^T = id$ implies $(A_{n1})^2 + \cdots + (A_{nn})^2 = 1$ so we also have $A_{n1} = \cdots = A_{n-1\ 1} = 0$. Thus, $A \in O(n-1)$ as required.

Exercise 1.6.13 Define an analogous transitive left action of $SO(n)$ on S^{n-1} and show that the isotropy subgroup of $e_n \in S^{n-1}$ is isomorphic to $SO(n-1)$.

This example generalizes to the complex and quaternionic cases as well. Specifically, we let $\mathbb{F}$ denote one of $\mathbb{C}$ or $\mathbb{H}$. As in Section 1.1 we regard $\mathbb{F}^n$ as a (right) vector space over $\mathbb{F}$ with bilinear form $< \xi, \zeta > = \bar{\xi}^1\zeta^1 + \cdots + \bar{\xi}^n\zeta^n$. The $\mathbb{F}$-unitary group $U(n, \mathbb{F})$ consists of all $n \times n$ matrices with entries in $\mathbb{F}$ that satisfy $\bar{A}^T A = A\bar{A}^T = id$ and we now show that it acts transitively on the left on the unit sphere $S = \{\xi \in \mathbb{F}^n : < \xi, \xi > = 1\}$ in $\mathbb{F}^n$. Regarding $A \in U(n, \mathbb{F})$ as the matrix relative to the standard basis $\{e_1, \ldots, e_n\}$ of an $\mathbb{F}$-linear map $A : \mathbb{F}^n \to \mathbb{F}^n$ that preserves $< \ , \ >$, we define $\rho : U(n, \mathbb{F}) \times S \to S$ by $\rho(A, \xi) = A \cdot \xi = A(\xi)$. Then ρ is a left action of $U(n, \mathbb{F})$ on S. Since we have seen (Exercise 1.1.22) that any element ξ_1 of S is an element of some orthonormal basis for $\mathbb{F}^n$ and (Exercise 1.1.21) that the elements of $U(n, \mathbb{F})$ are precisely the matrices of $\mathbb{F}$-linear transformations that carry one orthonormal basis onto another, the proof that this action is transitive on S is identical to that given above in the real case.

Exercise 1.6.14 Show that the isotropy subgroup of $e_n \in S$ relative to this action of $U(n, \mathbb{F})$ on S is isomorphic to $U(n-1, \mathbb{F})$.

We conclude then that $U(n)$ $(Sp(n))$ acts transitively on the left on S^{2n-1} (S^{4n-1}) with isotropy subgroup at the north pole isomorphic to $U(n-1)$ $(Sp(n-1))$. As in Exercise 1.6.13 one defines an analogous transitive left action of $SU(n)$ on S^{2n-1}. We will see that the existence of these actions has interesting things to say about the topologies of $O(n)$, $U(n)$, $SU(n)$ and $Sp(n)$. First, however, we consider a very general method of constructing transitive group actions.

Let G be a topological group, H a closed subgroup of G and $\mathcal{Q} : G \to G/H$ the quotient map onto the left coset space G/H. Notice that there is a natural left action of G on G/H obtained, in effect, by left translating the cosets of H in G. More precisely, let us fix an $x \in G/H$ and a $g \in G$ and define $g \cdot x \in G/H$ as follows: $\mathcal{Q}^{-1}(x)$ is a left coset of H in G and therefore so is $g\mathcal{Q}^{-1}(x)$. Consequently, $\mathcal{Q}(g\mathcal{Q}^{-1}(x))$ is a single point in G/H and can be computed by choosing any $g_0 \in \mathcal{Q}^{-1}(x)$, i.e., $\mathcal{Q}(g\mathcal{Q}^{-1}(x)) = \mathcal{Q}(g(g_0H)) = \mathcal{Q}(gg_0)$. We define

$$g \cdot x = \mathcal{Q}\left(g\mathcal{Q}^{-1}(x)\right) = \mathcal{Q}\left(g(g_0H)\right) = \mathcal{Q}\left((gg_0)H\right) = \mathcal{Q}(gg_0),$$

where g_0 is any element of $\mathcal{Q}^{-1}(x)$. Suppressing $\mathcal{Q}$ (i.e., identifying elements of G/H with cosets in G) our action is just

$$(g, g_0H) \longrightarrow (gg_0)H.$$

Observe that $e \cdot x = \mathcal{Q}(eg_0) = \mathcal{Q}(g_0) = x$ and $g_1 \cdot (g_2 \cdot x) = g_1 \cdot \mathcal{Q}(g_2g_0) = \mathcal{Q}(g_1(g_2g_0)) = \mathcal{Q}((g_1g_2)g_0) = (g_1g_2) \cdot x$. All that remains then is to show that the map $\rho : G \times G/H \to G/H$ given by $\rho(g, x) = g \cdot x = \mathcal{Q}(g\mathcal{Q}^{-1}(x))$ is continuous. To see this, let $U \subseteq G/H$ be an open set (i.e., $\mathcal{Q}^{-1}(U)$ is open in G). Then $\rho^{-1}(U) = \{(g, x) \in G \times G/H \colon \mathcal{Q}(g\mathcal{Q}^{-1}(x)) \in U\}$ and we must show that this is open in $G \times G/H$. Fix $(g, x) \in \rho^{-1}(U)$. Then $\mathcal{Q}(g\mathcal{Q}^{-1}(x)) \in U$. Select some $g_0 \in \mathcal{Q}^{-1}(x)$. Then $\mathcal{Q}(gg_0) \in U$ so $gg_0 \in \mathcal{Q}^{-1}(U)$. Let $W_1 \times W_2$ be a basic open set in $G \times G$ such that $(g, g_0) \in W_1 \times W_2$ and $W_1W_2 \subseteq \mathcal{Q}^{-1}(U)$. Since $\mathcal{Q}$ is an open map, $W_1 \times \mathcal{Q}(W_2)$ is an open nbd of (g, x) in $G \times G/H$. We complete the proof by showing that $W_1 \times \mathcal{Q}(W_2) \subseteq \rho^{-1}(U)$. Thus, let $(g', x') \in W_1 \times \mathcal{Q}(W_2)$. Then $\rho(g', x') = g' \cdot x' = \mathcal{Q}(g'\mathcal{Q}^{-1}(x'))$. Now, $g' \in W_1$ and $\mathcal{Q}^{-1}(x') \cap W_2 \neq \emptyset$ (since $x' \in \mathcal{Q}(W_2)$) so we may select $g'' \in \mathcal{Q}^{-1}(x') \cap W_2$. Then

$\mathcal{Q}(g'\mathcal{Q}^{-1}(x')) = \mathcal{Q}(g'g'')$. But $g'g'' \in W_1W_2 \subseteq \mathcal{Q}^{-1}(U)$ so $\mathcal{Q}(g'g'') \in U$ and therefore $\rho(g', x') \in U$, i.e., $(g', x') \in \rho^{-1}(U)$ as required.

Thus, we have shown that $\rho : G \times G/H \to G/H$ defined by $\rho(g, x) = g \cdot x = \mathcal{Q}(g\mathcal{Q}^{-1}(x))$ is a left action of G on G/H. Moreover, this action is transitive. To see this, let x_1 and x_2 be points in G/H. Select $g_1 \in \mathcal{Q}^{-1}(x_1)$ and $g_2 \in \mathcal{Q}^{-1}(x_2)$. Then $g = g_2g_1^{-1} \in G$ and

$$g \cdot x_1 = \mathcal{Q}\left(g\mathcal{Q}^{-1}(x_1)\right) = \mathcal{Q}\left((g_2g_1^{-1})g_1\right) = \mathcal{Q}(g_2) = x_2$$

as required.

The significance of this last construction is that, under certain conditions that are often met in practice, *any* transitive left group action can be thought of as having arisen in just this way. We consider an arbitrary topological group G, space X and transitive left action $\rho : G \times X \to X$, $\rho(g, x) = g \cdot x$, of G on X. Fix some arbitary point $x_0 \in X$ and consider its isotropy subgroup $H = \{g \in G : g \cdot x_0 = x_0\}$. We let $\mathcal{Q} : G \to G/H$ be the projection and define a map $\mathcal{Q}' : G \to X$ by $\mathcal{Q}'(g) = g \cdot x_0$ for each $g \in G$. Since the action is assumed continuous, $\mathcal{Q}'$, which is the composition $g \to (g, x_0) \to \rho(g, x_0) = g \cdot x_0$, is also continuous. We claim that the fibers $(\mathcal{Q}')^{-1}(x)$, $x \in X$, are precisely the left cosets of H in G. To see this, fix an $x \in X$. First note that any two elements $g_0, g_1 \in (\mathcal{Q}')^{-1}(x)$ are in the same coset of H. Indeed, $\mathcal{Q}'(g_0) = \mathcal{Q}'(g_1)$ gives $g_0 \cdot x_0 = g_1 \cdot x_0$ so $x_0 = (g_0^{-1}g_1) \cdot x_0$ and this, in turn, means that $g_0^{-1}g_1 \in H$, i.e., $g_1 \in g_0H$. Thus, $(\mathcal{Q}')^{-1}(x) \subseteq g_0H$. But, on the other hand, every element of g_0H has the same image under $\mathcal{Q}'$ since, for any $h \in H$, $\mathcal{Q}'(g_0h) = (g_0h) \cdot x_0 = g_0 \cdot (h \cdot x_0) = g_0 \cdot x_0 = \mathcal{Q}'(g_0)$ so $g_0H \subseteq (\mathcal{Q}')^{-1}(x)$ as well. Thus, $(\mathcal{Q}')^{-1}(x) = g_0H$ is a coset of H in G. Furthermore, any coset g_0H of H is $(\mathcal{Q}')^{-1}(x)$, where $x = \mathcal{Q}'(g_0)$. This defines a natural mapping $\varphi : G/H \to X$ for which the following diagram commutes, i.e., $\varphi \circ \mathcal{Q} = \mathcal{Q}'$:

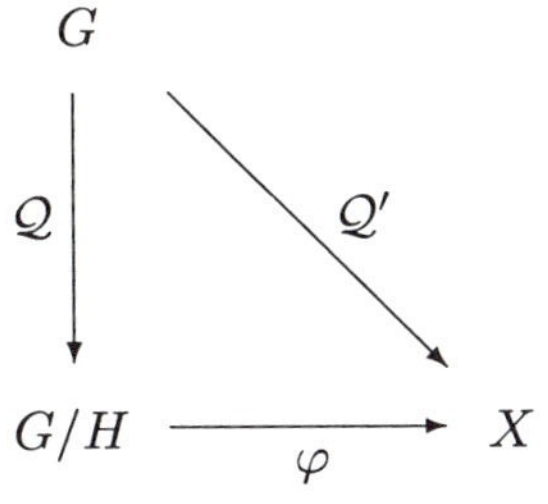

Since the action of G on X is assumed transitive, φ is surjective (for any $x \in X$, there exists a $g \in G$ with $g \cdot x_0 = x$ so that $\mathcal{Q}'(g) = x$ and therefore $\varphi(\mathcal{Q}(g)) = x$). Moreover, φ is one-to-one since the fibers of $\mathcal{Q}'$ are the cosets of H in G. To see that φ is continuous, let U be open in X. Since G/H has the quotient topology determined by $\mathcal{Q}$, it will suffice to show that $\mathcal{Q}^{-1}(\varphi^{-1}(U))$ is open in G. But $\mathcal{Q}^{-1}(\varphi^{-1}(U)) = (\varphi \circ \mathcal{Q})^{-1}(U) = (\mathcal{Q}')^{-1}(U)$ so this follows from the continuity of $\mathcal{Q}'$. Unfortunately, $\varphi^{-1}: X \to G/H$ is not always continuous so φ need not be a homeomorphism. However, under certain (frequently encountered) conditions, φ^{-1} will be continuous. For example, if G is compact, then so is G/H so the result follows from Theorem 1.4.4. On the other hand, if $\mathcal{Q}'$ is an open map, then, for any open set V in G/H, $\mathcal{Q}'(\mathcal{Q}^{-1}(V))$ is open in X so $\varphi(V)$ is open and therefore φ is an open map, i.e., φ^{-1} is a continuous. We summarize these considerations in the following result.

Theorem 1.6.6 *Let G be a topological group, X a topological space and $(g, x) \to g \cdot x$ a transitive left action of G on X. Fix an $x_0 \in X$, let $H = \{g \in G: g \cdot x_0 = x_0\}$ be its isotropy subgroup and define $\mathcal{Q}': G \to X$ by $\mathcal{Q}'(g) = g \cdot x_0$. Then H is a closed subgroup of G and, if $\mathcal{Q}: G \to G/H$ is the canonical projection, then there exists a unique continuous bijection $\varphi: G/H \to X$ for which the diagram*

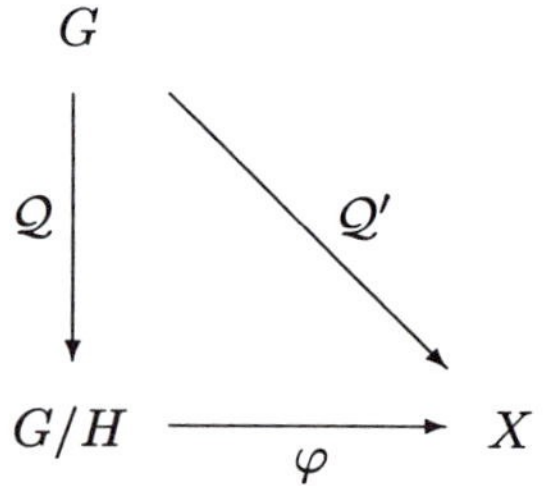

commutes, i.e., $\varphi \circ \mathcal{Q} = \mathcal{Q}'$. Moreover, if either (i) G is compact or (ii) $\mathcal{Q}'$ is an open map, then φ is a homeomorphism.

Here are some applications. $O(n)$ acts transitively on S^{n-1} on the left and has isotropy subgroup at the north pole isomorphic to $O(n-1)$. Since $O(n)$ is compact, we conclude that S^{n-1} is homeomorphic to the quotient group

$$S^{n-1} \cong O(n)/O(n-1)\,. \tag{1.6.7}$$

Similarly, using Exercise 1.6.13 and the fact that $SO(n)$ is closed in $O(n)$, we obtain

$$S^{n-1} \cong SO(n)/SO(n-1)\,. \tag{1.6.8}$$

In the same way, Exercise 1.6.14 gives the following homeomorphisms:

$$S^{2n-1} \cong U(n)/U(n-1) \tag{1.6.9}$$

$$S^{2n-1} \cong SU(n)/SU(n-1) \tag{1.6.10}$$

$$S^{4n-1} \cong Sp(n)/Sp(n-1)\,. \tag{1.6.11}$$

These homeomorphisms, together with Proposition 1.6.5 and a simple inductive argument, yield some important connectivity results. First note that $SO(1)$ and $SU(1)$ are connected since they both consist of a single point. $U(1) \cong S^1$ and $Sp(1) \cong SU(2) \cong S^3$ so these too are connected. Since any sphere of dimension greater than zero is connected, the homeomorphisms $S^1 \cong SO(2)/SO(1)$, $S^3 \cong U(2)/U(1)$, $S^3 \cong SU(2)/SU(1)$ and $S^7 \cong Sp(2)/Sp(1)$, together with Proposition 1.6.5, imply that $SO(2)$, $U(2)$, $SU(2)$ and $Sp(2)$ are all connected (there is nothing new here in the case of $SU(2)$).

Exercise 1.6.15 Show, by induction, that $SO(n)$, $U(n)$, $SU(n)$ and $Sp(n)$ are all connected for $n \geq 1$.

Note that this procedure fails for $O(n)$ since one cannot get the induction off the ground ($O(1)$ is homeomorphic to the two-point discrete space $\mathbb{Z}_2 = \{-1, 1\}$ and so is not connected). Indeed, it is now clear that every $O(n)$, $n \geq 1$, has two components since

$$O(n) = SO(n) \cup g \cdot SO(n),$$

where $g \in O(n)$ is given by

$$g = \begin{pmatrix} -1 & 0 & \cdots & 0 \\ 0 & 1 & \cdots & 0 \\ \vdots & \vdots & & \vdots \\ 0 & 0 & \cdots & 1 \end{pmatrix}.$$

It will be instructive, and useful in Chapter 4, to write out explicitly the homeomorphism (1.6.11) when $n = 2$. Thus, we regard S^7 as the set of all $x = \left(\begin{smallmatrix} q^1 \\ q^2 \end{smallmatrix}\right) \in \mathbb{H}^2$ with $|q^1|^2 + |q^2|^2 = 1$ (column vectors will be more convenient for these calculations) and $Sp(2)$ as the set of 2×2 quaternionic matrices $g = \left(\begin{smallmatrix} \alpha & \beta \\ \gamma & \delta \end{smallmatrix}\right)$ with $g\bar{g}^T = \bar{g}^T g = id$. In particular, $\bar{\alpha}\beta + \bar{\gamma}\delta = 0$ and $\gamma\bar{\gamma} + \delta\bar{\delta} = 1$. The transitive left action of $Sp(2)$ on S^7 is given by

$$(g, x) \longrightarrow g \cdot x = \begin{pmatrix} \alpha & \beta \\ \gamma & \delta \end{pmatrix} \begin{pmatrix} q^1 \\ q^2 \end{pmatrix} = \begin{pmatrix} \alpha q^1 + \beta q^2 \\ \gamma q^1 + \delta q^2 \end{pmatrix}.$$

Now, fix $x_0 = \left(\begin{smallmatrix} 1 \\ 0 \end{smallmatrix}\right) \in S^7$. Its isotropy subgroup consists of all $\left(\begin{smallmatrix} \alpha & \beta \\ \gamma & \delta \end{smallmatrix}\right)$ in $Sp(2)$ for which $\left(\begin{smallmatrix} \alpha & \beta \\ \gamma & \delta \end{smallmatrix}\right)\left(\begin{smallmatrix} 1 \\ 0 \end{smallmatrix}\right) = \left(\begin{smallmatrix} 1 \\ 0 \end{smallmatrix}\right)$, i.e., for which $\alpha = 1$ and $\gamma = 0$. But then $\bar{\alpha}\beta + \bar{\gamma}\delta = 0$ implies $\beta = 0$, while $\gamma\bar{\gamma} + \delta\bar{\delta} = 1$ gives $\delta\bar{\delta} = 1$. Thus, the isotropy subgroup consists of all $\left(\begin{smallmatrix} 1 & 0 \\ 0 & a \end{smallmatrix}\right)$ with $|a|^2 = 1$. Since $\left(\begin{smallmatrix} 1 & 0 \\ 0 & a_1 \end{smallmatrix}\right)\left(\begin{smallmatrix} 1 & 0 \\ 0 & a_2 \end{smallmatrix}\right) = \left(\begin{smallmatrix} 1 & 0 \\ 0 & a_1 a_2 \end{smallmatrix}\right)$, this subgroup is, as expected, isomorphic to the group $Sp(1)$ of unit quaternions and we will identify them:

$$Sp(1) = \left\{ \begin{pmatrix} 1 & 0 \\ 0 & a \end{pmatrix} \in Sp(2) : \ |a|^2 = 1 \right\}.$$

Now, fix a $g = \left(\begin{smallmatrix} \alpha & \beta \\ \gamma & \delta \end{smallmatrix}\right) \in Sp(2)$. The left coset of g modulo the subgroup $Sp(1)$ is

$$\begin{aligned} [g] = g\,Sp(1) &= \left\{ \begin{pmatrix} \alpha & \beta \\ \gamma & \delta \end{pmatrix} \begin{pmatrix} 1 & 0 \\ 0 & a \end{pmatrix} : \ |a|^2 = 1 \right\} \\ &= \left\{ \begin{pmatrix} \alpha & \beta a \\ \gamma & \delta a \end{pmatrix} : \ |a|^2 = 1 \right\}. \end{aligned}$$

The homeomorphism φ of $Sp(2)/Sp(1)$ onto S^7 described in Theorem 1.6.6 is then given by $\varphi([g]) = g \cdot \left(\begin{smallmatrix} 1 \\ 0 \end{smallmatrix}\right) = \left(\begin{smallmatrix} \alpha \\ \gamma \end{smallmatrix}\right)$:

$$\varphi : \ Sp(2)/Sp(1) \longrightarrow S^7 \ : \ \varphi([g]) = \begin{pmatrix} \alpha \\ \gamma \end{pmatrix}, \tag{1.6.12}$$

where $g = \left(\begin{smallmatrix} \alpha & \beta \\ \gamma & \delta \end{smallmatrix}\right)$ (any representative of the coset $[g]$ has first column $\left(\begin{smallmatrix} \alpha \\ \gamma \end{smallmatrix}\right)$).

Exercise 1.6.16 Show that $Sp(2)$ is a subgroup of $SL(2, \mathbb{H})$. **Hint:** For any $P \in Sp(2)$, let $\phi(P)$ be defined as in (1.1.26). Show that $\det \phi(P) = \pm 1$ and then use the connectivity of $Sp(2)$ (Exercise 1.6.15) and the continuity of $\det \circ \phi$ to conclude that $\det \phi(P) = 1$ for all $P \in Sp(2)$.

2

Homotopy Groups

2.1 Introduction

The real line $\mathbb{R}$ is *not* homeomorphic to the plane $\mathbb{R}^2$, but this fact is not quite the triviality one might hope. Perhaps the most elementary proof goes as follows: Suppose there were a homeomorphism h of $\mathbb{R}$ onto $\mathbb{R}^2$. Select some point $x_0 \in \mathbb{R}$. The restriction of h to $\mathbb{R} - \{x_0\}$ would then carry it homeomorphically onto $\mathbb{R}^2 - \{h(x_0)\}$. However, $\mathbb{R} - \{x_0\} = (-\infty, x_0) \cup (x_0, \infty)$ is not connected, whereas $\mathbb{R}^2 - \{h(x_0)\}$ certainly is connected (indeed, pathwise connected). Since connectedness is a topological property, this cannot be and we have our contradiction.

Notice that this argument would fail to distinguish $\mathbb{R}^2$ from $\mathbb{R}^3$ topologically since deleting a point from either of these yields a connected space. There is, however, a notion, familiar from vector calculus and that we will define precisely quite soon, which, if substituted for "connected" in the argument, will do the job. If $h : \mathbb{R}^2 \to \mathbb{R}^3$ were a homeomorphism, then, for any x_0 in $\mathbb{R}^2$, $\mathbb{R}^2 - \{x_0\}$ would be homeomorphic to $\mathbb{R}^3 - \{h(x_0)\}$. However, $\mathbb{R}^3 - \{h(x_0)\}$ is "simply connected" (intuitively, any closed curve in $\mathbb{R}^3 - \{h(x_0)\}$ can be continuously shrunk to a point in $\mathbb{R}^3 - \{h(x_0)\}$), but $\mathbb{R}^2 - \{x_0\}$ clearly is not. It seems plausible (and we will soon prove) that simple connectivity is a topological property so, again, we have a contradiction. Analogously, the difference between $\mathbb{R}^3$ and $\mathbb{R}^4$ is that, when a point is deleted from each, one obtains from the former a space in which 2-spheres cannot necessarily be shrunk to a point, whereas, with the extra dimension available in $\mathbb{R}^4$, the missing point presents no obstruction to collapsing 2-spheres.

All of these ideas, and their obvious higher dimensional generalizations, are made precise with the introduction of the so-called "homotopy groups" $\pi_n(X)$, $n = 1, 2, \ldots$, of an arbitrary (pathwise connected) topological space X. These are groups (Abelian if $n \geq 2$) which keep track of the number of essentially distinct ways in which spheres can be continuously mapped into X ("essentially distinct" means that the images cannot be "continuously deformed" into each other in X). They are powerful invariants for distinguishing topological spaces, but they also play a prominent role in contemporary mathematical physics. In Chapter 0 we intimated that a magnetic monopole is best viewed as a creature living in some principal $U(1)$-bundle over S^2. In Chapter 3 we will define such bundles precisely and prove the remarkable fact that they are in one-to-one correspondence with the elements of $\pi_1(U(1))$. In this chapter we will compute $\pi_1(U(1))$ and show that it is isomorphic to the group $\mathbb{Z}$ of integers. The resulting one-to-one correspondence between monopoles and integers will then emerge as a topological manifestation of the Dirac quantization condition (0.2.9). In the same way we will find, in Chapter 5, that the Yang-Mills instantons on S^4 are in one-to-one correspondence with the principal $SU(2)$-bundles over S^4 and that these, in turn, are classified by $\pi_3(SU(2))$. Now, $\pi_3(SU(2))$ is also isomorphic to $\mathbb{Z}$ and, in this context, each integer is essentially the so-called "instanton number" of the bundle (or of the corresponding Yang-Mills instanton).

2.2 Path Homotopy and the Fundamental Group

The first of the homotopy groups, also called the fundamental group, is rather special and we will examine it in some detail before introducing the so-called "higher homotopy groups." To do so we recall (Section 1.5) that a **path** in the topological space X is a continuous map $\alpha : [0,1] \to X$ and that $x_0 = \alpha(0)$ and $x_1 = \alpha(1)$ are, respectively, the **initial** and **terminal points** of α. One says that α is a path in X **from** x_0 **to** x_1. We wish to make precise the notion that one path in X from x_0 to x_1 can be "continuously deformed" into some other path in X from x_0 to x_1 without leaving X. Let $\alpha, \beta : [0,1] \to X$ be two paths in X with $\alpha(0) = \beta(0) = x_0$ and $\alpha(1) = \beta(1) = x_1$. We will say that α is **path homotopic to** β (or α is **homotopic to** β **relative to** $\{0,1\}$) and write $\alpha \simeq \beta \operatorname{rel}\{0,1\}$ if there exists a continuous map $F : [0,1] \times [0,1] \to X$, called a **path homotopy from** α **to** β in X satisfying

$$F(0,t) = x_0 \text{ and } F(1,t) = x_1 \text{ for all } t \in [0,1]$$

and

$$F(s,0) = \alpha(s) \text{ and } F(s,1) = \beta(s) \text{ for all } s \in [0,1]$$

(see Figure 2.2.1). For each t in $[0,1]$ one defines a path $F_t : [0,1] \to X$ from x_0 to x_1 by $F_t(s) = F(s,t)$. Then $F_0 = \alpha$, $F_1 = \beta$ and one regards F_t as the tth stage in the deformation of α into β. Alternatively, one thinks of $\{F_t : 0 \leq t \leq 1\}$ as a continuous sequence of paths in X from x_0 to x_1, beginning with α and ending with β. Here is a simple example: Suppose $X = \mathbb{R}^n$, x_0 and x_1 are any two points in $\mathbb{R}^n$ and α and β are any two paths in $\mathbb{R}^n$ from x_0 to x_1. Define $F : [0,1] \times [0,1] \to \mathbb{R}^n$ by

$$F(s,t) = (1-t)\alpha(s) + t\beta(s) \tag{2.2.1}$$

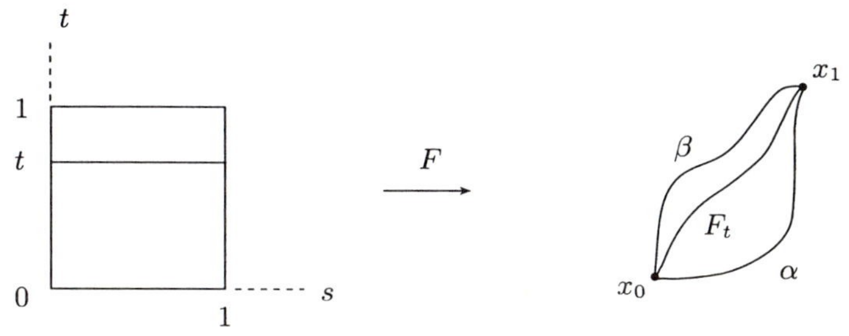

Figure 2.2.1

for all s and t in $[0,1]$. Then F is clearly a homotopy from α to β. On the other hand, proving that two paths in a space X are *not* homotopic can be rather formidable since it requires detailed information about the topological obstructions to deforming paths that are present in X. For example, define paths $\alpha, \beta : [0,1] \to \mathbb{R}^2$ by $\alpha(s) = (\cos \pi s, \sin \pi s)$ and $\beta(s) = (\cos \pi s, -\sin \pi s)$. Then α follows the top half of the unit circle from $(1,0)$ to $(-1,0)$, while β follows the bottom half and, as we have just shown, α and β are path homotopic in $\mathbb{R}^2$. However, we will see that the same two paths, regarded as maps into $X = \mathbb{R}^2 - \{(0,0)\}$ are *not* homotopic (intuitively, some stage of the deformation would have to pass through $(0,0)$, which isn't in X). Then again, thinking of $\mathbb{R}^2$ as a subspace of $\mathbb{R}^3$ (say, the xy-plane) and regarding α and β as maps into $\mathbb{R}^3 - \{(0,0,0)\}$, they once again become homotopic (intuitively, we can now use the extra dimension available to loop a deformation of α around the missing point $(0,0,0)$). Whether or not two paths $\alpha, \beta : [0,1] \to X$ from x_0 to x_1 are homotopic is a question about the topology of X.

The single most important fact about the notion of path homotopy is that it defines an equivalence relation on the set of all paths in X from x_0 to x_1.

Lemma 2.2.1 *Let X be a topological space and $\alpha, \beta, \gamma : [0,1] \to X$ paths in X from x_0 to x_1. Then*

(a) $\alpha \simeq \alpha \operatorname{rel} \{0,1\}$.

(b) $\alpha \simeq \beta \operatorname{rel} \{0,1\}$ *implies* $\beta \simeq \alpha \operatorname{rel} \{0,1\}$.

(c) $\alpha \simeq \beta \operatorname{rel} \{0,1\}$ *and* $\beta \simeq \gamma \operatorname{rel} \{0,1\}$ *imply* $\alpha \simeq \gamma \operatorname{rel} \{0,1\}$.

Proof: To prove (a) we need only produce a homotopy from α to α and this we accomplish by letting each stage of the deformation be α, i.e., we define $F: [0,1] \times [0,1] \to X$ by $F(s,t) = \alpha(s)$ for all (s,t) in $[0,1] \times [0,1]$. For (b) we simply reverse a deformation of α into β. More precisely, $\alpha \simeq \beta \operatorname{rel}\{0,1\}$ implies that there exists a continuous $F: [0,1] \times [0,1] \to X$ with $F(0,t) = x_0$, $F(1,t) = x_1$, $F(s,0) = \alpha(s)$ and $F(s,1) = \beta(s)$. Define $G: [0,1] \times [0,1] \to X$ by $G(s,t) = F(s,1-t)$. Then G is clearly a homotopy that begins at β and ends at α so $\beta \simeq \alpha \operatorname{rel}\{0,1\}$. Finally, if $\alpha \simeq \beta \operatorname{rel}\{0,1\}$ and $\beta \simeq \gamma \operatorname{rel}\{0,1\}$ we may select homotopies F from α to β and G from β to γ. To produce a homotopy H from α to γ we simply accomplish the deformations F and G, one after the other, but each in half the time. More precisely, we define $H: [0,1] \times [0,1] \to X$ by

$$H(s,t) = \begin{cases} F(s,2t), & 0 \le s \le 1, \quad 0 \le t \le \frac{1}{2} \\ G(s,2t-1), & 0 \le s \le 1, \quad \frac{1}{2} \le t \le 1 \end{cases}.$$

Notice that H is continuous by the Glueing Lemma 1.2.3 since $F(s,2t)$ and $G(s,2t-1)$ agree when $t = \frac{1}{2}$: $F(s,1) = \beta(s) = G(s,0)$. Thus, H is a homotopy from α to γ. ∎

It follows from Lemma 2.2.1 that the path homotopy relation divides the set of all paths in X from x_0 to x_1 into equivalence classes and these we will call **(path) homotopy classes**. If α is such a path, its homotopy class will be denoted $[\alpha]$.

We return now to two ideas that we first encountered in the proof of Lemma 1.5.4 and that are the very heart and soul of homotopy theory. First, suppose α is a path in X from x_0 to x_1. Define $\alpha^{\leftarrow}: [0,1] \to X$ ("α backwards") by $\alpha^{\leftarrow}(s) = \alpha(1-s)$ for all $s \in [0,1]$.

Exercise 2.2.1 Show that if $\alpha, \alpha': [0,1] \to X$ are paths in X from x_0 to x_1 and $\alpha' \simeq \alpha \operatorname{rel}\{0,1\}$, then $(\alpha')^{\leftarrow} \simeq \alpha^{\leftarrow} \operatorname{rel}\{0,1\}$.

Next suppose $\alpha: [0,1] \to X$ is a path in X from x_0 to x_1 and $\beta: [0,1] \to X$ is a path in X from x_1 to x_2. Define $\alpha\beta: [0,1] \to X$ ("α followed by β") by

$$\alpha\beta(s) = \begin{cases} \alpha(2s), & 0 \le s \le \frac{1}{2} \\ \beta(2s-1), & \frac{1}{2} \le s \le 1 \end{cases}.$$

Exercise 2.2.2 Show that if $\alpha, \alpha': [0,1] \to X$ are paths in X from x_0 to x_1, $\beta, \beta': [0,1] \to X$ are paths in X from x_1 to x_2, $\alpha' \simeq \alpha \operatorname{rel}\{0,1\}$ and $\beta' \simeq \beta \operatorname{rel}\{0,1\}$, then $\alpha'\beta' \simeq \alpha\beta \operatorname{rel}\{0,1\}$.

According to Exercise 2.2.1 one can unambiguously define the operation $\leftarrow$ of going backwards on homotopy classes of paths in X from x_0 to x_1.

More precisely, if $[\alpha]$ is any such homotopy class we may define $[\alpha]^{\leftarrow} = [\alpha^{\leftarrow}]$ and be assured that the definition does not depend on the choice of which representative α of the class one happens to turn around. In the same way, Exercise 2.2.2 guarantees that $[\alpha][\beta]$ is well-defined by $[\alpha][\beta] = [\alpha\beta]$, where $[\alpha]$ is a homotopy class of paths in X from x_0 to x_1 and $[\beta]$ is a homotopy class of paths in X from x_1 to x_2.

A path $\alpha : [0,1] \to X$ for which $\alpha(0) = \alpha(1) = x_0$ is called a **loop at** x_0 **in** X. Observe that if α and β are any two loops at x_0 in X, then $\alpha\beta$ is necessarily defined (the terminal point of α and the initial point of β are both x_0). A particularly simple loop at x_0 is the constant map on $[0,1]$ whose value at any s is x_0. We will abuse notation a bit and denote by x_0 also this **trivial loop** at x_0. Thus, $[x_0]$ will designate the homotopy class of the trivial loop at x_0. The set of all homotopy classes of loops at x_0 in X is denoted $\pi_1(X, x_0)$ and our major result of this section (Theorem 2.2.2) asserts that it has a natural group structure. With this structure $\pi_1(X, x_0)$ is called the **fundamental group**, or **first homotopy group, of** X **at** x_0. The point x_0 is called the **base point** of the group. We will eventually show (Theorem 2.2.3) that, if X is pathwise connected, different base points give rise to isomorphic groups so that one may drop all reference to them and speak simply of the "fundamental group of X".

Theorem 2.2.2 *Let X be a topological space and x_0 a point in X. Let $\pi_1(X, x_0)$ be the set of all homotopy classes of loops at x_0 in X. For $[\alpha], [\beta] \in \pi_1(X, x_0)$, define $[\alpha][\beta] = [\alpha\beta]$. Then, with this operation, $\pi_1(X, x_0)$ is a group in which the identity element is $[x_0]$ and the inverse of any $[\alpha]$ is given by $[\alpha]^{-1} = [\alpha]^{\leftarrow} = [\alpha^{\leftarrow}]$.*

Proof: We have already shown that our binary operation is well-defined by $[\alpha][\beta] = [\alpha\beta]$ and we must now show that it is associative, i.e., that for $[\alpha]$, $[\beta]$ and $[\gamma]$ in $\pi_1(X, x_0)$,

$$\begin{aligned} ([\alpha][\beta])\,[\gamma] &= [\alpha]\,([\beta][\gamma]) \\ ([\alpha\beta])\,[\gamma] &= [\alpha]\,([\beta\gamma]) \\ [(\alpha\beta)\gamma] &= [\alpha(\beta\gamma)]\ . \end{aligned}$$

Thus, we must show that $(\alpha\beta)\gamma \simeq \alpha(\beta\gamma) \operatorname{rel}\{0,1\}$. (We will see that $(\alpha\beta)\gamma$ and $\alpha(\beta\gamma)$ are, in general, *not* equal, but only path homotopic so that, in order to manufacture a group, one must consider homotopy classes of loops rather than the loops themselves.) Let us write out explicitly the path $(\alpha\beta)\gamma$.

$$(\,(\alpha\beta)\gamma\,)\,(s) = \begin{cases} (\alpha\beta)(2s), & 0 \le s \le \frac{1}{2} \\ \gamma(2s-1), & \frac{1}{2} \le s \le 1 \end{cases}\ .$$

But for $0 \le s \le \frac{1}{2}$,

$$(\alpha\beta)(2s) = \begin{cases} \alpha(2(2s)), & 0 \le 2s \le \frac{1}{2} \\ \beta(2(2s)-1), & \frac{1}{2} \le 2s \le 1 \end{cases} = \begin{cases} \alpha(4s), & 0 \le s \le \frac{1}{4} \\ \beta(4s-1), & \frac{1}{4} \le s \le \frac{1}{2} \end{cases}$$

so

$$((\alpha\beta)\gamma)(s) = \begin{cases} \alpha(4s), & 0 \le s \le \frac{1}{4} \\ \beta(4s-1), & \frac{1}{4} \le s \le \frac{1}{2} \\ \gamma(2s-1), & \frac{1}{2} \le s \le 1 \end{cases}.$$

Exercise 2.2.3 Show that

$$(\alpha(\beta\gamma))(s) = \begin{cases} \alpha(2s), & 0 \le s \le \frac{1}{2} \\ \beta(4s-1), & \frac{1}{2} \le s \le \frac{3}{4} \\ \gamma(4s-3), & \frac{3}{4} \le s \le 1 \end{cases}.$$

To construct the required homotopy we simply interpolate between these loops in the manner indicated in Figure 2.2.2.

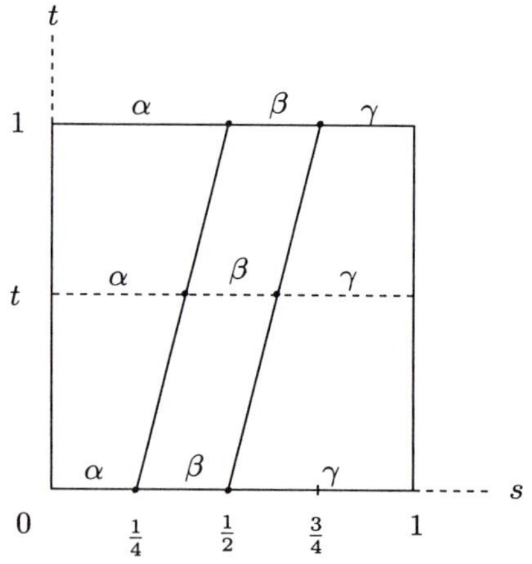

Figure 2.2.2

Thus, for each fixed $t \in [0,1]$ we traverse the entire loops α, β and γ on the indicated s-intervals. The equations of the two straight lines in Figure 2.2.2 are $s = \frac{t+1}{4}$ and $s = \frac{t+2}{4}$. To complete the action of α on $0 \le s \le \frac{t+1}{4}$, choose an increasing linear function of $[0, \frac{t+1}{4}]$ onto $[0,1]$, i.e., $s' = \frac{4s}{t+1}$, and take $F_t(s) = \alpha(s') = \alpha(\frac{4s}{t+1})$ for $0 \le s \le \frac{t+1}{4}$. Similarly, we traverse β on $\frac{t+1}{4} \le s \le \frac{t+2}{4}$ and γ on $\frac{t+2}{4} \le s \le 1$. The resulting map $F : [0,1]\times[0,1] \to X$ is given by

$$F(s,t) = \begin{cases} \alpha\left(\frac{4s}{t+1}\right), & 0 \le s \le \frac{t+1}{4}, & 0 \le t \le 1 \\ \beta(4s-1-t), & \frac{t+1}{4} \le s \le \frac{t+2}{4}, & 0 \le t \le 1 \\ \gamma\left(1 - \frac{4(1-s)}{2-t}\right), & \frac{t+2}{4} \le s \le 1, & 0 \le t \le 1 \end{cases}.$$

Exercise 2.2.4 Check the continuity of F with the Glueing Lemma 1.2.3 and show that it is the required homotopy from $(\alpha\beta)\gamma$ to $\alpha(\beta\gamma)$.

To show that $[x_0]$ acts as an identity we must prove that, for every $[\alpha] \in \pi_1(X, x_0)$, $[x_0][\alpha] = [\alpha][x_0] = [\alpha]$. Since both equalities are proved in the same way we show only that $[x_0][\alpha] = [\alpha]$, i.e., that $[x_0\alpha] = [\alpha]$, and for this we need a path homotopy from $x_0\alpha$ to α. Now,

$$(x_0\alpha)(s) = \begin{cases} x_0, & 0 \le s \le \frac{1}{2} \\ \alpha(2s-1), & \frac{1}{2} \le s \le 1 \end{cases}$$

and we construct a homotopy from this to α in the manner indicated in Figure 2.2.3.

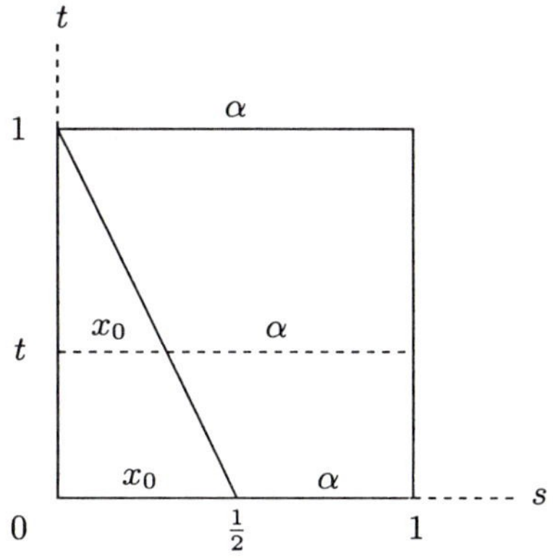

Figure 2.2.3

This time the equation of the straight line is $s = \frac{1-t}{2}$ so, for each fixed t, we take F_t to be x_0 for $0 \le s \le \frac{1-t}{2}$ and then complete the action of

α on $\frac{1-t}{2} \leq s \leq 1$. This is again accomplished by choosing an increasing linear function of $[\frac{1-t}{2}, 1]$ onto $[0,1]$, i.e., $s' = \frac{2s+t-1}{t+1}$, and evaluating $\alpha(s') = \alpha(\frac{2s+t-1}{t+1})$. Thus, we define $F : [0,1] \times [0,1] \to X$ by

$$F(s,t) = \begin{cases} x_0\,, & 0 \leq s \leq \frac{1-t}{2}\,, \quad 0 \leq t \leq 1 \\ \alpha\left(\frac{2s+t-1}{t+1}\right), & \frac{1-t}{2} \leq s \leq 1\,, \quad 0 \leq t \leq 1 \end{cases}.$$

Exercise 2.2.5 Show that this is a homotopy from $x_0\alpha$ to α.

Finally, we must show that, for each $[\alpha] \in \pi_1(X, x_0)$, $[\alpha]^{\leftarrow} = [\alpha^{\leftarrow}]$ acts as an inverse for $[\alpha]$, i.e., $[\alpha][\alpha]^{\leftarrow} = [\alpha]^{\leftarrow}[\alpha] = [x_0]$. Since these are similar we show only that

$$[\alpha][\alpha]^{\leftarrow} = [x_0]$$

$$[\alpha][\alpha^{\leftarrow}] = [x_0]$$

$$[\alpha\alpha^{\leftarrow}] = [x_0]\,.$$

For this we must prove that $\alpha\alpha^{\leftarrow} \simeq x_0 \operatorname{rel}\{0,1\}$ and one procedure for this is indicated in Figure 2.2.4.

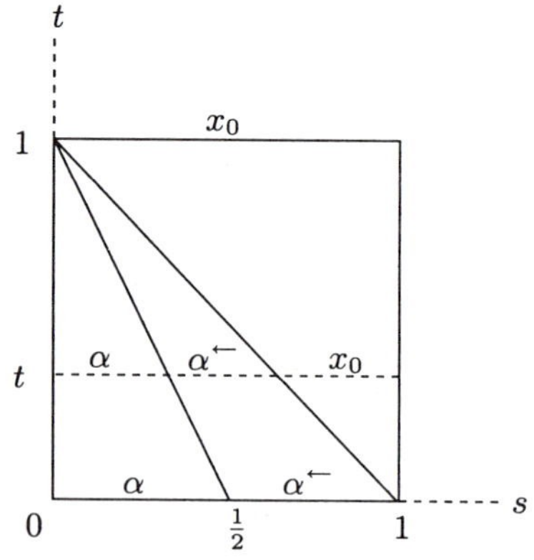

Figure 2.2.4

Exercise 2.2.6 Use Figure 2.2.4 to construct the required homotopy from $\alpha\alpha^{\leftarrow}$ to the trivial loop x_0. ∎

Exercise 2.2.7 There is nothing unique about the homotopies constructed in the proof of Theorem 2.2.2. Show, for example, that the following is a homotopy from $\alpha\alpha^{\leftarrow}$ to x_0 and describe intuitively how the deformation is being accomplished:

$$F(s,t) = \begin{cases} \alpha\,(2s(1-t))\,, & 0 \le s \le \frac{1}{2}\,, \quad 0 \le t \le 1 \\ \alpha\,(2(1-s)(1-t))\,, & \frac{1}{2} \le s \le 1\,, \quad 0 \le t \le 1 \end{cases}.$$

Thus, we have associated with every space X and every point $x_0 \in X$ a group $\pi_1(X, x_0)$ whose elements are the homotopy classes of loops in X at x_0. If x_0 and x_1 are distinct points in X, then, in general, there need not be any relationship between $\pi_1(X, x_0)$ and $\pi_1(X, x_1)$. However, if X is pathwise connected, then all such groups are isomorphic. More generally, we have the following result.

Theorem 2.2.3 *Let x_0 and x_1 be two points in an arbitrary space X and suppose there exists a path $\sigma : [0,1] \to X$ in X from $\sigma(0) = x_0$ to $\sigma(1) = x_1$. Then the map $\sigma_\# : \pi_1(X, x_1) \to \pi_1(X, x_0)$ defined by $\sigma_\#([\alpha]) = [\sigma\alpha\sigma^{\leftarrow}]$ for each $[\alpha] \in \pi_1(X, x_1)$ is an isomorphism* (see Figure 2.2.5).

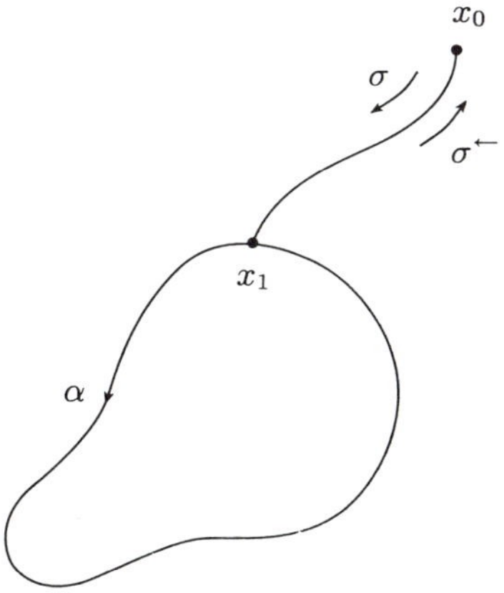

Figure 2.2.5

Proof: For each loop α at x_1, $\sigma\alpha\sigma^{\leftarrow}$ is clearly a loop at x_0. Moreover, by Exercise 2.2.2, $\alpha' \simeq \alpha \operatorname{rel}\{0,1\}$ implies $\sigma\alpha'\sigma^{\leftarrow} \simeq \sigma\alpha\sigma^{\leftarrow} \operatorname{rel}\{0,1\}$ so $\sigma_\#$ is well-defined. To show that it is a homomorphism we compute $\sigma_\#([\alpha])\sigma_\#([\beta]) = [\sigma\alpha\sigma^{\leftarrow}][\sigma\beta\sigma^{\leftarrow}] = [\sigma\alpha\sigma^{\leftarrow}\sigma\beta\sigma^{\leftarrow}] = [\sigma\alpha][\sigma^{\leftarrow}\sigma][\beta\sigma^{\leftarrow}] = [\sigma\alpha][x_0][\beta\sigma^{\leftarrow}] = [\sigma\alpha][\beta\sigma^{\leftarrow}] = [\sigma(\alpha\beta)\sigma^{\leftarrow}] = \sigma_\#([\alpha\beta])$. Switching the roles of x_0 and x_1 and of σ and $\sigma^{\leftarrow}$ we find that $(\sigma^{\leftarrow})_{\#} : \pi_1(X, x_0) \to \pi_1(X, x_1)$ is also a homomorphism. Indeed, we claim that it is the inverse of $\sigma_{\#}$ so, in particular, $\sigma_{\#}$ is an isomorphism. To see this we compute $(\sigma^{\leftarrow})_{\#} \circ \sigma_{\#}([\alpha]) = (\sigma^{\leftarrow})_{\#}([\sigma\alpha\sigma^{\leftarrow}]) = [\sigma^{\leftarrow}(\sigma\alpha\sigma^{\leftarrow})(\sigma^{\leftarrow})^{\leftarrow}] = [\sigma^{\leftarrow}(\sigma\alpha\sigma^{\leftarrow})\sigma] = [\sigma^{\leftarrow}\sigma][\alpha][\sigma^{\leftarrow}\sigma] = [x_0][\alpha][x_0] = [\alpha]$. In the same way, $\sigma_{\#} \circ (\sigma^{\leftarrow})_{\#}([\alpha]) = [\alpha]$ so the proof is complete. ∎

Corollary 2.2.4 *If X is pathwise connected, then, for any two points x_0 and x_1 in X, $\pi_1(X, x_0) \cong \pi_1(X, x_1)$.*

Exercise 2.2.8 Show that the isomorphism in Theorem 2.2.3 depends only on the homotopy class of the path σ, i.e., that if $\sigma' = \sigma \operatorname{rel}\{0, 1\}$, then $\sigma'_{\#} = \sigma_{\#}$.

If X is pathwise connected we may, by Corollary 2.2.4, speak of **the fundamental group of** X and write $\pi_1(X)$ without reference to any particular base point. Indeed, we shall often adopt this policy as a matter of convenience, but it is nevertheless important to keep in mind that, while all of the groups $\pi_1(X, x_0)$, $x_0 \in X$, are isomorphic, they are not, in general, "naturally" isomorphic in the sense that there is no canonical way to identify them. One must choose a homotopy class of paths from x_0 to x_1 in order to determine an isomorphism of $\pi_1(X, x_1)$ onto $\pi_1(X, x_0)$. Thus, it is often best to retain references to the base point even in the pathwise connected case. For this reason we introduce a bit of terminology. If X is a topological space and x_0 is a point in X we refer to the pair (X, x_0) as a **pointed space** with **base point** x_0. Thus, one may think of π_1 as an operator that assigns to every pointed space (X, x_0) a group $\pi_1(X, x_0)$. This operator π_1 does much more, however. If (X, x_0) and (Y, y_0) are two pointed spaces and f is a continuous map of X to Y that "preserves base points" in the sense that $f(x_0) = y_0$, then we will refer to f as a **map** of the pointed space (X, x_0) to the pointed space (Y, y_0) and write $f : (X, x_0) \to (Y, y_0)$. Note that, if f is such a map and α is a loop at x_0 in X, then $f \circ \alpha$ is a loop at y_0 in Y. We show next that the homotopy class $[f \circ \alpha]$ in $\pi_1(Y, y_0)$ depends only on $[\alpha]$ so that f actually determines a mapping $[\alpha] \to [f \circ \alpha]$ from $\pi_1(X, x_0)$ to $\pi_1(Y, y_0)$. Furthermore, this map has all sorts of terrific properties.

Theorem 2.2.5 *Let (X, x_0) and (Y, y_0) be pointed topological spaces and $f : (X, x_0) \to (Y, y_0)$ a map. Then f induces a homomorphism $f_* : \pi_1(X, x_0) \to \pi_1(Y, y_0)$ defined by $f_*([\alpha]) = [f \circ \alpha]$ for each $[\alpha] \in \pi_1(X, x_0)$. Furthermore,*

1. *If $(Y, y_0) = (X, x_0)$ and $f = id_X$, then $f_* = (id_X)_* = id_{\pi_1(X, x_0)}$.*
2. *If (Z, z_0) is another pointed space and $g : (Y, y_0) \to (Z, z_0)$ is another map, then $(g \circ f)_* = g_* \circ f_*$.*

Proof: We must show that f_* is well-defined, that it is a homomorphism and that properties #1 and #2 above are satisfied. f_* will be well-defined if $[\alpha'] = [\alpha]$ implies $[f \circ \alpha'] = [f \circ \alpha]$. Thus, suppose F is a homotopy from α to α'. We define $G: [0,1] \times [0,1] \to Y$ by $G = f \circ F$ (see Figure 2.2.6).

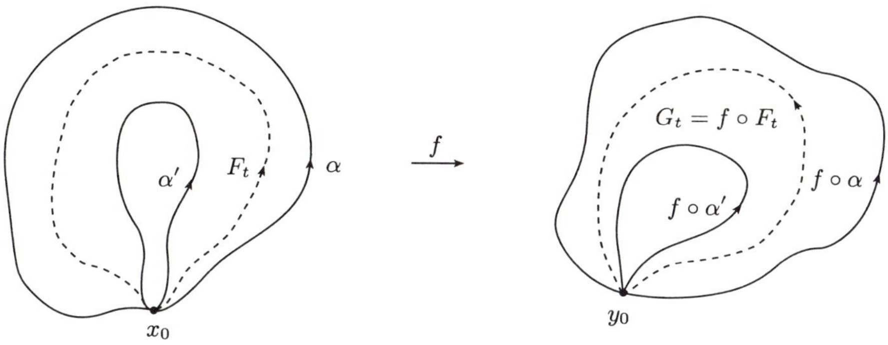

Figure 2.2.6

Then G is continuous, $G(0,t) = f(F(0,t)) = f(x_0) = y_0 = G(1,t)$ for all t, $G(s,0) = f(F(s,0)) = f(\alpha(s)) = (f \circ \alpha)(s)$ and $G(s,1) = f(F(s,1)) = f(\alpha'(s)) = (f \circ \alpha')(s)$ so G is a homotopy from $f \circ \alpha$ to $f \circ \alpha'$ as required.

To see that f_* is a homomorphism we compute $f_*([\alpha][\beta]) = f_*([\alpha\beta]) = [f \circ (\alpha\beta)] = [(f \circ \alpha)(f \circ \beta)] = [f \circ \alpha][f \circ \beta] = f_*([\alpha])f_*([\beta])$. Since $(id_X)_*([\alpha]) = [id_X \circ \alpha] = [\alpha]$, property #1 above is clear. Finally, to prove #2 we compute $(g \circ f)_*([\alpha]) = [(g \circ f) \circ \alpha] = [g \circ (f \circ \alpha)] = g_*([f \circ \alpha]) = g_*(f_*([\alpha])) = (g_* \circ f_*)([\alpha])$ and so $(g \circ f)_* = g_* \circ f_*$. ∎

Remark: We mention in passing some terminology that is often used to describe the sort of phenomenon with which we are now dealing. In mathematics one is often confronted with a particular collection of objects and a distinguished family of maps between these objects (vector spaces and linear maps; topological spaces and continuous maps; groups and homomorphisms; pointed spaces and base point preserving maps between them). In the current jargon, such a collection of objects together with its distinguished family of maps is referred to as a **category**. An operator which assigns to every object in one category a corresponding object in another category and to every map in the first a map in the second in such a way that compositions of maps are preserved and the identity map is taken to the identity map is called a **functor**. Thus, we may summarize our activities thus far by saying that we have constructed a functor (the **fundamental group functor**) from the category of pointed spaces and maps to the category of groups and homomorphisms. Such functors are the vehicles by which one translates topological problems into (hopefully more tractable) algebraic problems. We will see a particularly beautiful example of this technique when we prove the Brouwer Fixed Point Theorem.

It will not have escaped the reader's attention that, although we now know a fair amount about fundamental groups in general, we have yet to see our first example. It is sad, but true, that calculating $\pi_1(X, x_0)$ can be horrendously difficult and we will have to content ourselves with just those few examples that are specifically required for our purposes. The first one, at least, is easy. Since $\mathbb{R}^n$ is pathwise connected, all of the fundamental groups $\pi_1(\mathbb{R}^n, x_0)$, $x_0 \in \mathbb{R}^n$, are isomorphic. Moreover, we have already observed ((2.2.1)) that any two paths in $\mathbb{R}^n$ with the same initial and terminal points are path homotopic. In particular, any loop at any x_0 in $\mathbb{R}^n$ is homotopic to the trivial loop at x_0. Thus, there is just one homotopy class of loops at x_0 so $\pi_1(\mathbb{R}^n, x_0)$ is the trivial group. Dropping reference to x_0 we say that $\pi_1(\mathbb{R}^n)$ is trivial and write this symbolically as

$$\pi_1(\mathbb{R}^n) = 0 .$$

A pathwise connected space X whose fundamental group is the trivial group is said to be **simply connected** and, for such spaces, we will write $\pi_1(X) = 0$. Thus, we have shown that $\mathbb{R}^n$ is simply connected and we will find many more examples in the next section.

2.3 Contractible and Simply Connected Spaces

We begin by generalizing the notion of homotopy (continuous deformation) to maps other than paths. Thus, we suppose that X and Y are topological spaces, A is a subset of X (perhaps the empty set) and $f, g : X \to Y$ are two continuous maps with $f \mid A = g \mid A$. We say that f is **homotopic to g relative to A** and write $f \simeq g \operatorname{rel} A$ if there exists a continuous map $F : X \times [0,1] \to Y$ with

$$F(x,0) = f(x), \quad F(x,1) = g(x) \text{ for all } x \in X, \quad \text{and}$$
$$F(a,t) = f(a) = g(a) \quad \text{for all } a \in A \text{ and } t \in [0,1] .$$

F is called a **homotopy (relative to A) from f to g** in X. Defining, for each t in $[0,1]$, a map $F_t : X \to Y$ by $F_t(x) = F(x,t)$ for each $x \in X$ one thinks of $\{F_t : 0 \le t \le 1\}$ as a continuous sequence of maps, beginning with $F_0 = f$, ending with $F_1 = g$ and each agreeing with both f and

g everywhere on A. For maps on $[0,1]$ with $A = \{0,1\}$ this agrees with our definition of path homotopy. If $A = \emptyset$, then we simply say that f is **homotopic to** g, write $f \simeq g$, and call F a **free homotopy** from f to g.

Lemma 2.3.1 *Let X and Y be topological spaces, A a subset of X and $f, g, h : X \to Y$ continuous maps with $f \mid A = g \mid A = h \mid A$. Then*

(a) $f \simeq f \operatorname{rel} A$.

(b) $f \simeq g \operatorname{rel} A$ *implies* $g \simeq f \operatorname{rel} A$.

(c) $f \simeq g \operatorname{rel} A$ *and* $g \simeq h \operatorname{rel} A$ *imply* $f \simeq h \operatorname{rel} A$.

Exercise 2.3.1 Prove Lemma 2.3.1. ■

Exercise 2.3.2 Suppose $f, g : X \to Y$ are continuous and $f \simeq g$. Let $u : Y \to U$ and $v : V \to X$ be continuous. Show that $u \circ f \simeq u \circ g$ and $f \circ v \simeq g \circ v$. Show also that if $A \subseteq X$ and $f \simeq g \operatorname{rel} A$, then $u \circ f \simeq u \circ g \operatorname{rel} A$.

Thus, homotopy relative to A is an equivalence relation on the set of all mappings from X to Y that agree on A and so partitions this set into equivalence classes (called **homotopy classes relative to** A). If $A = \emptyset$, then the set of all homotopy classes of maps from X to Y is denoted $[X, Y]$. Similarly, if (X, x_0) and (Y, y_0) are pointed spaces, then we denote by $[(X, x_0), (Y, y_0)]$ the set of homotopy classes relative to $\{x_0\}$ of maps $f : (X, x_0) \to (Y, y_0)$.

For a simple example we return again to $\mathbb{R}^n$. Let X be any topological space and $f, g : X \to \mathbb{R}^n$ any two continuous maps. Define $F : X \times [0,1] \to \mathbb{R}^n$ by $F(x,t) = (1-t)f(x) + tg(x)$ for all $x \in X$ and all $t \in [0,1]$. Then F is a continuous map with $F(x,0) = f(x)$ and $F(x,1) = g(x)$ so it is a free homotopy from f to g. Thus, *any* two maps into $\mathbb{R}^n$ are homotopic, i.e., $[X, \mathbb{R}^n]$ consists of a single element.

Exercise 2.3.3 A subset Y of $\mathbb{R}^n$ is said to be **convex** if it contains the line segment joining any two of its points, i.e., if $(1-t)p_0 + tp_1$ is in Y whenever p_0 and p_1 are in Y and $0 \le t \le 1$. Show that if Y is a convex subspace of $\mathbb{R}^n$ and X is any topological space, then $[X, Y]$ consists of a single element.

Now, the n-sphere S^n is certainly not convex, but a naive attempt to generalize our proof that any two maps into $\mathbb{R}^n$ are homotopic very nearly works for S^n. More precisely, suppose X is any space and $f, g : X \to S^n$ are any two continuous maps. Then $(1-t)f(x) + t\,g(x)$ will not, in general, be in S^n, but we can project it radially out onto S^n by dividing it by its norm $\|(1-t)f(x) + t\,g(x)\|$. This, of course, is legal only if $(1-t)f(x) + t\,g(x)$ is never $0 \in \mathbb{R}^{n+1}$ and this is the case iff $f(x)$ and $g(x)$ are never antipodal points on S^n.

Exercise 2.3.4 Let X be any topological space and $f, g : X \to S^n$ two continuous maps such that, for every $x \in X$, $f(x) \neq -g(x)$ (i.e., f and g are never antipodal). Show that $f \simeq g$. In particular, if $f : X \to S^n$ is a nonsurjective map into S^n and $p \in S^n - f(X)$, then f is homotopic to the constant map on X whose value is $-p \in S^n$.

A particular consequence of what we have just seen is that any map into $\mathbb{R}^n$ (or a convex subspace of $\mathbb{R}^n$) is homotopic to a constant map into that space. A mapping $f : X \to Y$ which is homotopic to some constant map of X into Y is said to be **nullhomotopic**. Thus, any map into a convex subspace of $\mathbb{R}^n$ is nullhomotopic. In particular, the identity map on a convex subspace of $\mathbb{R}^n$ is nullhomotopic. A topological space Y for which the identity map $id_Y : Y \to Y$ is homotopic to some constant map of Y into Y is said to be **contractible**. The intuition here is that a contractible space Y, which is the image of id_Y, can be continuously deformed within itself to a point (the image of the constant map homotopic to id_Y). For example, the homotopy $F : \mathbb{R}^n \times [0,1] \to \mathbb{R}^n$ given by $F(x,t) = (1-t)x$ begins at $id_{\mathbb{R}^n}$ and ends at the map that is identically zero and each stage of the deformation (fixed t) is a radial contraction of $\mathbb{R}^n$ toward the origin.

Lemma 2.3.2 *A topological space Y is contractible iff, for any space X, any two maps $f, g : X \to Y$ are homotopic ($f \simeq g$).*

Proof: If any two maps into Y are homotopic, then $id_Y : Y \to Y$ is homotopic to a (indeed, to any) constant map of Y so id_Y is nullhomotopic and Y is contractible. For the converse, suppose Y is contractible. Then there exists a y_0 in Y such that id_Y is homotopic to the constant map on Y whose value is y_0 (we denote this map y_0 also). Let $F : Y \times [0,1] \to Y$ be a homotopy with $F(y,0) = y$ and $F(y,1) = y_0$ for all $y \in Y$. It will suffice to prove that any map $f : X \to Y$ is homotopic to the constant map y_0 for we may then appeal to Lemma 2.3.1. Define $G : X \times [0,1] \to Y \times [0,1]$ by $G(x,t) = (f(x), t)$ (G is "f at each t-level"). Now compose with F to get

$$H : X \times [0,1] \xrightarrow{G} Y \times [0,1] \xrightarrow{F} Y$$

so that

$$H(x,t) = F(\, f(x), t)\,).$$

Then H is continuous, $H(x,0) = F(f(x),0) = f(x)$ and $H(x,1) = F(f(x),1) = y_0$ as required. ∎

Lemma 2.3.3 *A contractible space is pathwise connected.*

Exercise 2.3.5 Prove Lemma 2.3.3. **Hint:** Fix two points y_0 and y_1 in the contractible space Y and consider a homotopy between the constant maps y_0 and y_1.

Exercise 2.3.6 Show that if Y is contractible, then any continuous map $f : Y \to X$ is nullhomotopic. **Hint:** If F is a homotopy from id_Y to some constant map, consider $H(y,t) = f(F(y,t))$.

Exercise 2.3.7 Two constant maps of X into Y are homotopic iff their images lie in the same path component of Y.

Our next objective is to show that a contractible space is simply connected. By Lemma 2.3.3, we need only show that any contractible space X has trivial fundamental group. Thus, we fix some base point $x_0 \in X$ and let α be a loop at x_0 in X. We must show that $[\alpha] = [x_0]$, i.e., that $\alpha \simeq x_0 \operatorname{rel}\{0,1\}$. Of course, Lemma 2.3.2 implies $\alpha \simeq x_0$, but this is not enough. We require a *path homotopy* from α to x_0, i.e., a *fixed-endpoint* deformation of α into x_0. Arranging this requires a bit of work, but along the way we will introduce some ideas that are of considerable independent interest. The first step is to construct a homotopy $F : [0,1] \times [0,1] \to X$ with $F(s,0) = x_0$, $F(s,1) = \alpha(s)$ and $F(0,t) = F(1,t)$ for all s, $t \in [0,1]$ (so that each stage F_t of the deformation is a loop in X, albeit not at x_0, in general).

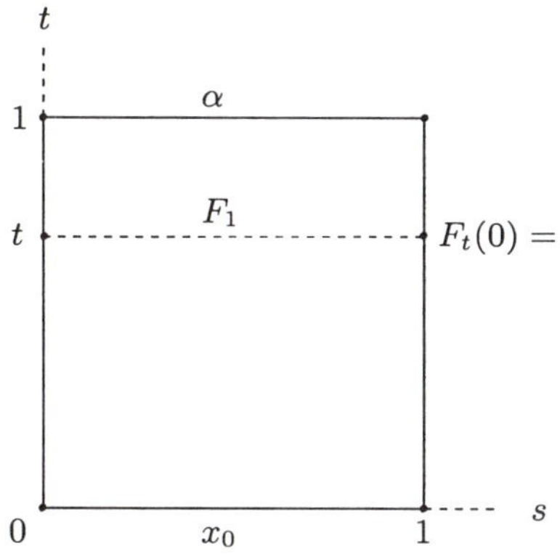

Figure 2.3.1

The idea behind the construction is that loops in X can also be regarded as base point preserving maps of S^1 into X. In fact one can prove quite a bit more. Define $\mathcal{Q} : [0,1] \to S^1$ by $\mathcal{Q}(s) = e^{2\pi s\mathbf{i}}$ and note that $\mathcal{Q}(0) = \mathcal{Q}(1) = 1 \in S^1$ (what we are doing here is identifying S^1 with the quotient space of $[0,1]$ that identifies the boundary points 0 and 1). Given a loop $\alpha : [0,1] \to X$ at x_0 in X one can define $\tilde{\alpha} : (S^1, 1) \to (X, x_0)$ by $\tilde{\alpha}(\mathcal{Q}(s)) = \alpha(s)$. Conversely, given an $\tilde{\alpha} : (S^1, 1) \to (X, x_0)$ one can define a loop $\alpha : [0,1] \to X$ at x_0 in X by $\alpha(s) = \tilde{\alpha}(\mathcal{Q}(s))$. We claim that if α and α' are two loops at x_0 in X, then $\alpha' \simeq \alpha \operatorname{rel}\{0,1\}$ iff $\tilde{\alpha}' \simeq \tilde{\alpha} \operatorname{rel}\{1\}$. Suppose $F : [0,1] \times [0,1] \to X$ satisfies $F(s,0) = \alpha'(s)$, $F(s,1) = \alpha(s)$ and $F(0,t) = F(1,t) = x_0$ for all s and t in $[0,1]$. Define $G : S^1 \times [0,1] \to X$ by $G(\mathcal{Q}(s), t) = F(s,t)$. Then $G(\mathcal{Q}(s), 0) = F(s,0) = \alpha'(s) = \tilde{\alpha}'(\mathcal{Q}(s))$ and $G(\mathcal{Q}(s), 1) = \tilde{\alpha}(\mathcal{Q}(s))$ so $G_0 = \tilde{\alpha}'$ and $G_1 = \tilde{\alpha}$. Moreover, $G(1,t) = G(e^{2\pi 0 \mathbf{i}}, t) = F(0,t) = x_0$ for all $t \in [0,1]$. Thus, G is a homotopy, relative to $\{1\}$, from $\tilde{\alpha}'$ to $\tilde{\alpha}$.

Exercise 2.3.8 Show, conversely, that $\tilde{\alpha}' \simeq \tilde{\alpha} \operatorname{rel}\{1\}$ implies $\alpha' \simeq \alpha \operatorname{rel}\{0,1\}$.

It follows that we may associate with each element $[\alpha]$ of $\pi_1(X, x_0)$ a unique element $[\tilde{\alpha}]$ of $[(S^1, 1), (X, x_0)]$ and that this correspondence is one-to-one and onto. Leaving it to the reader to show that any point p_0 in S^1 would serve just as well as 1, we have the following result.

Lemma 2.3.4 *Let X be an arbitrary topological space and $x_0 \in X$. Let p_0 be a fixed point in S^1. Then there is a one-to-one correspondence between $\pi_1(X, x_0)$ and $[(S^1, p_0), (X, x_0)]$.*

Now we return to our proof that a contractible space X is simply connected. We have fixed $x_0 \in X$ and a loop α at x_0 in X. As in the argument above we define $\tilde{\alpha} : S^1 \to X$ by $\tilde{\alpha}(\mathcal{Q}(s)) = \alpha(s)$. By Lemma 2.3.2, $\tilde{\alpha}$ is homotopic to the constant map of S^1 to X that sends everything to x_0. Let $\tilde{F} : S^1 \times [0,1] \to S^1$ be a homotopy with $\tilde{F}(p, 0) = x_0$ and $\tilde{F}(p,1) = \tilde{\alpha}(p)$ for all $p \in S^1$. Define $F : [0,1] \times [0,1] \to X$ by $F(s,t) = \tilde{F}(\mathcal{Q}(s), t)$. Then F is continuous, $F(s,0) = \tilde{F}(\mathcal{Q}(s), 0) = x_0$, $F(s,1) = \tilde{F}(\mathcal{Q}(s), 1) = \tilde{\alpha}(\mathcal{Q}(s)) = \alpha(s)$ and $F(0,t) = \tilde{F}(\mathcal{Q}(0), t) = \tilde{F}(\mathcal{Q}(1), t) = F(1,t)$ and this is what we were after (see Figure 2.3.1). What we have at this point then is a deformation of x_0 into α through a sequence of loops in X (not necessarily based at x_0, however). The final step (getting the intermediate loops based at x_0) requires a lemma that appears rather technical, but will prove its worth on numerous occasions.

Lemma 2.3.5 *Let X be a topological space and $F : [0,1] \times [0,1] \to X$ a continuous map. If $\alpha, \beta, \gamma, \delta : [0,1] \to X$ are the paths in X defined by $\alpha(s) = F(s,1)$, $\beta(s) = F(s,0)$, $\gamma(t) = F(0,t)$ and $\delta(t) = F(1,t)$ for $s, t \in [0,1]$ (see Figure 2.3.2), then $\alpha \simeq \gamma^{\leftarrow} \beta \delta \operatorname{rel}\{0,1\}$.*

Before embarking on the proof of this we remark that, with it, we can show that a contractible space must be simply connected. Indeed, referring to the homotopy (Figure 2.3.1) constructed above, we conclude from

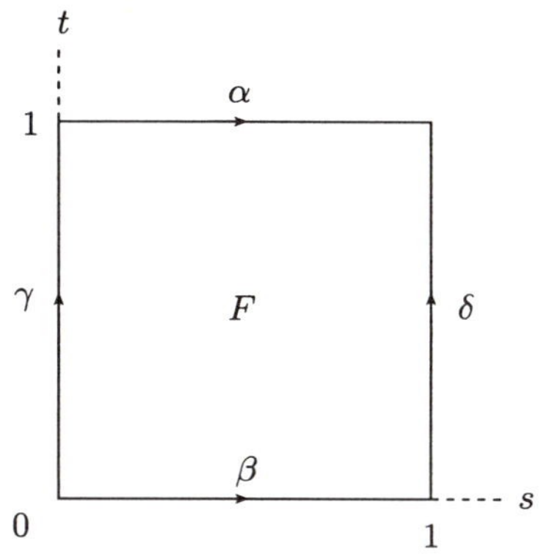

Figure 2.3.2

Lemma 2.3.5 (with $\gamma(t) = F(0,t)$, $\delta(t) = F(1,t) = F(0,t) = \gamma(t)$ and $\beta(s) = x_0$) that $\alpha \simeq \gamma^{\leftarrow}\beta\gamma \operatorname{rel}\{0,1\}$ so $[\alpha] = [\gamma^{\leftarrow}\beta\gamma] = [\gamma^{\leftarrow}][\beta][\gamma] = [\gamma]^{-1}[x_0][\gamma] = [\gamma]^{-1}[\gamma] = [x_0]$, i.e., $\alpha \simeq x_0 \operatorname{rel}\{0,1\}$.

Proof of Lemma 2.3.5 Let $x_0 = \alpha(0) = \gamma(1)$ and $x_1 = \alpha(1) = \delta(1)$ and note that $x_0\alpha x_1 \simeq \alpha \operatorname{rel}\{0,1\}$ (see Figure 2.3.3).

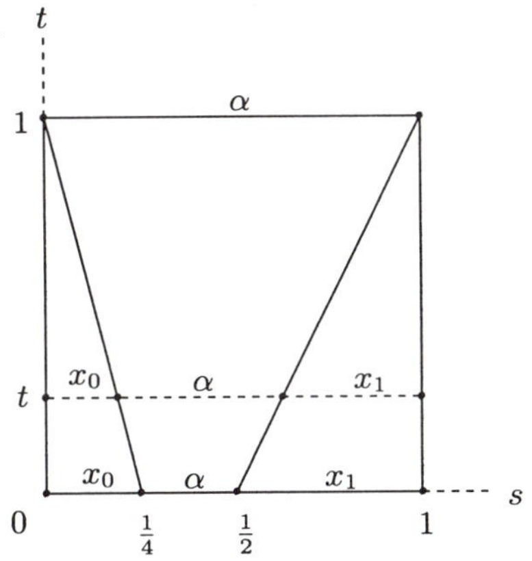

Figure 2.3.3

Exercise 2.3.9 Use Figure 2.3.3 to write out a homotopy from $x_0\alpha x_1$ to α.

Thus, it will suffice to show that $\gamma^{\leftarrow}\beta\delta \simeq x_0\alpha x_1 \operatorname{rel}\{0,1\}$. Our proof of this is based on Figure 2.3.4.

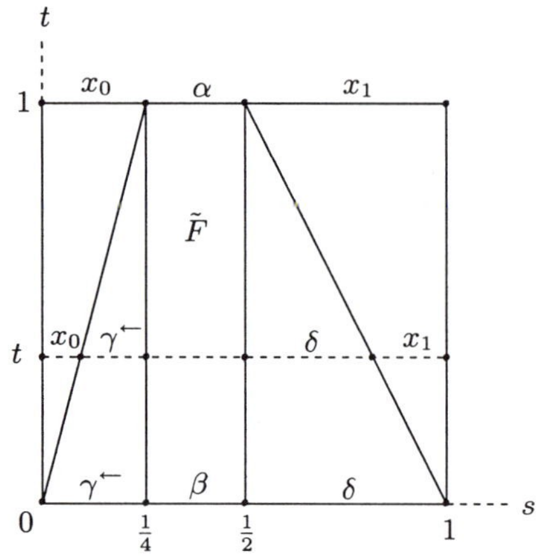

Figure 2.3.4

Here $\tilde{F}$ means "F accomplished over $[\frac{1}{4}, \frac{1}{2}] \times [0,1]$", i.e., $\tilde{F}(s,t) = F(4s-1,t)$ for $\frac{1}{4} \leq s \leq \frac{1}{2}$ and $0 \leq t \leq 1$. To construct the homotopy suggested by Figure 2.3.4 we examine in somewhat more detail what is going on at "height t" (see Figure 2.3.5).

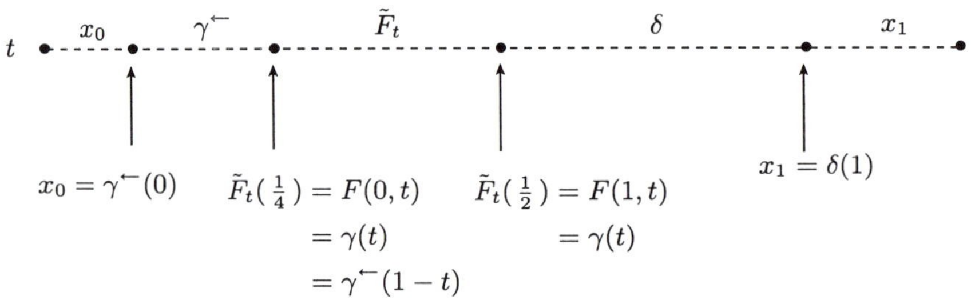

Figure 2.3.5

The equations of the two tilted lines in Figure 2.3.4 are $s = \frac{1}{4}t$ and $s =$

$\frac{-t+2}{2}$. Thus, at height t our homotopy should be x_0 for $0 \le s \le \frac{1}{4}t$, then complete $\gamma^{\leftarrow}$ from $\gamma^{\leftarrow}(0)$ to $\gamma^{\leftarrow}(1-t)$ over $\frac{1}{4}t \le s \le \frac{1}{4}$. Next, it will be $\tilde{F}_t$ for $\frac{1}{4} \le s \le \frac{1}{2}$, then it will complete δ from $\delta(t)$ to $\delta(1)$ over $\frac{1}{2} \le s \le \frac{-t+2}{2}$ and then finally it will be x_1 for $\frac{-t+2}{2} \le s \le 1$. To see how to complete $\gamma^{\leftarrow}$ from $\gamma^{\leftarrow}(0)$ to $\gamma^{\leftarrow}(1-t)$ over $\frac{1}{4}t \le s \le \frac{1}{4}$ we consider Figure 2.3.6.

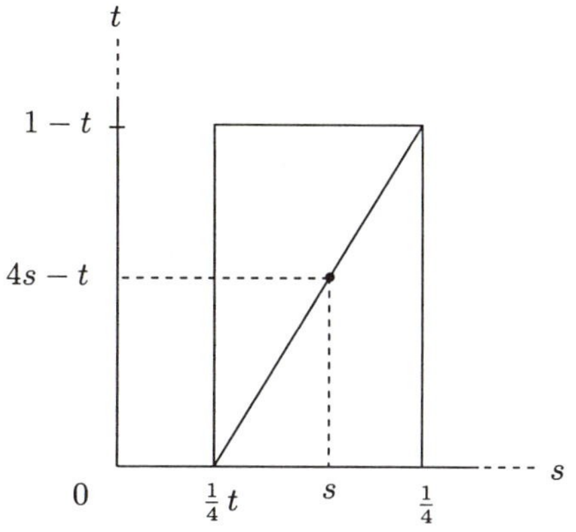

Figure 2.3.6

Thus, $s \to 4s - t$ is an increasing linear map of $[\frac{1}{4}t, \frac{1}{4}]$ onto $[0, 1-t]$ so $\gamma^{\leftarrow}(4s-t)$ will complete the action of $\gamma^{\leftarrow}$ from $\gamma^{\leftarrow}(0)$ to $\gamma^{\leftarrow}(1-t)$ over the interval $\frac{1}{4}t \le s \le \frac{1}{4}$.

Exercise 2.3.10 Draw a diagram similar to Figure 2.3.6 and show that $\delta(2s+t-1)$ completes the action of δ from $\delta(t)$ to $\delta(1)$ over the interval $\frac{1}{2} \le s \le \frac{-t+2}{2}$.

Thus, we define $H : [0,1] \times [0,1] \to X$ by

$$H(s,t) = \begin{cases} x_0, & 0 \le s \le \frac{1}{4}t, \quad 0 \le t \le 1 \\ \gamma^{\leftarrow}(4s-t), & \frac{1}{4}t \le s \le \frac{1}{4}, \quad 0 \le t \le 1 \\ F(4s-1,t), & \frac{1}{4} \le s \le \frac{1}{2}, \quad 0 \le t \le 1 \\ \delta(2s+t-1), & \frac{1}{2} \le s \le \frac{-t+2}{2}, \quad 0 \le t \le 1 \\ x_1, & -\frac{t+2}{2} \le s \le 1, \quad 0 \le t \le 1 \end{cases} .$$

Exercise 2.3.11 Verify that H is a path homotopy from $\gamma^{\leftarrow}\beta\delta$ to $x_0\,\alpha\,x_1$. ■

As we observed before the proof of Lemma 2.3.5 this establishes our major result.

Theorem 2.3.6 *A contractible space is simply connected.*

In particular, if X is any convex subspace of some $\mathbb{R}^n$ (e.g., an open or closed ball), then $\pi_1(X) = 0$.

It is easy to see that the fundamental group is a topological invariant, i.e., that homeomorphic pathwise connected spaces have isomorphic fundamental groups. Indeed, we have the following consequence of Theorem 2.2.5.

Exercise 2.3.12 Let X be an arbitrary topological space, $x_0 \in X$ and $h : X \to Y$ a homeomorphism of X onto Y that carries x_0 to $h(x_0) = y_0$. Show that $h_* : \pi_1(X, x_0) \to \pi_1(Y, y_0)$ is an isomorphism.

A very great deal more is true, however. Let us say that a continuous map $h : X \to Y$ is a **homotopy equivalence** if there exists a continuous map $h' : Y \to X$ such that $h' \circ h \simeq id_X$ and $h \circ h' \simeq id_Y$. If such maps exist we say that X and Y are **homotopically equivalent** (or of the **same homotopy type**) and write $X \simeq Y$.

Exercise 2.3.13 Justify our use of the term "equivalent" by showing that (a) $X \simeq X$, (b) $X \simeq Y$ implies $Y \simeq X$, and (c) $X \simeq Y$ and $Y \simeq Z$ imply $X \simeq Z$. **Hint:** For (c), use Exercise 2.3.2.

Of course, a homeomorphism is a homotopy equivalence, but the converse is very far from being true, as the following result amply demonstrates.

Theorem 2.3.7 *A space X is contractible iff it is homotopically equivalent to a point (i.e., to a one point discrete space).*

Proof: First suppose that X is contractible and select some $x_0 \in X$. Let Y be the (discrete) subspace of X consisting of the single point $\{x_0\}$. Let $h : X \to Y$ be the constant map $h(x) = x_0$ for all $x \in X$ and let $h' : Y \hookrightarrow X$ be the inclusion map. Both are continuous and $h \circ h' = id_Y$ so certainly $h \circ h' \simeq id_Y$. Furthermore, $h' \circ h : X \to X$ and $id_X : X \to X$ are both maps into a contractible space so they are homotopic by Lemma 2.3.2. Thus, X is homotopically equivalent to $Y = \{x_0\}$.

Conversely, suppose X is homotopically equivalent to a one point discrete space Y. Since all one point discrete spaces are homeomorphic we may assume $Y = \{x_0\}$, where $x_0 \in X$. Let $h : X \to Y$ and $h' : Y \to X$ be such that $h' \circ h \simeq id_X$. Since $h' \circ h$ is a constant map on X, id_X is homotopic to a constant map so X is contractible. ∎

Our goal here is to show that the fundamental group is actually a **homotopy invariant** in the sense that if $h : X \to Y$ is a homotopy equivalence and $x_0 \in X$, then $h_* : \pi_1(X, x_0) \to \pi_1(Y, h(x_0))$ is an isomorphism. This fact can enormously simplify the task of computing fundamental groups. Indeed, the underlying reason that a contractible space is simply connected is now particularly transparent (Theorem 2.3.7). We remark, however, that our proof of homotopy invariance will use Lemma 2.3.5 so that this approach does not significantly simplify the proof that contractible implies simply connected.

We begin by considering the following situation. Let $f : (X, x_0) \to (Y, y_0)$ and $g : (X, x_0) \to (Y, y_1)$ be maps that are homotopic. Let $F : X \times [0,1] \to Y$ be a homotopy with $F(x,0) = f(x)$ and $F(x,1) = g(x)$. Observe that $\sigma(t) = F(x_0, t)$ is then a path in Y from $\sigma(0) = y_0$ to $\sigma(1) = y_1$. Now consider the induced maps

$$f_* : \pi_1(X, x_0) \to \pi_1(Y, y_0) : f_*([\alpha]) = [f \circ \alpha]$$
$$g_* : \pi_1(X, x_0) \to \pi_1(Y, y_1) : g_*([\alpha]) = [g \circ \alpha]$$
$$\sigma_\# : \pi_1(Y, y_1) \to \pi_1(Y, y_0) : \sigma_\#([\tau]) = [\sigma\tau\sigma^\leftarrow] .$$

We claim that

$$f_* = \sigma_\# \circ g_* . \tag{2.3.1}$$

To prove this we must show that, for each $[\alpha] \in \pi_1(X, x_0)$, $[\sigma(g \circ \alpha)\sigma^\leftarrow] = [f \circ \alpha]$, i.e.,

$$\sigma(g \circ \alpha)\sigma^\leftarrow \simeq f \circ \alpha \operatorname{rel} \{0,1\} . \tag{2.3.2}$$

Define $\tilde{F} : [0,1] \times [0,1] \to Y$ by $\tilde{F}(s,t) = F(\alpha(s), t)$. Then $\tilde{F}(s,0) = F(\alpha(s), 0) = f(\alpha(s)) = (f \circ \alpha)(s)$, $\tilde{F}(s,1) = F(\alpha(s),1) = g(\alpha(s)) = (g \circ \alpha)(s)$, $\tilde{F}(0,t) = F(\alpha(0),t) = F(x_0,t) = \sigma(t)$ and $\tilde{F}(1,t) = F(\alpha(1),t) = F(x_0,t) = \sigma(t)$. Lemma 2.3.5 applied to $\tilde{F}$ now yields (2.3.2) and therefore (2.3.1). Notice also that, since we have already shown that $\sigma_\#$ is an isomorphism (Theorem 2.2.3), it follows from (2.3.1) that f_* is an isomorphism iff g_* is an isomorphism. Thus, we have proved the following result.

Theorem 2.3.8 *Let $f : (X, x_0) \to (Y, y_0)$ and $g : (X, x_0) \to (Y, y_1)$ be homotopic maps with $F : X \times [0,1] \to X$ a homotopy satisfying $F(x,0) = f(x)$ and $F(x,1) = g(x)$ for all $x \in X$. Let $\sigma : [0,1] \to Y$ be the path from y_0 to y_1 given by $\sigma(t) = F(x_0, t)$. Then the diagram*

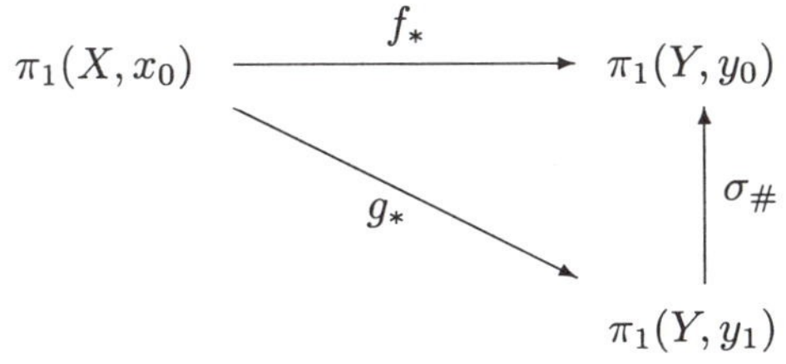

commutes, i.e., $f_* = \sigma_\# \circ g_*$. Moreover, f_* is an isomorphism iff g_* is an *isomorphism.*

Remark: An important special case of Theorem 2.3.8 arises when $y_1 = y_0$ and F is a homotopy relative to x_0. Then $\sigma_\#$ is the identity map so $f_* = g_*$.

Corollary 2.3.9 *Let $h : X \to Y$ be a homotopy equivalence. Then, for every $x_0 \in X$, $h_* : \pi_1(X, x_0) \to \pi_1(Y, h(x_0))$ is an isomorphism.*

Proof: Let $h' : Y \to X$ be such that $h \circ h' \simeq id_Y$ and $h' \circ h \simeq id_X$ and fix an $x_0 \in X$. We regard id_X as a map of (X, x_0) to (X, x_0) and $h' \circ h$ as a map from (X, x_0) to $(X, h'(h(x_0)))$. Since they are homotopic and $(id_X)_*$ is an isomorphism (indeed, it is $id_{\pi_1(X,x_0)}$, by Theorem 2.2.5), Theorem 2.3.8 implies that $(h' \circ h)_*$ is also an isomorphism. Similarly, $(h \circ h')_*$ is an isomorphism. But Theorem 2.2.5 also gives $(h' \circ h)_* = h'_* \circ h_*$ and $(h \circ h')_* = h_* \circ h'_*$. Since $h'_* \circ h_*$ is one-to-one, so is h_*. Since $h_* \circ h'_*$ is onto, so is h_*. Thus, h_* is a bijective homomorphism and therefore an isomorphism. ■

As we mentioned earlier, Corollary 2.3.9 and Theorem 2.3.7 together make clear the reason that a contractible space is simply connected (a point obviously has trivial fundamental group). What may not be so clear at the moment is that there are simply connected spaces that are *not* contractible. The spheres S^n for $n \geq 2$ are examples of this sort, but proving so is no mean feat. In this section we will content ourselves with showing that S^n is simply connected for $n \geq 2$ (S^1 is not simply connected and we will compute its fundamental group in Section 2.4). We do this by proving a special case of a very powerful result known as the Seifert-Van Kampen Theorem (see [**Gra**] for a detailed discussion of the general result).

Theorem 2.3.10 *Suppose $X = U \cup V$, where U and V are simply connected open subspaces of X with $U \cap V$ nonempty and pathwise connected. Then X is simply connected.*

Proof: We ask the reader to get the ball rolling.

Exercise 2.3.14 Show that X is pathwise connected.

Now we select an $x_0 \in U \cap V$ and show that $\pi_1(X, x_0)$ is the trivial group. Let $\alpha : [0,1] \to X$ be a loop at x_0 in X. We must show that $\alpha \simeq x_0 \operatorname{rel} \{0,1\}$. Now, $\alpha^{-1}(U)$ and $\alpha^{-1}(V)$ are nonempty open sets that cover $[0,1]$ so, by Theorem 1.4.6, we may select a Lebesgue number λ for the open cover $\{\alpha^{-1}(U), \alpha^{-1}(V)\}$ of $[0,1]$. Next we partition $[0,1]$ by selecting points $0 = t_0 < t_1 < \cdots < t_{k-1} < t_k = 1$ with $|\,t_{i+1} - t_i\,| < \lambda$ for each $i = 0, \ldots, k-1$. Thus, each $\alpha([t_i, t_{i+1}])$, $i = 0, \ldots, k-1$, is entirely contained in either U or V. Observe that if $\alpha([t_{i-1}, t_i])$ and $\alpha([t_i, t_{i+1}])$ are both contained in U (or both contained in V), then $\alpha([t_{i-1}, t_{i+1}])$ is contained in U (or, V, respectively). On the other hand, if $\alpha([t_{i-1}, t_i]) \subseteq U$ and $\alpha([t_i, t_{i+1}]) \subseteq V$ (or vice versa), then $\alpha(t_i) \in U \cap V$. We combine into a single interval $[t_{i-1}, t_{i+1}]$ all the intervals $[t_{i-1}, t_i]$ and $[t_i, t_{i+1}]$ which map into a single set (U or V). There are two possibilities. First suppose *all* of the intervals are combined. The result will be the single interval $[0, 1]$ which must then be mapped by α entirely into either U or V. Since U and V are simply connected there is a path homotopy from α to x_0 in either U or V and this is a path homotopy from α to x_0 in X by Lemma 1.1.2. In this case then $\alpha \simeq x_0 \operatorname{rel}\{0, 1\}$ as required.

The second possibility is that, in combining the t-intervals, one obtains a partition

$$0 = s_0 < s_1 < \cdots < s_{n-1} < s_n = 1$$

of $[0, 1]$ with each $[s_i, s_{i+1}]$ mapping into one of U or V and each s_i in $U \cap V$. For each $i = 0, \ldots, n-1$, let $\alpha_i = \alpha | \; [s_i, s_{i+1}]$. To obtain a path (defined on $[0, 1]$) that accomplishes α_i we reparametrize in the usual way: Define $\tilde{\alpha}_i : [0, 1] \to X$ by $\tilde{\alpha}_i(s) = \alpha_i((s_{i+1} - s_i)s + s_i)$. Then each $\tilde{\alpha}_i$ is a path in either U or V from $\tilde{\alpha}_i(0) = \alpha_i(s_i) = \alpha(s_i)$ to $\tilde{\alpha}_i(1) = \alpha_i(s_{i+1}) = \alpha(s_{i+1})$.

Exercise 2.3.15 Show that $\tilde{\alpha}_0 \tilde{\alpha}_1 \cdots \tilde{\alpha}_{n-1} \simeq \alpha \operatorname{rel}\{0, 1\}$.

Since $U \cap V$ is pathwise connected we may select, for each $i = 0, \ldots, n-1$, a path $\beta_i : [0, 1] \to U \cap V \subseteq X$ from $\beta_i(0) = x_0$ to $\beta_i(1) = \alpha(s_i)$ (see Figure 2.3.7).

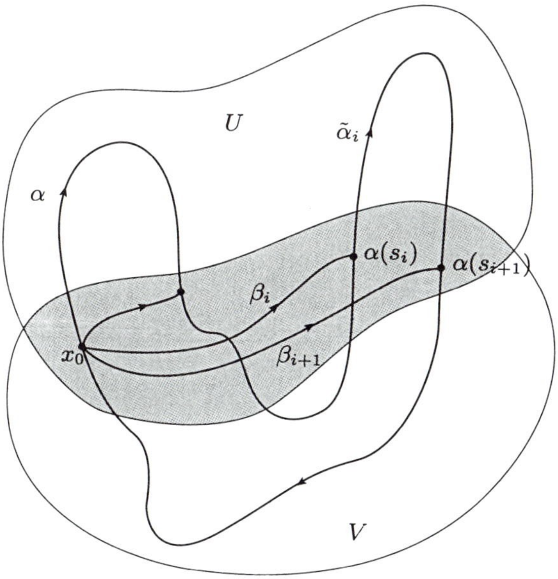

Figure 2.3.7

Thus, each $\beta_i \tilde{\alpha}_i \beta_{i+1}^{\leftarrow}$, $i = 0, \ldots, n-1$, is a loop at x_0 contained entirely in either U or V so $\beta_i \tilde{\alpha}_i \beta_{i+1}^{\leftarrow} \simeq x_0 \operatorname{rel}\{0,1\}$ in U or V and therefore in X.

Exercise 2.3.16 Justify the following sequence of homotopies:

$$\begin{aligned} x_0 &\simeq (\beta_0 \tilde{\alpha}_0 \beta_1^{\leftarrow})(\beta_1 \tilde{\alpha}_1 \beta_2^{\leftarrow}) \cdots (\beta_{n-1} \tilde{\alpha}_{n-1} \beta_n^{\leftarrow}) \operatorname{rel}\{0,1\} \\ &\simeq \beta_0 (\tilde{\alpha}_0 \tilde{\alpha}_1 \cdots \tilde{\alpha}_{n-1}) \beta_n^{\leftarrow} \operatorname{rel}\{0,1\} \\ &\simeq x_0 (\tilde{\alpha}_0 \tilde{\alpha}_1 \cdots \tilde{\alpha}_{n-1}) x_0 \operatorname{rel}\{0,1\} \\ &\simeq x_0 \alpha \, x_0 \operatorname{rel}\{0,1\} \\ &\simeq \alpha \operatorname{rel}\{0,1\}. \end{aligned}$$

■

As an application of Theorem 2.3.10 we show that, for $n \geq 2$, the n-sphere S^n is simply connected. As in Exercise 1.1.8 we denote by N and S the north and south poles of S^n. Observe that $S^n = (S^n - \{N\}) \cup (S^n - \{S\})$ and that, by stereographic projection, $S^n - \{N\}$ and $S^n - \{S\}$ are both homeomorphic to $\mathbb{R}^n$ and therefore are simply connected. Moreover, $(S^n - \{N\}) \cap (S^n - \{S\}) = S^n - \{N, S\}$ which, again by stereographic projection, is homeomorphic to $\mathbb{R}^n - \{0\}$ and this, for $n \geq 2$, is pathwise connected. Theorem 2.3.10 therefore implies that S^n is simply connected for $n \geq 2$.

Before moving on to some spaces whose fundamental groups are *not* trivial we must obtain two useful characterizations of simple connectivity. The first is commonly taken to be obvious in vector calculus, but is surprisingly tricky to prove.

Theorem 2.3.11 *Let X be pathwise connected. Then X is simply connected iff any two paths in X with the same initial and terminal points are path homotopic, i.e., iff, whenever x_0 and x_1 are in X and $\alpha, \beta : [0,1] \to X$ are paths from $x_0 = \alpha(0) = \beta(0)$ to $x_1 = \alpha(1) = \beta(1)$, then $\alpha \simeq \beta \operatorname{rel}\{0,1\}$.*

Proof: Since the sufficiency is obvious we prove only the necessity, i.e., we assume X is simply connected and let α and β be two paths in X from x_0 to x_1. Then $\gamma = \alpha\beta^{\leftarrow}$ is a loop at x_0 so $\gamma \simeq x_0 \operatorname{rel}\{0,1\}$. Let $F : [0,1] \times [0,1] \to X$ be a homotopy with

$$F(s,0) = \gamma(s) = \begin{cases} \alpha(2s), & 0 \leq s \leq \frac{1}{2} \\ \beta^{\leftarrow}(2s-1), & \frac{1}{2} \leq s \leq 1 \end{cases},$$

$F(s,1) = x_0$, $F(0,t) = F(1,t) = x_0$. Now we wish to define a map $h : [0,1] \times [0,1] \to [0,1] \times [0,1]$ as indicated schematically in Figure 2.3.8, where the ratio of the lengths of the segments determined by the dot is $\frac{s}{1-s}$. To gain some feel for what this map is intended to do we sketch a few specific images in Figure 2.3.9.

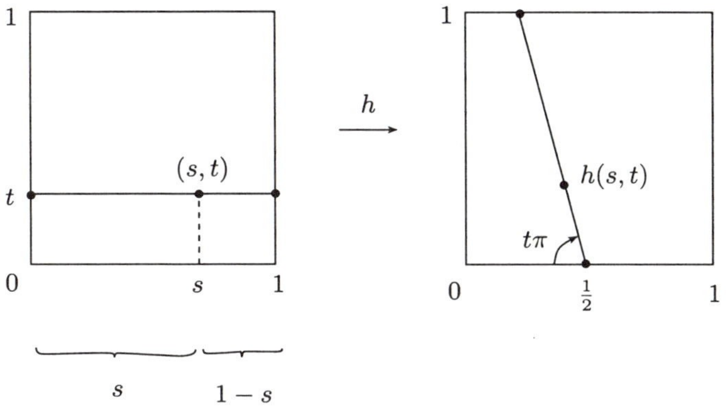

Figure 2.3.8

Exercise 2.3.17 Use Figure 2.3.8 to write out the map h explicitly and show that it is continuous.

Now define $H : [0,1] \times [0,1] \to X$ by $H = F \circ h$. We claim that H is a homotopy relative to $\{0,1\}$ from α to β. First note that, for $0 \le s \le 1$, $H(s,0) = F(h(s,0))$ and that $h(s,0) = (\frac{1}{2}s, 0)$ (because $h(s,0) = (s',0)$, where $\frac{s'}{\frac{1}{2}-s'} = \frac{s}{1-s}$). Thus, $H(s,0) = F(\frac{1}{2}s,0) = \gamma(\frac{1}{2}s) = \alpha(2(\frac{1}{2}s)) = \alpha(s)$. Similarly, $H(s,1) = F(h(s,1)) = F(-\frac{1}{2}s+1,0) = \gamma(-\frac{1}{2}s+1) = \beta^{\leftarrow}(2(-\frac{1}{2}s+1)-1) = \beta^{\leftarrow}(1-s) = \beta(s)$. Next, $H(0,t) = F(h(0,t))$ for $0 \le t \le 1$. But $h(0,t)$ is some point on the upper part of the boundary of $[0,1] \times [0,1]$, i.e., it has one of the forms $(0,t')$ or $(1,t')$ for some $0 \le t' \le 1$ or $(s',1)$ for some $0 \le s' \le 1$. Since F takes the value x_0 at all such points, $H(0,t) = x_0$ for all t in $[0,1]$. Finally, $H(1,t) = F(h(1,t)) = F(\frac{1}{2},0) = \gamma(\frac{1}{2}) = \alpha(1) = \beta^{\leftarrow}(0) = x_1$, as required. ∎

Simple connectivity is a condition phrased in terms of *fixed endpoint* homotopies (path homotopies or, by Lemma 2.3.4, homotopies relative to some fixed point p_0 in S^1). We show next that this notion can, nevertheless, be characterized in terms of *free* homotopies.

Exercise 2.3.18 Let X be any topological space, $n \ge 1$ and $f : S^n \to X$ a continuous map. Show that f is nullhomotopic iff it has a continuous extension to the disc D^{n+1}, i.e., iff there exists a continuous map $\tilde{f} : D^{n+1} \to X$ whose restriction to the boundary $\partial D^{n+1} = S^n$ is f. **Hint:** In each direction the relationship between $\tilde{f}$ and $F : S^n \times [0,1] \to X$ is given by $F(p,t) = \tilde{f}((1-t)p)$.

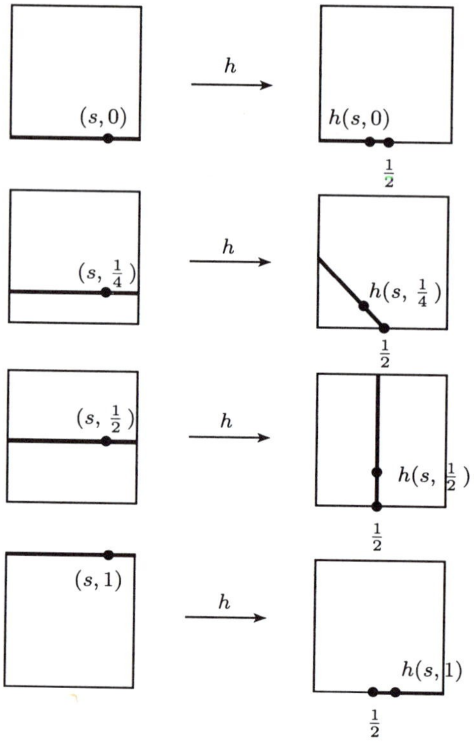

Figure 2.3.9

Theorem 2.3.12 *Let X be pathwise connected. Then the following are equivalent:*

(a) X is simply connected.

(b) Every continuous map of S^1 into X is nullhomotopic.

(c) Every continuous map of S^1 into X has a continuous extension to the disc D^2.

Proof: The equivalence of (b) and (c) follows from the $n = 1$ case of Exercise 2.3.18. For the remainder of the proof we will make use of the map $\mathcal{Q} : [0,1] \times [0,1] \to D^2$ defined by $\mathcal{Q}(s,t) = te^{2\pi s \mathbf{i}}$. Observe that $\mathcal{Q}$ is continuous, surjective and carries $[0,1] \times \{1\}$ onto $\partial D^2 = S^1$. Moreover,

$[0,1] \times [0,1]$ is compact so $\mathcal{Q}$ is a closed mapping and therefore D^2 has the quotient topology determined by $\mathcal{Q}$ (Exercise 1.3.13). In particular, a map out of D^2 is continuous iff its composition with $\mathcal{Q}$ is continuous (Lemma 1.2.1).

Now, suppose that X is simply connected and let f be an arbitrary continuous map of $S^1 = \partial D^2$ into X. We show that f has a continuous extension to D^2. Define $\alpha : [0,1] \to X$ by $\alpha(s) = f(\mathcal{Q}(s,1))$. Then α is continuous, $\alpha(0) = f(\mathcal{Q}(0,1)) = f(1 \cdot e^{2\pi 0 \mathbf{i}}) = f(1)$ and $\alpha(1) = f(\mathcal{Q}(1,1)) = f(1 \cdot e^{2\pi 1 \mathbf{i}}) = f(1) = \alpha(0)$ so α is a loop at $x_0 = f(1)$ in X. Since $\pi_1(X, x_0)$ is trivial, there exists a path homotopy $F : [0,1] \times [0,1] \to X$ with $F(s,1) = \alpha(s)$ and $F(s,0) = F(0,t) = F(1,t) = x_0$ for all $s, t \in [0,1]$. Now define $\tilde{f} : D^2 \to X$ by $\tilde{f}(\mathcal{Q}(s,t)) = F(s,t)$ for all $s, t \in [0,1]$. Thus, $\tilde{f} \circ \mathcal{Q} = F$ so, since F is continuous and D^2 has the quotient topology determined by $\mathcal{Q}$, $\tilde{f}$ is continuous. Moreover, $\tilde{f} \mid S^1$ is given by $\tilde{f}(\mathcal{Q}(s,1)) = F(s,1) = \alpha(s) = f(\mathcal{Q}(s,1))$ so $\tilde{f} \mid S^1 = f$ as required.

Next we suppose that every continuous map from S^1 to X has a continuous extension to D^2. Let $\alpha : [0,1] \to X$ be a loop in X and set $x_0 = \alpha(0) = \alpha(1)$. We show that $\alpha \simeq x_0 \operatorname{rel} \{0,1\}$. Define $f : S^1 \to X$ by $f(e^{2\pi s \mathbf{i}}) = f(\mathcal{Q}(s,1)) = \alpha(s)$.

Exercise 2.3.19 Show that f is continuous.

By assumption then f has a continuous extension $\tilde{f} : D^2 \to X$ to D^2. Define $F : [0,1] \times [0,1] \to X$ by $F(s,t) = \tilde{f}(\mathcal{Q}(s,t))$, i.e., $F = \tilde{f} \circ \mathcal{Q}$. Then F is continuous and, moreover,

$$\begin{aligned} F(s,1) &= \tilde{f}\left(\mathcal{Q}(s,1)\right) = \tilde{f}(1 \cdot e^{2\pi s \mathbf{i}}) = f(e^{2\pi s \mathbf{i}}) = \alpha(s) \\ F(s,0) &= \tilde{f}\left(\mathcal{Q}(s,0)\right) = \tilde{f}(0 \cdot e^{2\pi s \mathbf{i}}) = \tilde{f}(0) \\ F(0,t) &= \tilde{f}\left(\mathcal{Q}(0,t)\right) = \tilde{f}(te^{2\pi 0 \mathbf{i}}) = \tilde{f}(te^{2\pi 1 \mathbf{i}}) = F(1,t)\,. \end{aligned}$$

Letting $\gamma(t) = F(0,t) = F(1,t)$ for $0 \le t \le 1$ and denoting by $\tilde{f}(0)$ the constant path in X at $\tilde{f}(0)$ we conclude from Lemma 2.3.5 that $\alpha \simeq \gamma^{\leftarrow}(\tilde{f}(0))\gamma \operatorname{rel} \{0,1\}$.

Exercise 2.3.20 Complete the proof by showing that $\gamma^{\leftarrow}(\tilde{f}(0))\gamma \simeq x_0 \operatorname{rel} \{0,1\}$. ∎

2.4 The Covering Homotopy Theorem

Calculating fundamental groups that are not trivial is, well, not trivial. A number of powerful techniques have been devised, one of which (the Seifert-Van Kampen Theorem) we have already seen a special case of in Theorem

2.3.10. The use of covering spaces is also quite efficacious, primarily because these have what is known as a "covering homotopy property". Since this property will also figure heavily in our study of higher homotopy groups and principal bundles, we intend to derive at once a result general enough to serve all of our needs.

We consider a locally trivial bundle $(P, X, \mathcal{P}, Y)$ and a continuous map $f : Z \to X$ from some space Z into the base X. We have already observed (Section 1.5) that f may or may not have a lift to P, i.e., there may or may not exist a continuous map $\tilde{f} : Z \to B$ for which $\mathcal{P} \circ \tilde{f} = f$. Suppose f does, in fact, lift to P. Generally speaking, we are interested in whether or not the same must be true of maps homotopic to f. More specifically, we pose the following question: Suppose $F : Z \times [0,1] \to Z$ is a homotopy beginning at $F(z,0) = f(z)$. Does F "lift" to a homotopy $\tilde{F} : Z \times [0,1] \to P$ that begins at $\tilde{F}(z,0) = \tilde{f}(z)$ and satisfies $\mathcal{P} \circ \tilde{F} = F$? If this is the case for every map of Z into X that lifts to P, then $(P, X, \mathcal{P}, Y)$ is said to have the **homotopy lifting property** for the space Z. Although a great deal more can be proved (see Corollary 14, Section 7, Chapter 2 of [**Gra**]), it will suffice for our purposes to show that any locally trivial bundle has the homotopy lifting property for $I^n = [0,1]^n = [0,1] \times \cdots \times [0,1]$, $n = 1, 2, \ldots$ In contemporary parlance, what we prove in our next result is that a locally trivial bundle is a **Serre fibration**.

Theorem 2.4.1 (Homotopy Lifting Theorem) *Let $(P, X, \mathcal{P}, Y)$ be a locally trivial bundle and n a positive integer. Suppose $f : I^n \to X$ is a continuous map that lifts to $\tilde{f} : I^n \to P$ and $F : I^n \times [0,1] \to X$ is a homotopy with $F(x,0) = f(x)$ for each $x \in I^n$. Then there exists a homotopy $\tilde{F} : I^n \times [0,1] \to P$ such that $\mathcal{P} \circ \tilde{F} = F$ and $\tilde{F}(x,0) = \tilde{f}(x)$ for each $x \in I^n$.*

Proof: In order to simplify the notation we will, throughout the proof, regard $\tilde{f}$ as a map on $I^n \times \{0\} \subseteq I^n \times [0,1]$ in the obvious way. Thus, we have a commutative diagram

$$\begin{array}{ccc} I^n \times \{0\} & \xrightarrow{\ \tilde{f}\ } & P \\ \big\downarrow & & \big\downarrow \mathcal{P} \\ I^n \times [0,1] & \xrightarrow[\ F\ ]{} & X \end{array}$$

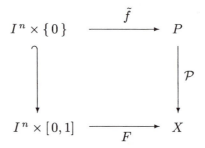

(where the first vertical mapping is inclusion) and our objective is to construct a continuous map $\tilde{F} : I^n \times [0,1] \to P$ such that the following diagram also commutes:

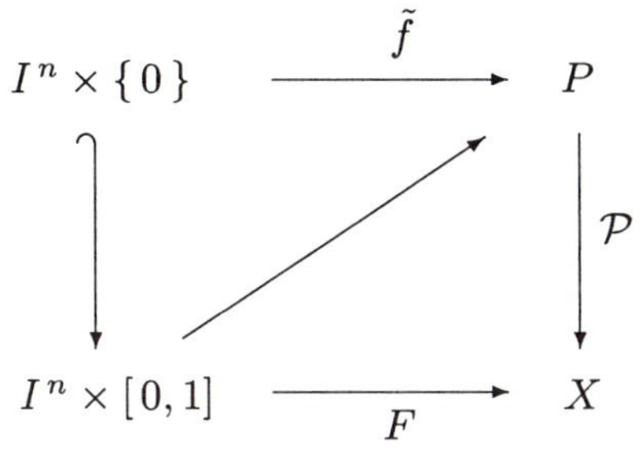

Cover X by trivializing nbds V_j with trivializations $\Phi_j : V_j \times Y \to \mathcal{P}^{-1}(V_j)$. Then $\{F^{-1}(V_j)\}$ is an open cover of $I^n \times [0,1]$, which is compact.

Exercise 2.4.1 Show that one can select a finite open cover $\{U_\lambda \times I_\nu\}$ of $I^n \times [0,1]$ with each $U_\lambda \times I_\nu$ contained in some $F^{-1}(V_j)$ and where $\{U_\lambda\}$ is an open cover of I^n and $\{I_\nu\}_{\nu=1}^r$ is a finite sequence of (relatively) open intervals in $[0,1]$ that cover $[0,1]$ and such that each I_ν intersects only $I_{\nu-1}$ and $I_{\nu+1}$ for $\nu = 2, \ldots, r-1$.

Choose numbers $0 = t_0 < t_1 < \cdots < t_r = 1$ such that $t_\nu \in I_\nu \cap I_{\nu+1}$ for each $\nu = 1, \ldots, r-1$. We will inductively define, for each $\nu = 0, \ldots, r$, a map $\tilde{F}_\nu : I^n \times [0, t_\nu] \to P$ such that $\tilde{F}_\nu \mid I^n \times \{0\} = \tilde{f}$ and $\mathcal{P} \circ \tilde{F}_\nu = F \mid I^n \times [0, t_\nu]$. This done, $\tilde{F} = \tilde{F}_r$ will be the required homotopy. Since $t_0 = 0$, so that $I^n \times [0, t_0] = I^n \times \{0\}$, we may start the induction at $\nu = 0$ by setting $\tilde{F}_0 = \tilde{f}$. Now suppose that $\tilde{F}_\nu$ has been defined. To define $\tilde{F}_{\nu+1}$ we must extend $\tilde{F}_\nu$ continuously to $I^n \times [t_\nu, t_{\nu+1}]$ in such a way that, on this set, $\mathcal{P} \circ \tilde{F}_{\nu+1} = F$.

Exercise 2.4.2 Show that, for each $x \in I^n$, one can select a pair of open sets W and W' in I^n such that

$$x \in W \subseteq \overline{W} \subseteq W' \subseteq \overline{W'} \subseteq U_\lambda$$

for some λ.

By compactness of I^n one can choose a finite number of such pairs W_α, W'_α, $\alpha = 1, \ldots, s$, with the W_α covering I^n. Now, for each $\alpha = 1, \ldots, s$, we wish to select a continuous function

$$\mu_\alpha : I^n \longrightarrow [t_\nu, t_{\nu+1}]$$

such that

$$\overline{W}_\alpha = \mu_\alpha^{-1}(t_{\nu+1})$$
$$I^n - W'_\alpha = \mu_\alpha^{-1}(t_\nu)$$

for each α. That such functions exist is not obvious. Although we could easily prove what we need at this point, we will eventually obtain a much stronger (C^∞) result (Exercise 4.1.5) so we prefer to defer the argument. The reader who wishes to see the proof now should proceed directly to Section 4.1 (it does not depend on any of the intervening material). Now define functions $\tau_0, \tau_1, \ldots, \tau_s$ on I^n as follows:

$$\tau_0(x) = t_\nu \quad \text{for all} \quad x \in I^n\,,$$

and, for $\alpha = 1, \ldots, s$,

$$\tau_\alpha(x) = \max\{\mu_1(x), \ldots, \mu_\alpha(x)\}$$

for all $x \in I^n$.

Exercise 2.4.3 Show that $\tau_0, \tau_1, \ldots, \tau_s$ are all continuous and that, for each $x \in I^n$,

$$t_\nu = \tau_0(x) \le \tau_1(x) \le \cdots \le \tau_{s-1}(x) \le \tau_s(x) = t_{\nu+1}\,.$$

Next, define subsets $X_0, X_1, \ldots, X_s$ of $I^n \times [t_\nu, t_{\nu+1}]$ by

$$\begin{aligned} X_\alpha &= \{\,(x,t) \in I^n \times [0,1] :\ t_\nu \le t \le \tau_\alpha(x)\,\} \\ &= \{\,(x,t) \in I^n \times [t_\nu, t_{\nu+1}] :\ t_\nu \le t \le \tau_\alpha(x)\,\}\,. \end{aligned}$$

Note that $X_0 = I^n \times \{t_\nu\}$, $X_s = I^n \times [t_\nu, t_{\nu+1}]$ and, for each $\alpha = 1, \ldots, s$,

$$X_\alpha - X_{\alpha-1} = \{\,(x,t) \in I^n \times [t_\nu, t_{\nu+1}] :\ \tau_{\alpha-1}(x) < t \le \tau_\alpha(x)\,\}\,.$$

We claim that, for each $\alpha = 1, \ldots, s$,

$$X_\alpha - X_{\alpha-1} \subseteq W'_\alpha \times [t_\nu, t_{\nu+1}]\,. \tag{2.4.1}$$

Indeed, $(x,t) \in X_\alpha - X_{\alpha-1}$ implies

$$\max\{\mu_1(x), \ldots, \mu_{\alpha-1}(x)\} < t \le \max\{\mu_1(x), \ldots, \mu_{\alpha-1}(x), \mu_\alpha(x)\}$$

which implies $t_\nu < t \le \mu_\alpha(x)$ so $\mu_\alpha(x) \ne t_\nu$. Thus, x is not in $\mu_\alpha^{-1}(t_\nu) = I^n - W'_\alpha$, i.e., $x \in W'_\alpha$ so $(x,t) \in W'_\alpha \times [t_\nu, t_{\nu+1}]$.

Now, since $\overline{W'_\alpha}$ is contained in some U_λ and $[t_\nu, t_{\nu+1}]$ is contained in I_ν, (2.4.1) implies that $\overline{X_\alpha - X_{\alpha-1}} \subseteq \overline{W'_\alpha} \times [t_\nu, t_{\nu+1}] \subseteq U_\lambda \times I_\nu$. But $U_\lambda \times I_\nu$ is contained in some $F^{-1}(V_j)$. Thus, for each $\alpha = 1, \ldots, s$, there exists a j such that

$$F\left(\,\overline{X_\alpha - X_{\alpha-1}}\,\right) \subseteq V_j\,. \tag{2.4.2}$$

At this point we have $I^n \times [t_\nu, t_{\nu+1}]$ carved up into a finite sequence of closed subsets

$$X_0 = I^n \times \{t_\nu\} \subseteq X_1 \subseteq \cdots \subseteq X_{s-1} \subseteq X_s = I^n \times [t_\nu, t_{\nu+1}]$$

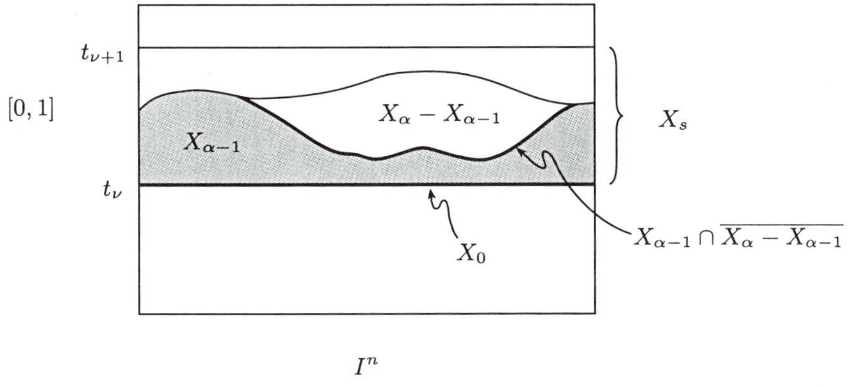

Figure 2.4.1

which satisfy (2.4.2) (see Figure 2.4.1). We define $\tilde{F}_{\nu+1}$ by inductively extending $\tilde{F}_\nu$ over this sequence of subsets. For this we let $X = I^n \times [0, t_\nu]$ (the domain of $\tilde{F}_\nu$) and define $\tilde{F}_{\nu+1} \,|\, X \cup X_0, \ldots, \tilde{F}_{\nu+1} \,|\, X \cup X_s$ inductively. Since $X \cup X_s = I^n \times [0, t_{\nu+1}]$, the last map will be the one we require. To start the induction note that $X \cup X_0 = X$ so we may take $\tilde{F}_{\nu+1} \,|\, X \cup X_0 = \tilde{F}_\nu$. Now suppose $\tilde{F}_{\nu+1} \,|\, X \cup X_{\alpha-1}$ has been defined so that $\tilde{F}_{\nu+1} \,|\, I^n \times \{0\} = \tilde{f}$ and, on $X \cup X_{\alpha-1}$, $\mathcal{P} \circ \tilde{F}_{\nu+1} = F$. To define $\tilde{F}_{\nu+1}$ on X_α observe that, since X_α and $X_{\alpha-1}$ are closed sets,

$$X_\alpha = X_{\alpha-1} \cup \overline{X_\alpha - X_{\alpha-1}}\,.$$

Thus, we need only define $\tilde{F}_{\nu+1}$ on $\overline{X_\alpha - X_{\alpha-1}}$ in such a way as to agree with $\tilde{F}_{\nu+1} \mid X \cup X_{\alpha-1}$ on $X_{\alpha-1} \cap \overline{X_\alpha - X_{\alpha-1}}$ (again, see Figure 2.4.1). This we will accomplish by using the fact that $\overline{X_\alpha - X_{\alpha-1}}$ maps into some trivializing nbd V_j under F (this will permit the lifting of F) and by constructing a continuous mapping of $\overline{X_\alpha - X_{\alpha-1}}$ onto $X_{\alpha-1} \cap \overline{X_\alpha - X_{\alpha-1}}$ that is the identity on $X_{\alpha-1} \cap \overline{X_\alpha - X_{\alpha-1}}$ (this will enable us to force agreement with $\tilde{F}_{\nu+1} \,|\, X \cup X_{\alpha-1}$ on this set).

To carry out the plan just outlined we need first a continuous map $r : \overline{X_\alpha - X_{\alpha-1}} \to X_{\alpha-1} \cap \overline{X_\alpha - X_{\alpha-1}}$ that is the identity on $X_{\alpha-1} \cap \overline{X_\alpha - X_{\alpha-1}}$ (such a map is called a "retraction" of $\overline{X_\alpha - X_{\alpha-1}}$ onto $X_{\alpha-1} \cap \overline{X_\alpha - X_{\alpha-1}}$ and we will have much to say about this type of mapping shortly). For each $(x, t) \in \overline{X_\alpha - X_{\alpha-1}}$ we define $r(x, t) = (x, \tau_{\alpha-1}(x))$.

Then $r(x,t)$ is clearly in $X_{\alpha-1}$ and we claim that it is in $\overline{X_\alpha - X_{\alpha-1}}$ also. By continuity of r it will suffice to show that if $(x,t) \in X_\alpha - X_{\alpha-1}$, then $r(x,t) \in \overline{X_\alpha - X_{\alpha-1}}$. But $(x,t) \in X_\alpha - X_{\alpha-1}$ implies $\tau_{\alpha-1}(x) < t$. For each $n = 1, 2, \ldots$ define $t_n = \tau_{\alpha-1}(x) + \frac{1}{2^n}(t - \tau_{\alpha-1}(x))$. Then $(x, t_n) \in X_\alpha - X_{\alpha-1}$ for each n and $(x, t_n) \to (x, \tau_{\alpha-1}(x)) = r(x,t)$ so $r(x,t) \in \overline{X_\alpha - X_{\alpha-1}}$ as required. Thus,

$$r : \overline{X_\alpha - X_{\alpha-1}} \longrightarrow X_{\alpha-1} \cap \overline{X_\alpha - X_{\alpha-1}}$$

is a continuous map and, moreover, if $(x,t) \in X_{\alpha-1} \cap \overline{X_\alpha - X_{\alpha-1}}$, then $r(x,t) = (x,t)$.

Exercise 2.4.4 Prove this last assertion.

Now we can finish off the proof by defining $\tilde{F}_{\nu+1}$ on $\overline{X_\alpha - X_{\alpha-1}}$ as follows: Choose a trivializing nbd V_j in X containing $F(\overline{X_\alpha - X_{\alpha-1}})$ (this is possible by (2.4.2)). Let $\Phi_j : V_j \times Y \to \mathcal{P}^{-1}(V_j)$ be a local trivialization on V_j and let $r : \overline{X_\alpha - X_{\alpha-1}} \to X_{\alpha-1} \cap \overline{X_\alpha - X_{\alpha-1}}$ be as above. Fix $(x,t) \in \overline{X_\alpha - X_{\alpha-1}}$. Then $F(x,t) \in V_j$ and $\tilde{F}_{\nu+1}(r(x,t))$ is in $\mathcal{P}^{-1}(V_j)$ because $\mathcal{P} \circ \tilde{F}_{\nu+1}(r(x,t)) = F(r(x,t)) \subseteq F(\overline{X_\alpha - X_{\alpha-1}}) \subseteq V_j$. Thus, $\Phi_j^{-1} \circ (\tilde{F}_{\nu+1} \mid X \cup X_{\alpha-1})(r(x,t))$ is in $V_j \times Y$ so its projection $\mathcal{P}_Y \circ \Phi_j^{-1} \circ (\tilde{F}_{\nu+1} | X \cup X_{\alpha-1})(r(x,t))$ is in Y. Define

$$\begin{aligned} &\tilde{F}_{\nu+1}(x,t) = \\ &\Phi_j\left(F(x,t),\ \mathcal{P}_Y \circ \Phi_j^{-1} \circ (\tilde{F}_{\nu+1} | X \cup X_{\alpha-1})\,(r(x,t))\right). \end{aligned} \tag{2.4.3}$$

Observe that for $(x,t) \in X_{\alpha-1} \cap \overline{X_\alpha - X_{\alpha-1}}$, $r(x,t) = (x,t)$ and $F(x,t) = \mathcal{P} \circ (\tilde{F}_{\nu+1} | X \cup X_{\alpha-1})(x,t)$ so the right-hand side of (2.4.3) becomes

$$\begin{aligned} &\Phi_j\left(\mathcal{P}((\tilde{F}_{\nu+1} | X \cup X_{\alpha-1})(x,t)),\ (\mathcal{P}_Y \circ \Phi_j^{-1})((\tilde{F}_{\nu+1} | X \cup X_{\alpha-1})(x,t))\right) \\ &\quad = \Phi_j\left(\Phi_j^{-1}((\tilde{F}_{\nu+1} | X \cup X_{\alpha-1})(x,t))\right) \\ &\quad = (\tilde{F}_{\nu+1} | X \cup X_{\alpha-1})(x,t) \end{aligned}$$

so (2.4.3) reduces to $\tilde{F}_{\nu+1} | X \cup X_{\alpha-1}$ on $X_{\alpha-1} \cap \overline{X_\alpha - X_{\alpha-1}}$ and this (mercifully) completes the proof. ∎

Since paths $\alpha : [0,1] \to X$ into the base of a locally trivial bundle always lift to the total space (Theorem 1.5.10), the Homotopy Lifting Theorem implies that any path homotopy beginning at α also lifts. This is particularly significant in the case of covering spaces where paths lift uniquely once an initial point is specified.

Corollary 2.4.2 *Let $\mathcal{P} : \tilde{X} \to X$ be a covering space, x_0 a point in X and $\tilde{x}_0$ a point in $\mathcal{P}^{-1}(x_0)$. Let $\alpha, \beta : [0,1] \to X$ be paths in X with $\alpha(0) = \beta(0) = x_0$ and suppose $F : [0,1] \times [0,1] \to X$ is a path homotopy from α to β (so that, in particular, $\alpha(1) = \beta(1)$). If $\tilde{\alpha}, \tilde{\beta} : [0,1] \to \tilde{X}$ are the unique lifts of α and β with $\tilde{\alpha}(0) = \tilde{\beta}(0) = \tilde{x}_0$, then there exists a path homotopy $\tilde{F} : [0,1] \times [0,1] \to \tilde{X}$ from $\tilde{\alpha}$ to $\tilde{\beta}$ with $\mathcal{P} \circ \tilde{F} = F$. In particular, $\tilde{\alpha}(1) = \tilde{\beta}(1)$.*

Proof: According to Theorem 2.4.1 (with $n = 1$) there exists a homotopy $\tilde{F} : [0,1] \times [0,1] \to \tilde{X}$ such that $\mathcal{P} \circ \tilde{F} = F$ and $\tilde{F}(s,0) = \tilde{\alpha}(s)$. We show that $\tilde{F}$ is actually a path homotopy (i.e., relative to $\{0,1\}$) and ends at $\tilde{\beta}$. First note that $\mathcal{P} \circ \tilde{F}(0,t) = F(0,t) = x_0$ for all t so $t \to \tilde{F}(0,t)$ is a path in $\mathcal{P}^{-1}(x_0)$. But $\mathcal{P}^{-1}(x_0)$ is discrete so this path must be constant. Since $\tilde{F}(0,0) = \tilde{\alpha}(0) = \tilde{x}_0$, we have $\tilde{F}(0,t) = \tilde{x}_0$ for all $t \in [0,1]$. Next observe that $\mathcal{P} \circ \tilde{F}(s,1) = F(s,1) = \beta(s)$ so $s \to \tilde{F}(s,1)$ is a path in $\tilde{X}$ that lifts β. Since we have just shown that $\tilde{F}(0,1) = \tilde{x}_0$, this lift of β begins at $\tilde{x}_0$ and so, by uniqueness, must coincide with $\tilde{\beta}$, i.e., $\tilde{F}(s,1) = \tilde{\beta}(s)$ for every $s \in [0,1]$.

Exercise 2.4.5 Complete the proof by showing that $\tilde{F}(1,t) = \tilde{\alpha}(1) = \tilde{\beta}(1)$ for all $t \in [0,1]$. ∎

It is interesting that the Homotopy Lifting Theorem 2.4.1 can be used to show that one of its own hypotheses is superfluous.

Corollary 2.4.3 *Let $(P, X, \mathcal{P}, Y)$ be a locally trivial bundle and n a positive integer. Then any continuous map $f : I^n \to X$ lifts to a continuous map $\tilde{f} : I^n \to P$ with $\mathcal{P} \circ \tilde{f} = f$.*

Proof: I^n is convex and therefore contractible. By Exercise 2.3.6, f is nullhomotopic. Let $F : I^n \times [0,1] \to X$ be a homotopy with $F(x,0) = x_0$ and $F(x,1) = f(x)$ for all $x \in I^n$. Now, the constant map x_0 of I^n to X obviously lifts to P so Theorem 2.4.1 implies that F also lifts to $\tilde{F} : I^n \times [0,1] \to P$. Defining $\tilde{f} : I^n \to P$ by $\tilde{f}(x) = \tilde{F}(x,1)$, one obtains the required lift of f. ∎

Corollary 2.4.2 is the key to calculating $\pi_1(S^1)$. We fix the base point $1 = e^{2\pi 0 \mathbf{i}} = e^{2\pi 1 \mathbf{i}}$ in S^1 and compute $\pi_1(S^1, 1)$. Recall (Section 1.5) that the map $\mathcal{P} : \mathbb{R} \to S^1$ given by $\mathcal{P}(s) = e^{2\pi s \mathbf{i}}$ is a covering space. Thus, each *loop* α at 1 in S^1 has a unique lift to a *path* $\tilde{\alpha}$ in $\mathbb{R}$ beginning at $\tilde{\alpha}(0) = 0 \in \mathbb{R}$. Now, $\tilde{\alpha}$ need not be a loop in $\mathbb{R}$, i.e., $\tilde{\alpha}(1)$ need not be $0 \in \mathbb{R}$. However, $\mathcal{P} \circ \tilde{\alpha}(1) = \alpha(1) = 1 \in S^1$ so $\tilde{\alpha}(1)$ is in $\mathcal{P}^{-1}(1)$. But $\mathcal{P}^{-1}(1)$ is just the set $\mathbb{Z}$ of integers in $\mathbb{R}$ so every loop α at 1 in S^1 lifts uniquely to a path $\tilde{\alpha}$ in $\mathbb{R}$ from 0 to some integer $\tilde{\alpha}(1)$. Since $\mathcal{P} : \mathbb{R} \to S^1$ essentially wraps $\mathbb{R}$

around itself to produce S^1, this integer $\tilde{\alpha}(1)$ effectively measures the net number of times α wraps around S^1 (a complete revolution being positive if the wrapping is counterclockwise and negative if clockwise). The essential point is that this integer is a *homotopy invariant.* That is, if α' is a loop at 1 in S^1 that is path homotopic to α and $\tilde{\alpha}'$ is its lift to $\mathbb{R}$ with $\tilde{\alpha}'(0) = 0$, then $\tilde{\alpha}'(1) = \tilde{\alpha}(1)$ (Corollary 2.4.2). Thus, we may define a mapping

$$\deg : \pi_1(S^1, 1) \longrightarrow \mathbb{Z}$$

by

$$\deg([\alpha]) = \tilde{\alpha}(1), \tag{2.4.4}$$

where $\tilde{\alpha}$ is the unique lift of α to $\mathbb{R}$ that begins at $\tilde{\alpha}(0) = 0 \in \mathbb{R}$. Now, $\mathbb{Z}$ is a group under addition and we propose to show next that deg is, in fact, an isomorphism onto this group.

Theorem 2.4.4 $\pi_1(S^1) \cong \mathbb{Z}$.

Proof: We show that the map deg defined by (2.4.4) is an isomorphism. First let $[\alpha], [\beta] \in \pi_1(S^1, 1)$. Let $\deg([\alpha]) = \tilde{\alpha}(1) = m$ and $\deg([\beta]) = \tilde{\beta}(1) = n$. Then $\deg([\alpha]) + \deg([\beta]) = m + n$ and we must show that $\deg([\alpha][\beta]) = m + n$ also. But $\deg([\alpha][\beta]) = \deg([\alpha\beta]) = \widetilde{\alpha\beta}(1)$ so we must determine the lift of $\alpha\beta$ to $\mathbb{R}$ starting at $0 \in \mathbb{R}$. Now, $\tilde{\alpha}$ is a path in $\mathbb{R}$ from 0 to m and $\tilde{\beta}$ is a path in $\mathbb{R}$ from 0 to n. Define $\gamma : [0,1] \to \mathbb{R}$ by $\gamma(s) = m + \tilde{\beta}(s)$. Then γ is a path in $\mathbb{R}$ from m to $m + n$. Moreover, $\mathcal{P} \circ \gamma(s) = \mathcal{P}(m + \tilde{\beta}(s)) = e^{2\pi(m+\tilde{\beta}(s))\mathbf{i}} = e^{2\pi m\mathbf{i}} e^{2\pi\tilde{\beta}(s)\mathbf{i}} = e^{2\pi\tilde{\beta}(s)\mathbf{i}} = \mathcal{P}(\tilde{\beta}(s)) = \mathcal{P} \circ \tilde{\beta}(s) = \beta(s)$. Thus,

$$\begin{aligned}
\mathcal{P} \circ (\tilde{\alpha}\gamma)(s) &= \begin{cases} \mathcal{P}(\tilde{\alpha}(2s)), & 0 \le s \le \frac{1}{2} \\ \mathcal{P}(\gamma(2s-1)), & \frac{1}{2} \le s \le 1 \end{cases} \\
&= \begin{cases} \alpha(2s), & 0 \le s \le \frac{1}{2} \\ \beta(2s-1), & \frac{1}{2} \le s \le 1 \end{cases} \\
&= (\alpha\beta)(s).
\end{aligned}$$

Consequently, $\tilde{\alpha}\gamma$ is a lift of $\alpha\beta$ to $\mathbb{R}$. Moreover, $\tilde{\alpha}\gamma(0) = \tilde{\alpha}(0) = 0$ so $\tilde{\alpha}\gamma$ begins at $0 \in \mathbb{R}$. By uniqueness, $\tilde{\alpha}\gamma = \widetilde{\alpha\beta}$. Thus, $\deg([\alpha][\beta]) = \widetilde{\alpha\beta}(1) = \tilde{\alpha}\gamma(1) = \gamma(1) = m + \tilde{\beta}(1) = m + n$ as required so we have shown that deg is a homomorphism.

All that remains is to show that deg is a bijection. It is onto since, given any $n \in \mathbb{Z}$, we may define $\tilde{\alpha} : [0,1] \to \mathbb{R}$ by $\tilde{\alpha}(s) = ns$ and then $\mathcal{P} \circ \tilde{\alpha}(s) = \mathcal{P}(ns) = e^{2\pi(ns)\mathbf{i}}$ is a loop at $1 \in S^1$ with $\deg([\mathcal{P} \circ \tilde{\alpha}]) = \tilde{\alpha}(1) = n$. Finally, we show that deg is injective by showing that it has trivial kernel, i.e., that $\deg([\alpha]) = 0$ implies $[\alpha]$ is the homotopy class of the constant loop at 1 in S^1. Thus, suppose $\deg([\alpha]) = 0$. Then $\tilde{\alpha}(1) = 0$ so $\tilde{\alpha}$ is a loop at 0 in $\mathbb{R}$. Now, $\mathbb{R}$ is contractible and therefore simply connected so $\tilde{\alpha} \simeq 0 \operatorname{rel}\{0,1\}$, i.e., $[\tilde{\alpha}] = [0]$ in $\pi_1(\mathbb{R}, 0)$. Thus, applying the induced homomorphism $\mathcal{P}_* : \pi_1(\mathbb{R}, 0) \to \pi_1(S^1, 1)$ gives $\mathcal{P}_*([\tilde{\alpha}]) = \mathcal{P}_*([0])$, i.e., $[\mathcal{P} \circ \tilde{\alpha}] = [\mathcal{P} \circ 0]$ in $\pi_1(S^1, 1)$. But $[\mathcal{P} \circ \tilde{\alpha}] = [\alpha]$ and $[\mathcal{P} \circ 0] = [1]$ so the result follows. ∎

One can generalize this discussion of $\pi_1(S^1)$ just a bit to obtain another example. Let $\mathcal{P} : \tilde{X} \to X$ be any covering space, $x_0 \in X$ and $\tilde{x}_0 \in \mathcal{P}^{-1}(x_0)$. Define a map $\varphi : \pi_1(X, x_0) \to \mathcal{P}^{-1}(x_0)$ by $\varphi([\alpha]) = \tilde{\alpha}(1)$, where $\tilde{\alpha}$ is the unique lift of α to $\tilde{X}$ satisfying $\tilde{\alpha}(0) = \tilde{x}_0$. This map is well-defined by Corollary 2.4.2 and we will show that, if $\tilde{X}$ is simply connected, then it is one-to-one. To do so we produce an inverse for φ. Let p be any point in $\mathcal{P}^{-1}(x_0)$. Select a path σ in $\tilde{X}$ from $\tilde{x}_0$ to p. Then $\mathcal{P} \circ \sigma$ is a loop at x_0 in X. Moreover, if σ' is any other path in $\tilde{X}$ from $\tilde{x}_0$ to p, then, by simple connectivity of $\tilde{X}$, $\sigma' \simeq \sigma \operatorname{rel}\{0,1\}$ (Theorem 2.3.11) so $\mathcal{P} \circ \sigma' \simeq \mathcal{P} \circ \sigma \operatorname{rel}\{0,1\}$, i.e., $[\mathcal{P} \circ \sigma'] = [\mathcal{P} \circ \sigma]$. Thus, we may define a map $\phi : \mathcal{P}^{-1}(x_0) \to \pi_1(X, x_0)$ by $\phi(p) = [\mathcal{P} \circ \sigma]$, where σ is any path in $\tilde{X}$ from $\tilde{x}_0$ to p.

Exercise 2.4.6 Show that $\varphi \circ \phi$ and $\phi \circ \varphi$ are the identities on $\mathcal{P}^{-1}(x_0)$ and $\pi_1(X, x_0)$ respectively.

Thus, when $\tilde{X}$ is simply connected, $\pi_1(X, x_0)$ is in one-to-one correspondence with $\mathcal{P}^{-1}(x_0)$. Consider, for example, the covering space $\mathcal{P} : S^{n-1} \to \mathbb{R}\mathbb{P}^{n-1}$ (Sections 1.3 and 1.5). For $n \geq 3$, S^{n-1} is simply connected so, for any $x_0 \in \mathbb{R}\mathbb{P}^{n-1}$, $\pi_1(\mathbb{R}\mathbb{P}^{n-1}, x_0)$ is in one-to-one correspondence with $\mathcal{P}^{-1}(x_0)$. But $\mathcal{P}^{-1}(x_0)$ consists of just two points and there is only one 2-element group so we have proved the following result.

Theorem 2.4.5 $\pi_1(\mathbb{R}\mathbb{P}^{n-1}) \cong \mathbb{Z}_2$ *for every* $n \geq 3$.

Thus, we (finally) have our first examples of nontrivial fundamental groups. Our patience and persistence are now rewarded with a sequence of beautiful applications that illustrate in its purest form the power of attaching algebraic invariants to topological spaces. First, a definition. Let A be a subset of a topological space X. Then A is said to be a **retract** of X if there exists a continuous map $r : X \to A$, called a **retraction** of X onto A, such that $r \,|\, A = id_A$. For example, S^1 is a retract of $\mathbb{R}^2 - \{(0,0)\}$ since $r : \mathbb{R}^2 - \{(0,0)\} \to S^1$ defined by $r(x) = \frac{x}{\|x\|}$ clearly has the required properties. If $D^2 = \{x \in \mathbb{R}^2 : \|\, x \,\| \leq 1\}$ is the 2-disc in $\mathbb{R}^2$, then $S^1 \subseteq D^2 - \{(0,0)\}$ and the same mapping r (restricted to $D^2 - \{(0,0)\}$) is a retraction of $D^2 - \{(0,0)\}$ onto S^1. We now use what we know about fundamental groups to prove the seemingly innocuous fact that, unless some point (like (0,0)) is deleted, one cannot retract D^2 onto S^1.

Lemma 2.4.6 *If A is a retract of X, then, for any $a \in A$, $\pi_1(A,a)$ is isomorphic to a subgroup of $\pi_1(X,a)$.*

Proof: Suppose $r : X \to A$ is a retraction of X onto A. Then the diagram

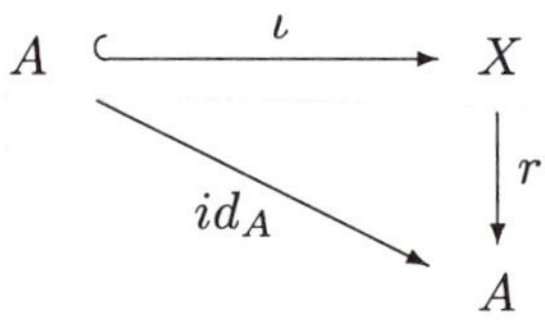

commutes. For any $a \in A$ the diagram of induced maps

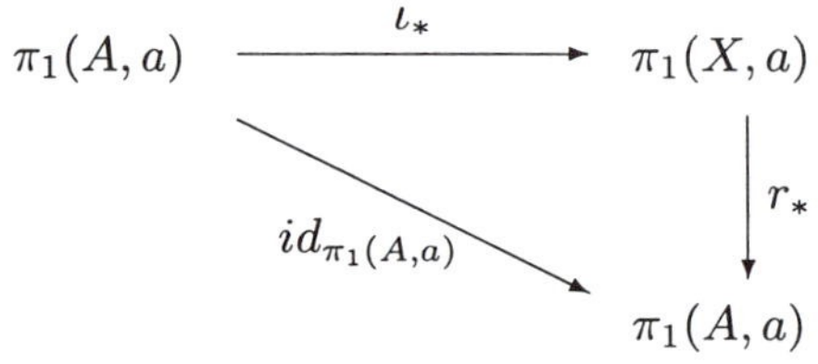

therefore also commutes (Theorem 2.2.5), i.e., $r_* \circ \iota_* = id_{\pi_1(A,a)}$. But then ι_* must be one-to-one so it carries $\pi_1(A,a)$ isomorphically onto a subgroup of $\pi_1(X,a)$. ∎

Since $\pi_1(S^1) \cong \mathbb{Z}$ and $\pi_1(D^2) = 0$ (D^2 is convex) it follows at once from Lemma 2.4.6 that S^1 is not a retract of D^2.

Remark: It is important to observe that Lemma 2.4.6 depends only on the functorial nature of the fundamental group (i.e., Theorem 2.2.5). Also note that, while it is true that S^n is not a retract of D^{n+1} for any $n \geq 1$, the argument just given with the fundamental group fails to establish this fact because, for $n \geq 2$, $\pi_1(S^n) = 0$ and the contradiction evaporates. The higher homotopy group functors (Section 2.5) or homology functors (see [**Gre**]) are required to prove, in the same way, that S^n is not a retract of D^{n+1} for $n \geq 2$.

Exercise 2.4.7 Show that A is a retract of X iff, for every space Y, any continuous map $g : A \to Y$ has a continuous extension $f : X \to Y$ to X.

The fact that the circle S^1 is not a retract of the disc D^2 is closely related to an issue of great practical significance to the applications of topology to analysis. If $f : X \to X$ is a continuous map of X into itself, then a point $x_0 \in X$ is called a **fixed point of** f if $f(x_0) = x_0$. A topological space X is said to have the **fixed point property** if *every* continuous map of X into itself has a fixed point.

Exercise 2.4.8 Use the Intermediate Value Theorem from calculus to show that $[0, 1]$ has the fixed point property.

Exercise 2.4.9 Show that the fixed point property is a topological property, i.e., show that if X has the fixed point property and Y is homeomorphic to X, then Y also has the fixed point property.

Exercise 2.4.10 Show that a retract of a space with the fixed point property also has the fixed point property.

Next we prove a theorem that asserts the logical equivalence of two statements about the sphere S^n and the disc D^{n+1}. The first of these statements we have proved to be valid when $n = 1$ (see Exercise 2.5.23 for larger n).

Theorem 2.4.7 *For any integer $n \geq 1$, the following statements are equivalent:*

> *(a) S^n is not a retract of D^{n+1}.*
>
> *(b) D^{n+1} has the fixed point property.*

Proof: We show first that (b) implies (a). We assume then that D^{n+1} has the fixed point property. Observe that S^n certainly does not have the fixed point property since, for example, the antipodal map $-id_{S^n}$ is continuous, but has no fixed points. According to Exercise 2.4.10, S^n can therefore not be a retract of D^{n+1}.

To show that (a) implies (b) we prove the contrapositive. Thus, we assume the existence of a continuous map $f : D^{n+1} \to D^{n+1}$ without fixed points and use it to construct a retraction $r : D^{n+1} \to S^n$. Geometrically, the idea is as follows: Since $f(x) \neq x$ for each $x \in D^{n+1}$, the points x and $f(x)$ determine a unique line in $\mathbb{R}^{n+1}$. We intend to let $r(x)$ be the point where the ray from $f(x)$ to x intersects S^n (see Figure 2.4.2). More precisely, let us fix an $x \in D^{n+1}$. Since $f(x) \neq x$, x and $f(x)$ determine a straight line in $\mathbb{R}^{n+1}$, every point y of which has the form $y = tx + (1-t)f(x)$ for some $t \in \mathbb{R}$. We claim that there is exactly one such y with $\| y \| = 1$ and $t \geq 0$. To see this, compute the inner product $\langle y, y \rangle = \| y \|^2$ and set it equal to 1 to obtain

$$t^2\langle x, x\rangle + 2(t - t^2)\langle x, f(x)\rangle + (1-t)^2\langle f(x), f(x)\rangle = 1\,, \qquad (2.4.5)$$

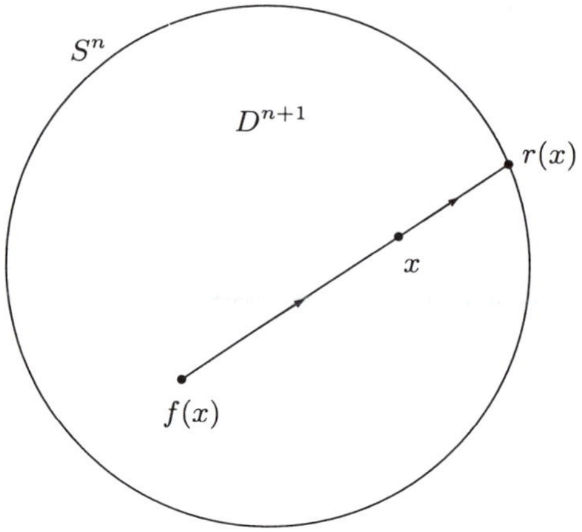

Figure 2.4.2

or, equivalently,

$$t^2 \,\|\, x - f(x) \,\|^2 + 2t \,\langle\, f(x), x - f(x)\,\rangle + (\,\|\, f(x)\,\|^2 - 1) = 0\,. \qquad (2.4.6)$$

Now, (2.4.6) is a quadratic in t and an elementary calculation reveals that it has precisely one root greater than or equal to 1. We take $r(x)$ to be the unique $y = tx + (1-t)f(x)$ with this root as the value of t. Thus, we have a map $r : D^{n+1} \to S^n$. To see that it is continuous one applies the quadratic formula to (2.4.6), selecting the root greater than or equal to 1. This gives t as a continuous function of x (the denominator is $2 \,\|\, x - f(x)\,\|^2 \neq 0$). Substituting this into $r(x) = tx + (1-t)f(x)$ gives r as a continuous function of x. Finally, if $x \in S^n$, then $\langle x, x\rangle = 1$ and it follows that $t = 1$ is a solution to (2.4.5) and therefore $r(x) = x$. Thus, $r \,|\, S^n = id_{S^n}$ and the proof is complete. ∎

Corollary 2.4.8 (Brouwer Fixed Point Theorem in Dimension 2) *The 2-disc D^2 has the fixed point property.*

The general Brouwer Fixed Point Theorem asserts that any disc D^{n+1} has the fixed point property (see Exercise 2.5.23).

We include two more examples of retractions that are of use in relation to the higher homotopy groups. Let n be a positive integer, D^n the n-disc in $\mathbb{R}^{n+1}$ and S^{n-1} its boundary sphere (Exercise 1.2.7). We consider the solid cylinder $D^n \times [0,1]$ and the subset $D^n \times \{0\} \cup S^{n-1} \times [0,1]$ consisting of its base and sides (see Figure 2.4.3). We retract $D^n \times [0,1]$ onto $D^n \times$

$\{0\} \cup S^{n-1} \times [0,1]$ by projecting from the point $(0,\ldots,0,2)$ in $\mathbb{R}^{n+1}$.

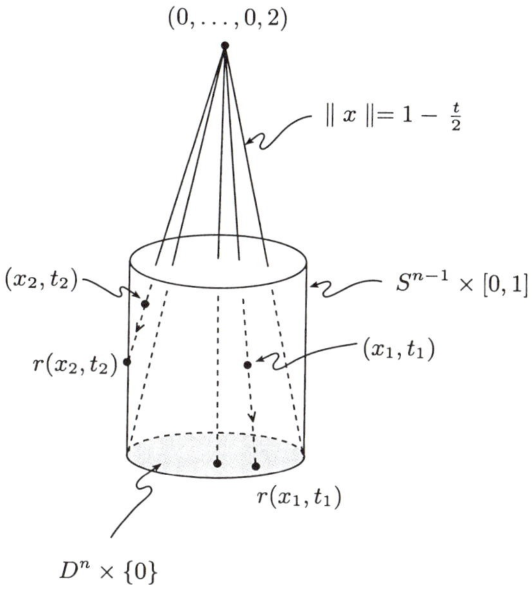

Figure 2.4.3

Exercise 2.4.11 Write out the equations suggested by Figure 2.4.3 and show that $r : D^n \times [0,1] \to D^n \times \{0\} \cup S^{n-1} \times [0,1]$ given by

$$r(x,t) = \begin{cases} \left(\frac{2x}{2-t}, 0\right), & \|x\| \le 1 - \frac{t}{2} \\ \left(\frac{x}{\|x\|}, 2 - \frac{2-t}{\|x\|}\right), & \|x\| \ge 1 - \frac{t}{2} \end{cases}$$

is a retraction.

Exercise 2.4.12 Show that $D^n \times [0,1] \times \{0\} \cup S^{n-1} \times [0,1] \times [0,1]$ is a retract of $D^n \times [0,1] \times [0,1]$.

A bit earlier we showed that S^1 *is* a retract of the punctured disc $D^2 - \{(0,0)\}$ (the map $r(x) = \frac{x}{\|x\|}$ is a retraction). In fact, it is much more. A subset A of a space X is said to be a **deformation retract** of X if there exists a homotopy $F : X \times [0,1] \to X$ such that $F_0 = id_X$ and $F_1 : X \to X$ satisfies $F_1(x) \in A$ for every $x \in X$ and $F_1(a) = a$ for every $a \in A$ (intuitively, there is a retraction of X onto A which, thought of as a map into X, is homotopic to the identity on X). Such an F is called a **deformation retraction** of X onto A. For example, $F : (D^2 - \{(0,0)\}) \times [0,1] \to D^2 - \{(0,0)\}$ given by

$$F(x,t) = (1-t)x + t\left(\frac{x}{\| x \|}\right) \tag{2.4.7}$$

in a deformation retraction of $D^2 - \{(0,0)\}$ onto S^1. Here's why you should care:

Lemma 2.4.9 *If $A \subseteq X$ is a deformation retract of X, then A and X have the same homotopy type.*

Proof: Let F be a deformation retraction of X onto A and consider the maps $F_1 \circ id_A : A \to X$ and $id_A \circ F_1 : X \to A$. Then $F_1 \circ id_A = id_A$ and $id_A \circ F_1 = F_1 \simeq id_X$ so F_1 and id_A are homotopy equivalences. ∎

Exercise 2.4.13 What are $\pi_1(\mathbb{R}^n - \{0,\ldots,0\})$ and $\pi_1(D^n - \{0,\ldots,0\})$?

With a few applications of Theorem 2.4.4 under our belts we now turn briefly to Theorem 2.4.5 and, in particular, to the $n = 4$ case ($\pi_1(\mathbb{R}\mathbb{P}^3) \cong \mathbb{Z}_2$). A space X with $\pi_1(X) \cong \mathbb{Z}_2$ contains a loop α which, although not itself homotopically trivial, has the property that $\alpha^2 = \alpha\alpha$ *is* homotopic to a constant loop. Since α^2 is just "α traversed twice", this is a rather counter-intuitive phenomenon and so it seems advisable to write out at least one concrete example explicitly. For this we regard $\mathbb{R}\mathbb{P}^3$ as the quotient of S^3 obtained by identifying antipodal points (Section 1.2). Furthermore, we identify S^3 with $SU(2)$ as in Theorem 1.1.4. Begin by defining $\tilde{\alpha} : [0,1] \to SU(2)$ by

$$\tilde{\alpha}(s) = \begin{pmatrix} e^{-\pi s\mathbf{i}} & 0 \\ 0 & e^{\pi s\mathbf{i}} \end{pmatrix}$$

for each $s \in [0,1]$. Note that $\tilde{\alpha}$ is a path from $\tilde{\alpha}(0) = id$ to $\tilde{\alpha}(1) = -id$ in $SU(2)$, whereas $\tilde{\alpha}^2$, given by

$$\tilde{\alpha}^2(s) = \begin{pmatrix} e^{-2\pi s\mathbf{i}} & 0 \\ 0 & e^{2\pi s\mathbf{i}} \end{pmatrix},$$

is a loop at id in $SU(2)$. Now, if $\mathcal{P}$ is the projection onto $\mathbb{R}\mathbb{P}^3$ and if we define $\alpha = \mathcal{P} \circ \tilde{\alpha}$, then both α and $\alpha^2 = \mathcal{P} \circ \tilde{\alpha}^2$ are loops in $\mathbb{R}\mathbb{P}^3$

at $\mathcal{P}(id) = \mathcal{P}(-id)$. Now, $SU(2)$, being homeomorphic to S^3, is simply connected so there is a homotopy $F : [0,1] \times [0,1] \to SU(2)$ with $F(s,0) = \tilde{\alpha}^2(s)$ and $F(s,1) = F(0,t) = F(1,t) = id$ for all $s,t \in [0,1]$. Thus, $\mathcal{P} \circ F$ is a homotopy in $\mathbb{R}\mathbb{P}^3$ from α^2 to the trivial loop at $\mathcal{P}(id)$.

Exercise 2.4.14 Use Corollary 2.4.2 to show that α itself is *not* homotopically trivial.

Some interesting consequences of this phenomenon in physics, based on the fact that $\mathbb{R}\mathbb{P}^3$ is homeomorphic to the rotation group $SO(3)$, are discussed in the Appendix.

Next we consider a simple device for producing new examples of fundamental groups from those we already know.

Theorem 2.4.10 *Let X and Y be topological spaces with $x_0 \in X$ and $y_0 \in Y$. Then*

$$\pi_1(X \times Y, (x_0, y_0)) \cong \pi_1(X, x_0) \times \pi_1(Y, y_0).$$

Proof: Let $\mathcal{P}_X$ and $\mathcal{P}_Y$ be the projections of $X \times Y$ onto X and Y, respectively. Then each induces a homomorphism $(\mathcal{P}_X)_* : \pi_1(X \times Y, (x_0, y_0)) \to \pi_1(X, x_0)$ and $(\mathcal{P}_Y)_* : \pi_1(X \times Y, (x_0, y_0)) \to \pi_1(Y, y_0)$. Define a homorphism $h : \pi_1(X \times Y, (x_0, y_0)) \to \pi_1(X, x_0) \times \pi_1(Y, y_0)$ by

$$h([\tau]) = ((\mathcal{P}_X)_*, (\mathcal{P}_Y)_*)([\tau]) = ((\mathcal{P}_X)_*([\tau]), (\mathcal{P}_Y)_*([\tau]))$$

for each $[\tau] \in \pi_1(X \times Y, (x_0, y_0))$. To show that h is an isomorphism we exhibit its inverse. Let $([\alpha], [\beta]) \in \pi_1(X, x_0) \times \pi_1(Y, y_0)$. Then α is a loop at x_0 in X and β is a loop at y_0 in Y. Define a loop $\tau = (\alpha, \beta)$ at (x_0, y_0) in $X \times Y$ by $\tau(s) = (\alpha, \beta)(s) = (\alpha(s), \beta(s))$ for each s in $[0,1]$.

Exercise 2.4.15 Show that if $\alpha' \simeq \alpha \operatorname{rel}\{0,1\}$ and $\beta' \simeq \beta \operatorname{rel}\{0,1\}$, then $(\alpha', \beta') \simeq (\alpha, \beta) \operatorname{rel}\{0,1\}$.

By virtue of Exercise 2.4.15, the mapping g that carries $([\alpha], [\beta])$ onto $[\tau] = [(\alpha, \beta)]$ is well-defined. We claim that it is, in fact, the inverse of h. First note that g carries $\pi_1(X, x_0) \times \pi_1(Y, y_0)$ onto $\pi_1(X \times Y, (x_0, y_0))$. Indeed, let σ be an arbitrary loop in $X \times Y$ at (x_0, y_0). Then, for each s in $[0,1]$, $\sigma(s) = (\mathcal{P}_X \circ \sigma(s), \mathcal{P}_Y \circ \sigma(s)) = (\mathcal{P}_X \circ \sigma, \mathcal{P}_Y \circ \sigma)(s)$ so $g([\mathcal{P}_X \circ \sigma], [\mathcal{P}_Y \circ \sigma]) = [(\mathcal{P}_X \circ \sigma, \mathcal{P}_Y \circ \sigma)] = [\sigma]$.

Exercise 2.4.16 Complete the proof by showing that $g \circ h$ and $h \circ g$ are the identity maps on $\pi_1(X \times Y, (x_0, y_0))$ and $\pi_1(X, x_0) \times \pi_1(Y, y_0)$, respectively. ∎

In particular, it follows from Theorem 2.4.10 that any product of simply connected spaces is simply connected. The torus has fundamental group $\pi_1(S^1 \times S^1) \cong \mathbb{Z} \times \mathbb{Z}$ and, by induction, $\pi_1(S^1 \times \cdots \times S^1) \cong \mathbb{Z} \times \cdots \times \mathbb{Z}$. Also note that, if X is simply connected, then $\pi_1(X \times Y) \cong \pi_1(Y)$ so, for example, $\pi_1(S^2 \times S^1) \cong \mathbb{Z}$. Since $\pi_1(S^3) = 0$ we have shown that, although S^3 and $S^2 \times S^1$ are locally homeomorphic (Hopf bundle), they cannot be globally homeomorphic, or even homotopically equivalent.

We conclude this section with an amusing result on the fundamental group of a topological group whose generalization to the higher homotopy groups in Section 2.5 will come in handy. Thus, we let G be a topological group with identity e and temporarily denote the group operation in G with a dot $\cdot$. Let $\alpha, \beta : [0,1] \to G$ be two loops at e in G. As usual, $\alpha\beta$ is the loop "α followed by β" at e and now we define $\alpha \cdot \beta : [0,1] \to G$ by $(\alpha \cdot \beta)(s) = \alpha(s) \cdot \beta(s)$ for every $s \in [0,1]$. Then $\alpha \cdot \beta$ is clearly also a loop at e in G.

Lemma 2.4.11 $\alpha \cdot \beta \simeq \alpha\beta \,\mathrm{rel}\, \{0,1\}$, *i.e.*, $[\alpha \cdot \beta] = [\alpha\beta] = [\alpha][\beta]$.

Proof: Denoting by e also the constant loop at $e \in G$ we have $[\alpha][e] = [\alpha]$ and $[e][\beta] = [\beta]$. Let H and K be homotopies, relative to $\{0,1\}$, from αe to α and $e\beta$ to β, respectively. Now define $H \cdot K : [0,1] \times [0,1] \to G$ by $H \cdot K(s,t) = H(s,t) \cdot K(s,t)$. Then $H \cdot K(s,0) = H(s,0) \cdot K(s,0) = (\alpha e)(s) \cdot (e\beta)(s)$. But

$$(\alpha e)(s) = \begin{cases} \alpha(2s), & 0 \le s \le \frac{1}{2} \\ e, & \frac{1}{2} \le s \le 1 \end{cases} \quad \text{and} \quad (e\beta)(s) = \begin{cases} e, & 0 \le s \le \frac{1}{2} \\ \beta(2s-1), & \frac{1}{2} \le s \le 1 \end{cases}$$

so $H \cdot K(s,0) = \begin{cases} \alpha(2s), & 0 \le s \le \frac{1}{2} \\ \beta(2s-1), & \frac{1}{2} \le s \le 1 \end{cases} = (\alpha\beta)(s)$. Furthermore, $H \cdot K(s,1) = H(s,1) \cdot K(s,1) = \alpha(s) \cdot \beta(s) = (\alpha \cdot \beta)(s)$, $H \cdot K(0,t) = H \cdot K(1,t) = e \cdot e = e$. Thus, $H \cdot K$ is a homotopy, relative to $\{0,1\}$, from $\alpha\beta$ to $\alpha \cdot \beta$. ∎

Exercise 2.4.17 Show that, if G is a topological group with identity e, then $\pi_1(G, e)$ is Abelian. **Hint:** Consider $K \cdot H$ as in the proof of Lemma 2.4.11.

2.5 Higher Homotopy Groups

A loop at x_0 in X is a map of $I = [0,1]$ into X that carries the boundary $\partial I = \{0,1\}$ onto x_0. $\pi_1(X, x_0)$ consists of all homotopy classes, relative to ∂I, of such loops and admits a natural, and very useful, group structure. We generalize as follows. For each positive integer n, $I^n = [0,1]^n =$

$\{(s^1, \ldots, s^n) \in \mathbb{R}^n : 0 \leq s^i \leq 1,\ i = 1, \ldots, n\}$ is called the **n-cube** in $\mathbb{R}^n$. The **boundary** ∂I^n of I^n consists of all $(s^1, \ldots, s^n) \in I^n$ for which $s^i = 0$ or $s^i = 1$ for at least one value of i. If X is a topological space and $x_0 \in X$, then an **n-loop at** x_0 is a continuous map $\alpha : I^n \to X$ such that $\alpha(\partial I^n) = \{x_0\}$ (see Figure 2.5.1).

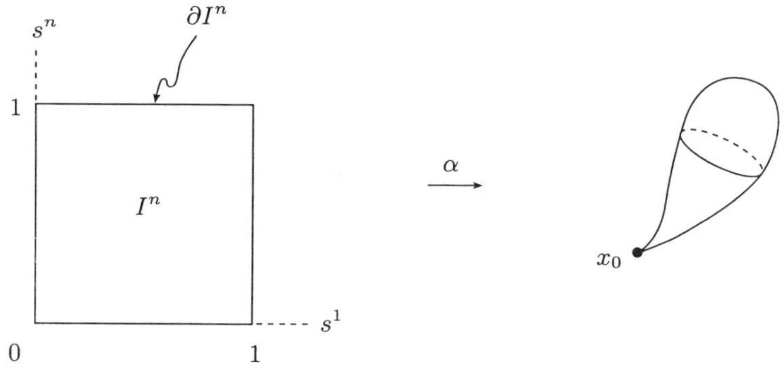

Figure 2.5.1

By Lemma 2.3.1, the collection of all n-loops at x_0 in X is partitioned into equivalence classes by homotopy relative to ∂I^n. The equivalence class containing an n-loop α is denoted $[\alpha]$.

If α is an n-loop at x_0 in X, we define $\alpha^{\leftarrow} : I^n \to X$ by $\alpha^{\leftarrow}(s^1, s^2, \ldots, s^n) = \alpha(1 - s^1, s^2, \ldots, s^n)$ for all $(s^1, s^2, \ldots, s^n) \in I^n$.

Exercise 2.5.1 Show that if $\alpha, \alpha' : I^n \to X$ are two n-loops at x_0 in X and $\alpha' \simeq \alpha \operatorname{rel} \partial I^n$, then $(\alpha')^{\leftarrow} \simeq \alpha^{\leftarrow} \operatorname{rel} \partial I^n$.

Next suppose that $\alpha, \beta : I^n \to X$ are two n-loops at x_0 in X. We define an n-loop $\alpha + \beta : I^n \to X$ by

$$(\alpha + \beta)(s^1, s^2, \ldots, s^n) = \begin{cases} \alpha(2s^1, s^2, \ldots, s^n), & 0 \leq s^1 \leq \frac{1}{2} \\ \beta(2s^1 - 1, s^2, \ldots, s^n), & \frac{1}{2} \leq s^1 \leq 1 \end{cases}$$

for all $(s^1, s^2, \ldots, s^n) \in I^n$ (see Figure 2.5.2). It is traditional to opt for additive rather than multiplicative notation in this context because the higher homotopy groups turn out to be Abelian (Theorem 2.5.3).

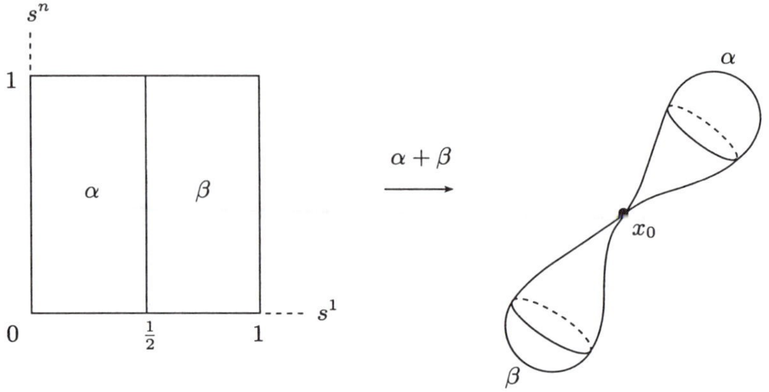

Figure 2.5.2

Exercise 2.5.2 Show that if $\alpha, \alpha', \beta, \beta' : I^n \to X$ are n-loops at x_0 with $\alpha' \simeq \alpha \operatorname{rel} \partial I^n$ and $\beta' \simeq \beta \operatorname{rel} \partial I^n$, then $\alpha' + \beta' \simeq \alpha + \beta \operatorname{rel} \partial I^n$.

Let $\pi_n(X, x_0)$ denote the set of all homotopy equivalences classes, relative to ∂I^n, of n-loops at x_0 in X. According to Exercises 2.5.1 and 2.5.2 we may unambiguously define, for $[\alpha], [\beta] \in \pi_n(X, x_0)$,

$$[\alpha]^{\leftarrow} = [\alpha^{\leftarrow}]$$

and

$$[\alpha] + [\beta] = [\alpha + \beta].$$

In order to prove that $\pi_n(X, x_0)$ thereby acquires a group structure and that, for $n \geq 2$, this group is Abelian, we require the following very useful lemma.

Lemma 2.5.1 *Let $\alpha : I^n \to X$ be an n-loop at x_0 in X and let $I_1^n = \{(s^1, \ldots, s^n) \in \mathbb{R}^n : a^i \leq s^i \leq b^i,\ i = 1, \ldots, n\}$, where $0 \leq a^i < b^i \leq 1$ for each $i = 1, \ldots, n$ (I_1^n is called a* **subcube** *of I^n). Then there exists an n-loop $\alpha' : I^n \to X$ such that $\alpha' \simeq \alpha \operatorname{rel} \partial I^n$ and $\alpha'(I^n - I_1^n) = \{x_0\}$.*

Proof: Define $\alpha' : I^n \to X$ by

$$\alpha'(s^1, \ldots, s^n) = \begin{cases} \alpha\left(\frac{s^1 - a^1}{b^1 - a^1}, \ldots, \frac{s^n - a^n}{b^n - a^n}\right), & a^i \leq s^i \leq b^i, \\ & i = 1, \ldots, n \\ x_0, & \text{otherwise} \end{cases} . \qquad (2.5.1)$$

Since α carries ∂I^n to x_0, the Glueing Lemma 1.2.3 implies that α' is continuous and it surely satisfies $\alpha'(I^n - I_1^n) = \{x_0\}$. To show that $\alpha' \simeq \alpha \operatorname{rel} \partial I^n$ we interpolate between α and α' as follows: Define $F : I^n \times [0,1] \to X$ by

$$F\left((s^1,\ldots,s^n),t\right) = \begin{cases} \alpha\left(\frac{s^1-a^1t}{(1-(1-b^1)t)-a^1t},\ldots,\frac{s^n-a^nt}{(1-(1-b^n)t)-a^nt}\right), & a^it \le s^i \le 1-(1-b^i)t, \\ & i=1,\ldots,n \\ x_0, & \text{otherwise} \end{cases}.$$

Then $F((s^1,\ldots,s^n),0) = \alpha(s^1,\ldots,s^n)$ and $F((s^1,\ldots,s^n),1) = \alpha'(s^1,\ldots,s^n)$ for all $(s^1,\ldots,s^n) \in I^n$. Next suppose $(s^1,\ldots,s^n) \in \partial I^n$. We show that $F((s^1,\ldots,s^n),t) = x_0$ for any $t \in [0,1]$. Some s^i is either 0 or 1. Suppose $s^i = 0$. If $a^it \le s^i \le 1-(1-b^i)t$ is satisfied, then $a^it = 0$ so $s^i - a^it = 0$ and $F((s^1,\ldots,s^n),t) = x_0$ because $\alpha(\partial I^n) = \{x_0\}$. If $a^it \le s^i \le 1-(1-b^i)t$ is not satisfied, then $F((s^1,\ldots,s^n),t) = x_0$ by definition. In the same way one shows that $s^i = 1$ implies $F((s^1,\ldots,s^n),t) = x_0$.

Thus, if we can show that F is continuous it will be a homotopy, relative to ∂I^n, from α to α'. First observe that, for each i, $(1-(1-b^i)t) - a^it = 1 + ((b^i - a^i) - 1)t \ge b^i - a^i > 0$ so none of the denominators in the definition of F can be zero. Since α is continuous on I^n it follows that F is continuous on $\{((s^1,\ldots,s^n),t) : a^it \le s^i \le 1-(1-b^i)t \text{ for } i = 1,\ldots,n\}$. Moreover, if either $s^i = a^it$ or $s^i = 1-(1-b^i)t$, F takes the value x_0 so, by the Glueing Lemma 1.2.3, F is continuous on all of $I^n \times [0,1]$. ∎

Given an n-loop α at x_0 in X, any other n-loop γ at x_0 in X with $\gamma \simeq \alpha \operatorname{rel} \partial I^n$ and $\gamma(I^n - I_1^n) = \{x_0\}$ is called a **concentration of α on the subcube I_1^n** and we say that γ is **concentrated on I_1^n**. The particular concentration of α on I_1^n given by (2.5.1) will be denoted α' (when the identity of the subcube I_1^n is clear from the context). Two special cases are of particular interest. We denote by I_L^n and I_R^n the left and right halves of I^n. That is,

$$I_L^n = \left\{(s^1,\ldots,s^n) \in \mathbb{R}^n : 0 \le s^1 \le \frac{1}{2},\ 0 \le s^i \le 1,\ i = 2,\ldots,n\right\}$$

and

$$I_R^n = \left\{(s^1,\ldots,s^n) \in \mathbb{R}^n : \frac{1}{2} \le s^1 \le 1,\ 0 \le s^i \le 1,\ i = 2,\ldots,n\right\}.$$

For any n-loop α at x_0 in X we will denote by α'_L and α'_R the concentrations of α on I_L^n and I_R^n, respectively, given by (2.5.1). As usual, we denote by x_0 also the constant n-loop at x_0 in X.

Exercise 2.5.3 Show that, for any n-loop α at x_0 in X, $\alpha'_L = \alpha + x_0$ and $\alpha'_R = x_0 + \alpha$ and conclude that

$$[\alpha] + [x_0] = [\alpha] = [x_0] + [\alpha]. \tag{2.5.2}$$

Exercise 2.5.4 Let G be a topological group with identity e and let α and β be two n-loops at e in G. Denoting the group operation in G with a dot $\cdot$, define $\alpha \cdot \beta : I^n \to G$ by $(\alpha \cdot \beta)(s^1, \ldots, s^n) = \alpha(s^1, \ldots, s^n) \cdot \beta(s^1, \ldots, s^n)$. Show that $\alpha \cdot \beta$ is an n-loop at e in G and is, moreover, homotopic to $\alpha + \beta$ relative to ∂I^n, i.e., $[\alpha \cdot \beta] = [\alpha + \beta] = [\alpha] + [\beta]$. **Hint:** Mimic the proof of Lemma 2.4.11.

Lemma 2.5.2 *Let $\alpha, \beta : I^n \to X$ be two n-loops at $x_0 \in X$. Let I_1^n and I_2^n be two subcubes of I^n given by $a_1^i \leq s^i \leq b_1^i$ and $a_2^i \leq s^i \leq b_2^i$, respectively, for $i = 1, \ldots, n$, where $0 \leq a_1^i < b_1^i \leq 1$ and $0 \leq a_2^i < b_2^i \leq 1$ for each $i = 1, \ldots, n$. Suppose that $b_1^1 \leq a_2^1$ (so that I_1^n is "to the left of" I_2^n as in Figure 2.5.3). Let γ be a concentration of α on I_1^n and δ a concentration of β on I_2^n and define $\omega : I^n \to X$ by $\omega \,|\, I_1^n = \gamma$, $\omega \,|\, I_2^n = \delta$ and $\omega \,|\, I^n - (I_1^n \cup I_2^n) = x_0$. Then*

$$[\omega] = [\alpha + \beta] = [\alpha] + [\beta].$$

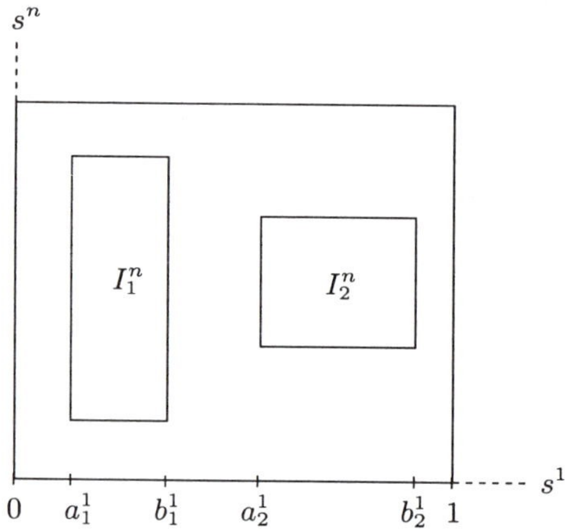

Figure 2.5.3

Proof: We first consider the special case in which $I_1^n = I_L^n$ and $I_2^n = I_R^n$.

Note first that if γ and δ happen to be α'_L and β'_R, respectively, then

$$\begin{aligned}\omega(s^1,s^2,\ldots,s^n) &= \begin{cases} \alpha'_L(s^1,s^2,\ldots,s^n), & 0\leq s^1\leq \frac{1}{2}\\ \beta'_R(s^1,s^2,\ldots,s^n), & \frac{1}{2}\leq s^1\leq 1\end{cases}\\ &= \begin{cases} \alpha(2s^1,s^2,\ldots,s^n), & 0\leq s^1\leq \frac{1}{2}\\ \beta(2s^1-1,s^2,\ldots,s^n), & \frac{1}{2}\leq s^1\leq 1\end{cases}\\ &= (\alpha+\beta)(s^1,s^2,\ldots,s^n)\end{aligned}$$

so $\omega = \alpha+\beta$ and therefore certainly $[\omega] = [\alpha+\beta]$. Now suppose γ and δ are arbitrary concentrations of α and β on I^n_L and I^n_R. Define $\Gamma, \Delta : I^n \to X$ by

$$\Gamma(s^1,s^2,\ldots,s^n) = \gamma(\frac{1}{2}s^1,s^2,\ldots,s^n)$$

and

$$\Delta(s^1,s^2,\ldots,s^n) = \delta(\frac{1}{2}s^1+\frac{1}{2},s^2,\ldots,s^n).$$

These are n-loops at x_0 in X and, moreover,

$$\begin{aligned}\Gamma'_L(s^1,s^2,\ldots,s^n) &= \begin{cases} \Gamma(2s^1,s^2,\ldots,s^n), & 0\leq s^1\leq \frac{1}{2}\\ x_0, & \text{otherwise}\end{cases}\\ &= \begin{cases} \gamma(s^1,s^2,\ldots,s^n), & 0\leq s^1\leq \frac{1}{2}\\ x_0, & \frac{1}{2}\leq s^1\leq 1\end{cases}\\ &= \gamma(s^1,s^2,\ldots,s^n)\end{aligned}$$

and

$$\begin{aligned}\Delta'_R(s^1,s^2,\ldots,s^n) &= \begin{cases} \Delta(2s^1-1,s^2,\ldots,s^n), & \frac{1}{2}\leq s^1\leq 1\\ x_0, & \text{otherwise}\end{cases}\\ &= \begin{cases} \delta(s^1,s^2,\ldots,s^n), & \frac{1}{2}\leq s^1\leq 1\\ x_0, & 0\leq s^1\leq \frac{1}{2}\end{cases}\\ &= \delta(s^1,s^2,\ldots,s^n).\end{aligned}$$

Thus, $\Gamma'_L = \gamma$ and $\Delta'_R = \delta$. As we have shown above, $\omega = \Gamma+\Delta$ so $[\omega] = [\Gamma+\Delta] = [\Gamma]+[\Delta] = [\Gamma'_L]+[\Delta'_R] = [\gamma]+[\delta] = [\alpha]+[\beta] = [\alpha+\beta]$ as required. The lemma is therefore proved when $I^n_1 = I^n_L$ and $I^n_2 = I^n_R$ and we turn now to the general case.

We claim that it suffices to consider the case in which $a^1_1 = 0$, $b^1_1 = a^1_2$, $b^1_2 = 1$, $a^i_1 = 0$ and $b^i_1 = 1$ for $i = 2,\ldots,n$ (see Figure 2.5.4). Indeed,

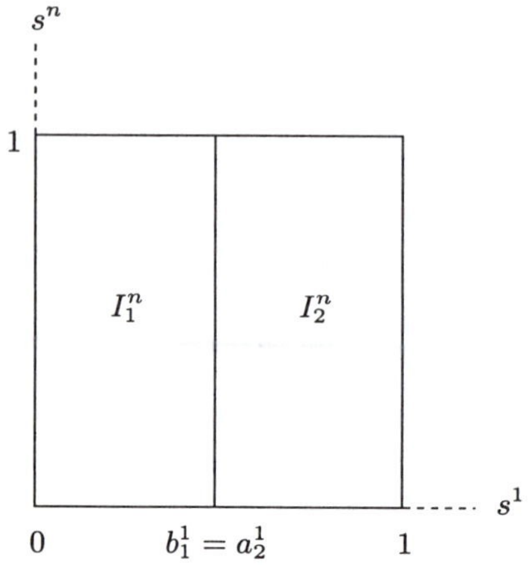

Figure 2.5.4

if γ and δ are concentrations of α and β to subcubes as shown in Figure 2.5.3, then they are surely also concentrations on the enlargements of these subcubes shown in Figure 2.5.4 and the map ω is the same in both cases. Letting $b_1^2 = a_2^1 = a$ we intend to continuously deform Figure 2.5.4 into $I^n = I_L^n \cup I_R^n$, carrying the concentrations along as we go, and then appeal to the special case of the lemma already proved. Begin by considering Figure 2.5.5.

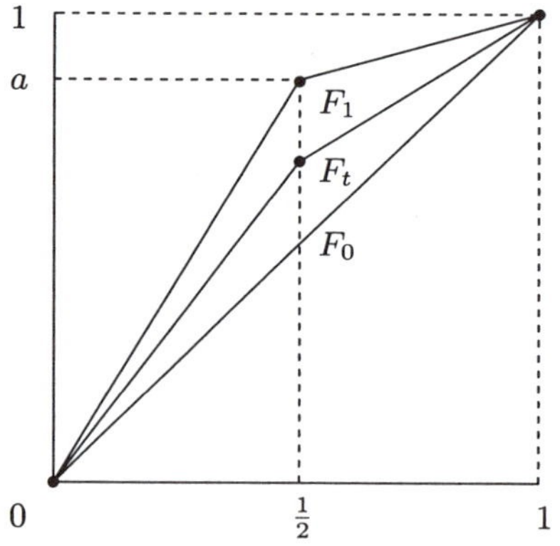

Figure 2.5.5

Define $F_0, F_1 : [0,1] \to [0,1]$ by $F_0(s^1) = s^1$ and

$$F_1(s^1) = \begin{cases} 2as^1, & 0 \le s^1 \le \frac{1}{2} \\ 1 - 2(1-a)(1-s^1), & \frac{1}{2} \le s^1 \le 1 \end{cases} .$$

We define a homotopy $F : [0,1] \times [0,1] \to [0,1]$, relative to $\{0,1\}$, from $F_0 = id$ to F_1 as indicated in Figure 2.5.5, that is,

$$F(s^1, t) = \begin{cases} (1 + t(2a-1))s^1, & 0 \le s^1 \le \frac{1}{2} \\ 1 - (1 - t(2a-1))(1-s^1), & \frac{1}{2} \le s^1 \le 1 \end{cases} .$$

Now define $G : I^n \times [0,1] \to I^n$ by

$$G\left((s^1, s^2, \ldots, s^n), t \right) = \left(F(s^1, t), s^2, \ldots, s^n \right) .$$

For each t,

$$G_t(s^1, s^2, \ldots, s^n) = \left(F_t(s^1), s^2, \ldots, s^n \right)$$

so G_0 is the identity on I^n and therefore $\omega \circ G_0 = \omega$, $\gamma \circ G_0 = \gamma$ and $\delta \circ G_0 = \delta$. Since $G_1 \simeq G_0 \operatorname{rel} \partial I^n$ it follows from Exercise 2.3.2 that $[\omega \circ G_1] = [\omega]$, $[\gamma \circ G_1] = [\gamma] = [\alpha]$ and $[\delta \circ G_1] = [\delta] = [\beta]$. Thus, it will suffice to prove that

$$[\omega \circ G_1] = [\gamma \circ G_1 + \delta \circ G_1] . \tag{2.5.3}$$

Note first that $\gamma \circ G_1$ is a concentration of α on I^n_L. Indeed,

$$\begin{aligned} \gamma \circ G_1(s^1, s^2, \ldots, s^n) &= \gamma(F_1(s^1), s^2, \ldots, s^n) \\ &= \begin{cases} \gamma(2as^1, s^2, \ldots, s^n), & 0 \le s^1 \le \frac{1}{2} \\ \gamma(1 - 2(1-a)(1-s^1), s^2, \ldots, s^n), & \frac{1}{2} \le s^1 \le 1 \end{cases} \end{aligned}$$

and, for $\frac{1}{2} \le s^1 \le 1$, $a \le 1 - 2(1-a)(1-s^1) \le 1$ so $\gamma(1 - 2(1-a)(1 - s^1), s^2, \ldots, s^n) = x_0$.

Exercise 2.5.5 Show that $\delta \circ G_1$ is a concentration of β on I^n_R.

Exercise 2.5.6 Show that $\omega \circ G_1 \,|\, I^n_L = \gamma \circ G_1$ and $\omega \circ G_1 \,|\, I^n_R = \delta \circ G_1$.

The required equality (2.5.3) now follows at once from Exercise 2.5.6 and the special case of the lemma proved earlier. ∎

The essential content of Lemma 2.5.2 is indicated schematically in Figure 2.5.6, where we have used the symbols α and β also for the concentrations of these n-loops on various subcubes.

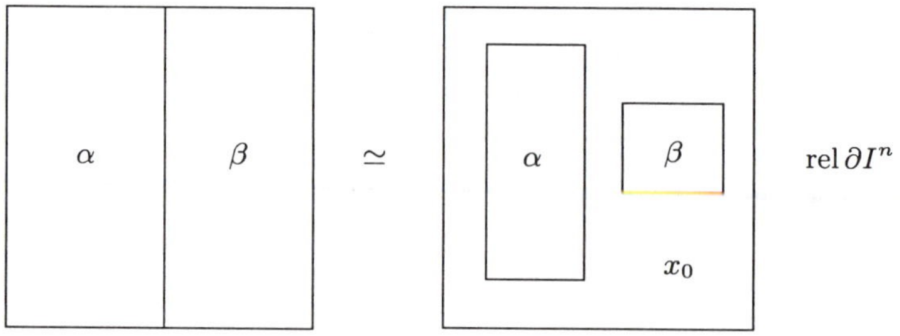

$$\alpha + \beta \simeq \omega \,\mathrm{rel}\, \partial I^n$$

Figure 2.5.6

Theorem 2.5.3 *Let X be a topological space, $x_0 \in X$ and $n \geq 2$ a positive integer. Let $\pi_n(X, x_0)$ be the set of all homotopy classes, relative to ∂I^n, of n-loops at x_0 in X. For $[\alpha], [\beta] \in \pi_n(X, x_0)$, define $[\alpha] + [\beta] = [\alpha + \beta]$. Then, with this operation, $\pi_n(X, x_0)$ is an Abelian group in which the identity element is $[x_0]$ and the inverse of any $[\alpha]$ is given by $-[\alpha] = [\alpha]^{\leftarrow} = [\alpha^{\leftarrow}]$.*

Proof: We have already shown that the operations described in the theorem are well-defined (Exercises 2.5.1 and 2.5.2) and that $[x_0]$ acts as an additive identity (Exercise 2.5.3).

Exercise 2.5.7 Prove that, for any n-loops α, β and γ at x_0 in X, $(\alpha + \beta) + \gamma \simeq \alpha + (\beta + \gamma) \,\mathrm{rel}\, \partial I^n$. **Hint:** Lemma 2.5.2.

To show that $[\alpha] + (-[\alpha]) = [x_0]$ we write out $\alpha + (-\alpha)$ explicitly.

$$(\alpha + (-\alpha))(s^1, s^2, \ldots, s^n) = \begin{cases} \alpha(2s^1, s^2, \ldots, s^n), & 0 \leq s^1 \leq \frac{1}{2} \\ \alpha^{\leftarrow}(2s^1 - 1, s^2, \ldots, s^n), & \frac{1}{2} \leq s^1 \leq 1 \end{cases}$$

$$= \begin{cases} \alpha(2s^1, s^2, \ldots, s^n), & 0 \leq s^1 \leq \frac{1}{2} \\ \alpha(2(1 - s^1), s^2, \ldots, s^n), & \frac{1}{2} \leq s^1 \leq 1 \end{cases}.$$

Define $F : I^n \times [0,1] \to X$ by

$$F\left((s^1, s^2, \ldots, s^n), t\right) = \begin{cases} \alpha(2(1-t)s^1, s^2, \ldots, s^n), & 0 \le s^1 \le \frac{1}{2} \\ \alpha(2(1-t)(1-s^1), s^2, \ldots, s^n), & \frac{1}{2} \le s^1 \le 1 \end{cases} .$$

Then F is continuous, $F_0 = \alpha + (-\alpha)$, $F_1 = x_0$ (because $\alpha(\partial I^n) = \{x_0\}$) and $F_t(\partial I^n) = \{x_0\}$ for each $t \in [0,1]$. Thus, $\alpha + (-\alpha) \simeq x_0 \operatorname{rel} \partial I^n$. One could prove that $-[\alpha] + [\alpha] = [x_0]$ in the same way, or one can appeal to commutativity, which we now prove.

The $n = 2$ case of Exercise 1.2.8 gives a homeomorphism $\varphi : I^2 \to D^2$ with the following properties: (1) $\varphi(\partial I^2) = S^1$, (2) $(s^1, s^2) \in I^2_L$ and $\varphi(s^1, s^2) = (y^1, y^2)$ implies $-1 \le y^1 \le 0$, and (3) $(s^1, s^2) \in I^2_R$ and $\varphi(s^1, s^2) = (y^1, y^2)$ implies $0 \le y^1 \le 1$. Let $R : D^2 \to D^2$ be the rotation through π given by $R(y^1, y^2) = -(y^1, y^2)$ and note that $R \simeq id_{D^2}$ since we may define, for each $t \in [0,1]$, $R_t : D^2 \to D^2$ (rotation through πt) by

$$R_t(y^1, y^2) = \left((\cos \pi t) y^1 - (\sin \pi t) y^2,\ (\sin \pi t) y^1 + (\cos \pi t) y^2\right)$$

and then $R_0 = id_{D^2}$ and $R_1 = R$. Now use φ to transfer the rotation R to I^2, i.e., define $\rho' : I^2 \to I^2$ by $\rho' = \varphi^{-1} \circ R \circ \varphi$ (see Figure 2.5.7). Observe that $\rho'(I^2_R) = I^2_L$, $\rho'(I^2_L) = I^2_R$ and $\rho'(\partial I^2) = \partial I^2$. Moreover, Exercise 2.3.2 implies that $\rho' \simeq id_{I^2}$. Now define $\rho : I^n \to I^n$ by $\rho(s^1, s^2, s^3, \ldots, s^n) = (\rho'(s^1, s^2), s^3, \ldots, s^n)$ (note that this requires $n \ge 2$). Observe that $\rho(I^n_R) = I^n_L$, $\rho(I^n_L) = I^n_R$ and $\rho(\partial I^n) = \partial I^n$.

Exercise 2.5.8 Show that $\rho \simeq id_{I^n}$.

We use ρ to show that $\pi_n(X, x_0)$ is Abelian as follows: Let $[\alpha], [\beta] \in \pi_n(X, x_0)$ and assume, without loss of generality, that α is concentrated on I^n_L and β is concentrated on I^n_R. Defining $\omega : I^n \to X$ by $\omega \mid I^n_L = \alpha$ and $\omega \mid I^n_R = \beta$ we have, by Lemma 2.5.2, $[\omega] = [\alpha + \beta] = [\alpha] + [\beta]$. Notice that $\alpha \circ \rho$, $\beta \circ \rho$ and $\omega \circ \rho$ are all n-loops at x_0 in X.

Exercise 2.5.9 Show that $[\alpha \circ \rho] = [\alpha]$, $[\beta \circ \rho] = [\beta]$ and $[\omega \circ \rho] = [\omega]$. **Hint:** $\rho \simeq id_{I^n}$, $\rho(\partial I^n) = \partial I^n$ and α, β and ω all carry ∂I^n to x_0.

Moreover, $\omega \circ \rho \mid I^n_L = \beta \circ \rho$ and $\omega \circ \rho \mid I^n_R = \alpha \circ \rho$ so Lemma 2.5.2 gives $[\omega \circ \rho] = [\beta \circ \rho + \alpha \circ \rho] = [\beta \circ \rho] + [\alpha \circ \rho]$. By Exercise 2.5.9, $[\omega] = [\beta] + [\alpha]$. Thus, $[\alpha] + [\beta] = [\beta] + [\alpha]$ as required. ■

The group $\pi_n(X, x_0)$ is called the **nth homotopy group of X at x_0** and $\pi_2(X, x_0)$, $\pi_3(X, x_0), \ldots$ are collectively referred to as the **higher homotopy groups** of X at x_0. While these latter are Abelian and the fundamental group $\pi_1(X, x_0)$ is generally not, we will find that the constructions

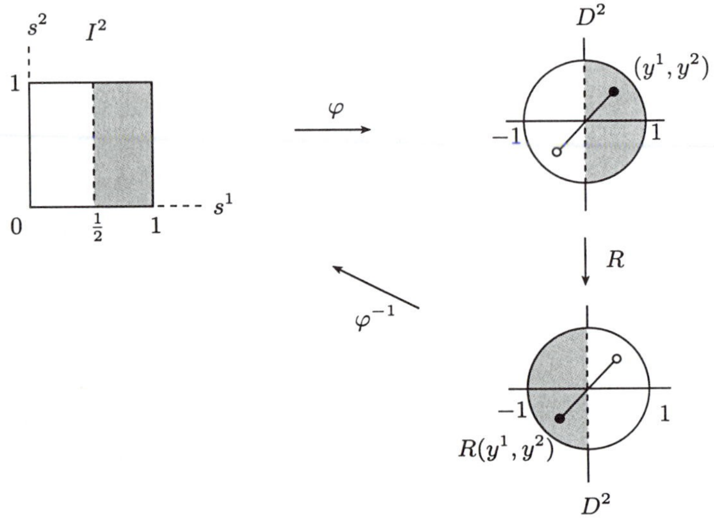

Figure 2.5.7

nevertheless share many common features. They are, for example, all "functorial" since we have the following analogue of Theorem 2.2.5.

Theorem 2.5.4 *Let (X, x_0) and (Y, y_0) be pointed spaces and $f : (X, x_0) \to (Y, y_0)$ a map. Then for each $n \geq 2$, f induces a homomorphism $f_* : \pi_n(X, x_0) \to \pi_n(Y, y_0)$ defined by $f_*([\alpha]) = [f \circ \alpha]$ for each $[\alpha] \in \pi_n(X, x_0)$. Furthermore,*

1. *If $(Y, y_0) = (X, x_0)$ and $f = id_X$, then $f_* = (id_X)_* = id_{\pi_n(X, x_0)}$.*
2. *If (Z, z_0) is another pointed space and $g : (Y, y_0) \to (Z, z_0)$ is another map, then $(g \circ f)_* = g_* \circ f_*$.*

Exercise 2.5.10 Prove Theorem 2.5.4. ∎

Next we wish to obtain a result analogous to Theorem 2.2.3 for the higher homotopy groups. Thus, we let x_0 and x_1 be points in X and suppose that there exists a path $\sigma : [0, 1] \to X$ in X from $\sigma(0) = x_0$ to $\sigma(1) = x_1$. Our objective is to show that, for any $n \geq 2$, σ induces an isomorphism

$\sigma_\# : \pi_n(X, x_1) \to \pi_n(X, x_0)$. Let $[\alpha] \in \pi_n(X, x_1)$.

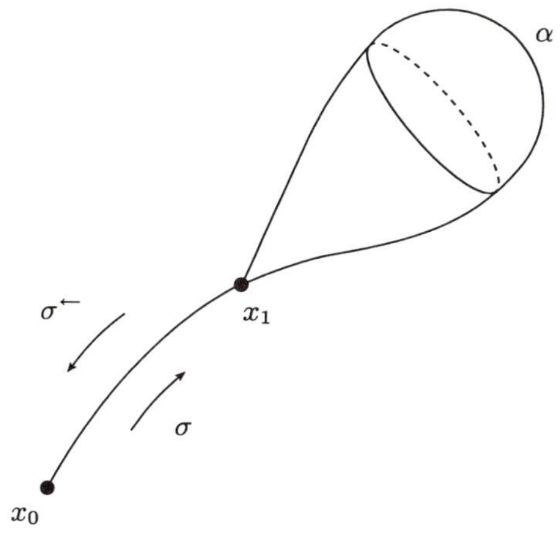

Figure 2.5.8

We define a map $A : I^n \times \{0\} \cup \partial I^n \times [0,1] \to X$ by

$$A(s,0) = \alpha(s)\,, \qquad s = (s^1, \ldots, s^n) \in I^n\,,$$
$$A(s,t) = \sigma^{\leftarrow}(t)\,, \qquad s \in \partial I^n,\ t \in [0,1]\,.$$

(see Figure 2.5.9). Since $\alpha(\partial I^n) = \{x_1\}$, A is continuous.

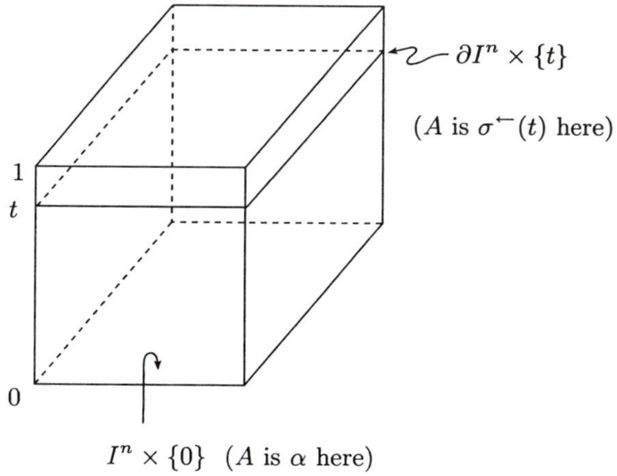

Figure 2.5.9

Thus, A is the n-loop α at height $t = 0$ and, at each height $0 < t \leq 1$, A collapses the entire $\partial I^n \times \{t\}$ to the point $\sigma^{\leftarrow}(t)$ which therefore moves toward x_0 along $\sigma^{\leftarrow}$ as t increases.

Exercise 2.5.11 Use Exercises 2.4.11 and 1.2.8 to show that there exists a retraction r of $I^n \times [0,1]$ onto $I^n \times \{0\} \cup \partial I^n \times [0,1]$.

Thus, $A \circ r : I^n \times [0,1] \to X$ is a continuous extension of A to $I^n \times [0,1]$ (intuitively, the empty box in Figure 2.5.9 is now filled in so that, at each height t, we now have an n-loop at $\sigma^{\leftarrow}(t)$ in X). Define an n-loop β at x_0 in X by $\beta(s) = A \circ r(s,1)$ for all $s \in I^n$. Then $[\beta] \in \pi_n(X, x_0)$ and we define $\sigma_\# : \pi_n(X, x_1) \to \pi_n(X, x_0)$ by

$$\sigma_\# \left([\alpha] \right) = [\beta] = [\, A \circ r(\,\cdot\,, 1) \,] \,. \tag{2.5.4}$$

We claim that $\sigma_\#$ is well-defined, depends only on the path homotopy class $[\sigma]$ of σ and is, in fact, an isomorphism. To prove all of this we begin by observing that $A \circ r$ is a special type of homotopy $F : I^n \times [0,1] \to X$ from $F_0 = \alpha$ to $F_1 = \beta$, i.e., one for which each F_t is a loop at $\sigma^{\leftarrow}(t)$ in X. We may view such an F as a continuous sequence of n-loops traveling along $\sigma^{\leftarrow}$ from x_1 to x_0 and beginning with α. Our first objective is to show that if $F : I^n \times [0,1] \to X$ is any continuous map with $F(s,0) = \alpha(s)$ for all $s \in I^n$ and $F(s,t) = (\sigma')^{\leftarrow}(t)$ for all $s \in \partial I^n$ and $t \in [0,1]$, where $\sigma' \simeq \sigma \operatorname{rel} \{0,1\}$, then $[F(\,\cdot\,,1)] = [\beta]$. This will prove simultaneously that our definition of $\sigma_\#([\alpha])$ does not depend on the choice of the retraction r and depends only on the path homotopy class of σ. We begin with a special case ($x_0 = x_1$ and $t \to F_t(s)$ a nullhomotopic loop for any $s \in \partial I^n$) that will be useful in another context as well.

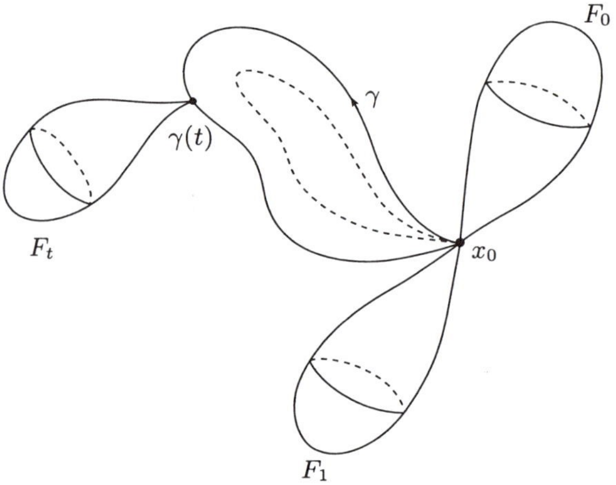

Figure 2.5.10

Lemma 2.5.5 *Let X be an arbitrary space, $x_0 \in X$, n a positive integer and $\gamma : [0,1] \to X$ a loop at x_0 in X with $\gamma \simeq x_0 \operatorname{rel}\{0,1\}$. Let $F : I^n \times [0,1] \to X$ be a homotopy with $F(s,t) = \gamma(t)$ for every $s \in \partial I^n$ and every $t \in [0,1]$. Then $F_0 \simeq F_1 \operatorname{rel} \partial I^n$, i.e., $[F_0] = [F_1]$ in $\pi_n(X, x_0)$. (See Figure 2.5.10)*

Proof: What is required here is to replace the continuous deformation F of F_0 to F_1 through a sequence of loops with varying base points by a continuous deformation of F_0 to F_1 through loops at x_0. Since γ is null-homotopic there exists a path homotopy $G : [0,1] \times [0,1] \to X$ from γ to x_0:

$$G(t,0) = \gamma(t)\,,\; G(t,1) = x_0, \quad t \in [0,1]$$

$$G(0,u) = G(1,u) = x_0, \quad u \in [0,1]\,.$$

Define a mapping $j' : I^n \times [0,1] \times \{0\} \cup \partial I^n \times [0,1] \times [0,1] \to X$ by

$$j'(s,t,0) = F(s,t), \quad s \in I^n,\; t \in [0,1]$$

$$j'(s,t,u) = G(t,u), \quad s \in \partial I^n,\; (t,u) \in [0,1] \times [0,1]$$

(see Figure 2.5.11, where we have represented I^n as being 1-dimensional).

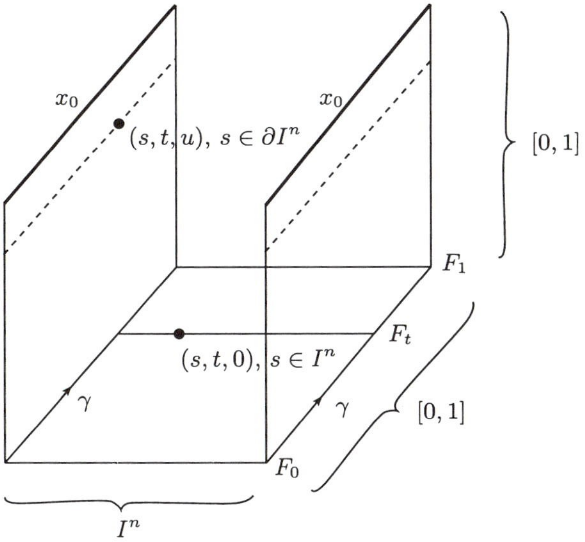

Figure 2.5.11

Exercise 2.5.12 Use Exercises 2.4.12 and 1.2.8 to show that $I^n \times [0,1] \times \{0\} \cup \partial I^n \times [0,1] \times [0,1]$ is a retract of $I^n \times [0,1] \times [0,1]$ and conclude that j' has a continuous extension $\jmath\colon I^n \times [0,1] \times [0,1] \to X$.

Intuitively, Exercise 2.5.12 has filled in the frame (base and sides) in Figure 2.5.11 with a solid box and extended j' to this box. Now we replace the homotopy F (along the base) from F_0 to F_1 by the homotopy from F_0 to F_1 that goes up the front, across the top and down the back of this box (see Figure 2.5.12).

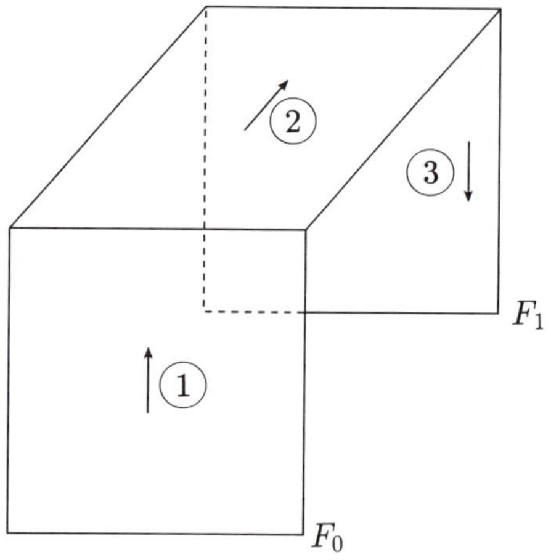

Figure 2.5.12

More precisely, we define $H_1, H_2, H_3 : I^n \times [0,1] \to X$ by $H_1(s,t) = \jmath(s,0,t)$, $H_2(s,t) = \jmath(s,t,1)$ and $H_3(s,t) = \jmath(s,1,1-t)$.

Exercise 2.5.13 Show that H_1, H_2 and H_3 are homotopies relative to ∂I^n from F_0 to x_0, x_0 to x_0 and x_0 to F_1, respectively.

Lemma 2.3.1 now gives $[F_0] = [F_1]$ so the proof of Lemma 2.5.5 is complete. ∎

Now we return to the map $\sigma_\#$ defined by (2.5.4). We consider a continuous map $F : I^n \times [0,1] \to X$ with $F(s,0) = \alpha(s)$ for all $s \in I^n$ and

$F(s,t) = (\sigma')^{\leftarrow}(t)$ for all $s \in \partial I^n$ and $t \in [0,1]$, where $\sigma' \simeq \sigma \operatorname{rel}\{0,1\}$. We claim that $[F_1] = [A \circ r(\,\cdot\,,1)]$. Define $K : I^n \times [0,1] \to X$ by

$$K(s,t) = \begin{cases} A \circ r(s, 1-2t), & 0 \le t \le \frac{1}{2} \\ F(s, 2t-1), & \frac{1}{2} \le t \le 1 \end{cases}.$$

Then K is a homotopy from $A \circ r(s,1)$ to $F_1(s)$. Moreover, if $s \in \partial I^n$, $A \circ r(s, 1-2t) = A(s, 1-2t) = \sigma^{\leftarrow}(1-2t) = \sigma(2t)$ and $F(s, 2t-1) = (\sigma')^{\leftarrow}(2t-1)$. Thus, $s \in \partial I^n$ implies $K(s,t) = \sigma(\sigma')^{\leftarrow}(t)$. Since $\sigma' \simeq \sigma \operatorname{rel}\{0,1\}$, $\sigma(\sigma')^{\leftarrow}$ is a nullhomotopic loop at x_0 in X. By Lemma 2.5.5, $[K_0] = [K_1]$, i.e., $[A \circ r(\,\cdot\,,1)] = [F_1]$ as required.

At this point we have proved the following: Suppose we are given $x_0, x_1 \in X$ and *any* representative σ of some path homotopy class $[\sigma]$ of paths from x_0 to x_1 in X. Let α be an n-loop at x_1 in X. Then one can uniquely define an element $[\beta]$ of $\pi_n(X, x_0)$ as follows: Select some homotopy $F : I^n \times [0,1] \to X$ with $F(s,0) = \alpha(s)$ for all $s \in I^n$ and $F(s,t) = \sigma^{\leftarrow}(t)$ for all $s \in \partial I^n$ and $t \in [0,1]$ (e.g., $A \circ r$ as described above). Then $[\beta] = [F_1]$. To show that our map $\sigma_\# : \pi_n(X, x_1) \to \pi_n(X, x_0)$ is well-defined we need only show that if $\alpha' \simeq \alpha \operatorname{rel} \partial I^n$, then $[\beta'] = [\beta]$. Let F' be a homotopy relative to ∂I^n from α' to α and define $F' + F : I^n \times [0,1] \to X$ by

$$(F'+F)(s,t) = \begin{cases} F'(s, 2t), & 0 \le t \le \frac{1}{2} \\ F(s, 2t-1), & \frac{1}{2} \le t \le 1 \end{cases}.$$

Then $F' + F$ is a homotopy from $(F'+F)(s,0) = F'(s,0) = \alpha'(s)$ to $(F'+F)(s,1) = F(s,1) = F_1(s)$ which, for $s \in \partial I^n$, satisfies

$$(F'+F)(s,t) = \begin{cases} x_1, & 0 \le t \le \frac{1}{2} \\ \sigma^{\leftarrow}(2t-1), & \frac{1}{2} \le t \le 1 \end{cases} = x_1 \sigma^{\leftarrow} = (\sigma x_1)^{\leftarrow}.$$

But $\sigma x_1 \simeq \sigma \operatorname{rel}\{0,1\}$ so, by what we have shown above, $[\beta'] = [F_1] = [\beta]$. Thus, $\sigma_\# : \pi_n(X, x_1) \to \pi_n(X, x_0)$ is well-defined.

Next we show that $\sigma_\#$ is a homomorphism. Let $[\alpha_1], [\alpha_2] \in \pi_n(X, x_1)$. Let $F, G : I^n \times [0,1] \to X$ be homotopies with $F(s,0) = \alpha_1(s)$ and $G(s,0) = \alpha_2(s)$ for all $s \in I^n$ and, for $s \in \partial I^n$, $F(s,t) = G(s,t) = \sigma^{\leftarrow}(t)$ for all $t \in [0,1]$. Define $H : I^n \times [0,1] \to X$ by

$$H(s^1, s^2, \ldots, s^n, t) = \begin{cases} F(2s^1, s^2, \ldots, s^n, t), & 0 \le s^1 \le \frac{1}{2} \\ G(2s^1 - 1, s^2, \ldots, s^n, t), & \frac{1}{2} \le s^1 \le 1 \end{cases}.$$

Then H is continuous since $F(1, s^2, \ldots, s^n, t) = \sigma^{\leftarrow}(t) = G(0, s^2, \ldots, s^n, t)$. Moreover, $H(s,0) = (\alpha_1 + \alpha_2)(s)$ and, for $s \in \partial I^n$, $H(s,t) = \sigma^{\leftarrow}(t)$. Thus, $\sigma_\#([\alpha_1] + [\alpha_2]) = \sigma_\#([\alpha_1 + \alpha_2]) = [H_1]$. But $\sigma_\#([\alpha_1]) + \sigma_\#([\alpha_2]) =$

$[F_1]+[G_1]=[F_1+G_1]=[H_1]=\sigma_\#([\alpha_1]+[\alpha_2])$ so $\sigma_\#$ is, indeed, a homomorphism.

Exercise 2.5.14 Show that $(\sigma^\leftarrow)_\# : \pi_n(X,x_0)\to\pi_n(X,x_1)$ is the inverse of $\sigma_\#$ and conclude that $\sigma_\#$ is an isomorphism.

Thus, we have proved the following result.

Theorem 2.5.6 *Let x_0 and x_1 be two points in an arbitrary space X and suppose there exists a path $\sigma:[0,1]\to X$ in X from $\sigma(0)=x_0$ to $\sigma(1)=x_1$. For each $n\geq 2$, define a map $\sigma_\# : \pi_n(X,x_1)\to\pi_n(X,x_0)$ as follows: Fix $[\alpha]\in\pi_n(X,x_1)$. Select a homotopy $F: I^n\times[0,1]\to X$ with $F(s,0)=\alpha(s)$ for $s\in I^n$ and, for $s\in\partial I^n$, $F(s,t)=\sigma^\leftarrow(t)$ for all $t\in[0,1]$ (e.g., $A\circ r$ as in (2.5.4)). Set $\sigma_\#([\alpha])=[F_1]$, where $F_1(s)=F(s,1)$ for all $s\in I^n$. Then $\sigma_\#$ is a well-defined isomorphism that depends only on the path homotopy class $[\sigma]$ of σ, i.e., if $\sigma'\simeq\sigma\operatorname{rel}\{0,1\}$, then $(\sigma')_\#=\sigma_\#$.*

In particular, if X happens to be pathwise connected, then any two of its nth homotopy groups $\pi_n(X,x_0)$, $\pi_n(X,x_1),\ldots$ are isomorphic and one often omits reference to the base point altogether, writing $\pi_n(X)$ and referring to this as the **$\mathbf{n}^{\text{th}}$ homotopy group of X**. If $\pi_n(X)$ is the trivial group we will simply write $\pi_n(X)=0$. At this point, of course, we still suffer from a rather embarrassing dearth of examples (none, as it were). These are, in fact, notoriously difficult to come by and when all is said and done we will still have precious few. The search is made somewhat less arduous by the fact that, like π_1, π_n is a homotopy invariant. To prove this we first obtain an analogue of Theorem 2.3.8.

Theorem 2.5.7 *Let $f:(X,x_0)\to(Y,y_0)$ and $g:(X,x_0)\to(Y,y_1)$ be homotopic maps with $F: X\times[0,1]\to Y$ a homotopy satisfying $F(x,0)=f(x)$ and $F(x,1)=g(x)$ for all $x\in X$. Let $\sigma:[0,1]\to Y$ be the path in Y from y_0 to y_1 given by $\sigma(t)=F(x_0,t)$. Then, for each $n\geq 2$, the diagram*

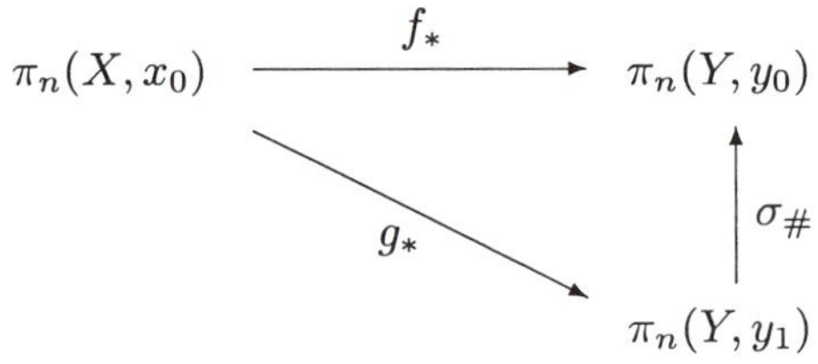

commutes, i.e., $f_*=\sigma_\#\circ g_*$. Moreover, f_* is an isomorphism iff g_* is an isomorphism.

Proof: Let $[\alpha]\in\pi_n(X,x_0)$. We must show that $\sigma_\#(g_*([\alpha]))=f_*([\alpha])$, i.e., that $\sigma_\#([g\circ\alpha])=[f\circ\alpha]$. Define $F': I^n\times[0,1]\to X$ by $F'(s,t)=$

$F(\alpha(s), 1-t)$. Then F' is a homotopy with $F'(s,0) = F(\alpha(s),1) = g(\alpha(s)) = (g \circ \alpha)(s)$ and $F'(s,1) = F(\alpha(s),0) = f(\alpha(s)) = (f \circ \alpha)(s)$. Moreover, for $s \in \partial I^n$, $F'(s,t) = F(\alpha(s), 1-t) = F(x_0, 1-t) = \sigma(1-t) = \sigma^{\leftarrow}(t)$. According to Theorem 2.5.6, $\sigma_{\#}([g \circ \alpha]) = [F_1'] = [f \circ \alpha]$ as required. Since $\sigma_{\#}$ is an isomorphism, it is clear that f_* is an isomorphism iff g_* is an isomorphism. ∎

Remark: Observe that if $y_1 = y_0$ and F is a homotopy relative to x_0, then $\sigma_{\#}$ is the identity map and $f_* = g_*$.

Corollary 2.5.8 *Let $h : X \to Y$ be a homotopy equivalence. Then, for every $x_0 \in X$ and every $n \geq 2$, $h_* : \pi_n(X, x_0) \to \pi_n(Y, h(x_0))$ is an isomorphism.*

Exercise 2.5.15 Prove Corollary 2.5.8. **Hint:** Follow the proof of Corollary 2.3.9. ∎

Since a contractible space is homotopically equivalent to a point (Theorem 2.3.7) and since any n-loop in a one-point space is obviously constant we obtain the following (admittedly, rather dull) class of examples.

Corollary 2.5.9 *Let X be a contractible space. Then, for any $n \geq 1$, $\pi_n(X) = 0$.*

Exercise 2.5.16 Show that, for any (X, x_0) and (Y, y_0) and any $n \geq 2$,

$$\pi_n\left(X \times Y,\ (x_0, y_0)\right) \cong \pi_n(X, x_0) \times \pi_n(Y, y_0).$$

Hint: Follow the proof of Theorem 2.4.10.

We wish next to compute all of the higher homotopy groups of the circle S^1. Recalling our experience with $\pi_1(S^1)$ it should come as no surprise that covering spaces will have a role to play in the calculation. Indeed, when $n \geq 2$, the relationship between $\pi_n(X)$ and $\pi_n(\tilde{X})$, where $\mathcal{P} : \tilde{X} \to X$ is a covering space, is even more direct.

Theorem 2.5.10 *Let $\mathcal{P} : \tilde{X} \to X$ be a covering space, x_0 a point in X and $\tilde{x}_0$ a point in $\mathcal{P}^{-1}(x_0)$. Then, for each $n \geq 2$, $\mathcal{P}_* : \pi_n(\tilde{X}, \tilde{x}_0) \to \pi_n(X, x_0)$ is an isomorphism.*

Proof: $\mathcal{P}_*$ is a homomorphism by Theorem 2.5.4. We show that it is surjective and injective. For surjectivity, we let $[\alpha] \in \pi_n(X, x_0)$ and seek a $[\tilde{\alpha}] \in \pi_n(\tilde{X}, \tilde{x}_0)$ with $\mathcal{P}_*([\tilde{\alpha}]) = [\alpha]$, i.e., $[\mathcal{P} \circ \tilde{\alpha}] = [\alpha]$. Thus, it will suffice to show that any n-loop α at x_0 in X lifts to an n-loop $\tilde{\alpha}$ at $\tilde{x}_0$ in $\tilde{X}$. Now, I^n is convex and therefore contractible so, by Exercise 2.3.6, α

is nullhomotopic. Since α must map into the path component of X containing x_0, Exercise 2.3.7 guarantees that α is homotopic to the constant n-loop x_0. Let $F : I^n \times [0,1] \to X$ be a homotopy with $F(s,0) = x_0$ and $F(s,1) = \alpha(s)$. Now, the constant n-loop at $\tilde{x}_0$ in $\tilde{X}$ clearly lifts x_0 to $\tilde{X}$. By the Homotopy Lifting Theorem 2.4.1, F lifts to a homotopy $\tilde{F} : I^n \times [0,1] \to \tilde{X}$ with $\tilde{F}(s,0) = \tilde{x}_0$. Define $\tilde{\alpha}$ by $\tilde{\alpha}(s) = \tilde{F}(s,1)$. Then $\mathcal{P} \circ \tilde{\alpha}(s) = \mathcal{P} \circ \tilde{F}(s,1) = F(s,1) = \alpha(s)$ so $\tilde{\alpha}$ lifts α. Moreover, $\tilde{\alpha}$ is an n-loop at $\tilde{x}_0$. Indeed, $s \in \partial I^n$ implies $\mathcal{P} \circ \tilde{\alpha}(s) = \mathcal{P} \circ \tilde{F}(s,1) = F(s,1) = \alpha(s) = x_0$ so $\tilde{\alpha}(s) \in \mathcal{P}^{-1}(x_0)$. But $\mathcal{P}^{-1}(x_0)$ is discrete and ∂I^n is connected so $\tilde{\alpha}(\partial I^n)$ must be a single point. Since $\tilde{F}$ must carry $I^n \times [0,1]$ into a single sheet of the covering and $\tilde{F}(s,0) = \tilde{x}_0$, we conclude that $\tilde{\alpha}(\partial I^n) = \{\tilde{x}_0\}$. Thus, $\tilde{\alpha}$ is the required n-loop at $\tilde{x}_0$ in $\tilde{X}$.

To prove that $\mathcal{P}_*$ is injective it will suffice to show that if $\tilde{\alpha}$ and $\tilde{\beta}$ are two n-loops at $\tilde{x}_0$ in $\tilde{X}$ with $\tilde{\alpha} \not\simeq \tilde{\beta} \operatorname{rel} \partial I^n$, then $\mathcal{P} \circ \tilde{\alpha} \not\simeq \mathcal{P} \circ \tilde{\beta} \operatorname{rel} \partial I^n$. But if F were a homotopy, relative to ∂I^n, of $\mathcal{P} \circ \tilde{\alpha}$ to $\mathcal{P} \circ \tilde{\beta}$, the Homotopy Lifting Theorem 2.4.1 would imply that F lifts to a homotopy $\tilde{F}$ of $\tilde{\alpha}$ to $\tilde{\beta}$. Moreover, we ask the reader to show that, because $\mathcal{P} : \tilde{X} \to X$ is a covering space, this homotopy must be relative to ∂I^n.

Exercise 2.5.17 Show that $s \in \partial I^n$ implies $\tilde{F}(s,t) = \tilde{x}_0$ for all $t \in [0,1]$.

Thus, $\mathcal{P} \circ \tilde{\alpha} \simeq \mathcal{P} \circ \tilde{\beta} \text{ rel } \partial I^n$ implies $\tilde{\alpha} \simeq \tilde{\beta} \text{ rel } \partial I^n$ so the proof is complete. ∎

Corollary 2.5.11 *For any $n \geq 2$, $\pi_n(S^1) = 0$.*

Proof: S^1 is the base space of the covering space $\mathcal{P} : \mathbb{R} \to S^1$ given by $\mathcal{P}(s) = e^{2\pi s \mathbf{i}}$. By Theorem 2.5.10, $\pi_n(S^1) \cong \pi_n(\mathbb{R})$. But $\mathbb{R}$ is contractible so Corollary 2.5.9 implies that $\pi_n(\mathbb{R}) = 0$. ∎

Corollary 2.5.12 *For any $n \geq 2$ and any $m \geq 1$, $\pi_n(\mathbb{R}\mathbb{P}^m) \cong \pi_n(S^m)$.*

Exercise 2.5.18 Prove Corollary 2.5.12. ∎

Corollary 2.5.12 is something of a triviality when $m = 1$ (see (1.2.7)) and would no doubt be rather more impressive when $m > 1$ if only we knew a few of the groups $\pi_n(S^m)$ for $n \geq 2$. As it happens, however, the problem of calculating the higher homotopy groups of the spheres is one of the deepest and most difficult in topology. The subject has a vast literature, virtually all of which is inaccessible with the crude tools we have developed thus far (see [**Rav**]). The "trivial case" (by the standards of homotopy theory) is that in which $n \leq m$. Here the result is at least easy to state:

$$\pi_n(S^m) \cong \begin{cases} 0, & \text{if } n < m \\ \mathbb{Z}, & \text{if } n = m \end{cases} . \tag{2.5.5}$$

The proof, however, is by no means "trivial". In particular, $\pi_n(S^n) \cong \mathbb{Z}$ follows from a deep theorem of Heinz Hopf concerning the "degree" of a map from S^n to S^n. To understand what is involved here (and for our classification of principal bundles over S^n in Chapter 3 as well) we must obtain a generalization of Lemma 2.3.4 for the higher homotopy groups. For the proof of Lemma 2.3.4 we identified S^1 with the quotient of $I = [0,1]$ obtained by identifying $\partial I = \{0,1\}$ to a point. Analogously, we will now identify S^n with the quotient of I^n obtained by identifying ∂I^n to a point. To construct a concrete realization of the quotient map we begin with the homeomorphism $\varphi : I^n \to D^n$ of Exercise 1.2.8. This carries ∂I^n onto the boundary sphere S^{n-1} of D^n.

Exercise 2.5.19 Show that the map of $D^n - S^{n-1}$ to $\mathbb{R}^n$ defined by

$$x \longrightarrow y = \frac{x}{\sqrt{1 - \| x \|^2}}$$

is a homeomorphism.

Now, the inverse of the stereographic projection $\varphi_S : S^n - \{N\} \to \mathbb{R}^n$ (Exercise 1.1.8) carries $\mathbb{R}^n$ onto S^n minus the north pole $N = (0, \ldots, 0, 1)$.

Exercise 2.5.20 Show that the composition $x \to y \to \varphi_S^{-1}(y)$ of the homeomorphism in Exercise 2.5.19 and φ_S^{-1} is given by

$$x = (x^1, \ldots, x^n) \longrightarrow \left(2x^1\sqrt{1 - \| x \|^2}, \ldots, 2x^n\sqrt{1 - \| x \|^2}, 2 \| x \|^2 - 1\right).$$

Next observe that the map in Exercise 2.5.20 is actually defined and continuous on all of D^n and carries S^{n-1} onto $\{N\}$. Thus, we have defined a continuous map $k : D^n \to S^n$ by

$$\begin{aligned} k(x) &= k(x^1, \ldots, x^n) \\ &= \left(2x^1\sqrt{1 - \| x \|^2}, \ldots, 2x^n\sqrt{1 - \| x \|^2},\ 2 \| x \|^2 - 1\right) \end{aligned} \tag{2.5.6}$$

with $k(\partial D^n) = k(S^{n-1}) = \{N\}$. Consequently, $\mathcal{Q} = k \circ \varphi : I^n \to S^n$ is a continuous map. Moreover, it maps $I^n - \partial I^n$ homeomorphically onto $S^n - \{N\}$ and sends all of ∂I^n onto $\{N\}$. With this we are in a position to establish a one-to-one correspondence between $\pi_n(X, x_0)$ and $[(S^n, N), (X, x_0)]$ exactly as in Lemma 2.3.4. Specifically, given an n-loop $\alpha : I^n \to X$ at x_0 in X one defines a map $\tilde{\alpha} : (S^n, N) \to (X, x_0)$ by $\tilde{\alpha}(\mathcal{Q}(s)) = \alpha(s)$. Since I^n is compact, $\mathcal{Q}$ is a closed map and therefore a quotient map (Exercise 1.3.13) so the continuity of $\tilde{\alpha}$ is guaranteed by Lemma 1.2.1. Moreover, $\tilde{\alpha}(N) = \tilde{\alpha}(\mathcal{Q}(\partial I^n)) = \alpha(\partial I^n) = x_0$. Conversely, given $\tilde{\alpha} : (S^n, N) \to (X, x_0)$ one defines an n-loop α at x_0 in X by

$\alpha(s) = \tilde{\alpha}(\mathcal{Q}(s))$.

Exercise 2.5.21 Let α and α' be two n-loops at x_0 in X and let $\tilde{\alpha}$ and $\tilde{\alpha}'$ be the corresponding maps of (S^n, N) to (X, x_0) just defined. Show that $\alpha' \simeq \alpha \operatorname{rel} \partial I^n$ iff $\tilde{\alpha}' \simeq \tilde{\alpha} \operatorname{rel} \{N\}$.

Lemma 2.5.13 *Let X be an arbitrary topological space and $x_0 \in X$. Let $n \geq 2$ be an integer and p_0 some fixed point in S^n. Then there is a one-to-one correspondence between $\pi_n(X, x_0)$ and $[(S^n, p_0), (X, x_0)]$.*

Exercise 2.5.22 Prove Lemma 2.5.13. **Hint:** For $p_0 = N$ the result is immediate from Exercise 2.5.21. Then show that one can equally well define a map $\mathcal{Q} : I^n \to S^n$ that maps $I^n - \partial I^n$ homeomorphically onto $S^n - \{p_0\}$ and sends all of ∂I^n onto $\{p_0\}$ and repeat the arguments leading up to Exercise 2.5.21.

It will be of considerable significance to us that Lemma 2.5.5 can be refashioned into a statement about maps on the n-sphere.

Lemma 2.5.14 *Let X be an arbitrary topological space, $x_0 \in X$ and $\gamma : [0,1] \to X$ a loop at x_0 in X with $\gamma \simeq x_0 \operatorname{rel} \{0,1\}$. Let n be a positive integer, $p_0 \in S^n$ and $G : S^n \times [0,1] \to X$ a homotopy with $G(p_0, t) = \gamma(t)$ for every $t \in [0,1]$. Then $G_0 \simeq G_1 \operatorname{rel}\{p_0\}$.*

Proof: It will suffice to consider the case in which $p_0 = N$ and $\mathcal{Q} : I^n \to S^n$ is the quotient map constructed above (so that $\mathcal{Q}(\partial I^n) = \{N\}$). Define $F : I^n \times [0,1] \to X$ by $F(s,t) = G(\mathcal{Q}(s), t)$. Then $s \in \partial I^n$ implies $F(s,t) = G(N,t) = \gamma(t)$ so Lemma 2.5.5 gives $F_0 \simeq F_1 \operatorname{rel} \partial I^n$. Thus, $\tilde{F}_0 \simeq \tilde{F}_1 \operatorname{rel} \{N\}$ by Exercise 2.5.21. But $\tilde{F}_0(\mathcal{Q}(s)) = F_0(s) = F(s,0) = G(\mathcal{Q}(s), 0) = G_0(\mathcal{Q}(s))$ so $\tilde{F}_0 = G_0$. Similarly, $\tilde{F}_1 = G_1$ so the result follows. ∎

We are now in a position to prove the remarkable, and very useful, fact that, when X is simply connected, the relative homotopies that go into the definition of $[(S^n, p_0), (X, x_0)]$ can be replaced by free homotopies.

Theorem 2.5.15 *Let X be a simply connected space and $x_0 \in X$. Let n be a positive integer, $p_0 \in S^n$ and $f, g : (S^n, p_0) \to (X, x_0)$. Then f and g are homotopic relative to p_0 ($f \simeq g \operatorname{rel}\{p_0\}$) iff they are homotopic ($f \simeq g$).*

Proof: Since $f \simeq g \operatorname{rel} \{p_0\}$ always implies $f \simeq g$, we need only prove the converse. Suppose $G : S^n \times [0,1] \to X$ satisfies $G_0 = f$ and $G_1 = g$. Then $\gamma(t) = G(p_0, t)$ is a path in X from $\gamma(0) = G(p_0, 0) = x_0$ to $\gamma(1) = G(p_0, 1) = g(p_0) = x_0$, i.e., γ is a loop at x_0 in X. Since X is simply connected, $\gamma \simeq x_0 \operatorname{rel}\{0,1\}$ so Lemma 2.5.14 gives $f \simeq g \operatorname{rel} \{p_0\}$. ∎

Now, with the one-to-one correspondence we have established between $[(S^n, p_0), (X, x_0)]$ and $\pi_n(X, x_0)$, one can transfer the algebraic structure of $\pi_n(X, x_0)$ to $[(S^n, p_0), (X, x_0)]$ and thereby obtain a group isomorphic to $\pi_n(X, x_0)$. In particular, $\pi_n(S^n)$ may be thought of as $[(S^n, N), (S^n, N)]$. Furthermore, two maps $(S^n, N) \to (S^n, N)$ are in the same equivalence class (i.e., are homotopic relative to $\{N\}$) iff they are (freely) homotopic. The usual proof that $\pi_n(S^n) \cong \mathbb{Z}$ proceeds by assigning to each continuous map $f : S^n \to S^n$ an integer, called the *(Brouwer) degree* of f and denoted $\deg f$, which satisfies

i. $f \simeq g \Rightarrow \deg f = \deg g$.
ii. $\deg f = \deg g \Rightarrow f \simeq g$ *(Hopf's Theorem)*.
iii. $\deg(id_{S^n}) = 1$.

Condition (i) implies, in particular, that the Brouwer degree provides a well-defined map $\deg : [(S^n, N), (S^n, N)] \to \mathbb{Z}$ and one shows that, when $[(S^n, N), (S^n, N)]$ has the group structure of $\pi_n(S^n, N)$ and $\mathbb{Z}$ has its customary additive group structure, this map is a homomorphism. Condition (ii) and Theorem 2.5.15 imply that this homomorphism is one-to-one. Since a homomorphism to $\mathbb{Z}$ that takes the value 1 must clearly be surjective, condition (iii) implies that deg is onto and therefore an isomorphism.

There are various ways of defining the Brouwer degree and establishing the required properties, none of which is easy, however. The most common approach is to apply the machinery of homology theory (see [**Hu**]), although it is possible to prune away much of the excess baggage and jargon and be left with an essentially "elementary" treatment (this is done in Sections 1 and 7, Chapter $\overline{\text{XVI}}$, of [**Dug**], which the reader who has followed us thus far is now fully prepared to assimilate). Another approach, about which we will have more to say in Chapter 4, utilizes the differentiable structure of the sphere and properties of smooth maps on it. Whatever the approach, however, the details of the argument, especially for Hopf's Theorem (property (ii) above) are formidable and, except for the discussion in Section 4.9, we will forego them here.

Exercise 2.5.23 Assume that $\pi_n(S^n) \cong \mathbb{Z}$. Use this fact to prove that S^n is not a retract of D^{n+1} and deduce from this the general Brouwer Fixed Point Theorem (D^{n+1} has the fixed point property). **Hint:** Mimic the proof of Lemma 2.4.6.

Remarks: Although we will have no occasion to utilize it, we would feel remiss if we failed to make the reader aware of a beautiful "interlacing" of the homotopy groups of the base, fiber and total space of a locally trivial bundle $(P, X, \mathcal{P}, Y)$. It goes like this: Select $x_0 \in X$, let $Y_0 = \mathcal{P}^{-1}(x_0)$ be the fiber above x_0 and select $y_0 \in Y_0$. Also let $\iota : Y_0 \hookrightarrow P$ be the inclusion map. For each $n \geq 1$, we have homomorphisms $\iota_* : \pi_n(Y_0, y_0) \to \pi_n(P, y_0)$ and $\mathcal{P}_* : \pi_n(P, y_0) \to \pi_n(X, x_0)$. One can also define a homomorphism

$\partial : \pi_n(X, x_0) \to \pi_{n-1}(Y_0, y_0)$ for each $n \geq 2$ in such a way that the following sequence is *exact* (i.e., the kernel of each map is the image of the map that precedes it):

$$\cdots \longrightarrow \pi_n(Y_0, y_0) \xrightarrow{\iota_*} \pi_n(P, y_0) \xrightarrow{\mathcal{P}_*} \pi_n(X, x_0) \xrightarrow{\partial} \pi_{n-1}(Y_0, y_0) \longrightarrow \cdots$$

To see where the maps ∂ come from we proceed as follows: Fix an $[\alpha]$ in $\pi_n(X, x_0)$. Then α is an n-loop at x_0 in X so $\alpha : I^n \to X$ with $\alpha(\partial I^n) = \{x_0\}$. Now regard I^n as $I^{n-1} \times [0,1]$ so that α can be thought of as a homotopy beginning at the constant map of I^{n-1} to X whose value is x_0. The constant map of I^{n-1} to P whose value is y_0 is clearly a lift of this to P. Thus, the Homotopy Lifting Theorem 2.4.1 implies that α lifts to a homotopy $\tilde{\alpha} : I^{n-1} \times [0,1] \to P$ beginning at the constant map y_0. Now, $\tilde{\alpha}(s^1, \ldots, s^{n-1}, 1)$ is an $(n-1)$-loop at y_0 in P whose image is in Y_0 since $\mathcal{P} \circ \tilde{\alpha} = \alpha$ and $\alpha(s^1, \ldots, s^{n-1}, 1) = x_0$. The homotopy class of this $(n-1)$-loop is $\partial([\alpha])$. One then verifies that ∂ is well-defined, a homomorphism and that the sequence above is exact (it is called the *Homotopy Exact Sequence* of the bundle). This sequence provides a very powerful tool. For example, applied to the Hopf bundle $S^1 \to S^3 \to S^2$ it yields (for $n = 3$) the exact sequence

$$\pi_3(S^1) \longrightarrow \pi_3(S^3) \longrightarrow \pi_3(S^2) \longrightarrow \pi_2(S^1)\,,$$

i.e.,

$$0 \longrightarrow \mathbb{Z} \longrightarrow \pi_3(S^2) \longrightarrow 0\,.$$

Since the first map is trivially one-to-one and the last is trivially onto, the exactness of the sequence implies at once that the map in the center is an isomorphism, i.e., $\pi_3(S^2) \cong \mathbb{Z}$. Indeed, this calculation was the motivation behind Hopf's construction of the bundle.

Exercise 2.5.24 Show that $\pi_n(S^3) \cong \pi_n(S^2)$ for all $n \geq 3$.

In order that the reader might gain some sense of the extraordinary unpredictability of the homotopy groups of spheres we close with a little table. There is a classical theorem of Freudenthal according to which $\pi_{n+k}(S^n)$ depends only on k provided $n > k+1$. The group $\pi_{n+k}(S^n)$, $n > k+1$, is denoted π_k^S and called the **k th stable homotopy group of spheres**. We have already mentioned ((2.5.5)) that $\pi_0^S = \mathbb{Z}$. Here are a few more.

$\pi_1^S = \mathbb{Z}_2$	$\pi_6^S = \mathbb{Z}_2$	$\pi_{11}^S = \mathbb{Z}_{504}$
$\pi_2^S = \mathbb{Z}_2$	$\pi_7^S = \mathbb{Z}_{240}$	$\pi_{12}^S = 0$
$\pi_3^S = \mathbb{Z}_{24}$	$\pi_8^S = \mathbb{Z}_2$	$\pi_{13}^S = \mathbb{Z}_3$
$\pi_4^S = 0$	$\pi_9^S = \mathbb{Z}_2 \oplus \mathbb{Z}_2 \oplus \mathbb{Z}_2$	$\pi_{14}^S = \mathbb{Z}_2 \oplus \mathbb{Z}_2$
$\pi_5^S = 0$	$\pi_{10}^S = \mathbb{Z}_6$	$\pi_{15}^S = \mathbb{Z}_{480} \oplus \mathbb{Z}_2$

Those who need to find some order in this chaos are referred to [**Rav**].

3

Principal Bundles

3.1 C^0 Principal Bundles

In this chapter we meld together locally trivial bundles and group actions to arrive at the notion of a C^0 (continuous) principal bundle (smoothness hypotheses are added in Chapter 4). The source of our interest in these structures was discussed at some length in Chapter 0, where we also suggested that principal bundles over spheres were of particular significance. Our goal here is to use the homotopy-theoretic information assembled in Chapter 2 to classify the principal bundles over S^n.

Let X be a Hausdorff topological space and G a topological group. Then a **C^0** (or **continuous**) **principal bundle over X with structure group** G (or, simply, a **C^0 G-bundle over** X) is a triple $\mathcal{B} = (P, \mathcal{P}, \sigma)$, where P is a topological space, $\mathcal{P}$ is a continuous map of P onto X and $\sigma : P \times G \to P$, $\sigma(p, g) = p \cdot g$, is a right action of G on P such that the following conditions are satisfied:

1. σ preserves the fibers of $\mathcal{P}$, i.e.,
$$\mathcal{P}(p \cdot g) = \mathcal{P}(p) \tag{3.1.1}$$
for all $p \in P$ and $g \in G$.

2. (**Local Triviality**) For each $x_0 \in X$ there exists an open nbd V of x_0 in X and a homeomorphism $\Psi : \mathcal{P}^{-1}(V) \to V \times G$ of the form
$$\Psi(p) = \left(\mathcal{P}(p), \psi(p)\right) , \tag{3.1.2}$$
where $\psi : \mathcal{P}^{-1}(V) \to G$ satisfies
$$\psi(p \cdot g) = \psi(p)g \tag{3.1.3}$$
for all $p \in \mathcal{P}^{-1}(V)$ and $g \in G$ ($\psi(p)g$ is the product in G of $\psi(p)$ and g).

In particular, $(P, X, \mathcal{P}, G)$ is a locally trivial bundle with local trivializations (V, Φ), where $\Phi = \Psi^{-1}$ (although we will often refer to (V, Ψ) itself, or even just Ψ, as a local trivialization as well). Consequently, P is necessarily a Hausdorff space (Exercise 1.3.23). Very often the intended group action will be clear from the context and we will feel free to refer to $\mathcal{P} : P \to X$ as a principal G-bundle. If the projection map $\mathcal{P}$ is also understood from the context we will say simply that P is a principal G-bundle over X and indicate this diagramatically by writing $G \to P \to X$.

Condition (3.1.1) asserts that σ acts on the bundle space P fiberwise. The significance of (3.1.3) will emerge from the following lemma.

Lemma 3.1.1 *For each $p \in P$, the fiber above $\mathcal{P}(p)$ coincides with the orbit of p under σ, i.e.,*

$$\mathcal{P}^{-1}(\mathcal{P}(p)) = \{p \cdot g : g \in G\} = p \cdot G .$$

Proof: $\mathcal{P}^{-1}(\mathcal{P}(p)) \supseteq \{p \cdot g : g \in G\}$ is immediate from (3.1.1). For the reverse containment, let $p' \in \mathcal{P}^{-1}(\mathcal{P}(p))$. We show that there is a $g \in G$ such that $p' = p \cdot g$. Choose V and Ψ at $x_0 = \mathcal{P}(p) = \mathcal{P}(p')$ as in condition (2) of the definition. Then $\psi(p)$ and $\psi(p')$ are in G so there is a g in G (namely, $(\psi(p))^{-1}\psi(p')$) such that $\psi(p)g = \psi(p')$. Thus, $\psi(p \cdot g) = \psi(p')$ so $\Psi(p \cdot g) = (\mathcal{P}(p \cdot g), \psi(p \cdot g)) = (\mathcal{P}(p), \psi(p')) = (\mathcal{P}(p'), \psi(p')) = \Psi(p')$. Since Ψ is one-to-one, $p \cdot g = p'$ as required. ∎

Thus, identifying a fiber with G via ψ, condition (3.1.3) asserts that the action of σ on fibers is "right multiplication by elements of G".

Exercise 3.1.1 Show that the action of G on P is necessarily free, but generally not transitive.

The simplest example of a G-bundle over X is the trivial bundle $(X \times G, X, \mathcal{P}, G)$, where $\mathcal{P} : X \times G \to X$ is the projection onto the first factor and the action of G on $X \times G$ is defined by $(x, h) \cdot g = (x, hg)$. In this case one takes V in condition (2) to be all of X and Ψ to be the identity on $\mathcal{P}^{-1}(V) = X \times G$. This is the **trivial G-bundle over X**. We have seen much more interesting examples than this, however. Consider, for example, the locally trivial bundle $(S^{n-1}, \mathbb{R}\mathbb{P}^{n-1}, \mathcal{P}, \mathbb{Z}_2)$, where $\mathcal{P} : S^{n-1} \to \mathbb{R}\mathbb{P}^{n-1}$ identifies antipodal points (Section 1.3). We define a natural action of $\mathbb{Z}_2 = \{-1, 1\}$ (with the discrete topology) on S^{n-1} by $p \cdot g = (x^1, \ldots, x^n) \cdot g = (x^1 g, \ldots, x^n g)$ for all $p = (x^1, \ldots, x^n)$ in S^{n-1} and all g in $\mathbb{Z}_2$. Thus, $p \cdot 1 = p$ and $p \cdot (-1) = -p$ for all $p \in S^{n-1}$ so $\mathcal{P}(p \cdot g) = [p \cdot g] = [\pm p] = [p] = \mathcal{P}(p)$. Furthermore, if we let (V_k, Φ_k), $k = 1, \ldots, n$, be the local trivializations described in Section 1.3 (Exercise 1.3.19) and define $\Psi_k : \mathcal{P}^{-1}(V_k) \to V_k \times \mathbb{Z}_2$ by $\Psi_k = \Phi_k^{-1}$, then $\Psi_k(p) = (\mathcal{P}(p), \psi_k(p))$, where $\psi_k : \mathcal{P}^{-1}(V_k) \to \mathbb{Z}_2$ takes the value 1 on U_k^+ and -1 on U_k^-. In particular, $\psi_k(p \cdot 1) = \psi_k(p) = \psi_k(p)1$ and $\psi_k(p \cdot (-1)) = \psi_k(-p) = -\psi_k(p) = \psi_k(p)(-1)$ so (3.1.3) is satisfied. Thus, S^{n-1} is a principal $\mathbb{Z}_2$-bundle over $\mathbb{R}\mathbb{P}^{n-1}$.

Next consider the locally trivial bundle $(S^{2n-1}, \mathbb{C}\mathbb{P}^{n-1}, \mathcal{P}, S^1 = U(1))$, also described in Section 1.3. Exercise 1.6.11 gives a natural action of $U(1)$ on S^{2n-1} defined by $p \cdot g = (z^1, \ldots, z^n) \cdot g = (z^1 g, \ldots, z^n g)$ which, as above for S^{n-1}, satisfies $\mathcal{P}(p \cdot g) = \mathcal{P}(p)$. We have also defined open sets $V_k = \{[z^1, \ldots, z^n] \in \mathbb{C}\mathbb{P}^{n-1} : z^k \neq 0\}$, $k = 1, \ldots, n$, and homeomorphisms $\Psi_k : \mathcal{P}^{-1}(V_k) \to V_k \times U(1)$ given by $\Psi_k(p) = \Psi_k(z^1, \ldots, z^n) = (\mathcal{P}(p), \psi_k(p))$, where $\psi_k(p) = \psi_k(z^1, \ldots, z^n) = |\, z^k \,|^{-1}\, z^k$. For $g \in U(1)$, $\psi_k(p \cdot g) = \psi_k(z^1 g, \ldots, z^n g) = |\, z^k g \,|^{-1}\, (z^k g) = (|\, z^k \,|^{-1}\, z^k) g = \psi_k(p) g$ so S^{2n-1} is a principal $U(1)$-bundle over $\mathbb{C}\mathbb{P}^{n-1}$.

Exercise 3.1.2 Show that, with the action described in Exercise 1.6.12, S^{4n-1} is a principal $Sp(1)$-bundle over $\mathbb{H}\mathbb{P}^{n-1}$. **Hint:** Exercise 1.3.22.

These last few examples are just the Hopf bundles, of course, and, as we have mentioned repeatedly, the $n = 2$ cases are of particular interest to us. Here, by virtue of (1.2.8) and (1.2.9), we obtain S^3 as a principal $U(1)$-bundle over S^2 and S^7 as a principal $Sp(1)$-bundle over S^4. Our major objective in this chapter is a complete classification of all the principal G-bundles over any S^n. As a first step in this direction we now proceed to build a machine (Theorem 3.3.4) for the mass production of principal bundles.

Exercise 3.1.3 Let $\mathcal{B}' = (P', \mathcal{P}', \sigma')$ be any principal G-bundle over X' and suppose X is a topological subspace of X'. Let $P = (\mathcal{P}')^{-1}(X)$, $\mathcal{P} = \mathcal{P}' \mid P$ and $\sigma = \sigma' \mid P \times G$. For each local trivialization (V', Ψ') of $\mathcal{B}'$ with $V' \cap X \neq \emptyset$ set $V = V' \cap X$ and $\Psi = \Psi' \mid \mathcal{P}^{-1}(V)$. Show that, with these definitions, P becomes a principal G-bundle over X (called the **restriction of $\mathcal{B}'$ to X** and denoted $\mathcal{B}' | X$).

3.2 Transition Functions

We consider a principal G-bundle $\mathcal{P} : P \to X$ over X and fix a **trivializing cover** of X, i.e., a family $\{(V_j, \Psi_j)\}_{j \in J}$ of local trivializations with $\bigcup_{j \in J} V_j = X$. We write each Ψ_j as $(\mathcal{P}, \psi_j)$ as in (3.1.2). Now, suppose $i, j \in J$ and $V_i \cap V_j \neq \emptyset$. For each $x \in V_i \cap V_j$, ψ_i and ψ_j both carry $\mathcal{P}^{-1}(x)$ homeomorphically onto G so

$$\left(\psi_j | \mathcal{P}^{-1}(x)\right) \circ \left(\psi_i | \mathcal{P}^{-1}(x)\right)^{-1} : G \longrightarrow G \qquad (3.2.1)$$

is a homeomorphism.

Exercise 3.2.1 Show that $\psi_j(p)(\psi_i(p))^{-1}$ takes the same value for every p in the fiber $\mathcal{P}^{-1}(x)$ above x.

By virtue of Exercise 3.2.1 we may define a map

$$g_{ji} : V_i \cap V_j \longrightarrow G$$

by

$$g_{ji}(x) = \psi_j(p)\left(\psi_i(p)\right)^{-1}, \tag{3.2.2}$$

where p is *any* element of $\mathcal{P}^{-1}(x)$. Since ψ_j and ψ_i are continuous and G is a topological group, g_{ji} is also continuous.

Lemma 3.2.1 *For each $x \in V_i \cap V_j$,*

$$\left(\psi_j \,|\, \mathcal{P}^{-1}(x)\right) \circ \left(\psi_i \,|\, \mathcal{P}^{-1}(x)\right)^{-1}(g) = g_{ji}(x)g \tag{3.2.3}$$

for all $g \in G$.

Proof: Let $(\psi_i \,|\, \mathcal{P}^{-1}(x))^{-1}(g) = p$. Then $g = \psi_i(p)$ and $(\psi_j \,|\, \mathcal{P}^{-1}(x)) \circ (\psi_i \,|\, \mathcal{P}^{-1}(x))^{-1}(g) = \psi_j(p)$. But $p \in \mathcal{P}^{-1}(x)$ so $g_{ji}(x)g = \psi_j(p)(\psi_i(p))^{-1}g = \psi_j(p)(\psi_i(p))^{-1}(\psi_i(p)) = \psi_j(p)$ also and the result follows. ∎

Thus, the homeomorphism (3.2.1) is actually left multiplication by the element $g_{ji}(x)$ in G. The maps $g_{ji} : V_i \cap V_j \to G$, defined whenever $V_i \cap V_j \neq \emptyset$, are called the **transition functions** of the principal bundle associated with the trivializing cover $\{(V_j, \psi_j)\}_{j \in J}$ of X. As an example we consider the Hopf bundle $(S^{2n-1}, \mathbb{C}\mathbb{P}^{n-1}, \mathcal{P}, U(1))$ and the trivializing cover $\{(V_j, \psi_j)\}_{j=1,\ldots,n}$ described in Section 3.1. Thus, $\psi_j : \mathcal{P}^{-1}(V_j) \to U(1)$ is given by $\psi_j(p) = |\, z^j \,|^{-1} z^j$ for any $p = (z^1, \ldots, z^n) \in \mathcal{P}^{-1}(V_j)$. Thus, if $x \in V_i \cap V_j$ and $p = (z^1, \ldots, z^n)$ is any point in $\mathcal{P}^{-1}(x)$, $g_{ji}(x) = \psi_j(p)(\psi_i(p))^{-1} = |z^j|^{-1} z^j (|z^i|^{-1} z^i)^{-1} = |z^j|^{-1} z^j (z^i)^{-1} |z^i|$, or

$$g_{ji}(x) = \frac{z^j / |z^j|}{z^i / |z^i|}.$$

The calculation is the same in the quaternionic case, but, due to the failure of commutativity in $\mathbb{H}$, it is best to stick with the somewhat less aesthetic expression $g_{ji}(x) = |q^j|^{-1} q^j (q^i)^{-1} |q^i|$.

Exercise 3.2.2 Show that if $V_i \cap V_j \cap V_k \neq \emptyset$ and $x \in V_i \cap V_j \cap V_k$, then

$$g_{kj}(x)g_{ji}(x) = g_{ki}(x) \tag{3.2.4}$$

(this is called the **cocycle condition**). Show also that

$$g_{ii}(x) = e \quad \text{(the identity in } G\text{)}, \tag{3.2.5}$$

and

$$g_{ij}(x) = \left(g_{ji}(x)\right)^{-1}. \tag{3.2.6}$$

3.3 Bundle Maps and Equivalence

We fix a topological group G and consider two principal G-bundles $\mathcal{B}_1(\mathcal{P}_1 : P_1 \to X_1)$ and $\mathcal{B}_2(\mathcal{P}_2 : P_2 \to X_2)$; for convenience, we denote the actions of G on P_1 and P_2 by the same dot $\cdot$. A **(principal) bundle map** from $\mathcal{B}_1$ to $\mathcal{B}_2$ is a continuous map $\tilde{f} : P_1 \to P_2$ such that

$$\tilde{f}(p \cdot g) = \tilde{f}(p) \cdot g \tag{3.3.1}$$

for all $p \in P_1$ and $g \in G$. Since the fiber containing p in P_1 is, by Lemma 3.1.1, $\{p \cdot g : g \in G\}$ and the fiber containing $\tilde{f}(p)$ in P_2 is $\{\tilde{f}(p) \cdot g : g \in G\}$, (3.3.1) implies that $\tilde{f}$ preserves fibers, i.e., carries the fiber containing p in P_1 to the fiber containing $\tilde{f}(p)$ in P_2.

Exercise 3.3.1 Show that, in fact, $\tilde{f}$ carries each fiber of $\mathcal{P}_1$ homeomorphically onto a fiber of $\mathcal{P}_2$. **Hint:** Locally trivialize near $\mathcal{P}_1(p)$ and $\mathcal{P}_2(\tilde{f}(p))$.

In particular, $\tilde{f}$ determines a map $f : X_1 \to X_2$ defined by

$$\mathcal{P}_2 \circ \tilde{f} = f \circ \mathcal{P}_1 \tag{3.3.2}$$

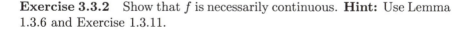

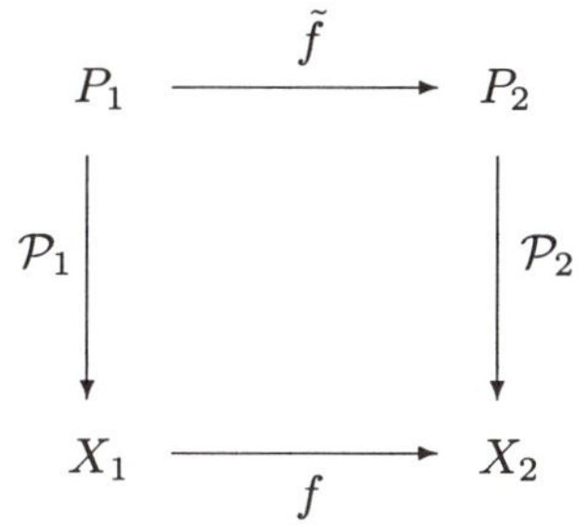

Exercise 3.3.2 Show that f is necessarily continuous. **Hint:** Use Lemma 1.3.6 and Exercise 1.3.11.

We say that the bundle map $\tilde{f}$ **induces** (or **covers**) f.

Exercise 3.3.3 Show that if $\tilde{f}$ induces a homeomorphism $f : X_1 \to X_2$, then $\tilde{f} : P_1 \to P_2$ is also a homeomorphism and $\tilde{f}^{-1} : P_2 \to P_1$ is a bundle map of $\mathcal{B}_2$ onto $\mathcal{B}_1$. **Hint:** It it enough to prove that $\tilde{f}^{-1}$ is continuous locally. Use the fact that inversion is continuous on G.

The case of most interest to us is described as follows. Suppose $\mathcal{B}_1(\mathcal{P}_1 : P_1 \to X)$ and $\mathcal{B}_2(\mathcal{P}_2 : P_2 \to X)$ are both principal G-bundles over the same base space X. Then a bundle map $\tilde{f} : P_1 \to P_2$ is called an

equivalence (and $\mathcal{B}_1$ and $\mathcal{B}_2$ are said to be **equivalent**) if the induced map $f : X \to X$ is the identity id_X. It follows from Exercise 3.3.3 that $\tilde{f}$ is necessarily a homeomorphism and its inverse $\tilde{f}^{-1} : P_2 \to P_1$ is also an equivalence. If $\mathcal{B}(\mathcal{P} : P \to X)$ is a fixed principal G-bundle, then an equivalence $\tilde{f} : P \to P$ is called an **automorphism** of the bundle. A principal G-bundle $\mathcal{B}$ over X is said to be **trivial** if it is equivalent to the trivial G-bundle $\mathcal{P} : X \times G \to X$ over X (Section 3.1).

Exercise 3.3.4 Show that a principal G-bundle $\mathcal{B}$ is trivial iff it has a **global trivialization**, i.e., iff one can take $V = X$ in condition (2) of the definition in Section 3.1.

Deciding whether or not a given principal G-bundle is trivial is generally not a simple matter. A useful test for triviality is based on the notion of a "cross section", which we have seen already in Section 1.5, but now generalize. If V is an open set in the base X of some locally trivial bundle (e.g., a principal bundle), then a (**local**) **cross-section** of the bundle defined on V is a continuous map $s : V \to P$ of V into the bundle space P such that $\mathcal{P} \circ s = id_V$, i.e., it is a continuous selection of an element in each fiber above V. We observe that if $\Psi : \mathcal{P}^{-1}(V) \to V \times G$ is a trivialization of a principal G-bundle, then one can define a local cross-section on V by transferring back to $\mathcal{P}^{-1}(V)$ the obvious "horizontal" cross-section of $V \times G$, i.e., by defining $s_V : V \to P$ by $s_V(x) = \Psi^{-1}(x, e)$ (see Figure 3.3.1).

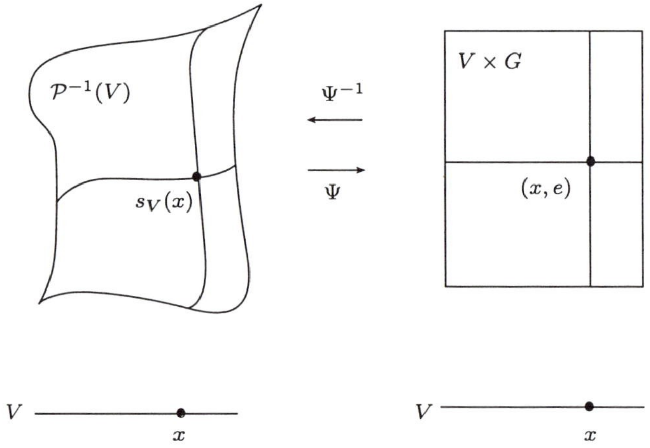

Figure 3.3.1

We call this s_V the **canonical cross-section** associated with the trivialization $\Psi : \mathcal{P}^{-1}(V) \to V \times G$. As an example we consider the Hopf bundle $(S^{2n-1}, \mathbb{C}\mathbb{P}^{n-1}, \mathcal{P}, U(1))$ and its standard trivializations (V_k, Ψ_k),

$k = 1, \ldots, n$. For each such k we define $s_k = s_{V_k} : V_k \to \mathcal{P}^{-1}(V_k)$ by $s_k(x) = \Psi_k^{-1}([z^1, \ldots, z^n], e) = (z^1(z^k)^{-1} \mid z^k \mid, \ldots, z^n(z^k)^{-1} \mid z^k \mid)$ for any $(z^1, \ldots, z^n) \in \mathcal{P}^{-1}(x)$ (see (1.3.1)). Thus,

$$s_k(x) = \left(\frac{z^1 \mid z^k \mid}{z^k}, \ldots, \mid z^k \mid, \ldots, \frac{z^n \mid z^k \mid}{z^k} \right),$$

where the $\mid z^k \mid$ is in the kth slot. For example, if $n = 2$, the canonical cross-sections for $(S^3, \mathbb{C}\mathbb{P}^1, \mathcal{P}, U(1))$ are

$$s_1(x) = \left(\mid z^1 \mid, \frac{z^2 \mid z^1 \mid}{z^1} \right) \quad \text{and} \quad s_2(x) = \left(\frac{z^1 \mid z^2 \mid}{z^2}, \mid z^2 \mid \right).$$

The calculations are the same in the quaternionic case, but it is again best not to risk inadvertent lapses into commutativity by writing the result as $s_k(x) = (\xi^1(\xi^k)^{-1} \, |\xi^k|, \ldots, |\xi^k|, \ldots, \xi^n(\xi^k)^{-1} \, |\xi^k|)$.

The group action on a principal bundle permits us to reverse this process in the following way. Suppose we are given a local cross-section $s : V \to \mathcal{P}^{-1}(V)$, $\mathcal{P} \circ s = id_V$, on some open set V in the base. Since

$$\mathcal{P}^{-1}(V) = \bigcup_{x \in V} \mathcal{P}^{-1}(x) = \bigcup_{x \in V} \{s(x) \cdot g : g \in G\}$$

(because $s(x) \in \mathcal{P}^{-1}(x)$) we can define $\Psi : \mathcal{P}^{-1}(V) \to V \times G$ by

$$\Psi(s(x) \cdot g) = (x, g).$$

We claim that (V, Ψ) is a local trivialization of our principal bundle. Ψ is clearly a bijection and $\Psi(s(x) \cdot g) = (\mathcal{P}(s(x) \cdot g), \psi(s(x) \cdot g))$, where $\psi(s(x) \cdot g) = g$. Thus, $\psi((s(x) \cdot g) \cdot g') = \psi(s(x) \cdot (gg')) = gg' = \psi(s(x) \cdot g)g'$, i.e., $\psi(p \cdot g') = \psi(p)g'$ for all p in $\mathcal{P}^{-1}(V)$ and all g' in G. All that remains is to show that Ψ and Ψ^{-1} are continuous. Now, $\Psi^{-1}(x, g) = s(x) \cdot g$, which is the composition $(x, g) \to (s(x), g) \to s(x) \cdot g$ and so is continuous. Finally, continuity of Ψ will follow from the continuity of $\psi : \mathcal{P}^{-1}(V) \to G$. For any $p = s(x) \cdot g$ in $\mathcal{P}^{-1}(V)$, $\psi(p) = \psi(s(x) \cdot g) = g$. Choose a trivialization (V', Ψ') at $\mathcal{P}(p)$ with $\Psi' = (\mathcal{P}, \psi')$. Then $\psi'(p) = \psi'((s \circ \mathcal{P})(p) \cdot g) = ((\psi' \circ s \circ \mathcal{P})(p))g$ so $g = \psi'(p)((\psi' \circ s \circ \mathcal{P})(p))^{-1} = \psi(p)$, from which the continuity of ψ follows. Observe also that, since $\Psi^{-1}(x, g) = s(x) \cdot g$, the canonical cross-section s_V associated with the trivialization (V, Ψ) just constructed is s.

Thus, we have established a one-to-one correspondence between local cross-sections and local trivializations of a principal bundle. In particular, one has the following consequence of Exercise 3.3.4.

Theorem 3.3.1 *A principal G-bundle $\mathcal{P} : P \to X$ is trivial iff it admits a global cross-section $s : X \to P$.*

Exercise 3.3.5 Let $\Psi_i : \mathcal{P}^{-1}(V_i) \to V_i \times G$ and $\Psi_j : \mathcal{P}^{-1}(V_j) \to V_j \times G$ be two local trivializations of a principal G-bundle with $V_i \cap V_j \neq \emptyset$. Let $s_i = s_{V_i}$ and $s_j = s_{V_j}$ be the associated canonical cross-sections. Show that, for each $x \in V_i \cap V_j$, $s_i(x) = s_j(x) \cdot g_{ji}(x)$, where $g_{ji} : V_i \cap V_j \to G$ is the transition function (defined by (3.2.2)).

Corollary 3.3.2 *Let G be a topological group and n a positive integer. Then any principal G-bundle over the n-cube I^n is trivial.*

Proof: Let $\mathcal{P} : P \to I^n$ be a principal G-bundle over I^n. In particular, $(P, I^n, \mathcal{P}, G)$ is a locally trivial bundle. Since I^n is convex and therefore contractible, the identity map $id : I^n \to I^n$ is nullhomotopic. Let $F : I^n \times [0,1] \to I^n$ be a homotopy with $F(x,0) = x_0$ and $F(x,1) = x$ for all $x \in I^n$. If $f : I^n \to I^n$ is the constant map $f(x) = x_0$ and p_0 is any point in $\mathcal{P}^{-1}(x_0)$, then the constant map $\tilde{f} : I^n \to P$, $\tilde{f}(x) = p_0$, lifts f. Since $F(x,0) = f(x)$, the Homotopy Lifting Theorem 2.4.1 implies that there is a homotopy $\tilde{F} : I^n \times [0,1] \to P$ such that $\mathcal{P} \circ \tilde{F} = F$. In particular, $\mathcal{P} \circ \tilde{F}(x,1) = F(x,1) = x$ for each $x \in I^n$ so the map $x \to \tilde{F}(x,1)$ is a global cross-section of the principal bundle and our result follows from Theorem 3.3.1. ∎

Remark: Using more general versions of the Homotopy Lifting Theorem one can prove that any principal G-bundle over a contractible, paracompact (see [**Dug**]) space is trivial. In particular, any principal bundle over $\mathbb{R}^n$ is trivial.

Exercise 3.3.6 Show that any principal G-bundle over the disc D^n is trivial.

The notion of equivalence for principal bundles has an important reformulation in terms of transition functions. First suppose that $\mathcal{B}(\mathcal{P} : P \to X)$ is a principal G-bundle over X and $\{V_j\}_{j \in J}$ is a cover of X by trivializing nbds with trivializations $\Psi_j : \mathcal{P}^{-1}(V_j) \to V_j \times G$. Suppose also that $\{V'_k\}_{k \in K}$ is an open cover of X with each V'_k contained in some V_j ($\{V'_k\}_{k \in K}$ is called a **refinement** of $\{V_j\}_{j \in J}$). For each $k \in K$ select a $j = j(k) \in J$ with $V'_k \subseteq V_{j(k)}$. Define $\Psi'_k : \mathcal{P}^{-1}(V'_k) \to V'_k \times G$ by $\Psi'_k = \Psi_j \mid \mathcal{P}^{-1}(V'_k)$. Then $\{(V'_k, \Psi'_k)\}_{k \in K}$ is a family of trivializations for $\mathcal{B}$ with trivializing nbds V'_k. The point here is that if we are given a cover of X by trivializing nbds and a refinement of that cover, then the refinement is also a family of trivializing nbds.

Now suppose that we are given two principal G-bundles $\mathcal{B}_1(\mathcal{P}_1 : P_1 \to X)$ and $\mathcal{B}_2(\mathcal{P}_2 : P_2 \to X)$ over the same base X. If $\{V^1_{j_1}\}_{j_1 \in J_1}$ and $\{V^2_{j_2}\}_{j_2 \in J_2}$ are covers of X by trivializing nbds for $\mathcal{B}_1$ and $\mathcal{B}_2$, respectively, then $\{V^1_{j_1} \cap V^2_{j_2}\}_{j_1 \in J_1, j_2 \in J_2}$ is a common refinement of both of these covers and so is a cover by trivializing nbds for *both* $\mathcal{B}_1$ and $\mathcal{B}_2$. As a result, we might just as well assume at the outset that $\mathcal{B}_1$ and $\mathcal{B}_2$ have the same trivializing nbds.

Lemma 3.3.3 *Let $\mathcal{B}_1(\mathcal{P}_1 : P_1 \to X)$ and $\mathcal{B}_2(\mathcal{P}_2 : P_2 \to X)$ be two principal G-bundles over X and suppose (without loss of generality) that $\{V_j\}_{j\in J}$ is a cover of X by trivializing nbds for both $\mathcal{B}_1$ and $\mathcal{B}_2$. Let $g^1_{ji}, g^2_{ji} : V_i \cap V_j \to G$ be the corresponding transition functions for $\mathcal{B}_1$ and $\mathcal{B}_2$, respectively. Then $\mathcal{B}_1$ and $\mathcal{B}_2$ are equivalent iff there exist continuous maps $\lambda_j : V_j \to G$, $j \in J$, such that*

$$g^2_{ji}(x) = (\lambda_j(x))^{-1}\, g^1_{ji}(x)\lambda_i(x)$$

for all $x \in V_i \cap V_j$.

Proof: Suppose first that $\mathcal{B}_1$ and $\mathcal{B}_2$ are equivalent and $\tilde{f} : P_1 \to P_2$ is a bundle map that induces $f = id_X$. Fix $x \in V_i \cap V_j$. Then, for any $p \in \mathcal{P}_1^{-1}(x)$, we have $\tilde{f}(p) \in \mathcal{P}_2^{-1}(x)$. Let $\Psi^1_i : \mathcal{P}_1^{-1}(V_i) \to V_i \times G$ and $\Psi^2_i : \mathcal{P}_2^{-1}(V_i) \to V_i \times G$ be trivializations on V_i for $\mathcal{B}_1$ and $\mathcal{B}_2$, respectively, and similarly for V_j.

Exercise 3.3.7 Show that $\psi^1_i(p)\,(\psi^2_i(\tilde{f}(p)))^{-1}$ takes the same value in G for every p in $\mathcal{P}_1^{-1}(x)$.

Thus, we may define $\lambda_i : V_i \to G$ by

$$\lambda_i(x) = \psi^1_i(p)\,\left(\psi^2_i(\tilde{f}(p))\right)^{-1},$$

where p is any point in $\mathcal{P}_1^{-1}(x)$. Similarly for $\lambda_j(x)$. Thus,

$$(\lambda_j(x))^{-1} = \psi^2_j(\tilde{f}(p))\,\left(\psi^1_j(p)\right)^{-1}.$$

Since $g^1_{ji}(x) = \psi^1_j(p)\,(\psi^1_i(p))^{-1}$ and $g^2_{ji}(x) = \psi^2_j(\tilde{f}(p))\,(\psi^2_i(\tilde{f}(p)))^{-1}$, it follows at once that $g^2_{ji}(x) = (\lambda_j(x))^{-1} g^1_{ji}(x)\lambda_i(x)$.

Exercise 3.3.8 Prove the converse. **Hint:** For each $j \in J$ define $f_j : \mathcal{P}_1^{-1}(V_j) \to \mathcal{P}_2^{-1}(V_j)$ by $f_j(p) = (\Psi^2_j)^{-1}(x, (\lambda_j(x))^{-1}\psi^1_j(p))$. ■

Notice that it follows, in particular, from Lemma 3.3.3 that two principal G-bundles with the same trivializing nbds and the same associated transition functions ($g^2_{ji} = g^1_{ji}$) are equivalent (take $\lambda_i(x) = \lambda_j(x) = e$ for all i and j). We prove next the remarkable fact that, given only the $\{V_j\}_{j\in J}$ and a family $\{g_{ji}\}_{j,i\in J}$ of maps into G satisfying the cocycle condition, one can manufacture a principal bundle having these as its trivializing nbds and transition functions.

Theorem 3.3.4 **(The Reconstruction Theorem)** *Let X be a Hausdorff space, G a topological group and $\{V_j\}_{j\in J}$ an open cover of X. Suppose that, for each $i,j \in J$ with $V_i \cap V_j \neq \emptyset$, there is given a continuous map*

$$g_{ji} : V_i \cap V_j \longrightarrow G$$

and that these maps have the property that, if $V_i \cap V_j \cap V_k \neq \emptyset$, then

$$g_{kj}(x)g_{ji}(x) = g_{ki}(x) \tag{3.3.3}$$

for all $x \in V_i \cap V_j \cap V_k$. Then there exists a principal G-bundle $\mathcal{B}$ over X which has the V_j as trivializing nbds and the g_{ji} as corresponding transition functions. Furthermore, $\mathcal{B}$ is unique up to equivalence.

Proof: First note that, should such a bundle exist, its uniqueness up to equivalence is assured by Lemma 3.3.3. Also note that the following are immediate consequences of (3.3.3).

$$g_{ii}(x) = e\,, \qquad x \in V_i\,. \tag{3.3.4}$$

$$g_{ij}(x) = (g_{ji}(x))^{-1}\,, \quad x \in V_i \cap V_j\,. \tag{3.3.5}$$

Now we set about constructing a bundle space P and a projection $\mathcal{P} : P \to X$. First provide the index set J with the discrete topology and consider the space $X \times G \times J$ (a disjoint union of copies of $X \times G$, one for each $j \in J$). Now consider the subspace $T = \{(x,g,j) \in X \times G \times J : x \in V_j\}$ (pick out of the jth level of $X \times G \times J$ just those things that set above V_j). T is a disjoint union of the open sets $V_j \times G \times \{j\}$ and so is open in $X \times G \times J$. Now define a relation $\sim$ on T as follows:

$$(x,g,j) \sim (x',g',k) \Longleftrightarrow x' = x \ \text{ and } \ g' = g_{kj}(x)g$$

(so, in particular, $x \in V_j \cap V_k$).

Exercise 3.3.9 Show that $\sim$ is an equivalence relation on T.

For each $(x,g,j) \in T$, its equivalence class is

$$[x,g,j] = \{\,(x, g_{kj}(x)g, k) : k \in J \quad \text{and} \quad x \in V_j \cap V_k\,\}\,.$$

We let P denote the set of all such equivalence classes, $\mathcal{Q} : T \to P$ the quotient map ($\mathcal{Q}(x,g,j) = [x,g,j]$) and we provide P with the quotient topology determined by $\mathcal{Q}$.

Exercise 3.3.10 Describe P if the open cover of X consists of just one set $V_j = X$. Now describe P if the open cover consists of precisely two sets $\{V_j, V_k\}$ (see Figure 3.3.2).

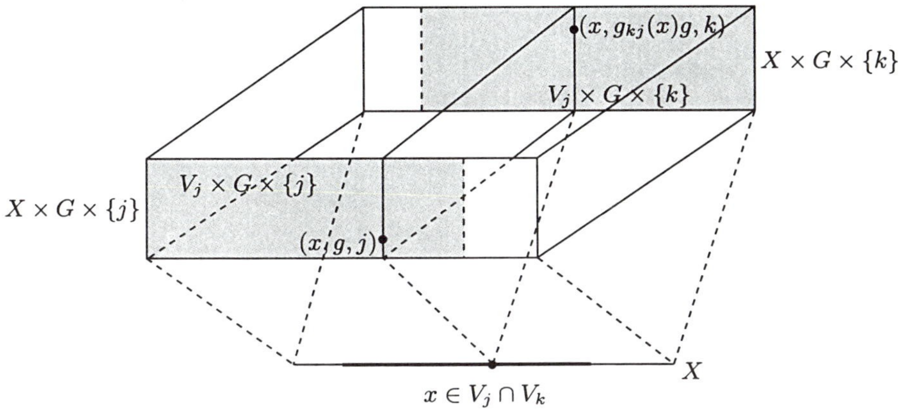

Figure 3.3.2

Now define $\mathcal{P} : P \to X$ by $\mathcal{P}([x,g,j]) = x$. This is well-defined by the definition of $\sim$ and is clearly surjective. To show that it is continuous we let W be an open set in X. Since P has the quotient topology determined be $\mathcal{Q}$, $\mathcal{P}^{-1}(W)$ will be open in P iff $\mathcal{Q}^{-1}(\mathcal{P}^{-1}(W))$ is open in T. Thus, it will suffice to prove that

$$\mathcal{Q}^{-1}\left(\mathcal{P}^{-1}(W)\right) = (W \times G \times J) \cap T. \tag{3.3.6}$$

Exercise 3.3.11 Prove (3.3.6).

At this point we have a continuous surjection $\mathcal{P} : P \to X$ and we must now define the required bundle structure on it. We begin by identifying the fibers of $\mathcal{P}$. Let $x \in X$ and select some (fixed) $j \in J$ with $x \in V_j$. We claim that

$$\mathcal{P}^{-1}(x) = \{\, [x,g,j] : g \in G \,\} \quad (j \text{ fixed with } x \in V_j). \tag{3.3.7}$$

To prove this we proceed as follows. Every $g \in G$ gives an $(x,g,j) \in T$ and therefore determines an equivalence class $[x,g,j]$ so $\{[x,g,j] : g \in G\} \subseteq \mathcal{P}^{-1}(x)$ is clear. Next, every element of $\mathcal{P}^{-1}(x)$ is an equivalence class $[x, g', k]$ for some $g' \in G$ and some $k \in J$ with $x \in V_k$. We must show that this equals $[x, g, j]$ for some $g \in G$ and for the specific j fixed above. But, if we let $g = g_{jk}(x)g'$, then $(x, g', k) = (x, g_{kj}(x)g, k) \sim (x, g, j)$ so $[x, g', k] = [x, g, j]$ as required. We conclude that

$$\mathcal{P}^{-1}(V_j) = \{ [x, g, j] : x \in V_j, g \in G \} . \tag{3.3.8}$$

Now we define mappings $\Psi_j : \mathcal{P}^{-1}(V_j) \to V_j \times G$ and $\Phi_j : V_j \times G \to \mathcal{P}^{-1}(V_j)$ by $\Psi_j([x, g, j]) = (x, g)$ and $\Phi_j(x, g) = [x, g, j] = \mathcal{Q}(x, g, j)$.

Exercise 3.3.12 Show that Ψ_j and Φ_j are inverse bijections.

Note that Φ_j is clearly continuous since it is the composition

$$(x, g) \hookrightarrow (x, g, j) \xrightarrow{\mathcal{Q}} [x, g, j] .$$

Thus, to show that Ψ_j and Φ_j are inverse homeomorphisms it will suffice to prove that Φ_j is an open mapping. Let W be open in $V_j \times G$. We must show that $\mathcal{Q}^{-1}(\Phi_j(W))$ is open in T. Now, the sets $V_k \times G \times \{k\}$ are open in T and cover T so it will suffice to show that, for each $k \in J$,

$$\mathcal{Q}^{-1} (\Phi_j(W)) \cap (V_k \times G \times \{k\}) \tag{3.3.9}$$

is open in $V_k \times G \times \{k\}$. First note that the set in (3.3.9) is contained in

$$(V_k \cap V_j) \times G \times \{k\} , \tag{3.3.10}$$

which, in turn, is an open subset of $V_k \times G \times \{k\}$. Indeed, let (x, g, k) be in (3.3.9). Then $x \in V_k$ and $\mathcal{Q}(x, g, k) \in \Phi_j(W)$ so $[x, g, k] \in \Phi_j(W)$. This implies that $[x, g, k] = [x', g', j]$ for some $(x', g') \in V_j \times G$. Consequently, $x = x' \in V_j$ so $x \in V_k \cap V_j$ and so (x, g, k) is in (3.3.10). Now consider the restriction of $\mathcal{Q}$ to $(V_k \cap V_j) \times G \times \{k\}$. This restriction can be written as the composition

$$(V_k \cap V_j) \times G \times \{k\} \xrightarrow{r} V_j \times G \xrightarrow{\Phi_j} P ,$$

where $r(x, g, k) = (x, g_{jk}(x)g)$ since

$$\begin{aligned} \Phi_j (r(x, g, k)) &= \Phi_j(x, g_{jk}(x)g) = \mathcal{Q}(x, g_{jk}(x)g, j) \\ &= [x, g_{jk}(x)g, j] = [x, g, k] = \mathcal{Q}(x, g, k) . \end{aligned}$$

Now, r is continuous since $g_{jk} : V_k \cap V_j \to G$ is continuous and right multiplication in G is continuous. Thus, $r^{-1}(W)$ is open in $(V_k \cap V_j) \times G \times \{k\}$ and therefore also in $V_k \times G \times \{k\}$. But

$$(\Phi_j \circ r)^{-1} (\Phi_j(W)) = r^{-1} \left(\Phi_j^{-1}(\Phi_j(W)) \right) = r^{-1}(W)$$

since Φ_j is a bijection so $(\Phi_j \circ r)^{-1}(\Phi_j(W))$ is open in $V_k \times G \times \{k\}$. Furthermore,

$$\mathcal{Q}^{-1}(\Phi_j(W)) \cap (V_k \times G \times \{k\}) = \mathcal{Q}^{-1}(\Phi_j(W)) \cap ((V_k \cap V_j) \times G \times \{k\})$$
$$= (\Phi_j \circ r)^{-1}(\Phi_j(W))$$

so the set in (3.3.9) is open in $V_k \times G \times \{k\}$ as required. Thus, Φ_j and Ψ_j are inverse homeomorphisms.

Noting that $\mathcal{P} \circ \Phi_j(x,g) = \mathcal{P}([x,g,j]) = x$ we now have that $(P, X, \mathcal{P}, G)$ is a locally trivial bundle (so, in particular, P must be Hausdorff by Exercise 1.3.23).

Exercise 3.3.13 Define $\sigma : P \times G \to P$ by $\sigma(p,h) = p \cdot h = [x,g,j] \cdot h = [x, gh, j]$ for all $p = [x,g,j]$ in P and all $h \in G$. Show that σ is a right action of G on P and satisfies $\mathcal{P}(p \cdot h) = \mathcal{P}(p)$.

Since $\Psi_j([x,g,j]) = (x,g) = (\mathcal{P}([x,g,j]), \psi_j([x,g,j]))$, where $\psi_j([x,g,j]) = g$ and $\psi_j([x,g,j]\cdot h) = \psi_j([x,gh,j]) = gh = \psi_j([x,g,j])h$, (3.1.3) is satisfied and $\mathcal{P} : P \to X$ is a principal G-bundle. Furthermore, if $x \in V_i \cap V_j$ and $p = [x,g,j]$ is any point in $\mathcal{P}^{-1}(x)$,

$$\psi_j(p)\,(\psi_i(p))^{-1} = g\,(g_{ij}(x)g)^{-1} = (g_{ij}(x))^{-1} = g_{ji}(x)$$

so the transition functions relative to the trivializations $\{(V_j, \Psi_j)\}_{j \in J}$ are precisely the g_{ji}. ∎

Given a space X and a topological group G the data that goes into the manufacture of a principal G-bundle over X via the Reconstruction Theorem 3.3.4 is an open cover of X and a family of maps into G satisfying the cocycle condition. In the next section we will proceed in this manner to build an $Sp(1)$-bundle over S^4 analogous, but not equivalent to the quaternionic Hopf bundle.

3.4 Principal G-Bundles Over Spheres

We are at last in a position to classify the principal G-bundles over S^n, $n \geq 2$, for pathwise connected groups G, in terms of the homotopy group $\pi_{n-1}(G)$. This classification will eventually play a pivotal role in our understanding of monopoles and instantons. Intuitively, the idea is quite simple. S^n consists of two copies of the n-dimensional disc (the upper and lower hemispheres) glued together along the equator, which is a copy of S^{n-1}. The restriction of any G-bundle over S^n to either of these discs is trivial by Exercise 3.3.6. This provides a trivializing cover of S^n consisting of just two trivializations and hence essentially one transition function. This one transition function determines the bundle up to equivalence and its restriction to the equator is a map of S^{n-1} into G. Fixing a couple of base points we show that the bundles determined by homotopic maps are equivalent and thereby arrive at a one-to-one correspondence between equivalence classes of G-bundles over S^n and, by Lemma 2.5.13, the elements of $\pi_{n-1}(G)$.

Let us begin by fixing some notation and terminology for the duration of this section. Throughout, G will denote a pathwise connected topological group and, as usual, we will regard S^n, $n \geq 2$, as the subset of $\mathbb{R}^{n+1}$ consisting of all $x = (x^1, \ldots, x^n, x^{n+1})$ with $< x, x > = (x^1)^2 + \cdots + (x^n)^2 + (x^{n+1})^2 = 1$. We identify S^{n-1} with the set of $x \in S^n$ with $x^{n+1} = 0$ (the "equator" in S^n). Select some fixed base point $x_0 \in S^{n-1}$ and an ε with $0 < \varepsilon < 1$. Now define the following subsets of S^n:

$$
\begin{aligned}
D_1 &= \{x = (x^1, \ldots, x^{n+1}) \in S^n : x^{n+1} \geq 0\} \\
D_2 &= \{x = (x^1, \ldots, x^{n+1}) \in S^n : x^{n+1} \leq 0\} \\
V_1 &= \{x = (x^1, \ldots, x^{n+1}) \in S^n : -\varepsilon < x^{n+1} \leq 1\} \supseteq D_1 \\
V_2 &= \{x = (x^1, \ldots, x^{n+1}) \in S^n : -1 \leq x^{n+1} < \varepsilon\} \supseteq D_2 .
\end{aligned}
$$

Thus, $V_1 \cap V_2$ is an open "band" on S^n containing the equator S^{n-1}.

Now let $\mathcal{B}(\mathcal{P} : P \to S^n)$ be an arbitrary principal G-bundle over S^n. We claim that V_1 and V_2 are necessarily locally trivializing nbds and that, moreover, one can choose trivializations $\Psi_i : \mathcal{P}^{-1}(V_i) \to V_i \times G$, $i = 1, 2$, for which the transition functions g_{ji} all carry our selected base point x_0 to the identity e in G. Consider first the restrictions $\mathcal{B} \,|\, \bar{V}_i$ (see Exercise 3.1.3). Since $\bar{V}_i$ is homeomorphic to D^n via stereographic projection, Exercise 3.3.6 implies that each $\mathcal{B} \,|\, \bar{V}_i$ is trivial and so admits a global cross-section by Theorem 3.3.1. Consequently, $\mathcal{B} \,|\, V_i$ also admits a global cross-section and so it too must be trivial. Choose equivalences $\tilde{f}_i : \mathcal{P}^{-1}(V_i) \to V_i \times G$, $i = 1, 2$, where $V_i \times G$ is regarded as the bundle space of the trivial G-bundle over V_i.

Exercise 3.4.1 Show that $(V_i, \tilde{f}_i)$ is a local trivialization of $\mathcal{B}$.

Remark: One could use our standard coordinate nbds U_S and U_N in place of V_1 and V_2, but then to get the triviality of $\mathcal{B} \,|\, U_S$ and $\mathcal{B} \,|\, U_N$ one would require the generalization of Corollary 3.3.2 mentioned in the Remark following that Corollary.

Now let $\tilde{g}_{12} : V_2 \cap V_1 \to G$ be the transition function corresponding to the local trivializations $\{(V_1, \tilde{f}_1), (V_2, \tilde{f}_2)\}$ and suppose $\tilde{g}_{12}(x_0) = a \in G$. Then, for any $p \in \mathcal{P}^{-1}(x_0)$, $\tilde{\psi}_1(p)(\tilde{\psi}_2(p))^{-1} = a$, where $\tilde{f}_i(p) = (\mathcal{P}(p), \tilde{\psi}_i(p))$. Define $\psi_2 : \mathcal{P}^{-1}(V_2) \to G$ by $\psi_2(p) = a\tilde{\psi}_2(p)$ for each $p \in \mathcal{P}^{-1}(V_2)$. Then

$$
\tilde{\psi}_1(p) \left(\psi_2(p)\right)^{-1} = \tilde{\psi}_1(p)\Big(a\tilde{\psi}_2(p)\Big)^{-1} = \tilde{\psi}_1(p)\Big(\tilde{\psi}_2(p)\Big)^{-1} a^{-1} = aa^{-1} = e.
$$

Exercise 3.4.2 Define $\Psi_i : \mathcal{P}^{-1}(V_i) \to V_i \times G$, $i = 1, 2$, by $\Psi_1 = \tilde{f}_1$ and $\Psi_2(p) = (\mathcal{P}(p), \psi_2(p))$. Show that these are local trivializations of $\mathcal{B}$ and that the corresponding transition functions $\{g_{11}, g_{12}, g_{21}, g_{22}\}$ all carry x_0 onto e.

Henceforth we assume that all of our principal G-bundles over S^n are trivialized over V_1 and V_2 in such a way that the transition functions $\{g_{11}, g_{12}, g_{21}, g_{22}\}$ all carry x_0 to e. Thus, we may define, for each such bundle $\mathcal{B}$, the **characteristic map**

$$T : \left(S^{n-1}, x_0\right) \longrightarrow (G, e)$$

by

$$T = g_{12} \,|\, S^{n-1}.$$

We will eventually show that the (relative) homotopy class of T completely characterizes the bundle $\mathcal{B}$.

Lemma 3.4.1 *Any continuous map $T : (S^{n-1}, x_0) \to (G, e)$ is the characteristic map of some principal G-bundle over S^n.*

Proof: We write down a set of continuous maps $g_{ji} : V_i \cap V_j \to G$, $i, j = 1, 2$, that satisfy the cocycle condition (3.3.3) and $g_{12} \,|\, S^{n-1} = T$ and then appeal to Theorem 3.3.4. Of course, we must take $g_{11}(x) = e$ for all $x \in V_1$ and $g_{22}(x) = e$ for all $x \in V_2$. Since $g_{21}(x)$ must be defined to be $(g_{12}(x))^{-1}$ for all $x \in V_1 \cap V_2$, we need only specify $g_{12} : V_2 \cap V_1 \to G$ with $g_{12} \,|\, S^{n-1} = T$.

Exercise 3.4.3 Use stereographic projection to define a retraction $r : V_2 \cap V_1 \to S^{n-1}$ of $V_2 \cap V_1$ onto the equator S^{n-1}.

Now we set $g_{12}(x) = T(r(x))$ for all $x \in V_2 \cap V_1$. Then $g_{12} \mid S^{n-1} = T$ as required. Since (3.3.3) is trivially satisfied by $\{g_{ji}\}_{j,i=1,2}$, the proof is complete. ∎

Theorem 3.4.2 *Let G be a pathwise connected topological group and $\mathcal{B}_1$ and $\mathcal{B}_2$ two principal G-bundles over S^n, $n \geq 2$. Let T_1 and T_2 be the characteristic maps of $\mathcal{B}_1$ and $\mathcal{B}_2$, respectively. Then $\mathcal{B}_1$ and $\mathcal{B}_2$ are equivalent iff $T_2 \simeq T_1 \operatorname{rel} \{x_0\}$.*

Proof: Suppose first that $\mathcal{B}_1$ and $\mathcal{B}_2$ are equivalent. We assume that both are trivialized in the manner described above so, in particular, they have the same trivializing neighborhoods V_1 and V_2. By Lemma 3.3.3, there exist continuous maps $\lambda_j : V_j \to G$, $j = 1, 2$, such that $g_{12}^2(x) = (\lambda_1(x))^{-1} g_{12}^1(x) \lambda_2(x)$ for all $x \in V_2 \cap V_1$. Let $\mu_j = \lambda_j \mid S^{n-1}$. Then $T_2(x) = (\mu_1(x))^{-1} T_1(x) \mu_2(x)$ for all $x \in S^{n-1}$. Since $T_1(x_0) = T_2(x_0) = e$,

$\mu_1(x_0) = \mu_2(x_0)$, and we denote this element of G by a.

Now, each D_i, $i = 1, 2$, is homeomorphic to D^n with boundary S^{n-1} and x_0 is in S^{n-1}. Thus, D_i is contractible so, by Corollary 2.5.9,

$$\pi_{n-1}(D_i, x_0) = 0, \quad i = 1, 2.$$

Thus, by Lemma 2.5.13, the set $[(S^{n-1}, x_0), (D_i, x_0)]$ of homotopy classes contains only one element. In particular, the inclusion maps $(S^{n-1}, x_0) \hookrightarrow (D_i, x_0)$ are both homotopic, relative to x_0, to the constant map of S^{n-1} onto x_0. For each $i = 1, 2$, let H_i be such a homotopy:

$$\begin{gathered} H_i : S^{n-1} \times [0,1] \longrightarrow D_i \\ H_i(x, 0) = x, \quad H_i(x, 1) = x_0, \quad x \in S^{n-1} \\ H_i(x_0, t) = x_0, \quad t \in [0,1]. \end{gathered}$$

Compose H_i with λ_i to get $K_i = \lambda_i \circ H_i : S^{n-1} \times [0,1] \to G$:

$$\begin{aligned} K_i(x,0) &= \lambda_i(x) = \mu_i(x), \quad x \in S^{n-1} \\ K_i(x,1) &= \lambda_i(x_0) = \mu_i(x_0) = a, \quad x \in S^{n-1} \\ K_i(x_0,t) &= \lambda_i(x_0) = \mu_i(x_0) = a, \quad t \in [0,1]. \end{aligned}$$

Thus, K_i is a homotopy, relative to x_0, from μ_i to the constant map of S^{n-1} to G whose value is a. Finally, define $K : S^{n-1} \times [0,1] \to G$ by $K(x,t) = (K_1(x,t))^{-1} T_1(x) K_2(x,t)$. Then

$$\begin{aligned} K(x,0) &= (K_1(x,0))^{-1} T_1(x) K_2(x,0) = (\mu_1(x))^{-1} T_1(x) \mu_2(x) = T_2(x) \\ K(x,1) &= (K_1(x,1))^{-1} T_1(x) K_2(x,1) = a^{-1} T_1(x) a \\ K(x_0,t) &= (K_1(x_0,t))^{-1} T_1(x_0) K_2(x_0,t) = a^{-1} e a = e \end{aligned}$$

so K is a homotopy, relative to x_0, of T_2 to $a^{-1}T_1 a$.

Exercise 3.4.4 Show that if $\alpha : [0,1] \to G$ is a path in G from $\alpha(0) = a$ to $\alpha(1) = e$, then $H : S^{n-1} \times [0,1] \to G$ defined by $H(x,t) = (\alpha(t))^{-1} T_1(x) \alpha(t)$ is a homotopy, relative to x_0, from $a^{-1}T_1 a$ to T_1.

Thus, $T_2 \simeq a^{-1}T_1 a \operatorname{rel}\{x_0\}$ implies $T_2 \simeq T_1 \operatorname{rel}\{x_0\}$. We conclude that $\mathcal{B}_2$ equivalent to $\mathcal{B}_1$ implies $T_2 \simeq T_1 \operatorname{rel}\{x_0\}$.

Now we suppose, conversely, that $T_2 \simeq T_1 \operatorname{rel}\{x_0\}$ and show that $\mathcal{B}_1$ and $\mathcal{B}_2$ are equivalent. First we wish to show that the map $T_1 T_2^{-1} : S^{n-1} \to G$ defined by $(T_1 T_2^{-1})(x) = T_1(x)(T_2(x))^{-1}$ is homotopic, relative to x_0, to the constant map of S^{n-1} onto $\{e\}$. This follows at once from our hypothesis and the following exercise.

Exercise 3.4.5 Suppose X is a topological space, G is a topological group and $x_0 \in X$. Suppose also that $f, g, f', g' : (X, x_0) \to (G, e)$ with $f \simeq f' \operatorname{rel}\{x_0\}$ and $g \simeq g' \operatorname{rel}\{x_0\}$. Define $fg : (X, x_0) \to (G, e)$ and $f'g' : (X, x_0) \to (G, e)$ by $(fg)(x) = f(x)g(x)$ and $(f'g')(x) = f'(x)g'(x)$. Show that $fg \simeq f'g' \operatorname{rel}\{x_0\}$.

Since $T_1 T_2^{-1} : S^{n-1} \to G$ is nullhomotopic, it has a continuous extension $\nu : D_1 \to G$ (extend to the disc $D^n = \{(x^1, \ldots, x^n, x^{n+1}) \in \mathbb{R}^{n+1} : (x^1)^2 + \cdots + (x^n)^2 \le 1,\ x^{n+1} = 0\}$ by Exercise 2.3.18 and then compose with the projection $(x^1, \ldots, x^n, x^{n+1}) \to (x^1, \ldots, x^n, 0)$ of D_1 onto D^n). Define $\lambda_1 : V_1 \to G$ by

$$\lambda_1(x) = \begin{cases} \nu(x)\,, & x \in D_1 \\ g_{12}^1(x)(\,g_{12}^2(x)\,)^{-1}\,, & x \in D_2 \cap V_1 \end{cases}.$$

Now, $D_1 \cap (D_2 \cap V_1) = S^{n-1}$ and, on S^{n-1}, $\nu(x) = T_1(x)(T_2(x))^{-1} = g_{12}^1(x)(g_{12}^2(x))^{-1}$ so $\lambda_1(x)$ is continuous by the Glueing Lemma 1.2.3.

Now let $\tilde{V}_2$ be the interior of D_2 (so that $\tilde{V}_2 \subseteq V_2$) and $\tilde{\Psi}_2^i = \Psi_2^i \,|\, \mathcal{P}_i^{-1}(\tilde{V}_2)$ for $i = 1, 2$. Then $\{(V_1, \Psi_1^1), (\tilde{V}_2, \tilde{\Psi}_2^1)\}$ and $\{(V_1, \Psi_1^2), (\tilde{V}_2, \tilde{\Psi}_2^2)\}$ trivialize $\mathcal{B}_1$ and $\mathcal{B}_2$, respectively, and the corresponding transition functions are just the appropriate restrictions of g_{ji}^1 and g_{ji}^2 (we will continue to use the same symbols for these restrictions). Now define $\lambda_2 : \tilde{V}_2 \to G$ by $\lambda_2(x) = e$ for each $x \in \tilde{V}_2$. Then, for $x \in V_1 \cap \tilde{V}_2$,

$$\begin{aligned}(\,\lambda_1(x)\,)^{-1}\, g_{12}^1(x)\lambda_2(x) &= (\,\lambda_1(x)\,)^{-1}\, g_{12}^1(x) \\ &= g_{12}^2(x)\,\left(\,g_{12}^1(x)\,\right)^{-1} g_{12}^1(x) = g_{12}^2(x)\,.\end{aligned}$$

Exercise 3.4.6 Show that it follows from this and Lemma 3.3.3 that $\mathcal{B}_1$ and $\mathcal{B}_2$ are equivalent. ∎

Before stating our major result we pause momentarily to use the Reconstruction Theorem 3.3.4 and Theorem 3.4.2 to produce a new example of an $Sp(1)$-bundle over the 4-sphere. We will want to have handy both of the usual descriptions of the base space.

Exercise 3.4.7 Re-examine the proof of (1.2.9) and show that, under the homeomorphism of $\mathbb{HP}^1$ onto S^4 constructed there, the south pole in S^4 corresponds to $[0,1]$ in $\mathbb{HP}^1$, the north pole corresponds to $[1,0]$, the equator S^3 to $\{[q^1, q^2] \in \mathbb{HP}^1 : |q^1| = |q^2| = \frac{\sqrt{2}}{2}\}$, U_N to U_1 and U_S to U_2.

Now, to apply the Reconstruction Theorem we take $X = S^4 = \mathbb{HP}^1$, $G = Sp(1)$ (identified with the unit quaternions) and $\{V_j\}_{j \in J} = \{U_S, U_N\} = \{U_1, U_2\}$. We then need only specify maps $g_{ji} : U_i \cap U_j \to Sp(1)$ for $i, j =$

1, 2 satisfying the cocycle condition. But the cocycle condition requires that $g_{21} = (g_{12})^{-1}$, $g_{11} = 1$ and $g_{22} = 1$ so this amounts to simply deciding on a choice for $g_{12} : U_2 \cap U_1 \to Sp\,(1)$. For the usual Hopf bundle $S^3 \to S^7 \to S^4$ one has $g'_{12}([q^1, q^2]) = (q^1/\mid q^1 \mid)(q^2/\mid q^2 \mid)^{-1}$. In the hope of getting something different we define

$$g_{12}\left([q^1, q^2]\right) = \left((q^1/|q^1|)(q^2/|q^2|)^{-1}\right)^{-1} = \left(q^2/|q^2|\right)\left(q^1/|q^1|\right)^{-1}$$

for all $[q^1, q^2] \in U_2 \cap U_1$. Now, Theorem 3.3.4 guarantees the existence of an $Sp\,(1)$-bundle over $S^4 = \mathbb{HP}^1$ with transition functions $\{g_{11}, g_{12}, g_{21}, g_{22}\}$, but it does not assure us that this bundle is genuinely new, i.e., not equivalent to the Hopf bundle.

To prove that we do have something new here we apply Theorem 3.4.2. Thus, we need the characteristic maps for both bundles. Notice that our trivializing nbds for both are U_N and U_S and that these contain the V_1 and V_2 specified at the beginning of this section (for any $0 < \varepsilon < 1$). Furthermore, if we take x_0 to be any point $[q^1, q^2]$ in the equator S^3 with $q^1 = q^2$, then $g_{ji}(x_0) = g'_{ji}(x_0) = 1 \in Sp\,(1)$ for all $i, j = 1, 2$. Thus, the characteristic maps T' and T for the Hopf bundle and our new bundle, respectively, are $T' = g'_{12}\,|\,S^3$ and $T = g_{12}\,|\,S^3$. But then, for each $[q^1, q^2] \in S^3$,

$$T\left([q^1, q^2]\right) = q^2(q^1)^{-1} = \left(q^1(q^2)^{-1}\right)^{-1} = \left(T'([q^1, q^2])\right)^{-1}$$

so T and T' take values in $Sp\,(1)$ that are multiplicative inverses. We claim that it follows from this that $T \not\simeq T' \operatorname{rel}\{x_0\}$ so that the bundles are not equivalent by Theorem 3.4.2. Suppose on the contrary, that $T \simeq T' \operatorname{rel}\{x_0\}$. It follows from Exercise 3.4.5 that the map $T'T^{-1} : S^3 \to Sp\,(1)$ given by $(T'T^{-1})(x) = (T'(x))(T(x))^{-1}$ is homotopic, relative to x_0, to the constant map of S^3 onto $\{1\} \subseteq Sp\,(1)$. Now, recall from Section 2.5 that each of the maps $T', T : (S^3, x_0) \to (Sp\,(1), 1)$ corresponds to a 3-loop $\alpha, \beta : I^3 \to Sp\,(1)$ at $1 \in Sp\,(1)$ and that $T \simeq T' \operatorname{rel}\{x_0\}$ implies $[\beta] = [\alpha]$ in $\pi_3(Sp\,(1))$.

Exercise 3.4.8 Use Exercise 2.5.4 to show that $[\alpha] = -[\beta]$ in $\pi_3(Sp\,(1))$.

To arrive at a contradiction from Exercise 3.4.8 we must observe that $[\alpha]$ is not the zero element of $\pi_3(Sp\,(1)) \cong \mathbb{Z}$. But $Sp\,(1)$ is not contractible and $h : Sp(1) \to S^3$ given by $h(q) = [\frac{q}{\sqrt{2}}, \frac{1}{\sqrt{2}}]$ is a homeomorphism for which $T' \circ h = id_{Sp\,(1)}$. It follows that $T' \not\simeq 1 \operatorname{rel}\{x_0\}$ so $[\alpha] \neq 0 \in \pi_3(Sp\,(1))$.

Exercise 3.4.9 Show that the underlying locally trivial bundle for this new principal $Sp\,(1)$-bundle over S^4 is the same as that of the Hopf bundle, but that the action of $Sp(1)$ on S^7 is different. Describe the action.

With this brief detour behind us we now turn our attention once again to the major result of this chapter.

Theorem 3.4.3 *Let G be a pathwise connected topological group. Then the set of equivalence classes of principal G-bundles over S^n, $n \geq 2$, is in one-to-one correspondence with the elements of $\pi_{n-1}(G)$.*

Proof: Lemma 2.5.13 gives a one-to-one correspondence between $\pi_{n-1}(G)$ and $[(S^{n-1}, x_0), (G, e)]$. On the other hand, Lemma 3.4.1 and Theorem 3.4.2 establish a one-to-one correspondence between equivalence classes of principal G-bundles over S^n and elements of $[(S^{n-1}, x_0), (G, e)]$ so the result follows. ■

Two instances of particular interest to us should be pointed out at once. The equivalence classes of principal $U(1)$-bundles over S^2 are in one-to-one correspondence with the elements of $\pi_1(U(1)) \cong \pi_1(S^1) \cong \mathbb{Z}$ and the equivalence classes of principal $Sp(1)$-bundles over S^4 are in one-to-one correspondence with the elements of $\pi_3(Sp(1)) \cong \pi_3(S^3) \cong \mathbb{Z}$. As we have mentioned in Chapter 0, the integers associated with such bundles characterize Dirac magnetic monopoloes and BPST instantons, respectively. In particular, the fact that principal $U(1)$-bundles over S^2 are in one-to-one correspondence with the integers can be viewed as the topological manifestation of the Dirac quantization condition (see Section 0.2).

Exercise 3.4.10 Define transition functions that will generate the remaining principal $Sp(1)$-bundles over S^4 from the Reconstruction Theorem.

There is a generalization of Theorem 3.4.3 in which the group G need not be assumed pathwise connected and which, moreover, classifies principal G-bundles over S^1 as well (since we have not defined $\pi_0(G)$, Theorem 3.4.3 makes no statement about bundles over the circle). We shall not require the more general result and so will be content to refer those interested in the matter to Theorem 18.5 of [**St**]. It might amuse the reader, however, to obtain one particular consequence of this theorem independently.

Exercise 3.4.11 Show that if G is pathwise connected, then every principal G-bundle over S^1 is trivial.

In particular, there are no nontrivial $U(1)$- or $SU(2)$-bundles over S^1. If G is not pathwise connected, however, nontrivial G-bundles over S^1 do exist and we will close with a simple, but not insignificant example. We take G to be $\mathbb{Z}_2 = \{-1, 1\}$ with the discrete topology and define a principal $\mathbb{Z}_2$-bundle over S^1 whose total space is also S^1 (which, of course, is not homeomorphic to $S^1 \times \mathbb{Z}_2$ so that such a bundle cannot be trivial). According to Exercise 1.5.10, the map $\mathcal{P}_2 : S^1 \to S^1$ given by $\mathcal{P}_2(z) = z^2$ for all $z \in S^1 \subseteq \mathbb{C}$ is a covering space and therefore a locally trivial bundle (see Figure 3.4.1). Define $\sigma : S^1 \times \mathbb{Z}_2 \to S^1$ by $\sigma(z, g) = zg$ (i.e., $\sigma(z, 1) = z$ and $\sigma(z, -1) = -z$).

Exercise 3.4.12 Show that $(S^1, \mathcal{P}_2, \sigma)$ is a nontrivial principal $\mathbb{Z}_2$-bundle over S^1.

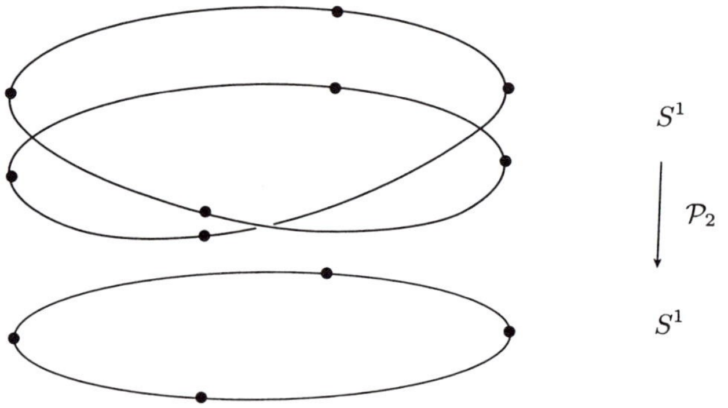

Figure 3.4.1

4

Differentiable Manifolds and Matrix Lie Groups

4.0 Introduction

If X is a topological manifold and (U_1, φ_1) and (U_2, φ_2) are two charts on X with $U_1 \cap U_2 \neq \emptyset$, then the overlap functions $\varphi_1 \circ \varphi_2^{-1} : \varphi_2(U_1 \cap U_2) \to \varphi_1(U_1 \cap U_2)$ and $\varphi_2 \circ \varphi_1^{-1} : \varphi_1(U_1 \cap U_2) \to \varphi_2(U_1 \cap U_2)$ are necessarily homeomorphisms between open sets in some Euclidean space. In the examples that we have encountered thus far (most notably, spheres and projective spaces) these maps actually satisfy the much stronger condition of being C^∞, that is, their coordinate functions have continuous partial derivatives of all orders and types (see Exercise 1.1.8 and (1.2.4)).

From the mathematical point-of-view, differentiable mappings are much more manageable creatures than those that are merely continuous and it seems foolish not to exploit this additional structure. On the other hand, physics, whose language is differential equations, cannot get off the ground without some sort of differentiability assumptions. From either vantage point, the time has come to introduce "smoothness" into our deliberations.

4.1 Smooth Maps on Euclidean Spaces

We remind the reader that our basic reference for real analysis is [**Sp1**] and, whenever feasible, we will utilize notation that conforms with that adopted there. Thus, if U is an open set in $\mathbb{R}^n$ and $f : U \to \mathbb{R}$ is a continuous real-valued function, we denote by $D_i f$ the partial derivative of f with respect to the ith coordinate. Higher order derivatives are written $D_{i,j} f = D_j(D_i f)$, $D_{i,j,k} f = D_k(D_{i,j} f)$, and so on. We note the traditional reversal of order in the indices and also the fact that, if the partial derivatives are continuous, the order is immaterial (Theorem 2-5 of [**Sp1**]).

If $f : U \to \mathbb{R}^m$ is a continuous map of U into $\mathbb{R}^m$, then f has m coordinate functions $f^1, \ldots, f^m : U \to \mathbb{R}$ $(f(x) = (f^1(x), \ldots, f^m(x))$ for each $x \in U)$, all of which are continuous. f is said to be $\mathbf{C^\infty}$ (or **smooth**) **on** U if each f^j, $j = 1, \ldots, m$, has continuous partial derivatives of all orders on U, i.e., if, for any $k = 1, 2, \ldots$ and any indices $i_1, \ldots, i_k$,

$D_{i_1,\ldots,i_k} f^j : U \to \mathbb{R}$ is continuous for each $j = 1, \ldots, m$. The **Jacobian** of f at any $a \in U$ is then the $m \times n$ matrix

$$f'(a) = \left(D_i f^j(a) \right)_{\substack{1 \le j \le m \\ 1 \le i \le n}} = \begin{pmatrix} D_1 f^1(a) & \cdots & D_n f^1(a) \\ \vdots & & \vdots \\ D_1 f^m(a) & \cdots & D_n f^m(a) \end{pmatrix}.$$

$f'(a)$ is a remarkable matrix, containing, as it does, essentially all of the local information about f near a. To make this more precise and for future reference we record the Inverse and Implicit Function Theorems (Theorems 2-11 and 2-12 of [**Sp1**]).

Theorem 4.1.1 **(Inverse Function Theorem)** *Let U be an open set in $\mathbb{R}^n$ and $f : U \to \mathbb{R}^n$ a C^∞ map. Suppose $a \in U$ and $f'(a)$ is nonsingular (i.e., $\det f'(a) \neq 0$). Then there exist open sets V and W in $\mathbb{R}^n$ with $a \in V \subseteq U$ and $f(a) \in W \subseteq \mathbb{R}^n$ such that $f \,|\, V : V \to W$ is a C^∞ bijection with $(f \,|\, V)^{-1} : W \to V$ also C^∞. Moreover, $(f^{-1})'(y) = (f'(f^{-1}(y)))^{-1}$ for all $y \in W$.*

Theorem 4.1.2 **(Implicit Function Theorem)** *Let U be an open set in $\mathbb{R}^n \times \mathbb{R}^m$ $(= \mathbb{R}^{n+m})$ containing (a, b) and let $f : U \to \mathbb{R}^m$, $f = (f^1, \ldots, f^m)$, be a C^∞ map with $f(a, b) = 0$. Suppose that the $m \times m$ matrix*

$$\left(D_{n+i} f^j(a, b) \right)_{\substack{1 \le j \le m \\ 1 \le i \le m}}$$

is nonsingular. Then there exist open sets V and W with $a \in V \subseteq \mathbb{R}^n$ and $b \in W \subseteq \mathbb{R}^m$ such that, for each $x \in V$, there exists a unique $g(x) \in W$ with $f(x, g(x)) = 0$. Moreover, the map $g : V \to W$ thus defined is C^∞.

We will require the existence of "lots" of C^∞ functions on $\mathbb{R}^n$. More precisely, we wish to prove that, for any two disjoint closed subsets A_0 and A_1 of $\mathbb{R}^n$, there exists a C^∞ function $\phi : \mathbb{R}^n \to \mathbb{R}$ with $0 \le \phi(x) \le 1$ for all $x \in \mathbb{R}^n$, $A_0 = \phi^{-1}(0)$ and $A_1 = \phi^{-1}(1)$. We begin by constructing some special C^∞ functions on $\mathbb{R}$. Define $\lambda : \mathbb{R} \to \mathbb{R}$ by

$$\lambda(t) = \begin{cases} 0, & t \le 0 \\ e^{-1/t}, & t > 0 \end{cases}.$$

Since $\lim_{t \to 0^\pm} \lambda(t) = 0$, λ is continuous at $t = 0$ and therefore everywhere. Moreover, $0 \le \lambda(t) < 1$ and $\lim_{t \to \infty} \lambda(t) = 1$. We claim that λ is C^∞ and this is clear everywhere except at $t = 0$.

Exercise 4.1.1 Show, by induction, that the nth derivative of $f(t) = e^{-1/t}$ on $t > 0$ is given by $f^{(n)}(t) = e^{-1/t} q(\frac{1}{t})$, where q is a polynomial of degree $2n$ and conclude, from l'Hôpital's Rule, that $\lim_{t \to 0^+} f^{(n)}(t) = 0$.

It follows from Exercise 4.1.1 that all of λ's derivatives exist and are continuous at $t = 0$ and so λ is C^∞.

Now, let $\epsilon > 0$ be given and define a function $\phi_\epsilon : \mathbb{R} \to \mathbb{R}$ by

$$\phi_\epsilon(t) = \frac{\lambda(t)}{\lambda(t) + \lambda(\epsilon - t)} .$$

Since $\lambda(t) > 0$ for $t > 0$ and $\lambda(\epsilon - t) = 0$ iff $t \geq \epsilon$, the denominator is never zero so ϕ_ϵ is C^∞. Moreover, $0 \leq \phi_\epsilon(t) \leq 1$ for all t, $\phi_\epsilon(t) = 0$ iff $t \leq 0$ and $\phi_\epsilon(t) = 1$ iff $t \geq \epsilon$. Next, define $\psi_\epsilon : \mathbb{R} \to \mathbb{R}$ by

$$\psi_\epsilon(t) = \phi_\epsilon(2\epsilon + t)\phi_\epsilon(2\epsilon - t) .$$

Exercise 4.1.2 Show that ψ_ϵ is C^∞ and satisfies $\psi_\epsilon(t) = 0$ iff $t \geq 2\epsilon$ or $t \leq -2\epsilon$, $\psi_\epsilon(t) = 1$ iff $-\epsilon \leq t \leq \epsilon$ and $0 < \psi_\epsilon(t) < 1$ for $-2\epsilon < t < -\epsilon$ and $\epsilon < t < 2\epsilon$. Sketch the graph of ψ_ϵ.

Now, for each $i = 1, \ldots, n$, define a function $\psi_\epsilon^i : \mathbb{R}^n \to \mathbb{R}$ by $\psi_\epsilon^i = \psi_\epsilon \circ \mathcal{P}^i$, where $\mathcal{P}^i : \mathbb{R}^n \to \mathbb{R}$ is the projection onto the ith coordinate. Then $\psi_\epsilon^i(x) = \psi_\epsilon^i(x^1, \ldots, x^n) = \psi_\epsilon(x^i)$, ψ_ϵ^i is C^∞, $\psi_\epsilon^i(x) = 0$ iff $x^i \geq 2\epsilon$ or $x^i \leq -2\epsilon$, $\psi_\epsilon^i(x) = 1$ iff $-\epsilon \leq x^i \leq \epsilon$ and $0 < \psi_\epsilon^i(x) < 1$ for $-2\epsilon < x^i < -\epsilon$ and $\epsilon < x^i < 2\epsilon$. Finally, define $\tau_\epsilon : \mathbb{R}^n \to \mathbb{R}$ by

$$\tau_\epsilon(x) = \psi_\epsilon^1(x)\psi_\epsilon^2(x) \cdots \psi_\epsilon^n(x) .$$

Thus, τ_ϵ is C^∞, $\tau_\epsilon(x) = 1$ iff x is in the closed cube $[-\epsilon, \epsilon] \times \cdots \times [-\epsilon, \epsilon]$, $\tau_\epsilon(x) = 0$ iff x is outside the open cube $(-2\epsilon, 2\epsilon) \times \cdots \times (-2\epsilon, 2\epsilon)$ and satisfies $0 \leq \tau_\epsilon(x) \leq 1$ everywhere.

Exercise 4.1.3 Describe (and sketch, if you're up to it) the graph of τ_ϵ in the $n = 2$ case.

Notice that if x_0 is any fixed point in $\mathbb{R}^n$, then $\tau_\epsilon(x - x_0)$ is a C^∞ function on $\mathbb{R}^n$ that is 1 on the closed cube $[x_0^1 - \epsilon, x_0^1 + \epsilon] \times \cdots \times [x_0^n - \epsilon, x_0^n + \epsilon]$ centered at x_0 and 0 outside the open cube $(x_0^1 - 2\epsilon, x_0^1 + 2\epsilon) \times \cdots \times (x_0^n - 2\epsilon, x_0^n + 2\epsilon)$ about x_0 and satisfies $0 < \tau_\epsilon(x - x_0) < 1$ for all other x.

Exercise 4.1.4 Show that the C^∞ functions on $\mathbb{R}^n$ separate points and closed sets in the following sense: If x_0 is a point in $\mathbb{R}^n$ and A is a closed subset of $\mathbb{R}^n$ not containing x_0, then there exists a C^∞ function $f : \mathbb{R}^n \to \mathbb{R}$ with $0 \leq f(x) \leq 1$ for all $x \in \mathbb{R}^n$, $f(x_0) = 1$ and $f(A) = \{0\}$.

Theorem 4.1.3 *Any closed set in $\mathbb{R}^n$ is the set of zeros of some non-negative C^∞ function on $\mathbb{R}^n$, i.e., if $A \subseteq \mathbb{R}^n$ is closed, then there exists a C^∞ function $f : \mathbb{R}^n \to \mathbb{R}$ with $f(x) \geq 0$ for all $x \in \mathbb{R}^n$ and $A = f^{-1}(0)$.*

Proof: If either $A = \mathbb{R}^n$ or $A = \emptyset$ the result is trivial so we assume that $U = \mathbb{R}^n - A$ is a nonempty, proper open subset of $\mathbb{R}^n$. Choosing an open cube about each point of U that is contained entirely in U and appealing to Theorems 1.3.1 and 1.4.1, we can write $U = \bigcup_{m=1}^{\infty} C_{r_m}(x_m)$, where $x_m \in U$ and $C_{r_m}(x_m)$ is an open cube $(x_m^1 - r_m, x_m^1 + r_m) \times \cdots \times (x_m^n - r_m, x_m^n + r_m)$ centered at x_m and contained in U. For each $m = 1, 2, \ldots$, let $f_m : \mathbb{R}^n \to \mathbb{R}$ be the C^∞ function defined by $f_m(x) = \tau_{r_m/2}(x - x_m)$. In particular, $f_m(x) \neq 0$ iff $x \in C_{r_m}(x_m)$ and $f_m(x) \geq 0$ everywhere.

Remark: What we would like to do now is just add up the f_m's since this sum would be zero precisely on A. Of course, this sum need not even converge, much less represent a C^∞ function. Our task then is to cut the f_m's and their derivatives down to size so as to ensure convergence, but without changing the sets on which they vanish.

Fix an $m \geq 1$. Since f_m is zero outside the compact set $\overline{C_{r_m}(x_m)} = [x_m^1 - r_m, x_m^1 + r_m] \times \cdots \times [x_m^n - r_m, x_m^n + r_m]$, the same is true of all of the partial derivatives of f_m. Consider the (finite) set of functions consisting of f_m and all of its partial derivatives of order $\leq m$. Each element of this set is continuous, and therefore bounded, on $\overline{C_{r_m}(x_m)}$. Consequently, there exists a constant $M_m > 0$ such that f_m and all of its derivatives of order $\leq m$ are bounded by M_m on all of $\mathbb{R}^n$. Let $\delta_m = (2^m M_m)^{-1}$ and $g_m = \delta_m f_m$. Then g_m is C^∞ on $\mathbb{R}^n$ and, moreover, g_m and all of its derivatives of order $\leq m$ are bounded by 2^{-m} on all of $\mathbb{R}^n$. Finally, define $f : \mathbb{R}^n \to \mathbb{R}$ by

$$f(x) = \sum_{m=1}^{\infty} g_m(x) .$$

Since $| g_m(x) | = g_m(x) \leq 2^{-m}$ for all x in $\mathbb{R}^n$, the Weierstrass M-test implies that the series converges uniformly and so represents a continuous function on $\mathbb{R}^n$ (consult Theorem 9.6 of [**Apos**] if you are unfamiliar with this test). Since f clearly vanishes precisely on A, all that remains is to show that f is C^∞. We prove the existence and continuity of the partial derivatives of f by induction on the order of the derivative.

To get the induction started we consider first order derivatives. Thus, we let i be some fixed integer from $1, \ldots, n$. By assumption, $| D_i g_m(x) | \leq 2^{-m}$ for each $x \in \mathbb{R}^n$ so $\sum_{m=1}^{\infty} D_i g_m$ converges uniformly on $\mathbb{R}^n$ to a (necessarily continuous) function that must be the ith derivative of f (if this last assertion is not clear to you, consult Theorem 9.13 of [**Apos**]).

Now, as an induction hypothesis, assume that, for all orders $\leq k$, the term-by-term derivatives of $f = \sum_{m=1}^{\infty} g_m$ converge uniformly on $\mathbb{R}^n$ to the corresponding derivative of f. Let $\mathcal{D}f = \sum_{m=1}^{\infty} \mathcal{D}g_m$ be some such term-by-term derivative of order k. Now let i be some fixed integer from

$1, \ldots, n$. Write $\mathcal{D}f = \sum_{m=1}^{k} \mathcal{D}g_m + \sum_{m=k+1}^{\infty} \mathcal{D}g_m$ so that

$$\mathcal{D}f - \sum_{m=1}^{k} \mathcal{D}g_m = \sum_{m=k+1}^{\infty} \mathcal{D}g_m \,. \tag{4.1.1}$$

By assumption, for $m \geq k+1$, the ith derivative of $\mathcal{D}g_m$ is bounded by 2^{-m} so $\sum_{m=k+1}^{\infty} D_i(\mathcal{D}g_m)$ converges uniformly on $\mathbb{R}^n$ to the ith derivative of the left-hand side of (4.1.1), which, being a finite sum, can be computed term-by-term. Putting the two sums back together gives $D_i(\mathcal{D}f)$ as the uniform sum of $\sum_{m=1}^{\infty} D_i(\mathcal{D}g)$ and the induction is complete. ∎

Corollary 4.1.4 *Let A_0 and A_1 be disjoint closed subsets of $\mathbb{R}^n$. Then there exists a C^∞ function $\phi : \mathbb{R}^n \to \mathbb{R}$ such that $0 \leq \phi(x) \leq 1$ for all $x \in \mathbb{R}^n$, $A_0 = \phi^{-1}(0)$ and $A_1 = \phi^{-1}(1)$.*

Proof: Let f_0 and f_1 be non-negative C^∞ functions on $\mathbb{R}^n$ with $A_0 = f_0^{-1}(0)$ and $A_1 = f_1^{-1}(0)$. Since $A_0 \cap A_1 = \emptyset$,

$$\phi(x) = \frac{f_0(x)}{f_0(x) + f_1(x)}$$

defines a C^∞ function on $\mathbb{R}^n$ and clearly has the required properties. ∎

Exercise 4.1.5 Show that if A_0 and A_1 are disjoint closed sets in $\mathbb{R}^n$ and $a < b$, then there exists a C^∞ function $\psi : \mathbb{R}^n \to \mathbb{R}$ with $a \leq \psi(x) \leq b$ for all $x \in \mathbb{R}^n$, $A_0 = \psi^{-1}(a)$ and $A_1 = \psi^{-1}(b)$.

4.2 Differentiable Manifolds

Let X be a topological manifold (Section 1.3) and let (U_1, φ_1) and (U_2, φ_2) be two charts on X. We will say that these charts are $\mathbf{C^\infty}$**-related** if either $U_1 \cap U_2 = \emptyset$, or $U_1 \cap U_2 \neq \emptyset$ and both overlap functions $\varphi_1 \circ \varphi_2^{-1} : \varphi_2(U_1 \cap U_2) \to \varphi_1(U_1 \cap U_2)$ and $\varphi_2 \circ \varphi_1^{-1} : \varphi_1(U_1 \cap U_2) \to \varphi_2(U_1 \cap U_2)$ are C^∞. This is the case, for example, for the two stereographic projection charts (U_S, φ_S) and (U_N, φ_N) on S^n (Exercise 1.1.8) and also for any two of the standard charts (U_k, φ_k), $k = 1, \ldots, n$, on the projective spaces $\mathbb{RP}^{n-1}$, $\mathbb{CP}^{n-1}$ and $\mathbb{HP}^{n-1}$ introduced in Section 1.2 (see (1.2.4)).

Exercise 4.2.1 Use the Inverse Function Theorem to show that if two charts on X with intersecting domains are C^∞-related, then they necessarily have the same dimension. **Remark:** This is true, but much more difficult to prove, even when the charts are not C^∞-related. The proof requires a deep theorem of Brouwer called **Invariance of Domain** (see

Theorem 3.1, Chapter XVII of [**Dug**]).

An **atlas** of **dimension** n on X is a collection $\{(U_\alpha, \varphi_\alpha)\}_{\alpha \in \mathcal{A}}$ of n-dimensional charts on X, any two of which are C^∞-related, and such that $\bigcup_{\alpha \in \mathcal{A}} U_\alpha = X$. Thus, $\{(U_S, \varphi_S), (U_N, \varphi_N)\}$ is an altas of dimension n on S^n and $\{(U_1, \varphi_1), \ldots, (U_n, \varphi_n)\}$, as defined in Section 1.2, is an atlas for $\mathbb{F}\mathbb{P}^{n-1}$ (the dimension being $n-1$ if $\mathbb{F} = \mathbb{R}$, $2n-2$ if $\mathbb{F} = \mathbb{C}$ and $4n-4$ if $\mathbb{F} = \mathbb{H}$). The single chart $\{(\mathbb{R}^n, id)\}$ is an atlas for $\mathbb{R}^n$ called the **standard atlas on** $\mathbb{R}^n$. A chart (U, φ) on X is said to be **admissible** to the atlas $\{(U_\alpha, \varphi_\alpha)\}_{\alpha \in \mathcal{A}}$ if it is C^∞-related to each $(U_\alpha, \varphi_\alpha)$. For example, letting $U = \{x = (x^1, \ldots, x^n, x^{n+1}) \in S^n : x^{n+1} > 0\}$ (the open upper hemisphere of S^n) and defining $\varphi : U \to \mathbb{R}^n$ by $\varphi(x) = \varphi(x^1, \ldots, x^n, x^{n+1}) = (x^1, \ldots, x^n)$ (the projection), we obtain a chart (U, φ) on S^n that is easily seen to be C^∞-related to both (U_S, φ_S) and (U_N, φ_N) and therefore admissible to the stereographic projection atlas for S^n.

Exercise 4.2.2 Verify this.

As another example we let U be an arbitrary open set in $\mathbb{R}^n$ and φ a homeomorphism of U onto an open subset $\varphi(U)$ in $\mathbb{R}^n$. Then (U, φ) is admissible to the standard atlas $\{(\mathbb{R}^n, id)\}$ for $\mathbb{R}^n$ iff both $\varphi : U \to \varphi(U)$ and $\varphi^{-1} : \varphi(U) \to U$ are C^∞. For example, when $n = 1$, $((-\frac{\pi}{2}, \frac{\pi}{2}), \tan)$ is admissible since $\tan : (-\frac{\pi}{2}, \frac{\pi}{2}) \to \mathbb{R}$ and $\arctan : \mathbb{R} \to (-\frac{\pi}{2}, \frac{\pi}{2})$ are both C^∞. However, $(\mathbb{R}, \varphi)$, with $\varphi(x) = x^3$, is not admissible because $\varphi^{-1}(x) = x^{\frac{1}{3}}$ is not C^∞ on all of $\mathbb{R}$. For a somewhat less trivial example we consider the standard spherical coordinate charts on $\mathbb{R}^3$ (see Figure 0.2.1). Every point (x, y, z) in $\mathbb{R}^3$ can be written as $(x, y, z) = (\rho \sin\phi \cos\theta, \rho \sin\phi \sin\theta, \rho \cos\phi)$ for some (ρ, ϕ, θ) in $[0, \infty) \times [0, \pi] \times [0, 2\pi]$.

Exercise 4.2.3 Show that the map $(\rho, \phi, \theta) \to (\rho \sin\phi \cos\theta, \rho \sin\phi \sin\theta, \rho \cos\phi)$ is a homeomorphism of $(0, \infty) \times (0, \pi) \times (0, 2\pi)$ onto the open set $U = \mathbb{R}^3 - \{(x, y, z) : x = 0 \text{ and } y \geq 0\}$. Show also that the Jacobian of the map is nonzero on $(0, \infty) \times (0, \pi) \times (0, 2\pi)$ and conclude that the inverse φ of the map gives a chart (U, φ) on $\mathbb{R}^n$. Show that all of this is true also for $(\rho, \phi, \theta) \in (0, \infty) \times (0, \pi) \times (-\pi, \pi)$.

Exercise 4.2.3 gives two charts on $\mathbb{R}^3$ that cover everything except the z-axis. Relabeling the axes in Figure 0.2.1 will give two more spherical coordinate charts that cover everything except, say, the x-axis. These four charts cover $\mathbb{R}^3 - \{(0, 0, 0)\}$ and this is the best one can do (Why?).

An atlas $\{(U_\alpha, \varphi_\alpha)\}_{\alpha \in \mathcal{A}}$ for X is said to be **maximal** if it contains every chart on X that is admissible to it, i.e., if, whenever (U, φ) is a chart on X that is C^∞-related to every $(U_\alpha, \varphi_\alpha)$, $\alpha \in \mathcal{A}$, there exists an $\alpha_0 \in \mathcal{A}$

such that $U = U_{\alpha_0}$ and $\varphi = \varphi_{\alpha_0}$. Maximal atlases are generally huge. For example, a maximal atlas on $\mathbb{R}^n$ containing the standard chart $(\mathbb{R}^n, id)$ contains every pair (U, φ), where U is open in $\mathbb{R}^n$ and $\varphi : U \to \varphi(U) \subseteq \mathbb{R}^n$ is C^∞ with a C^∞ inverse.

Theorem 4.2.1 *Let X be a topological manifold. Then every atlas for X is contained in a unique maximal atlas for X.*

Proof: Let $\{(U_\alpha, \varphi_\alpha)\}_{\alpha \in \mathcal{A}}$ be an atlas for X and let $\{U'_\beta, \varphi'_\beta)\}_{\beta \in \mathcal{B}}$ be the set of all charts on X that are admissible to this atlas. This latter set certainly contains the given atlas so we need only show that it too is an atlas for X, that it is maximal and that any other maximal atlas containing each $(U_\alpha, \varphi_\alpha)$ necessarily coincides with it. First observe that $\bigcup_{\beta \in \mathcal{B}} U'_\beta \supseteq \bigcup_{\alpha \in \mathcal{A}} U_\alpha = X$ so the U'_β cover X. Since any two of the $(U_\alpha, \varphi_\alpha)$ are C^∞-related and any $(U'_\beta, \varphi'_\beta)$ is C^∞-related to every $(U_\alpha, \varphi_\alpha)$, we need only prove that any two $(U'_{\beta_1}, \varphi'_{\beta_1})$ and $(U'_{\beta_2}, \varphi'_{\beta_2})$ are C^∞-related. Assume $U'_{\beta_1} \cap U'_{\beta_2} \neq \emptyset$, for otherwise they are C^∞-related by definition. By symmetry, it will suffice to show that $\varphi'_{\beta_1} \circ \varphi'^{-1}_{\beta_2}$ is C^∞ on $\varphi'_{\beta_2}(U'_{\beta_1} \cap U'_{\beta_2})$ and this we may prove locally, i.e., we need only show that, for each $x \in \varphi'_{\beta_2}(U'_{\beta_1} \cap U'_{\beta_2})$ there is a nbd of x in $\varphi'_{\beta_2}(U'_{\beta_1} \cap U'_{\beta_2})$ on which $\varphi'_{\beta_1} \circ \varphi'^{-1}_{\beta_2}$ is C^∞. Select a chart (U, φ) in our atlas with $\varphi'^{-1}_{\beta_2}(x) \in U$. Then, on $V = U'_{\beta_1} \cap U'_{\beta_2} \cap U$, φ'_{β_1}, φ'_{β_2} and φ are all homeomorphisms onto open subsets of $\mathbb{R}^n$. Moreover, on $\varphi'_{\beta_2}(V)$, $\varphi \circ \varphi'^{-1}_{\beta_2}$ is C^∞ and, on $\varphi(V)$, $\varphi'_{\beta_1} \circ \varphi^{-1}$ is C^∞ so the composition $(\varphi'_{\beta_1} \circ \varphi^{-1}) \circ (\varphi \circ \varphi'^{-1}_{\beta_2}) = \varphi'_{\beta_1} \circ \varphi'^{-1}_{\beta_2}$ is C^∞ on $\varphi'_{\beta_2}(V)$ as required. Thus, $\{(U'_\beta, \varphi'_\beta)\}_{\beta \in \mathcal{B}}$ is an atlas.

Exercise 4.2.4 Show that $\{(U'_\beta, \varphi'_\beta)\}_{\beta \in \mathcal{B}}$ is maximal and that any other maximal atlas for X containing $\{(U_\alpha, \varphi_\alpha)\}_{\alpha \in \mathcal{A}}$ necessarily coincides with it. ∎

A maximal n-dimensional atlas for a topological manifold X is called a **differentiable structure** on X and a topological manifold together with some differentiable structure is called a **differentiable** (or **smooth**, or $\mathbf{C^\infty}$) **manifold** (if the differentiable structure is understood we will often refer to X itself as a differentiable manifold). The dimension n of each chart in a differentiable structure on X is called the **dimension** of the manifold and is written $\dim X$. By Theorem 4.2.1, every atlas determines (we will say, **generates**) a unique differentiable structure.

Exercise 4.2.5 Show that to prove that a chart on X is admissible to some differentiable structure on X it is enough to show that it is C^∞-related to every chart in some atlas contained in the differentiable structure. **Hint:** Argue as in the proof of Theorem 4.2.1.

The charts in a differentiable structure are often called **(local) coordinate systems** on the manifold. The idea here is that, if (U, φ) is such a chart, then φ identifies $U \subseteq X$ with an open set $\varphi(U) \subseteq \mathbb{R}^n$ and every point in $\varphi(U)$ has standard Cartesian coordinates which one can then regard as coordinates labeling the corresponding point in U. If $\mathcal{P}^i : \mathbb{R}^n \to \mathbb{R}$ is the projection onto the ith coordinate, then the functions $\mathcal{P}^i \circ \varphi : U \to \mathbb{R}$ are called the **coordinate functions of** φ and are often given names such as x^i, y^i, u^i, v^i, etc. (the same names being used to label the usual coordinate axes in the image Euclidean space $\mathbb{R}^n$). For each $p \in U$ one thinks of $(x^1(p), \ldots, x^n(p)) = (\mathcal{P}^1(\varphi(p)), \ldots, \mathcal{P}^n(\varphi(p)))$ as the coordinates of p supplied by the chart φ. U itself is called a **coordinate neighborhood (nbd)** on X. From this point-of-view, the fact that two charts (U_1, φ_1) and (U_2, φ_2) in a differentiable structure are C^∞-related amounts to the assertion that the coordinates supplied by φ_1 on $U_1 \cap U_2$ are C^∞ functions of those supplied by φ_2: If $x^i = \mathcal{P}^i \circ \varphi_1$ and $y^i = \mathcal{P}^i \circ \varphi_2$, then $(y^1, \ldots, y^n) = \varphi_2 \circ \varphi_1^{-1}(x^1, \ldots, x^n)$ is a C^∞ coordinate transformation on $\varphi_1(U_1 \cap U_2)$.

We already have at our disposal a nice collection of examples of differentiable manifolds (Euclidean spaces, spheres and projective spaces) and we will add many more to the list before we are through. Some of the simpler ones are worth introducing at once. First we point out that a given topological manifold may admit many differentiable structures. $\mathbb{R}$ has its standard structure generated by the atlas $\{(\mathbb{R}, id)\}$, but, as we have already seen, $(\mathbb{R}, \varphi)$, where $\varphi(x) = x^3$, is a chart on $\mathbb{R}$ that is not in the standard structure. Thus, the atlas $\{(\mathbb{R}, \varphi)\}$ generates a (nonstandard) differentiable structure on $\mathbb{R}$ that is different from the standard one (not "too different", however, as we will see in Section 4.3).

Now consider an arbitrary manifold X with differentiable structure $\{(U_\alpha, \varphi_\alpha)\}_{\alpha \in \mathcal{A}}$ and let Y be an open subspace of X. Then Y is also a topological manifold. If $(U_\alpha, \varphi_\alpha)$ is any chart in the differentiable structure on X for which $Y \cap U_\alpha \neq \emptyset$, then maximality implies that $(Y \cap U_\alpha, \varphi_\alpha | Y \cap U_\alpha)$ is also in the differentiable structure. The collection of all such charts on Y is an atlas for Y and therefore generates a differentiable structure for Y. With this differentiable structure Y is called an **open submanifold of** X (we will introduce the general notion of a "submanifold" a bit later). Note that $\dim Y = \dim X$. As concrete examples we point out the open submanifolds $GL(n, \mathbb{R})$, $GL(n, \mathbb{C})$ and $GL(n, \mathbb{H})$ of $\mathbb{R}^{n^2}$, $\mathbb{R}^{2n^2}$ and $\mathbb{R}^{4n^2}$, respectively (Section 1.1).

Another simple way of manufacturing new examples from those we already have is by forming products. Let X and Y be differentiable manifolds of dimension n and m, respectively. Provide $X \times Y$ with the product topology. Now let (U, φ) and (V, ψ) be charts on X and Y, respectively, with $\mathcal{P}^i \circ \varphi = x^i$ for $i = 1, \ldots, n$ and $\mathcal{P}^j \circ \psi = y^j$ for $j = 1, \ldots, m$. Then $\varphi(U) \times \psi(V)$ is open in $\mathbb{R}^n \times \mathbb{R}^m \cong \mathbb{R}^{n+m}$. Define $\varphi \times \psi : U \times V \to \mathbb{R}^{n+m}$ by $(\varphi \times \psi)(p, q) = (x^1(p), \ldots, x^n(p), y^1(q), \ldots, y^m(q))$. Then $\varphi \times \psi$ is a homeomorphism of $U \times V$ onto $\varphi(U) \times \psi(V)$ (see Exercise 1.3.14) so $(U \times V, \varphi \times \psi)$ is a chart on $X \times Y$.

Exercise 4.2.6 Show that any two such charts are C^∞-related so that the collection of all such is an atlas for $X \times Y$.

The atlas described in Exercise 4.2.6 generates a differentiable structure on $X \times Y$ and with this structure $X \times Y$ is called the **product manifold** of X and Y. Its dimension is $n + m$. The process obviously extends to larger (finite) products by induction. Important examples are the tori $S^1 \times S^1$, $S^1 \times S^1 \times S^1, \ldots$ and various other products of spheres $S^n \times S^m$.

As a final item in our list of elementary examples we do for an arbitrary finite dimensional vector space over $\mathbb{R}$ what we have done for Euclidean spaces. Thus, let $\mathcal{V}$ be a real vector space of dimension n. Select a basis $\{e_1, \ldots, e_n\}$ for $\mathcal{V}$ and let $\{e^1, \ldots, e^n\}$ be its dual basis (thus, e^i is the real-valued linear functional defined on $\mathcal{V}$ by $e^i(e_j) = \delta^i_j$ for $i, j = 1, \ldots, n$). Define $\varphi : \mathcal{V} \to \mathbb{R}^n$ by $\varphi(v) = (e^1(v), \ldots, e^n(v)) = (v^1, \ldots, v^n)$, where $v = v^1 e_1 + \cdots + v^n e_n$. Then φ is a vector space isomorphism so we may define a topology on $\mathcal{V}$ by insisting that φ be a homeomorphism, i.e., $U \subseteq \mathcal{V}$ is open in $\mathcal{V}$ iff $\varphi(U)$ is open in $\mathbb{R}^n$.

Exercise 4.2.7 Show that if $\{\hat{e}_1, \ldots, \hat{e}_n\}$ is another basis for $\mathcal{V}$, $\{\hat{e}^1, \ldots, \hat{e}^n\}$ is its dual basis and $\hat{\varphi} : \mathcal{V} \to \mathbb{R}^n$ is defined by $\hat{\varphi}(v) = (\hat{e}^1(v), \ldots, \hat{e}^n(v))$, then $\hat{\varphi}(U)$ is open in $\mathbb{R}^n$ iff $\varphi(U)$ is open in $\mathbb{R}^n$. Conclude that the topology of $\mathcal{V}$ does not depend on the choice of the basis with which it is defined.

Thus, $(\mathcal{V}, \varphi)$ is a chart on $\mathcal{V}$ and so the atlas $\{(\mathcal{V}, \varphi)\}$ generates a unique differentiable structure on $\mathcal{V}$ which we call its **natural differentiable structure**.

Exercise 4.2.8 Show that if $\hat{\varphi}$ is as in Exercise 4.2.7, then $\{(\mathcal{V}, \hat{\varphi})\}$ generates the same differentiable structure on $\mathcal{V}$.

4.3 Smooth Maps on Manifolds

The charts in a differentiable structure supply local coordinates on a manifold and coordinates are things that one can differentiate with respect to. We will see that, because the charts are C^∞-related, the resulting notion of differentiability does not depend on the coordinate system in which the derivatives are computed.

Let X be a smooth manifold of dimension n and $f : X \to \mathbb{R}$ a real-valued function. Let (U, φ) be a chart in the differentiable structure for X. The

((U,φ)–) **coordinate expression for** f is the function $\tilde{f} : \varphi(U) \to \mathbb{R}$ defined by $\tilde{f} = f \circ \varphi^{-1}$. We say that f is $\mathbf{C}^\infty$ (or **smooth**) **on** X if, for every chart (U,φ) in the differentiable structure for X, the coordinate expression $\tilde{f}$ is a C^∞ real-valued function on the open subset $\varphi(U)$ of $\mathbb{R}^n$.

Exercise 4.3.1 Show that if f is C^∞ on X, then it is necessarily continuous on X.

Lemma 4.3.1 *Let X be a differentiable manifold and $f : X \to \mathbb{R}$ a real-valued function on X. Then f is C^∞ on X iff its coordinate expressions $\tilde{f} = f \circ \varphi^{-1} : \varphi(U) \to \mathbb{R}$ are C^∞ for all charts (U,φ) in some atlas for X.*

Proof: The necessity is clear. To prove the sufficiency we suppose $\tilde{f}$ is C^∞ for all charts (U,φ) in some atlas for X and let (V,ψ) be some arbitrary chart in the differentiable structure. We show that $f \circ \psi^{-1} : \psi(V) \to \mathbb{R}$ is C^∞ by proving that it is C^∞ on a nbd of each point $\psi(p)$ in $\psi(V)$. Let (U,φ) be a chart in the given atlas with $p \in U$. Then $U \cap V$ is open in V so $\psi(U \cap V)$ is open in $\psi(V)$ and contains $\psi(p)$. Moreover, since (V,ψ) and (U,φ) are C^∞-related, on $\psi(U \cap V)$, $\varphi \circ \psi^{-1}$ is C^∞. But, on $\psi(U \cap V)$, $f \circ \psi^{-1} = (f \circ \varphi^{-1}) \circ (\varphi \circ \psi^{-1})$ so this too is C^∞. ∎

Exercise 4.3.2 Regard S^n as the set of $x = (x^1, \ldots, x^n, x^{n+1})$ in $\mathbb{R}^{n+1}$ with $\|x\| = 1$ and define the **height function** $h : S^n \to \mathbb{R}$ by $h(x) = x^{n+1}$. Choose a convenient atlas for the standard differentiable structure on S^n, find the coordinate expressions for h relative to its charts and conclude that h is smooth on S^n.

If W is an open subset of X and f is a real-valued function on X, then we will say that f is $\mathbf{C}^\infty$ (or **smooth**) **on** W if $f \mid W$ is C^∞ on the open submanifold W of X.

Exercise 4.3.3 Show that $f : X \to \mathbb{R}$ is C^∞ on X iff it is C^∞ on an open nbd of every point in X (smoothness is a local property).

Exercise 4.3.4 Let U be a coordinate nbd for X and let A_0 and A_1 be closed subsets of U. Show that there exists a non-negative, C^∞, real-valued function f on U such that $A_0 = f^{-1}(0)$ and $A_1 = f^{-1}(1)$. **Hint:** Corollary 4.1.4.

Lemma 4.3.2 *Let W be an open subset of the smooth manifold X and p a point in W. Then there exists a real-valued, C^∞ function g on X that is 1 on an open nbd of p in W and 0 outside W (g is called a* **bump function** *at p in W).*

Proof: Select a chart (U,φ) for X with $p \in U \subseteq W$. There exist open sets V_1 and V_2 in U with $p \in V_1 \subseteq \bar{V}_1 \subseteq V_2 \subseteq \bar{V}_2 \subseteq U$ (find analogous sets in $\varphi(U)$ containing $\varphi(p)$). By Exercise 4.3.4 we may select a C^∞ function f on U with $\bar{V}_1 = f^{-1}(1)$ and $U - V_2 = f^{-1}(0)$. Now, define $g : X \to \mathbb{R}$ by $g \,|\, U = f$ and $g \,|\, X - \bar{V}_2 = 0$. Observe that g is well-defined because $U \cap (X - \bar{V}_2) \subseteq U \cap (X - V_2) = U - V_2 = f^{-1}(0)$. Furthermore, g is C^∞ on X by Exercise 4.3.3. Since g is 1 on V_1 and 0 on $X - V_2 \supseteq X - W$ the result follows. ∎

Now let X and Y be two smooth manifolds of dimension n and m, respectively, and let $F : X \to Y$ be a continuous map. Let (U, φ) be a chart on X and (V, ψ) a chart on Y with $U \cap F^{-1}(V) \neq \emptyset$. The **coordinate expression for** F relative to (U, φ) and (V, ψ) is the map $\tilde{F} : \varphi(U \cap F^{-1}(V)) \to \mathbb{R}^m$ defined by $\tilde{F}(p) = \psi \circ F \circ \varphi^{-1}(p)$ for each $p \in \varphi(U \cap F^{-1}(V))$. If $x^i = \mathcal{P}^i \circ \varphi$, $i = 1, \ldots, n$, and $y^j = \mathcal{P}^j \circ \psi$, $j = 1, \ldots, m$, and if the x^i and y^j are regarded also as names for the standard coordinate axes in $\mathbb{R}^n$ and $\mathbb{R}^m$, respectively, then $\tilde{F}$ is just an ordinary map from an open set in $\mathbb{R}^n$ to $\mathbb{R}^m$: $(y^1, \ldots, y^m) = \tilde{F}(x^1, \ldots, x^n) = (\tilde{F}^1(x^1, \ldots, x^n), \ldots, \tilde{F}^m(x^1, \ldots, x^n))$. If these coordinate expressions are C^∞ for all charts (U, φ) and (V, ψ) in the differentiable structures for X and Y with $U \cap F^{-1}(V) \neq \emptyset$, then we say that F itself is $\mathbf{C^\infty}$ (or **smooth**).

Exercise 4.3.5 Show that F is C^∞ iff its coordinate expressions are C^∞ for all charts in some atlases for X and Y.

Exercise 4.3.6 Show that compositions of C^∞ maps are C^∞.

Exercise 4.3.7 Show that smoothness is a local property, i.e., that $F : X \to Y$ is C^∞ iff, for each $x \in X$, there exists an open nbd W of x such that $F \,|\, W$ is C^∞ (as a map on the open submanifold W of X).

Exercise 4.3.8 Show that, if (U, φ) is a chart on X, then both $\varphi : U \to \varphi(U)$ and $\varphi^{-1} : \varphi(U) \to U$ are C^∞.

A bijection $F : X \to Y$ for which both F and $F^{-1} : Y \to X$ are C^∞ is called a **diffeomorphism** and, if such an F exists, we say that X and Y are **diffeomorphic**. According to Exercise 4.3.8, if (U, φ) is a chart, then $\varphi : U \to \varphi(U)$ is a diffeomorphism.

Exercise 4.3.9 Show, conversely, that if V is open in the differentiable manifold X and ψ is a diffeomorphism of (the open submanifold) V onto an open set $\psi(V)$ in $\mathbb{R}^n$, then (V, ψ) is an admissible chart on X.

Compositions and inverses of diffeomorphisms are clearly also diffeomorphisms. A diffeomorphism is necessarily a homeomorphism, but even a C^∞

homeomorphism need not be a diffeomorphism, e.g., $\varphi : \mathbb{R} \to \mathbb{R}$ given by $\varphi(x) = x^3$.

Exercise 4.3.10 Show that $\mathbb{RP}^1$, $\mathbb{CP}^1$ and $\mathbb{HP}^1$ are diffeomorphic to S^1, S^2 and S^4, respectively. **Hint:** Re-examine the proof of (1.2.7), (1.2.8) and (1.2.9) and show that the homeomorphism described there can be written $[\xi^1, \xi^2] \to (2\xi^1\bar{\xi}^2, |\xi^1|^2 - |\xi^2|^2)$.

Observe that it is possible for two different differentiable structures (maximal atlases) on the same set to yield manifolds that, while "different", are nevertheless diffeomorphic. Consider $\mathbb{R}$ with its standard differentiable structure (generated by $\{(\mathbb{R}, id)\}$) and the manifold $\mathbb{R}'$ whose underlying topological space is also $\mathbb{R}$, but whose differentiable structure is the nonstandard one generated by $\{(\mathbb{R}', \varphi)\}$, where $\varphi(x) = x^3$. We claim that φ, regarded as a map from $\mathbb{R}'$ to $\mathbb{R}$, is a diffeomorphism (even though φ^{-1}, regarded as a map from $\mathbb{R}$ to $\mathbb{R}$, is not C^∞ in the usual sense). The coordinate expression for $\varphi : \mathbb{R}' \to \mathbb{R}$ relative to the charts $(\mathbb{R}', \varphi)$ and $(\mathbb{R}, id)$ is $id \circ \varphi \circ \varphi^{-1} = id$, which is a C^∞ map from $\mathbb{R}$ to $\mathbb{R}$. Now consider $\varphi^{-1} : \mathbb{R} \to \mathbb{R}'$. Its coordinate expression relative to the same two charts is $\varphi \circ \varphi^{-1} \circ id^{-1} = id$ so it too is C^∞. Thus, φ is a diffeomorphism so $\mathbb{R}'$ and $\mathbb{R}$, although not identical as manifolds, are diffeomorphic and so not "too different".

Exercise 4.3.11 Show that diffeomorphic manifolds have precisely the same C^∞ functions (into and out of).

One can actually show that *any* two differentiable structures on the topological manifold $\mathbb{R}$ are necessarily diffeomorphic. Remarkably, the same is true of any $\mathbb{R}^n$, *except* $\mathbb{R}^4$. The profound work of Freedman on topological 4-manifolds and Donaldson on the implications of gauge theory for smooth 4-manifolds has culminated in the existence of what are called **fake** $\mathbb{R}^4$**'s**, i.e., 4-dimensional differentiable manifolds that are homeomorphic, but not diffeomorphic to $\mathbb{R}^4$ with its standard differentiable structure (see Section 5.5 for more on this).

Many of the purely topological notions introduced in earlier chapters have important smooth analogues. The smooth version of a topological group is a Lie group. More precisely, a **Lie group** is a differentiable manifold G that is also a group in which the operations of multiplication

$$(x, y) \longrightarrow xy : G \times G \longrightarrow G$$

and inversion

$$x \longrightarrow x^{-1} : G \longrightarrow G$$

are C^∞ (here $G \times G$ has the product manifold structure, defined in Section 4.2). With this one can define a **smooth right action** of the Lie group

G on the differentiable manifold P to be a C^∞ map $\sigma : P \times G \to P$, $\sigma(p, g) = p \cdot g$, which satisfies

1. $p \cdot e = p$ for all $p \in P$, and
2. $p \cdot (g_1 g_2) = (p \cdot g_1) \cdot g_2$ for all $p \in P$ and all $g_1, g_2 \in G$.

Then, if X is a differentiable manifold and G is a Lie group one defines a **$\mathbf{C}^\infty$ (smooth) principal bundle over X with structure group G** (or, simply, a **(smooth) G-bundle over X**) to be a triple $\mathcal{B} = (P, \mathcal{P}, \sigma)$, where P is a differentiable manifold, $\mathcal{P}$ is a C^∞ map of P onto X and $\sigma : P \times G \to P$, $\sigma(p, g) = p \cdot g$, is a smooth right action of G on P such that the following conditions are satisfied.

1. σ preserves the fibers of $\mathcal{P}$, i.e.,
$$\mathcal{P}(p \cdot g) = \mathcal{P}(p) \tag{4.3.1}$$
for all $p \in P$ and $g \in G$.
2. **(Local Triviality)** For each x_0 in X there exists an open set V in X containing x_0 and a diffeomorphism $\Psi : \mathcal{P}^{-1}(V) \to V \times G$ of the form
$$\Psi(p) = \left(\mathcal{P}(p), \psi(p)\right), \tag{4.3.2}$$
where $\psi : \mathcal{P}^{-1}(V) \to G$ satisfies
$$\psi(p \cdot g) = \psi(p)g \tag{4.3.3}$$
for all $p \in \mathcal{P}^{-1}(V)$ and $g \in G$.

Since we have explicit formulas recorded for all of the mappings involved it is a simple matter to check that the Hopf bundles are, in fact, smooth principal bundles. We leave this and a few related matters to the reader.

Exercise 4.3.12 Identifying $U(1)$ with S^1 (the unit complex numbers) and $Sp(1)$ with S^3 (the unit quaternions), show that these are Lie groups and that the Hopf bundles $(S^{2n-1}, \mathbb{CP}^{n-1}, \mathcal{P}, U(1))$ and $(S^{4n-1}, \mathbb{HP}^{n-1}, \mathcal{P}, Sp(1))$ are smooth principal bundles.

Exercise 4.3.13 Show that transition functions for smooth principal bundles are smooth maps from the open submanifolds on which they are defined to the Lie structure group of the bundle.

Exercise 4.3.14 The definition of a principal bundle map in Section 3.3 has an obvious smooth version. Prove the corresponding smooth analogues of the results in Exercises 3.3.1, 3.3.2 and 3.3.3.

Exercise 4.3.15 Exhibit a one-to-one correspondence between local trivializations of a smooth principal bundle and smooth local cross-sections of the bundle.

Lie groups and smooth principal bundles will occupy center stage from this point forward, but in order to effectively exploit their smooth structure we will require a number of tools that are not available in the purely topological setting. We turn to these now.

4.4 Tangent Vectors and Derivatives

One is accustomed, from calculus, to thinking of the tangent plane to a smooth surface in $\mathbb{R}^3$ as a linear subspace of $\mathbb{R}^3$ (or, perhaps, as the plane obtained by translating this subspace to the point of tangency). Since a general differentiable manifold (e.g., a projective space) need not live naturally in any ambient Euclidean space one is forced to seek another, intrinsic, characterization of tangent vectors if one wishes to define tangent spaces to such manifolds (and we do). Fortunately, one is at hand. A tangent vector $\mathbf{v}$ to a point p on a surface will assign to every smooth real-valued function f on the surface a "directional derivative" $\mathbf{v}(f) = \nabla f(p) \cdot \mathbf{v}$. Thought of as an operator on such functions, $\mathbf{v}$ is linear and satisfies a product rule. More importantly, the tangent vector $\mathbf{v}$ is uniquely determined if one knows the value it assigns to every such f. Consequently, one may identify a tangent vector with a certain type of real-valued operator on smooth functions and this view of things is entirely intrinsic.

If X is a differentiable manifold we denote by $C^\infty(X)$ the set of all smooth, real-valued functions on X and provide it with an algebraic structure as follows: Let $f, g \in C^\infty(X)$ and $a, b \in \mathbb{R}$. Define $af + bg$ and $f \cdot g$ (also denoted simply fg) by $(af + bg)(x) = af(x) + bg(x)$ and $(f \cdot g)(x) = f(x)g(x)$. These are clearly also in $C^\infty(X)$. Moreover, on $X - g^{-1}(0)$, we may define $\frac{f}{g}$ by $(\frac{f}{g})(x) = \frac{f(x)}{g(x)}$ and this too is C^∞ on its domain. Thus, $C^\infty(X)$ becomes a commutative algebra with identity $\mathbf{1} : X \to \mathbb{R}$ defined by $\mathbf{1}(x) = 1$ for each $x \in X$. If W is an open subset of X, then we may regard W as an open submanifold of X and thereby define $C^\infty(W)$. According to Exercise 4.3.3, if $f \in C^\infty(X)$, then $f\,|\,W \in C^\infty(W)$. If p is a point in X we will denote by $C^\infty(p)$ the set of all real-valued functions that are defined and C^∞ on *some* open nbd of p in X (different functions may have different domains). The algebraic structure of $C^\infty(p)$ is defined in exactly the same way as that of $C^\infty(X)$ except that one must "intersect domains", e.g., if f is defined and smooth on W_1 and g is defined and smooth on W_2, then $af + bg$ and fg are defined pointwise on $W_1 \cap W_2$.

Now let X be a differentiable manifold and $p \in X$. A **tangent vector to** X **at** p is a real-valued function $\mathbf{v} : C^\infty(X) \to \mathbb{R}$ that satisfies

1. **(Linearity)** $\mathbf{v}(af + bg) = a\mathbf{v}(f) + b\mathbf{v}(g)$, and

2. **(Leibnitz Product Rule)** $\mathbf{v}(fg) = f(p)\mathbf{v}(g) + \mathbf{v}(f)g(p)$,

for all $f, g \in C^\infty(X)$ and all $a, b \in \mathbb{R}$. Before producing examples we obtain a few preliminary results and show that one could replace $C^\infty(X)$ in the definition with $C^\infty(p)$.

Lemma 4.4.1 *Let* $\mathbf{v}$ *be a tangent vector to* X *at* p *and suppose that* f *and* g *are elements of* $C^\infty(X)$ *that agree on some nbd of* p *in* X. *Then* $\mathbf{v}(f) = \mathbf{v}(g)$.

Proof: By linearity (#1) it is enough to show that if $f(x) = 0$ for all x in some nbd W of p in X, then $\mathbf{v}(f) = 0$. By Lemma 4.3.2 we may select a bump function $g \in C^\infty(X)$ that is 1 on an open nbd of p in W and 0 outside W. Then $fg = 0$ on all of X. But $\mathbf{v}(0) = \mathbf{v}(0+0) = \mathbf{v}(0) + \mathbf{v}(0)$ implies $\mathbf{v}(0) = 0$ so $\mathbf{v}(fg) = 0$. Thus, $0 = \mathbf{v}(fg) = f(p)\mathbf{v}(g) + \mathbf{v}(f)g(p) = 0 \cdot \mathbf{v}(g) + 1 \cdot \mathbf{v}(f)$ as required. ∎

Lemma 4.4.2 *Let* $\mathbf{v}$ *be a tangent vector to* X *at* p. *If* $f \in C^\infty(X)$ *is constant on some nbd of* p, *then* $\mathbf{v}(f) = 0$.

Proof: By Lemma 4.4.1 we need only show that if $\mathbf{c}$ is a constant function on all of X, then $\mathbf{v}(\mathbf{c}) = 0$. First consider the function $\mathbf{1}$ that takes the value 1 everywhere on X. Then $\mathbf{v}(\mathbf{1}) = \mathbf{v}(\mathbf{1}\cdot\mathbf{1}) = 1\cdot\mathbf{v}(\mathbf{1}) + 1\cdot\mathbf{v}(\mathbf{1}) = 2\mathbf{v}(\mathbf{1})$ so $\mathbf{v}(\mathbf{1}) = 0$. But then $\mathbf{v}(\mathbf{c}) = \mathbf{v}(c\cdot\mathbf{1}) = c\mathbf{v}(\mathbf{1}) = 0$ (c is the value of the constant function $\mathbf{c}$). ∎

Now observe that, since $C^\infty(X) \subseteq C^\infty(p)$, any operator $\tilde{\mathbf{v}} : C^\infty(p) \to \mathbb{R}$ that is linear and satisfies the Leibnitz rule is necessarily a tangent vector to X at p. On the other hand, if $\mathbf{v}$ is a tangent vector to X at p, then we can uniquely extend $\mathbf{v}$ to a linear and Leibnitzian operator $\tilde{\mathbf{v}} : C^\infty(p) \to \mathbb{R}$ as follows: Let $f \in C^\infty(p)$, defined on a nbd W of p.

Exercise 4.4.1 Show that there exists an $F \in C^\infty(X)$ that agrees with f on some nbd of p in W. **Hint:** Lemma 4.3.2.

If $F, F' \in C^\infty(X)$ both agree with f on some nbd of p, then they will agree with each other on some nbd of p so, according to Lemma 4.4.1, $\mathbf{v}(F') = \mathbf{v}(F)$. Thus, we may unambiguously define $\tilde{\mathbf{v}}(f) = \mathbf{v}(F)$. The upshot of all this is that one can regard tangent vectors to X at p as linear, Leibnitzian operators on either $C^\infty(X)$ or $C^\infty(p)$, whichever is convenient.

The set of all tangent vectors to X at p is called the **tangent space to** X **at** p and denoted $T_p(X)$. It has a natural vector space structure defined as follows: If $\mathbf{v}, \mathbf{w} \in T_p(X)$ and $a \in \mathbb{R}$ we define $\mathbf{v} + \mathbf{w}$ and $a\mathbf{v}$ by $(\mathbf{v} + \mathbf{w})(f) = \mathbf{v}(f) + \mathbf{w}(f)$ and $(a\mathbf{v})(f) = a\mathbf{v}(f)$.

Exercise 4.4.2 Show that $\mathbf{v} + \mathbf{w}$ and $a\mathbf{v}$ are in $T_p(X)$.

We show in Theorem 4.4.3 that the dimension of $T_p(X)$ as a vector space coincides with the dimension of X as a manifold.

If $f \in C^\infty(\mathbb{R}^n)$ it would be pointless to insist on writing $f \circ id^{-1}$ for the coordinate expression of f in the standard chart on $\mathbb{R}^n$ and so we will do the sensible thing and denote this coordinate expression by f also. Then, for any $p \in \mathbb{R}^n$, the partial differentiation operators $D_1, \ldots, D_n$, evaluated at p, give elements of $T_p(\mathbb{R}^n)$ whose values on f we denote $D_i|_p(f)$, or $D_i f(p)$. We will soon see that these form a basis for $T_p(\mathbb{R}^n)$.

Now let $\mathcal{I}$ be an open interval in $\mathbb{R}$ (regarded as an open submanifold of $\mathbb{R}$). A C^∞ map $\alpha : \mathcal{I} \to X$ from $\mathcal{I}$ into the manifold X is called a **smooth curve** in X. Fix a $t_0 \in \mathcal{I}$ and let $p = \alpha(t_0)$. For each $f \in C^\infty(p)$, $f \circ \alpha$ is an element of $C^\infty(t_0)$. Again, we choose to write simply $f \circ \alpha$ also for the coordinate expression $id \circ (f \circ \alpha) \circ id^{-1}$. Now define $\alpha'(t_0) : C^\infty(p) \to \mathbb{R}$ by $\alpha'(t_0)(f) = D_1(f \circ \alpha)(t_0)$. Thus, $\alpha'(t_0)(f)$ is the derivative of f along α at p.

Exercise 4.4.3 Show that $\alpha'(t_0) \in T_p(X)$. **Hint:** This is just the linearity and product rule for ordinary differentiation.

$\alpha'(t_0)$ is called the **velocity vector of** α **at** t_0 and we will see shortly that every element of $T_p(X)$ is such a velocity vector.

Next consider a chart (U, φ) at p in X and denote its coordinate functions $\mathcal{P}^i \circ \varphi$ by x^i for $i = 1, \ldots, n$. Define operators

$$\left.\frac{\partial}{\partial x^i}\right|_p : C^\infty(p) \longrightarrow \mathbb{R}$$

by

$$\left.\frac{\partial}{\partial x^i}\right|_p (f) = D_i\left(f \circ \varphi^{-1}\right)\left((\varphi(p)\right).$$

Intuitively, we simply write f locally in terms of the coordinates supplied by (U, φ) and compute the ordinary partial derivative with respect to the ith one (at $\varphi(p)$). As an obvious notational convenience we will often denote this simply $\frac{\partial f}{\partial x^i}(p)$. Note that if $\varphi(p) = (x_0^1, \ldots, x_0^n)$, then $\frac{\partial}{\partial x^i}|_p$ is just $\alpha'(x_0^i)$, where $\alpha(t) = \varphi^{-1}(x_0^1, \ldots, t, \ldots, x_0^n)$ (t in the ith slot).

Exercise 4.4.4 Show that $\frac{\partial}{\partial x^i}|_p (x^j) = \delta_i^j$ (the Kronecker delta).

Exercise 4.4.5 Let W be an open submanifold of X and $p \in W$. For each $\mathbf{v} \in T_p(W)$ define $\tilde{\mathbf{v}} \in T_p(X)$ by $\tilde{\mathbf{v}}(f) = \mathbf{v}(f \mid W)$ for every $f \in C^\infty(X)$. Show that $\mathbf{v} \to \tilde{\mathbf{v}}$ is an isomorphism of $T_p(W)$ onto $T_p(X)$. Henceforth we will suppress the isomorphism altogether and simply identify $T_p(W)$ with $T_p(X)$.

Theorem 4.4.3 *Let X be a differentiable manifold and $p \in X$. If (U, φ) is a chart at p in X with coordinate functions $x^i = \mathcal{P}^i \circ \varphi$, $i = 1, \ldots, n$,* *then* $\{\frac{\partial}{\partial x^i}|_p\}_{i=1,\ldots,n}$ *is a basis for* $T_p(X)$. *Moreover, for each* $\mathbf{v} \in T_p(X)$,

$$\mathbf{v} = \sum_{i=1}^{n} \mathbf{v}(x^i) \left.\frac{\partial}{\partial x^i}\right|_p = \mathbf{v}(x^i) \left.\frac{\partial}{\partial x^i}\right|_p . \tag{4.4.1}$$

Note: As of the second equality in (4.4.1) we begin to employ the *Einstein summation convention*, according to which an index appearing twice in some expression (once as a superscript and once as a subscript) is to be summed over the range of values that the index can assume. An index appearing as a superscript in a denominator counts as a subscript.

Proof: We ask the reader to show first that we may assume, without loss of generality, that $\varphi(p) = (x^1(p), \ldots, x^n(p)) = (0, \ldots, 0)$.

Exercise 4.4.6 Let $T : \mathbb{R}^n \to \mathbb{R}^n$ be the translation defined by $T(x) = x + x_0$ for every $x \in \mathbb{R}^n$. Define $\varphi' : U \to \mathbb{R}^n$ by $\varphi' = T \circ \varphi$. Show that (U, φ') is a chart at p in X and that, if its coordinate functions are $y^i = \mathcal{P}^i \circ \varphi'$ for $i = 1, \ldots, n$, then $\frac{\partial}{\partial y^i}|_p = \frac{\partial}{\partial x^i}|_p$ and $\mathbf{v}(y^i) = \mathbf{v}(x^i)$ for every $\mathbf{v} \in T_p(X)$.

Thus, by taking $x_0 = -\varphi(p)$ in Exercise 4.4.6 we may assume at the outset that $x^i(p) = 0$ for $i = 1, \ldots, n$. By shrinking U if necessary we may also assume that $\varphi(U)$ is the open ε-ball about $0 \in \mathbb{R}^n$.

Next we need a preliminary result from calculus. Let g be a C^∞ real-valued function on the open ε-ball $U_\varepsilon(0)$ about 0 in $\mathbb{R}^n$. We show that, on $U_\varepsilon(0)$, g can be written in the form

$$g\left(x^1, \ldots, x^n\right) = g(0) + x^i g_i\left(x^1, \ldots, x^n\right), \tag{4.4.2}$$

where $x^1, \ldots, x^n$ are the standard coordinates in $\mathbb{R}^n$ and each g_i, $i = 1, \ldots, n$, is C^∞ (don't forget the summation convention in (4.4.2)). To prove this note first that, for each $x \in U_\varepsilon(0)$ and each t in $[0,1]$, tx is in $U_\varepsilon(0)$ and $\frac{d}{dt} g(tx) = D_i g(tx) x^i$ so the Fundamental Theorem of Calculus gives

$$g(x) - g(0) = \int_0^1 D_i g(tx) x^i \, dt = x^i \int_0^1 D_i(tx)\, dt\,.$$

Let $g_i(x) = \int_0^1 D_i(tx)\, dt$. Then (4.4.2) will be established if we can show that g_i is C^∞ on $U_\varepsilon(0)$. Since this is just a matter of repeatedly differentiating under the integral sign we ask the reader to do the calculus.

Exercise 4.4.7 Apply the following result from advanced calculus to show that $g_i(x)$ is C^∞ on $U_\varepsilon(0)$: Suppose $h(x^1, \ldots, x^n, t)$ has continuous partial derivatives on some open set $V \subseteq \mathbb{R}^{n+1} = \mathbb{R}^n \times \mathbb{R}$ that contains $\{(x^1, \ldots, x^n)\} \times [0,1]$ for each $(x^1, \ldots, x^n)$ in the projection V' of V into

$\mathbb{R}^n$. Define $k : V' \to \mathbb{R}$ by $k(x^1, \ldots, x^n) = \int_0^1 h(x^1, \ldots, x^n, t)\, dt$. Then k also has continuous partial derivatives and, for each $j = 1, \ldots, n$,

$$\frac{\partial k}{\partial x^j}\left(x^1, \ldots, x^n\right) = \int_0^1 \frac{\partial h}{\partial x^j}\left(x^1, \ldots, x^n, t\right)\, dt\,.$$

Now we return to the proof of (4.4.1). Let $f \in C^\infty(p)$ be arbitrary. Denote by $\tilde{f} : U \to \mathbb{R}$ the coordinate expression for f relative to (U, φ). We have just shown that we can write $\tilde{f}(x^1, \ldots, x^n) = \tilde{f}(0) + x^i \tilde{f}_i(x^1, \ldots, x^n) = f(p) + x^i \tilde{f}_i(x^1, \ldots, x^n)$, where the $\tilde{f}_i$ are C^∞ on $U_\varepsilon(0)$. Letting $f_i = \tilde{f}_i \circ \varphi$ for $i = 1, \ldots, n$, we obtain elements of $C^\infty(p)$ with $f = f(p) + x^i f_i$ (here we are regarding the x^i as elements of $C^\infty(p)$ and $f(p)$ as a constant function in $C^\infty(p)$). Now we compute

$$\begin{aligned}
\left.\frac{\partial}{\partial x^j}\right|_p (f) &= \left.\frac{\partial}{\partial x^j}\right|_p \left(f(p) + x^i f_i\right) \\
&= \left.\frac{\partial}{\partial x^j}\right|_p (f(p)) + x^i(p) \left.\frac{\partial}{\partial x^j}\right|_p (f_i) + f_i(p) \left.\frac{\partial}{\partial x^j}\right|_p (x^i) \\
&= 0 + 0 + f_i(p)\, \delta^i_j \\
&= f_j\,(p)
\end{aligned}$$

and

$$\begin{aligned}
\mathbf{v}(f) &= \mathbf{v}\left(f(p) + x^i f_i\right) = \mathbf{v}\left(f(p)\right) + x^i(p)\mathbf{v}(f_i) + f_i(p)\mathbf{v}(x^i) \\
&= 0 + 0 + \left.\frac{\partial}{\partial x^i}\right|_p (f)\mathbf{v}(x^i) \\
&= \mathbf{v}(x^i) \left.\frac{\partial}{\partial x^i}\right|_p (f)\,.
\end{aligned}$$

Since $f \in C^\infty(p)$ was arbitrary, (4.4.1) follows.

Exercise 4.4.8 Complete the proof by showing that $\frac{\partial}{\partial x^1}\big|_p, \ldots, \frac{\partial}{\partial x^n}\big|_p$ are linearly independent. **Hint:** Exercise 4.4.4. ■

Corollary 4.4.4 *If X is an n-dimensional C^∞ manifold and $p \in X$, then the dimension of the vector space $T_p(X)$ is n.*

Corollary 4.4.5 *Let (U, φ) and (V, ψ) be two charts on the C^∞ manifold X with $U \cap V \neq \emptyset$ and with coordinate functions $x^j = \mathcal{P}^j \circ \varphi$ and $y^i = \mathcal{P}^i \circ \psi$, respectively. Then, for any $p \in U \cap V$,*

$$\left.\frac{\partial}{\partial y^i}\right|_p = \frac{\partial x^j}{\partial y^i}(p) \left.\frac{\partial}{\partial x^j}\right|_p\,. \tag{4.4.3}$$

Corollary 4.4.6 *Let X be a C^∞ manifold, $p \in X$ and $\mathbf{v} \in T_p(X)$. Then there exists a smooth curve α in X, defined on some interval about 0 in $\mathbb{R}$, such that $\alpha'(0) = \mathbf{v}$.*

Proof: Select a chart (U, φ) at p in X with coordinate functions $x^i = \mathcal{P}^i \circ \varphi$ and write $\mathbf{v} = a^i \frac{\partial}{\partial x^i}|_p$, where $a^i = \mathbf{v}(x^i)$. Define a smooth curve α in X on some interval about 0 in $\mathbb{R}$ by $\alpha(t) = \varphi^{-1}(x^1(p)+ta^1, \ldots, x^n(p)+ta^n)$. Then the components of $\alpha'(0)$ relative to $\{\frac{\partial}{\partial x^i}|_p\}_{i=1,\ldots,n}$ are

$$\begin{aligned}\alpha'(0)(x^i) &= D_1\left(x^i \circ \alpha\right)(0) \\ &= D_1\left(\mathcal{P}^i \circ \varphi \circ \varphi^{-1}(x^1(p) + ta^1, \ldots, x^n(p) + ta^n)\right)(0) \\ &= \left(x^i(p) + ta^i\right)'(0) = a^i\end{aligned}$$

so $\alpha'(0) = \mathbf{v}$. ∎

Exercise 4.4.9 Let $\mathcal{V}$ be an n-dimensional real vector space with its natural differentiable structure (Section 4.2) and let p be a fixed point in $\mathcal{V}$. For each $v \in \mathcal{V}$ define $\mathbf{v}_p \in T_p(\mathcal{V})$ by $\mathbf{v}_p = \alpha'(0)$, where $\alpha : \mathbb{R} \to \mathcal{V}$ is given by $\alpha(t) = p + tv$. Show that $v \to \mathbf{v}_p$ is an isomorphism of $\mathcal{V}$ onto $T_p(\mathcal{V})$. Henceforth we use this **canonical isomorphism** to identify $T_p(\mathcal{V})$ with $\mathcal{V}$.

Now we consider two smooth manifolds X and Y of dimensions n and m, respectively, and a smooth map $f : X \to Y$. At each $p \in X$ we define a linear transformation

$$f_{*p} : T_p(X) \longrightarrow T_{f(p)}(Y),$$

called the **derivative of f at** p, which is intended to serve as a "linear approximation to f near p." We offer two independent definitions and show that they give the same result:

1. For each $\mathbf{v} \in T_p(X)$ we define $f_{*p}(\mathbf{v})$ to be the operator on $C^\infty(f(p))$ defined by $(f_{*p}(\mathbf{v}))(g) = \mathbf{v}(g \circ f)$ for all $g \in C^\infty(f(p))$.

2. For each $\mathbf{v} \in T_p(X)$ we select a smooth curve α in X with $\alpha'(0) = \mathbf{v}$. Then $f \circ \alpha$ is a smooth curve in Y through $f(p)$ at $t = 0$ and we define $f_{*p}(\mathbf{v}) = f_{*p}(\alpha'(0)) = (f \circ \alpha)'(0)$.

Observe that, in #1, f_{*p} is well-defined, but not obviously a tangent vector at $f(p)$, whereas, in #2, f_{*p} is clearly a tangent vector, but not obviously independent of the choice of α. We resolve all of these issues by noting that if α is any curve in X with $\alpha'(0) = \mathbf{v}$, then, for any $g \in C^\infty(f(p))$,

$$\begin{aligned}\mathbf{v}(g \circ f) &= \alpha'(0)(g \circ f) = D_1\left((g \circ f) \circ \alpha\right)(0) \\ &= D_1\left(g \circ (f \circ \alpha)\right)(0) = (f \circ \alpha)'(0)(g).\end{aligned}$$

Note also that definition #1 makes it clear that f_{*p} is a linear transformation.

Exercise 4.4.10 Let (U, φ) be a chart at p in X with coordinate functions $x^j = \mathcal{P}^j \circ \varphi$, $j = 1, \ldots, n$, and (V, ψ) a chart at $f(p)$ in Y with coordinate functions $y^i = \mathcal{P}^i \circ \psi$, $i = 1, \ldots, m$. Show that the matrix of f_{*p} relative to the bases $\left\{ \frac{\partial}{\partial x^j}\big|_p \right\}_{j=1,\ldots,n}$ and $\left\{ \frac{\partial}{\partial y^i}\big|_{f(p)} \right\}_{i=1,\ldots,m}$ is just the usual Jacobian of the coordinate expression $\psi \circ f \circ \varphi^{-1}$ at $\varphi(p)$.

Theorem 4.4.7 (Chain Rule) *Let $f : X \to Y$ and $g : Y \to Z$ be smooth maps between differentiable manifolds. Then $g \circ f : X \to Z$ is smooth and, for every $p \in X$,*

$$(g \circ f)_{*p} = g_{*f(p)} \circ f_{*p} \, .$$

Proof: Smoothness of $g \circ f$ is Exercise 4.3.6. Now let $\mathbf{v} \in T_p(X)$. Then, for every $h \in C^\infty(g(f(p)))$, $(g \circ f)_{*p}(\mathbf{v})(h) = \mathbf{v}(h \circ (g \circ f)) = \mathbf{v}((h \circ g) \circ f) = f_{*p}(\mathbf{v})(h \circ g) = g_{*f(p)}(f_{*p}(\mathbf{v}))(h) = (g_{*f(p)} \circ f_{*p})(\mathbf{v})(h)$. ∎

Exercise 4.4.11 By choosing charts at p, $f(p)$ and $g(f(p))$ and writing out the Jacobians, explain why Theorem 4.4.7 is called the Chain Rule.

Corollary 4.4.8 *Let $f : X \to Y$ be a smooth map between differentiable manifolds and let $p \in X$. Then $f_{*p} : T_p(X) \to T_{f(p)}(Y)$ is an isomorphism iff f is a local diffeomorphism at p, i.e., iff there exist open nbds V and W of p and $f(p)$, respectively, such that $f \mid V$ is a diffeomorphism of V onto W.*

Proof: First suppose f_{*p} is an isomorphism. Selecting charts (U, φ) and (V, ψ) at p and $f(p)$, respectively, the Jacobian of $\psi \circ f \circ \varphi^{-1}$ is nonsingular at $\varphi(p)$. By the Inverse Function Theorem 4.1.1, $\psi \circ f \circ \varphi^{-1}$ is a local diffeomorphism near $\varphi(p)$ and so f is a local diffeomorphism near p.

Next, suppose f is a local diffeomorphism near p. Then, on some nbd of p, $f^{-1} \circ f = id$. Theorem 4.4.7 then gives $(f^{-1})_{*f(p)} \circ f_{*p} = id_{*p}$. But id_{*p} is clearly the identity on $T_p(X)$. Similarly, $f_{*p} \circ (f^{-1})_{*f(p)}$ is the identity on $T_{f(p)}(Y)$ so f_{*p} is an isomorphism. ∎

Observe that the proof of Corollary 4.4.8 shows that if f_{*p} is an isomorphism, then $(f_{*p})^{-1} = (f^{-1})_{*f(p)}$.

Exercise 4.4.12 Show that diffeomorphic manifolds must have the same dimension. In particular, $\mathbb{R}^n$ is diffeomorphic to $\mathbb{R}^m$ iff $n = m$.

The standard chart on an open interval in $\mathbb{R}$ has a single coordinate function (generally denoted x, or t, or s, etc. rather than x^1) and so, at each point, a single coordinate vector spans the tangent space (this is usually written $\frac{d}{dx}|_{x_0}$, or $\frac{d}{dt}|_{t_0}$, or $\frac{d}{ds}|_{s_0}$, etc., rather than $\frac{\partial}{\partial x^1}|_p$). If α is a smooth curve in X defined on some open interval $\mathcal{I}$ about t_0, then $\alpha_{*t_0} : T_{t_0}(\mathcal{I}) \to T_{\alpha(t_0)}(X)$ and we claim that

$$\alpha_{*t_0}\left(\frac{d}{dt}\bigg|_{t_0}\right) = \alpha'(t_0)\,. \tag{4.4.4}$$

Indeed, if $f \in C^\infty(\alpha(t_0))$, then $\alpha_{*t_0}(\frac{d}{dt}|_{t_0})(f) = \frac{d}{dt}|_{t_0}(f \circ \alpha) = D_1(f \circ \alpha)(t_0) = \alpha'(t_0)(f)$, as required.

Exercise 4.4.13 Let $\mathcal{J}$ be another open interval in $\mathbb{R}$ and $h : \mathcal{J} \to \mathcal{I}$ a diffeomorphism. Then $\beta = \alpha \circ h$ is called a **reparametrization of** α. Fix an $s_0 \in \mathcal{J}$ and let $t_0 = h(s_0) \in \mathcal{I}$. Show that $\beta'(s_0) = h'(s_0)\alpha'(t_0)$, where h' is the ordinary derivative of the real-valued function h on the open interval $\mathcal{J}$.

Next we consider a differentiable manifold X, a $p \in X$ and an element f of $C^\infty(p)$. Then $f_{*p} : T_p(X) \to T_{f(p)}(\mathbb{R})$ (recall Exercise 4.4.5). But $T_{f(p)}(\mathbb{R})$ is spanned by the single coordinate vector $\frac{d}{dx}|_{f(p)}$ so, for every $\mathbf{v} \in T_p(X)$, $f_{*p}(\mathbf{v})$ is a multiple of $\frac{d}{dx}|_{f(p)}$, the coefficient being

$$f_{*p}(\mathbf{v})(x) = \mathbf{v}(x \circ f) = \mathbf{v}\left((\mathcal{P} \circ id) \circ f\right) = \mathbf{v}(f)\,.$$

Thus,

$$f_{*p}(\mathbf{v}) = \mathbf{v}(f)\,\frac{d}{dx}\bigg|_{f(p)} \qquad (f \in C^\infty(p)) \tag{4.4.5}$$

for all $\mathbf{v} \in T_p(X)$. Thus, $\mathbf{v}(f)$ completely determines $f_{*p}(\mathbf{v})$. For any $f \in C^\infty(p)$ we define an operator $df(p) = df_p : T_p(X) \to \mathbb{R}$, called the **differential of f at p**, by

$$df(p)(\mathbf{v}) = df_p(\mathbf{v}) = \mathbf{v}(f)$$

for every $\mathbf{v} \in T_p(X)$. Since df_p is clearly linear, it is an element of the dual space of $T_p(X)$, which we denote $T_p^*(X)$ and call the **cotangent space of X at p**. The elements of $T_p^*(X)$ are called **covectors** at p. On occasion (especially in the physics literature) one sees the elements of $T_p(X)$ referred to as **contravariant vectors**, while those of $T_p^*(X)$ are called **covariant vectors**.

If (U, φ) is a chart at p with coordinate functions $x^1, \ldots, x^n$, then each x^i is in $C^\infty(p)$ and so $dx^i{}_p$ is an element of $T_p^*(X)$. For any $\mathbf{v} \in T_p(X)$, $dx^i{}_p(\mathbf{v}) = \mathbf{v}(x^i)$. Since $\mathbf{v} = \mathbf{v}(x^i)\frac{\partial}{\partial x^i}|_p$ we find that $dx^i{}_p$ just picks out the ith component of $\mathbf{v}$ relative to the coordinate basis for $T_p(X)$. In particular,

$$dx^i{}_p\left(\frac{\partial}{\partial x^j}\bigg|_p\right) = \delta^i_j$$

so $\{dx^1{}_p, \ldots, dx^n{}_p\}$ is the basis for $T_p^*(X)$ dual to the coordinate basis $\{\frac{\partial}{\partial x^1}|_p, \ldots, \frac{\partial}{\partial x^n}|_p\}$ for $T_p(X)$.

Exercise 4.4.14 Show that if $\boldsymbol{\theta} \in T_p^*(X)$, then $\boldsymbol{\theta} = \boldsymbol{\theta}(\frac{\partial}{\partial x^i}|_p)dx^i{}_p$ for any chart at p. Conclude that, for any $f \in C^\infty(p)$, $df_p = \frac{\partial f}{\partial x^i}(p)dx^i{}_p$. **Remark:** One often drops the references to p to obtain the familiar looking formula $df = \frac{\partial f}{\partial x^i}dx^i$.

Exercise 4.4.15 Show that, if (U, φ) and (V, ψ) are two charts on X with coordinate functions x^i and y^j, respectively, and if $p \in U \cap V$, then $dy^j_p = \frac{\partial y^j}{\partial x^i}(p)dx^i{}_p$.

By virtue of our obsession with local products we will need a convenient description of the tangent space to a product manifold $X \times Y$. Fortunately, it is just what you think it should be. Let $(p, q) \in X \times Y$ and denote by $\mathcal{P}_X : X \times Y \to X$ and $\mathcal{P}_Y : X \times Y \to Y$ the projections. We define maps $e_q : X \to X \times Y$ and $e_p : Y \to X \times Y$ by $e_q(x) = (x, q)$ and $e_p(y) = (p, y)$.

Exercise 4.4.16 Show that $\mathcal{P}_X$, $\mathcal{P}_Y$, e_q and e_p are all smooth.

Now define $R : T_{(p,q)}(X \times Y) \to T_p(X) \times T_q(Y)$ by $R(\mathbf{v}) = ((\mathcal{P}_X)_{*(p,q)}(\mathbf{v}), (\mathcal{P}_Y)_{*(p,q)}(\mathbf{v}))$. Then R is clearly linear and we show that it is an isomorphism by explicitly exhibiting an inverse $S : T_p(X) \times T_q(Y) \to T_{(p,q)}(X \times Y)$. Thus, for $\mathbf{v}_X \in T_p(X)$ and $\mathbf{v}_Y \in T_q(Y)$ we set $S(\mathbf{v}_X, \mathbf{v}_Y) = (e_q)_{*p}(\mathbf{v}_X) + (e_p)_{*q}(\mathbf{v}_Y)$. Then S is also linear and, noting that $\mathcal{P}_X \circ e_q = id_X$, $\mathcal{P}_Y \circ e_p = id_Y$ and that both $\mathcal{P}_X \circ e_p$ and $\mathcal{P}_Y \circ e_q$ are constant maps, we have

$$\begin{aligned} R \circ S(\mathbf{v}_X, \mathbf{v}_Y) &= R\big((e_q)_{*p}(\mathbf{v}_X) + (e_p)_{*q}(\mathbf{v}_Y)\big) \\ &= \big((\mathcal{P}_X \circ e_q)_{*p}(\mathbf{v}_X) + (\mathcal{P}_X \circ e_p)_{*q}(\mathbf{v}_Y), \\ &\qquad (\mathcal{P}_Y \circ e_q)_{*p}(\mathbf{v}_X) + (\mathcal{P}_Y \circ e_p)_{*q}(\mathbf{v}_Y)\big) \\ &= (\mathbf{v}_X, \mathbf{v}_Y). \end{aligned}$$

Consequently, R and S are inverse isomorphisms and we may henceforth identify $T_{(p,q)}(X \times Y)$ with $T_p(X) \times T_q(Y)$. In particular, $(\mathcal{P}_X)_{*(p,q)}$ and $(\mathcal{P}_Y)_{*(p,q)}$ are now just the projections onto $T_p(X)$ and $T_q(Y)$, respectively. It follows, for example, that if $f : X \to X' \times Y'$ is of the form $f(x) = (f^1(x), f^2(x))$ and $f(p) = (p', q')$, then $f_{*p} : T_p(X) \to T_{p'}(X') \times T_{q'}(Y')$ is given by $f_{*p}(\mathbf{v}) = (f^1_{*p}(\mathbf{v}), f^2_{*p}(\mathbf{v}))$ since, e.g., $f^1_{*p} = (\mathcal{P}_{X'} \circ f)_{*p} = (\mathcal{P}_{X'})_{(p',q')} \circ f_{*p}$. For simplicity we will write

$$f_{*p} = \left(f^1, f^2\right)_{*p} = \left(f^1_{*p}, f^2_{*p}\right).$$

As another example we consider two smooth maps $f : X \to X'$ and $g : Y \to Y'$ and their product $f \times g : X \times Y \to X' \times Y'$ defined by $(f \times g)(x, y) = ((f(x), g(y))$ (see Exercise 1.3.14). Then $f \times g$ is clearly smooth and, if $(p, q) \in X \times Y$ and $\mathbf{v} = (\mathbf{v}_X, \mathbf{v}_Y) \in T_{(p,q)}(X \times Y)$, we claim that $(f \times g)_{*(p,q)}(\mathbf{v}) = (f_{*p}(\mathbf{v}_X), g_{*q}(\mathbf{v}_Y))$, i.e.,

$$(f \times g)_{*(p,q)} = f_{*p} \times g_{*q} \,.$$

To see this note that $(f \times g)(x, y) = (\bar{f}(x, y), \bar{g}(x, y))$, where $\bar{f}(x, y) = f(x)$ and $\bar{g}(x, y) = g(y)$ for all $(x, y) \in X \times Y$. Thus, by what we have just proved, $(f \times g)_{*(p,q)}(\mathbf{v}) = (\bar{f}_{*(p,q)}(\mathbf{v}), \bar{g}_{*(p,q)}(\mathbf{v}))$. Now, use Corollary 4.4.6 to write $\mathbf{v} = \alpha'(0) = ((\mathcal{P}_X)_{*(p,q)}(\alpha'(0)), (\mathcal{P}_Y)_{*(p,q)}(\alpha'(0))) = ((\mathcal{P}_X \circ \alpha)'(0), (\mathcal{P}_Y \circ \alpha)'(0)) = (\mathbf{v}_X, \mathbf{v}_Y)$. Then $\bar{f}_{*(p,q)}(\mathbf{v}) = \bar{f}_{*(p,q)}(\alpha'(0)) = (\bar{f} \circ \alpha)'(0)$. But $(\bar{f} \circ \alpha)(t) = \bar{f}(\alpha(t)) = \bar{f}((\mathcal{P}_X \circ \alpha)(t), (\mathcal{P}_Y \circ \alpha)(t)) = f((\mathcal{P}_X \circ \alpha)(t)) = (f \circ (\mathcal{P}_X \circ \alpha))(t)$ so $(\bar{f} \circ \alpha)'(0) = (f \circ (\mathcal{P}_X \circ \alpha))'(0) = f_{*p}((\mathcal{P}_X \circ \alpha)'(0)) = f_{*p}(\mathbf{v}_X)$. Similarly, $\bar{g}_{*(p,q)}(\mathbf{v}) = g_{*q}(\mathbf{v}_Y)$ so the result follows.

Exercise 4.4.17 Show that if $f : X \times Y \to Z$ is smooth and $(p, q) \in X \times Y$, then for every $\mathbf{v} = (\mathbf{v}_X, \mathbf{v}_Y) \in T_{(p,q)}(X \times Y)$, $f_{*(p,q)}(\mathbf{v}) = (f_1)_{*p}(\mathbf{v}_X) + (f_2)_{*q}(\mathbf{v}_Y)$, where $f_1 : X \to Z$ and $f_2 : Y \to Z$ are defined by $f_1(x) = f(x, q)$ and $f_2(y) = f(p, y)$ for all $x \in X$ and all $y \in Y$.

Exercise 4.4.18 Let $\mathcal{P} : P \to X$ be the projection of a smooth principal G-bundle. Show that, for each $x \in X$ and $p \in \mathcal{P}^{-1}(x)$, $\mathcal{P}_{*p} : T_p(P) \to T_x(X)$ is surjective. **Hint:** Prove this first for the projection of the product manifold $X \times G$ onto X.

In order to ease the typography we will now and then omit reference to the point of tangency and write, for example, f_* for $f_{*p} : T_p(X) \to T_{f(p)}(Y)$.

4.5 Submanifolds

In general, there is no natural way for an arbitrary subset X' of a C^∞ manifold X to "inherit" a differentiable structure from X. Those subsets for which this is possible are called "submanifolds" and the idea behind the definition is as follows: A subset X' of an n-dimensional smooth manifold X will be called a "k-dimensional submanifold of X" if, for each $p \in X'$, there exists a chart (U, φ) for X at p such that $\varphi \,|\, U \cap X'$ carries $U \cap X'$ onto an open set in some copy of $\mathbb{R}^k$ in $\mathbb{R}^n$. However, by composing φ with some orthogonal transformation of $\mathbb{R}^n$ we may clearly assume that this copy of $\mathbb{R}^k$ is $\mathbb{R}^k \times \{0\} = \{(x^1, \ldots, x^k, 0, \ldots, 0) \in \mathbb{R}^n\}$ and $\varphi(p) = 0$.

Thus, we let X be a C^∞ manifold of dimension n and $0 \leq k \leq n$ an integer. A topological subspace X' of X is called a **k-dimensional submanifold of** X if, for each $p \in X'$, there exists a chart (U, φ) in the differentiable structure for X with $\varphi(p) = 0$ and such that $\varphi(U \cap X') = \varphi(U) \cap (\mathbb{R}^k \times \{0\}) = \{x \in \varphi(U) : x^{k+1} = \cdots = x^n = 0\}$. Note that if $k = 0$ this set must be a point so that a 0-dimensional submanifold of X is just a discrete subspace of X. If (U, φ) is a chart of the type described with coordinate functions $x^1, \ldots, x^k, \ldots, x^n$, we define $\varphi' : U \cap X' \to \mathbb{R}^k$ by $\varphi'(q) = (x^1(q), \ldots, x^k(q))$ for every $q \in U \cap X'$, i.e., φ' is $\varphi \mid U \cap X'$ followed by the projection of $\mathbb{R}^n = \mathbb{R}^k \times \mathbb{R}^{n-k}$ onto $\mathbb{R}^k$.

Exercise 4.5.1 Show that the collection of all such $(U \cap X', \varphi')$ form an atlas for X' and so determine a differentiable structure on X' called the **submanifold differentiable structure for** X'.

Exercise 4.5.2 Show that being a submanifold of X is not the same as being a subset of X that is also a manifold. **Hint:** Keep in mind that a submanifold must be a topological subspace.

Exercise 4.5.3 Show that if $f : X \to Y$ is a smooth map and X' is a submanifold of X, then $f|X' : X' \to Y$ is smooth.

Exercise 4.5.4 Let (U, φ) be a chart on X with coordinate functions x^i, $i = 1, \ldots, n$. Let $x^{i_1}, \ldots, x^{i_{n-k}}$ be any $n - k$ of these coordinate functions and $c^1, \ldots, c^{n-k}$ constants. The set $\{p \in U : x^{i_1}(p) = c^1, \ldots, x^{i_{n-k}}(p) = c^{n-k}\}$ is called a **φ-coordinate slice** of U. Show that a topological subspace X' of X is a k-dimensional submanifold of X iff for each point $p' \in X'$ there exists a chart (U, φ) for X such that $U \cap X'$ is a φ-coordinate slice of U containing p'.

Setting $\rho = 1$ in the spherical coordinate charts for $\mathbb{R}^3$ (see Exercise 4.2.3 and the remarks that follow) and applying Exercise 4.5.4 we obtain standard spherical coordinate submanifold charts on S^2.

Exercise 4.5.5 Let $f : X \to Y$ be a smooth map and Y' a submanifold of Y with $f(X) \subseteq Y'$. Show that f, regarded as a map of X into Y', is smooth. **Hint:** $f : X \to Y'$ is continuous by Lemma 1.1.2.

Let $f : X \to Y$ be a smooth map and $p \in X$. f is said to be an **immersion at** p if $f_{*p} : T_p(X) \to T_{f(p)}(Y)$ is one-to-one. f is a **submersion at** p if f_{*p} is onto. f is an **immersion** (respectively, **submersion**) if, for every p in X, f is an immersion (respectively, submersion) at p. An immersion that is also a homeomorphism onto its image is an **imbedding**. A point $q \in Y$ is called a **regular value of** f if, for every $p \in f^{-1}(q)$, f is a submersion at p (this, in particular, is the case if $f^{-1}(q) = \emptyset$); otherwise, q is a **critical value of** f. Here are some examples.

Lemma 4.5.1 *If X' is a submanifold of X, then the inclusion map $\iota : X' \hookrightarrow X$ is an imbedding.*

Proof: ι is a homeomorphism onto its image because X' is assumed to have the subspace topology it inherits from X. Now, fix a $p \in X'$ and let (U, φ) be a chart at p in X with $(U \cap X', \varphi')$ a submanifold chart at p in X'. The coordinate expression for ι relative to these charts is $(x^1, \ldots, x^k) \to (x^1, \ldots, x^k, 0, \ldots, 0)$ on $U \cap X'$. The Jacobian of this coordinate expression has rank $k = \dim T_p(X')$ so, by Exercise 4.4.10, ι_{*p} is one-to-one. ∎

Remark: One consequence of Lemma 4.5.1 is worth pointing out at once. If Y' is a submanifold of Y, then for any $p \in Y'$, ι_{*p} carries $T_p(Y')$ isomorphically onto a linear subspace of $T_p(Y)$. We will not deny ourselves the convenience of suppressing this isomorphism and identifying $T_p(Y')$ with a subspace of $T_p(Y)$ (the velocity vectors to smooth curves in Y whose images happen to lie in Y'). This is particularly convenient for submanifolds of Euclidean spaces.

The converse of Lemma 4.5.1 is also true, but we will presently prove much more (see Corollary 4.5.6). A nice example of a submersion is the projection of $\mathbb{R}^n = \mathbb{R}^k \times \mathbb{R}^{n-k}$ onto $\mathbb{R}^k$. More generally, according to Exercise 4.4.18, any projection of a smooth principal bundle $\mathcal{P} : P \to X$ is a submersion. Of course, if $f : X \to Y$ is a submersion, then every $q \in Y$ is a regular value of f. For some less trivial examples we first consider an $f \in C^\infty(X)$. Then $f_{*p} : T_p(X) \to T_{f(p)}(\mathbb{R})$ is either surjective or identically zero. Thus, by (4.4.5), f is a submersion at p iff df_p is not the zero element of the cotangent space $T_p^*(X)$. By Exercise 4.4.14, this is the case iff for every chart (U, φ) at p in X, some of the $\frac{\partial f}{\partial x^i}(p)$ are nonzero. Thus, a regular value of f is an $r \in \mathbb{R}$ for which $f^{-1}(r)$ contains no points at which *all* of the $\frac{\partial f}{\partial x^i}$ vanish (in some and therefore, by (4.4.3), every chart).

Exercise 4.5.6 Let $X = \mathbb{R}^3$ and denote by x, y and z the coordinate functions for the standard chart $(\mathbb{R}^3, id)$. Let f be the element of $C^\infty(\mathbb{R}^3)$ whose standard coordinate expression is $f(x, y, z) = x^2 + y^2 - z^2$. Show that $0 \in \mathbb{R}$ is the only critical value of f and then describe the level sets $f^{-1}(-r)$, $f^{-1}(0)$ and $f^{-1}(r)$, for $r > 0$. What's "wrong" with $f^{-1}(0)$?

An even less trivial example that will be important to us quite soon is obtained as follows. Consider the collection M_n of all $n \times n$ real matrices. As in Section 1.1 we identify M_n with $\mathbb{R}^{n^2}$ by stringing out the entries in each matrix. Now, however, it will be more convenient to list the entries "lower triangle first" rather that "row after row", e.g., we identify M_3 with $\mathbb{R}^9$ via the map

$$\begin{pmatrix} a_{11} & a_{12} & a_{13} \\ a_{21} & a_{22} & a_{23} \\ a_{31} & a_{32} & a_{33} \end{pmatrix} \longrightarrow (a_{11}, a_{21}, a_{22}, a_{31}, a_{32}, a_{33}, a_{12}, a_{13}, a_{23}) .$$

Recall now that the orthogonal group $O(n)$ is just the set of $A \in M_n$ such that $AA^T = id$. Also note that, for any $A \in M_n$, AA^T is a symmetric matrix ($(AA^T)^T = (A^T)^T A^T = AA^T$). Denoting by S_n the set of symmetric elements in M_n we may therefore define a map $f : M_n \to S_n$ by $f(A) = AA^T$. Now, S_n is a linear subspace of M_n and each of its elements is uniquely determined by its lower triangle so the projection of $\mathbb{R}^{n^2}$ onto its first $\frac{n(n+1)}{2}$ coordinates is, when restricted to S_n, one-to-one and onto. It follows that S_n is a submanifold of M_n diffeomorphic to $\mathbb{R}^{n(n+1)/2}$.

Exercise 4.5.7 Show, more generally, that if $\mathcal{V}$ is an n-dimensional vector space with its natural differentiable structure and $\mathcal{W}$ is a k-dimensional linear subspace of $\mathcal{V}$, then $\mathcal{W}$ is a k-dimensional submanifold of $\mathcal{V}$ diffeomorphic to $\mathbb{R}^k$.

We claim now that $f : M_n \to S_n$ is a smooth map and that the identity matrix $id \in S_n$ is a regular value of f. (If you're wondering where all of this is going note that $O(n) = f^{-1}(id)$ and keep in mind your response to the last question in Exercise 4.5.6.)

To prove this we identify M_n with $\mathbb{R}^{n^2}$ and S_n with $\mathbb{R}^{n(n+1)/2}$ and regard f as a map from $\mathbb{R}^{n^2}$ to $\mathbb{R}^{n(n+1)/2}$. Smoothness is clear since the entries (coordinates) of AA^T are quadratic functions of the entries of A. Now we fix an $A \in M_n$ and compute $f_{*A} : T_A(M_n) \to T_{f(A)}(S_n)$. Let $x^1, \ldots, x^{n^2}$ be the standard coordinate functions on M_n. For any $B^i \frac{\partial}{\partial x^i}|_A \in T_A(M_n)$ we let $B = (B^1, \ldots, B^{n^2}) \in M_n$. A curve $\alpha : \mathbb{R} \to M_n$ whose velocity vector at $t = 0$ is $B^i \frac{\partial}{\partial x^i}|_A$ is $\alpha(t) = A + tB$. Thus, $f_{*A}(B^i \frac{\partial}{\partial x^i}|_A) = f_{*A}(\alpha'(0)) = (f \circ \alpha)'(0)$. Now,

$$\begin{aligned}(f \circ \alpha)(t) &= f(\alpha(t)) = f(A + tB) = (A + tB)(A + tB)^T \\ &= (A + tB)(A^T + tB^T) = AA^T + tAB^T + tBA^T + t^2 BB^T\end{aligned}$$

so $(f \circ \alpha)'(0) = AB^T + BA^T$. Thus, $f_{*A}(B^i \frac{\partial}{\partial x^i}|_A) = AB^T + BA^T$. We must show that if $A \in f^{-1}(id) = O(n)$, then f_{*A} is onto, i.e., given any C in $T_{f(A)}(S_n)$ there exists a $B^i \frac{\partial}{\partial x^i}|_A \in T_A(M_n)$ with $AB^T + BA^T = C$. But $T_{f(A)}(S_n)$ is, by Exercise 4.4.9, canonically identified with S_n so we need only show that, for $A \in O(n)$ and $C \in S_n$ we can find a $B \in M_n$ such that $AB^T + BA^T = C$.

Exercise 4.5.8 Show that $B = \frac{1}{2} CA$ does the job.

Thus, we have $O(n) \subseteq \mathbb{R}^{n^2}$ written as the inverse image of a regular value for a smooth map of $\mathbb{R}^{n^2}$ onto $\mathbb{R}^{n(n+1)/2}$. We will put this fact to use shortly.

Our objective now is to show that images of imbeddings and inverse images of regular values are always submanifolds. We begin with some results on Euclidean spaces.

Lemma 4.5.2 *Let U be an open set in $\mathbb{R}^n$ and $f : U \to \mathbb{R}^{n+k}$ a smooth map. Suppose $p \in U$ and $f_{*p} : T_p(U) \to T_{f(p)}(\mathbb{R}^{n+k})$ is one-to-one. Then there exists a nbd W of $f(p)$ in $\mathbb{R}^{n+k}$ and a diffeomorphism ψ of W onto an open set $\psi(W)$ in $\mathbb{R}^{n+k}$ such that, on some nbd of p in $\mathbb{R}^n$, $\psi \circ f$ is given by $\psi \circ f(x^1, \ldots, x^n) = (x^1, \ldots, x^n, 0, \ldots, 0)$.*

Remarks: More succinctly, the lemma says that if f is an immersion at p, then some coordinate expression for f near p is just the inclusion of the first n coordinates into $\mathbb{R}^{n+k}$. The idea of the proof is as follows: If $\psi \circ f$ is to have the required form it must, in particular, "undo" f in the first n slots. This we accomplish by identifying U with $U \times \{0\}$ in $\mathbb{R}^{n+k}$ and extending f to a map $\tilde{f}$ on an open set in $\mathbb{R}^{n+k}$ containing $U \times \{0\}$ in such a way that $\tilde{f}$ is still an immersion at $(p, 0)$. The Inverse Function Theorem then gives a local inverse for $\tilde{f}$ and this is our ψ.

Proof: Let $\widetilde{U} = \{(x^1, \ldots, x^n, y^1, \ldots, y^k) \in \mathbb{R}^{n+k} : (x^1, \ldots, x^n) \in U\}$. Then $\widetilde{U}$ is open in $\mathbb{R}^{n+k}$. Notice that, since f_{*p} is one-to-one, the Jacobian $f'(p) = (D_i f^j(p))$ has rank n and so contains an $n \times n$ submatrix that is nonsingular. Renumber the coordinates in $\mathbb{R}^{n+k}$ if necessary so that we may assume that this nonsingular submatrix is

$$\begin{pmatrix} D_1 f^1(p) & \cdots & D_n f^1(p) \\ \vdots & & \vdots \\ D_1 f^n(p) & \cdots & D_n f^n(p) \end{pmatrix}.$$

Now define $\tilde{f} : \widetilde{U} \to \mathbb{R}^{n+k}$ by $\tilde{f}(x^1, \ldots, x^n, y^1, \ldots, y^k) = (f^1(x), \ldots, f^n(x), f^{n+1}(x) + y^1, \ldots, f^{n+k}(x) + y^k)$, where $x = (x^1, \ldots, x^n) \in U$. $\tilde{f}$ is clearly smooth and its Jacobian at $(p, 0)$ is

$$\left(\begin{array}{ccc|c} D_1 f^1(p) & \cdots & D_n f^1(p) & \\ \vdots & & \vdots & 0 \\ D_1 f^n(p) & \cdots & D_n f^n(p) & \\ \hline D_1 f^{n+1}(p) & \cdots & D_n f^{n+1}(p) & \\ \vdots & & \vdots & id \\ D_1 f^{n+k}(p) & \cdots & D_n f^{n+k}(p) & \end{array}\right).$$

Since this is nonsingular, the Inverse Function Theorem implies that there exist open nbds V and W of $(p, 0)$ and $f(p)$, respectively, in $\mathbb{R}^{n+k}$ such that $\tilde{f}$ carries V diffeomorphically onto W. Let ψ be the inverse of $\tilde{f}|V$. Then $\psi : W \to \psi(W) = V$ is a diffeomorphism. Now let $W' = \{x \in \mathbb{R}^n : (x,0) \in W\}$. Then W' is an open nbd of p in $\mathbb{R}^n$ and f maps W' into V so $\psi \circ f$ is defined on W'. Moreover, for $x \in W'$, $\psi \circ f(x) = \psi(\tilde{f}(x,0)) = (x,0)$ as required. ■

Entirely analogous arguments yield the following.

Lemma 4.5.3 *Let U be an open set in $\mathbb{R}^{n+k}$ and $f : U \to \mathbb{R}^n$ a smooth map. Suppose $p \in U$ and $f_{*p} : T_p(U) \to T_{f(p)}(\mathbb{R}^n)$ is onto. Then there exists a nbd V of p in $\mathbb{R}^{n+k}$ and a diffeomorphism φ of V onto an open set $\varphi(V)$ in $\mathbb{R}^{n+k}$ such that, on some nbd of p in $\mathbb{R}^{n+k}$, $f \circ \varphi(x^1, \ldots, x^n, x^{n+1}, \ldots, x^{n+k}) = (x^1, \ldots, x^n)$.*

Exercise 4.5.9 Prove Lemma 4.5.3. **Hint:** This time consider the map $\tilde{f} : U \to \mathbb{R}^{n+k}$ defined by $\tilde{f}(x) = (f^1(x), \ldots, f^n(x), x^{n+1}, \ldots, x^{n+k})$, where $x = (x^1, \ldots, x^n, x^{n+1}, \ldots, x^{n+k}) \in U$. ■

Since manifolds are locally Euclidean both of these lemmas extend easily to the manifold setting.

Theorem 4.5.4 *Let X and Y be smooth manifolds of dimension n and $n+k$, respectively, $f : X \to Y$ a smooth map and p a point in X at which $f_{*p} : T_p(X) \to T_{f(p)}(Y)$ is one-to-one. Then there exist charts (U, φ) at p and (V, ψ) at $f(p)$ such that $\varphi(p) = 0 \in \mathbb{R}^n$, $\psi(f(p)) = 0 \in \mathbb{R}^{n+k}$, $f(U) \subseteq V$ and $\psi \circ f \circ \varphi^{-1}(x^1, \ldots, x^n) = (x^1, \ldots, x^n, 0, \ldots, 0)$.*

Proof: Select charts (U_1, φ_1) and (V_1, ψ_1) at p and $f(p)$, respectively, with $f(U_1) \subseteq V_1$ and $\varphi_1(p) = 0 \in \mathbb{R}^n$. Let $f_1 = \psi_1 \circ f \circ \varphi_1^{-1}$ be the corresponding coordinate expression for f. Since f_{*p} is one-to-one, so is $(f_1)_{*\varphi_1(p)}$. By Lemma 4.5.2, there is an open nbd W' of $f_1(\varphi_1(p))$ in $\psi_1(V_1)$ and a diffeomorphism ψ' of W' onto an open set $\psi'(W')$ in $\mathbb{R}^{n+k}$ such that, on some open nbd U' of $\varphi_1(p)$, $f_1(U') \subseteq W'$ and $\psi' \circ f_1(x^1, \ldots, x^n) = (x^1, \ldots, x^n, 0, \ldots, 0)$. Let $U = \varphi_1^{-1}(U') \subseteq U_1$, $V = \psi_1^{-1}(W') \subseteq V_1$, $\varphi = \varphi_1|U$ and $\psi = \psi' \circ (\psi_1|V)$. Then $f(U) \subseteq V$ since $f(U) = f \circ \varphi_1^{-1}(U') = \psi_1^{-1} \circ f_1(U') = \psi_1^{-1}(f_1(U')) \subseteq \psi_1^{-1}(W') = V$. Moreover, $\psi \circ f \circ \varphi^{-1}(x^1, \ldots, x^n) = (\psi' \circ \psi_1|V) \circ f \circ \varphi^{-1}(x^1, \ldots, x^n) = \psi' \circ f_1(x^1, \ldots, x^n) = (x^1, \ldots, x^n, 0, \ldots, 0)$. ■

Theorem 4.5.5 *Let X and Y be smooth manifolds of dimension $n+k$ and n, respectively, $f : X \to Y$ a smooth map and p a point in X at which $f_{*p} : T_p(X) \to T_{f(p)}(Y)$ is onto. Then there exist charts (U, φ) at p and (V, ψ) at $f(p)$ such that $\varphi(p) = 0 \in \mathbb{R}^{n+k}$, $\psi(f(p)) = 0 \in \mathbb{R}^n$, $\varphi(U) \subseteq V$*

and $\psi \circ f \circ \varphi^{-1}(x^1, \ldots, x^n, x^{n+1}, \ldots, x^{n+k}) = (x^1, \ldots, x^n)$.

Exercise 4.5.10 Use Lemma 4.5.3 to prove Theorem 4.5.5. ■

Thus, we find that if $f : X \to Y$ is an immersion (submersion) at $p \in X$, then, with a proper choice of coordinates, f is locally just an inclusion (projection) map. We derive two important consequences.

Corollary 4.5.6 *Let X and Y be smooth manifolds and $f : X \to Y$ an imbedding. Then $f(X)$ is a submanifold of Y and, regarded as a map of X onto $f(X)$, f is a diffeomorphism.*

Proof: Fix a $p \in X$. We must produce a chart (V, ψ) for Y at $f(p)$ such that $\psi(f(p)) = 0$ and $\psi(V \cap f(X)) = \psi(V) \cap (\mathbb{R}^n \times \{0\})$, where $n = \dim X$. Since f is an imbedding, f_{*p} is one-to-one so $\dim Y \geq \dim X = n$. Let $\dim Y = n + k$. Theorem 4.5.4 yields charts (U, φ) at p and (V', ψ') at $f(p)$ with $\varphi(p) = 0 \in \mathbb{R}^n$, $\psi'(f(p)) = 0 \in \mathbb{R}^{n+k}$, $\varphi(U) \subseteq V'$ and $\psi' \circ f \circ \varphi^{-1}(x^1, \ldots, x^n) = (x^1, \ldots, x^n, 0, \ldots, 0)$. Thus, identifying $\mathbb{R}^n$ with $\mathbb{R}^n \times \{0\} \subseteq \mathbb{R}^n \times \mathbb{R}^k = \mathbb{R}^{n+k}$, $\psi'(f(U)) = (\psi' \circ f \circ \varphi^{-1})(\varphi(U)) = \varphi(U) \times \{0\}$, which is open in $\mathbb{R}^n \times \{0\}$. Now, observe that $f(U) \subseteq V' \cap f(X)$, but the containment might well be proper. However, f is, by assumption, a homeomorphism onto its image $f(X)$ so $f(U)$ is open in $f(X)$. Thus, $f(U) \cap V'$ is open in $f(X)$. Replacing V' by $f(U) \cap V'$ and ψ' by $\psi' | f(U) \cap V'$ if necessary we may therefore assume without loss of generality that $f(U) = V' \cap f(X)$.

Now, since $\psi'(f(U))$ is open in $\mathbb{R}^n \times \{0\}$ we may select an open set W in $\mathbb{R}^{n+k}$ such that $\psi'(f(U)) = W \cap (\mathbb{R}^n \times \{0\})$. Let $V = (\psi')^{-1}(\psi'(V') \cap W)$ and $\psi = \psi' | V$. Note that $\psi(f(p)) = \psi'(f(p)) = 0$.

Exercise 4.5.11 Show that $f(U) \subseteq V$.

In order to complete the proof that $f(X)$ is a submanifold of Y we need only show that

$$\psi(V \cap f(X)) = \psi(V) \cap (\mathbb{R}^n \times \{0\}).$$

First note that $\psi(V \cap f(X)) \subseteq \psi(V' \cap f(X)) = \psi(f(U)) = \psi'(f(U)) = W \cap (\mathbb{R}^n \times \{0\})$ so $\psi(V \cap f(X)) \subseteq \mathbb{R}^n \times \{0\}$. But $\psi(V \cap f(X)) \subseteq \psi(V)$ is obvious so $\psi(V \cap f(X)) \subseteq \psi(V) \cap (\mathbb{R}^n \times \{0\})$. For the reverse containment we note that $\psi'(V) \subseteq \psi'(V') \cap W$ so $\psi(V) \subseteq \psi'(V') \cap W$ and therefore $\psi(V) \cap (\mathbb{R}^n \times \{0\}) \subseteq \psi'(V') \cap W \cap (\mathbb{R}^n \times \{0\}) = \psi'(V') \cap \psi'(f(U)) = \psi'(V' \cap f(U)) = \psi(V' \cap f(U))$ since $f(U) \subseteq V$. But also, $V' \cap f(U) = V' \cap f(U) \cap V = (V' \cap V) \cap f(U) = V \cap f(U) = V \cap f(X) \cap V' = V \cap f(X)$.

Thus, $\psi(V' \cap f(U)) = \psi(V \cap f(X))$ so $\psi(V) \cap (\mathbb{R}^n \times \{0\}) \subseteq \psi(V \cap f(X))$ as required.

Now we regard f as a map from X to the submanifold $f(X)$ of Y. Then f is a homeomorphism by assumption. To show that $f : X \to f(X)$ is a diffeomorphism we need only show that it and its inverse are C^∞.

Exercise 4.5.12 Let (U, φ) and (V, ψ) be the charts for X and Y constructed above. Then (V, ψ) determines a submanifold chart $(\tilde{V}, \tilde{\psi})$ for $f(X)$. Show that the coordinate expression $\tilde{\psi} \circ f \circ \varphi^{-1}$ for $f : X \to f(X)$ is the identity map $(x^1, \ldots, x^n) \to (x^1, \ldots, x^n)$. Conclude that $f : X \to f(X)$ is a diffeomorphism. ∎

Corollary 4.5.7 *Let X and Y be smooth manifolds of dimension n and m, respectively, with $n \geq m$, $f : X \to Y$ a smooth map and $q \in Y$ a regular value of f. Then $f^{-1}(q)$ is either empty or a submanifold of X of dimension $n - m$.*

Proof: Assume $f^{-1}(q) \neq \emptyset$ and let $p \in f^{-1}(q)$. We must find a chart (U, φ) for X at p with $\varphi(p) = 0$ and $\varphi(U \cap f^{-1}(q)) = \varphi(U) \cap (\mathbb{R}^{n-m} \times \{0\})$. Since f_{*p} is surjective, Theorem 4.5.5 yields charts (U, φ) at p and (V, ψ) at $f(p) = q$ such that $\varphi(p) = 0 \in \mathbb{R}^n$, $\psi(q) = 0 \in \mathbb{R}^m$, $\varphi(U) \subseteq V$ and (renumbering the coordinates, if necessary),

$$\psi \circ f \circ \varphi^{-1}\left(x^1, \ldots, x^{n-m}, x^{n-m+1}, \ldots, x^n\right) = \left(x^{n-m+1}, \ldots, x^n\right)$$

on $\varphi(U)$. Thus, $\psi \circ f \circ \varphi^{-1}(0) = 0$ and $\psi \circ f \circ \varphi^{-1}(x^1, \ldots, x^n) = (0, \ldots, 0)$ iff $(x^{n-m+1}, \ldots, x^n) = (0, \ldots, 0)$. Thus $\psi(f \circ \varphi^{-1}(x^1, \ldots, x^n)) = (0, \ldots, 0)$ iff $(x^1, \ldots, x^n) \in \varphi(U) \cap (\mathbb{R}^{n-m} \times \{0\})$. But, on $V \supseteq f(U)$, ψ is one-to-one so, since $\psi(q) = 0$, $f \circ \varphi^{-1}(x^1, \ldots, x^n) = q$ iff $(x^1, \ldots, x^n) \in \varphi(U) \cap (\mathbb{R}^{n-m} \times \{0\})$. Thus, $\varphi^{-1}(x^1, \ldots, x^n) \in f^{-1}(q)$ iff $(x^1, \ldots, x^n) \in \varphi(U) \cap (\mathbb{R}^{n-m} \times \{0\})$ so $\varphi(U \cap f^{-1}(q)) = \varphi(U) \cap (\mathbb{R}^{n-m} \times \{0\})$ as required. ∎

Corollary 4.5.7 provides us with a wealth of examples. Since any projection $\mathcal{P} : P \to X$ of a smooth principal bundle is a submersion, the fibers $\mathcal{P}^{-1}(x)$ partition the bundle space P into submanifolds of dimension $\dim P - \dim X$. But then, by local triviality, each such fiber is diffeomorphic to the structure group G so $\dim P - \dim X = \dim G$, i.e.,

$$\dim P = \dim X + \dim G\,.$$

If X is any n-dimensional smooth manifold and $f \in C^\infty(X)$, then a nonempty level set $f^{-1}(r)$, $r \in \mathbb{R}$, will be a submanifold of dimension $n - 1$ provided it contains no points p at which df_p is the zero element of $T_p^*(X)$.

Exercise 4.5.13 Rephrase your response to the last question in Exercise 4.5.6.

Exercise 4.5.14 Show that S^n is a submanifold of $\mathbb{R}^{n+1}$ by applying Corollary 4.5.7 to the map $f : \mathbb{R}^{n+1} \to \mathbb{R}$ given by $f(x^1, \ldots, x^{n+1}) = (x^1)^2 + \cdots + (x^{n+1})^2$.

Finally, recall that we have shown that the orthogonal group $O(n)$ can be written as the inverse image of a regular value for a smooth map from $\mathbb{R}^{n^2}$ to $\mathbb{R}^{n(n+1)/2}$. We may therefore conclude from Corollary 4.5.7 that $O(n)$ is a submanifold of $\mathbb{R}^{n^2}$ of dimension $n^2 - \frac{1}{2}n(n+1) = \frac{1}{2}n(n-1)$.

Exercise 4.5.15 Show that $SO(n)$ is an open submanifold of $O(n)$. **Hint:** The determinant function $\det : GL(n, \mathbb{R}) \to \mathbb{R}$ is continuous and $SO(n) = O(n) \cap \det^{-1}(0, \infty)$.

We will show somewhat later (Section 4.7) that $U(n)$, $SU(n)$ and $Sp(n)$ are all submanifolds of the Euclidean spaces in which they live.

4.6 Vector Fields and 1-Forms

A **vector field** on a smooth manifold X is a map $\mathbf{V}$ that assigns to each $p \in X$ a tangent vector $\mathbf{V}(p)$, also written $\mathbf{V}_p$, in $T_p(X)$. If (U, φ) is a chart on X with coordinate functions x^i, $i = 1, \ldots, n$, and $p \in U$, then, by Theorem 4.4.3, $\mathbf{V}(p)$ can be written $\mathbf{V}(p) = \mathbf{V}_p = \mathbf{V}_p(x^i) \frac{\partial}{\partial x^i}|_p$. The real-valued functions $V^i : U \to \mathbb{R}$ defined by $V^i(p) = \mathbf{V}_p(x^i)$ are called the **components of V relative to** (U, φ). If these component functions V^i are continuous, or C^∞ for all charts in (some atlas for) the differentiable structure of X, then we say that the vector field $\mathbf{V}$ itself is **continuous**, or $\mathbf{C}^\infty$ **(smooth)**.

The collection of all C^∞ vector fields on a manifold X is denoted $\boldsymbol{\mathcal{X}}(X)$ and we provide it with an algebraic structure as follows: Let $\mathbf{V}, \mathbf{W} \in \boldsymbol{\mathcal{X}}(X)$, $a \in \mathbb{R}$ and $f \in C^\infty(X)$. We define $\mathbf{V} + \mathbf{W}$, $a\mathbf{V}$ and $f\mathbf{V}$ by $(\mathbf{V} + \mathbf{W})(p) = \mathbf{V}(p) + \mathbf{W}(p)$, $(a\mathbf{V})(p) = a\mathbf{V}(p)$ and $(f\mathbf{V})(p) = f(p)\mathbf{V}(p)$. All of these are clearly also in $\boldsymbol{\mathcal{X}}(X)$ since they have smooth components in any chart. The first two operations therefore provide $\boldsymbol{\mathcal{X}}(X)$ with the structure of a real vector space. With the third operation, $\boldsymbol{\mathcal{X}}(X)$ acquires the structure of a module over (the commutative ring with identity) $C^\infty(X)$.

Exercise 4.6.1 Let x_0 be a point in X and (U, φ) a chart at x_0 with coordinate functions x^i, $i = 1, \ldots, n$. Regard U as an open submanifold of X and define vector fields $\frac{\partial}{\partial x^i}$ on U by $\frac{\partial}{\partial x^i}(p) = \frac{\partial}{\partial x^i}|_p$ for each $p \in U$.

Show that $\frac{\partial}{\partial x^i} \in \mathcal{X}(U)$ (these are called the **coordinate vector fields** of (U, φ)).

Exercise 4.6.2 Let x_0 be a point in X and W an open subset of X containing x_0. Show that, for any $\mathbf{V} \in \mathcal{X}(W)$, there exists a $\tilde{\mathbf{V}}$ in $\mathcal{X}(X)$ that agrees with $\mathbf{V}$ on some open nbd of x_0 contained in W. **Hint:** Exercise 4.4.1.

The upshot of the last two exercises is that locally defined smooth vector fields, e.g., $\frac{\partial}{\partial x^i}$, may, by shrinking the domain a bit if necessary, be regarded as elements of $\mathcal{X}(X)$.

If $\mathbf{V} \in \mathcal{X}(X)$, (U, φ) is a chart on X and $V^i : U \to \mathbb{R}$, $i = 1, \ldots, n$, are the components of $\mathbf{V}$ relative to (U, φ), then $\mathbf{V} = V^i \frac{\partial}{\partial x^i}$ on U. For any $f \in C^\infty(X)$ we define a real-valued function $\mathbf{V}f$, also denoted $\mathbf{V}(f)$, on X by

$$(\mathbf{V}f)(p) = \mathbf{V}_p(f)$$

for each $p \in X$. Relative to any chart, $(\mathbf{V}f)(p) = \mathbf{V}_p(f) = \mathbf{V}_p(x^i) \frac{\partial}{\partial x^i}|_p(f) = V^i(p) \frac{\partial f}{\partial x^i}(p)$ so $\mathbf{V}f$ is also in $C^\infty(X)$ and we often write simply $\mathbf{V}f = V^i \frac{\partial f}{\partial x^i}$. Note, in particular, that $V^i = \mathbf{V}(x^i)$ so, on the domain of the chart, $\mathbf{V} = \mathbf{V}(x^i) \frac{\partial}{\partial x^i}$.

Exercise 4.6.3 A **derivation on** $C^\infty(X)$ is a map $\mathcal{D} : C^\infty(X) \to C^\infty(X)$ that satisfies the following two conditions:

1. ($\mathbb{R}$**-linearity**) $\mathcal{D}(af + bg) = a\mathcal{D}(f) + b\mathcal{D}(g)$ for all $a, b \in \mathbb{R}$ and $f, g \in C^\infty(X)$.
2. (**Leibnitz Product Rule**) $\mathcal{D}(fg) = f\mathcal{D}(g) + \mathcal{D}(f)g$.

Show that, for each $\mathbf{V} \in \mathcal{X}(X)$, $\mathcal{D}(f) = \mathbf{V}f$ defines a derivation on $C^\infty(X)$. Show also that, conversely, every derivation $\mathcal{D}$ on $C^\infty(X)$ arises in this way from some $\mathbf{V} \in \mathcal{X}(X)$. **Hint:** For the converse, define $\mathbf{V}$ at each $p \in X$ by $\mathbf{V}_p(f) = \mathcal{D}(f)(p)$ for each $f \in C^\infty(X)$.

According to Exercise 4.6.3, one could identify C^∞ vector fields on X with derivations on $C^\infty(X)$ and this attitude is often convenient. For example, if $\mathbf{V}$ and $\mathbf{W}$ are fixed elements of $\mathcal{X}(X)$ we can define a derivation $[\mathbf{V}, \mathbf{W}]$ on $C^\infty(X)$, called the **Lie bracket** of $\mathbf{V}$ and $\mathbf{W}$, by

$$[\mathbf{V}, \mathbf{W}](f) = \mathbf{V}(\mathbf{W}f) - \mathbf{W}(\mathbf{V}f). \tag{4.6.1}$$

Exercise 4.6.4 Show that (4.6.1) does, indeed, define a derivation on $C^\infty(X)$.

Thus, $[\mathbf{V},\mathbf{W}]$ defines a smooth vector field on X whose value at any $p \in X$ is the tangent vector $[\mathbf{V},\mathbf{W}]_p \in T_p(X)$ given by

$$[\mathbf{V},\mathbf{W}]_p(f) = \mathbf{V}_p(\mathbf{W}f) - \mathbf{W}_p(\mathbf{V}f)$$

(we use the same symbol $[\mathbf{V},\mathbf{W}]$ for the derivation and the vector field). Intuitively, one thinks of $[\mathbf{V},\mathbf{W}](f)$ as the difference between two "mixed second order derivatives" (the $\mathbf{V}$ rate of change of $\mathbf{W}f$ minus the $\mathbf{W}$ rate of change of $\mathbf{V}f$).

Exercise 4.6.5 Show, for example, that the coordinate vector fields for any chart satisfy $[\frac{\partial}{\partial x^i}, \frac{\partial}{\partial x^j}](f) = 0$ for all $f \in C^\infty(X)$.

Two vector fields $\mathbf{V}$ and $\mathbf{W}$ for which $[\mathbf{V},\mathbf{W}] = 0$ (i.e., $[\mathbf{V},\mathbf{W}](f) = 0$ for all $f \in C^\infty(X)$) are said to **commute** (because, symbolically at least, $[\mathbf{V},\mathbf{W}] = \mathbf{VW} - \mathbf{WV}$). Next we list a number of properties of the Lie bracket that we will use repeatedly. For all $\mathbf{V},\mathbf{W} \in \boldsymbol{\mathcal{X}}(X)$, it follows at once from the definition (4.6.1) that

$$[\mathbf{W},\mathbf{V}] = -[\mathbf{V},\mathbf{W}] \quad \textbf{(skew–symmetry)}. \tag{4.6.2}$$

For $a, b \in \mathbb{R}$ and $\mathbf{V}_1, \mathbf{V}_2, \mathbf{V}_3 \in \boldsymbol{\mathcal{X}}(X)$,

$$[a\mathbf{V}_1 + b\mathbf{V}_2, \mathbf{V}_3] = a[\mathbf{V}_1,\mathbf{V}_3] + b[\mathbf{V}_2,\mathbf{V}_3], \quad \text{and} \tag{4.6.3}$$

$$[\mathbf{V}_1, a\mathbf{V}_2 + b\mathbf{V}_3] = a[\mathbf{V}_1,\mathbf{V}_2] + b[\mathbf{V}_1,\mathbf{V}_3] \tag{4.6.4}$$

(**$\mathbb{R}$-bilinearity**). Again, these are immediate consequences of the definition. The next two are, perhaps, less obvious. First, for $\mathbf{V},\mathbf{W} \in \boldsymbol{\mathcal{X}}(X)$ and $f, g \in C^\infty(X)$,

$$[f\mathbf{V}, g\mathbf{W}] = fg[\mathbf{V},\mathbf{W}] + f(\mathbf{V}g)\mathbf{W} - g(\mathbf{W}f)\mathbf{V}. \tag{4.6.5}$$

For the proof of (4.6.5) we let h be an arbitrary element of $C^\infty(X)$ and compute

$$\begin{aligned}
[f\mathbf{V}, g\mathbf{W}](h) &= (f\mathbf{V})\,(\,(g\mathbf{W})h\,) - (g\mathbf{W})\,(\,(f\mathbf{V})h\,) \\
&= (f\mathbf{V})\,(\,g(\mathbf{W}h)\,) - (g\mathbf{W})\,(\,f(\mathbf{V}h)\,) \\
&= f\,[\mathbf{V}\,(\,g(\mathbf{W}h)\,)\,] - g\,[\mathbf{W}\,(\,f(\mathbf{V}h)\,)\,] \\
&= f\,[\,g\mathbf{V}(\mathbf{W}h) + (\mathbf{V}g)(\mathbf{W}h)\,] - g\,[\,f\mathbf{W}(\mathbf{V}h) + (\mathbf{W}f)(\mathbf{V}h)\,] \\
&= fg\,[\,\mathbf{V}(\mathbf{W}h) - \mathbf{W}(\mathbf{V}h)\,] + f(\mathbf{V}g)(\mathbf{W}h) - g(\mathbf{W}f)(\mathbf{V}h) \\
&= fg\,[\,\mathbf{V},\mathbf{W}\,]\,(h) + (\,f(\mathbf{V}g)\mathbf{W}\,)\,(h) - (\,g(\mathbf{W}f)\mathbf{V}\,)\,(h) \\
&= [\,fg[\mathbf{V},\mathbf{W}] + f(\mathbf{V}g)\mathbf{W} - g(\mathbf{W}f)\mathbf{V}\,]\,(h)
\end{aligned}$$

as required. Next, we let $\mathbf{V}_1, \mathbf{V}_2, \mathbf{V}_3 \in \boldsymbol{\mathcal{X}}(X)$. Then

$$[\mathbf{V}_1, [\mathbf{V}_2, \mathbf{V}_3]] + [\mathbf{V}_3, [\mathbf{V}_1, \mathbf{V}_2]] + [\mathbf{V}_2, [\mathbf{V}_3, \mathbf{V}_1]] = 0 \qquad (4.6.6)$$

is known as the **Jacobi identity**.

Exercise 4.6.6 Prove (4.6.6).

Finally, we record a local, component expression for the Lie bracket. Let (U, φ) be a chart on X with coordinate functions x^i, $i = 1, \ldots, n$, and let $\mathbf{V}, \mathbf{W} \in \boldsymbol{\mathcal{X}}(X)$. On U we write $\mathbf{V} = V^j \frac{\partial}{\partial x^j}$ and $\mathbf{W} = W^j \frac{\partial}{\partial x^j}$. Then

$$[\mathbf{V}, \mathbf{W}] = \left(V^j \frac{\partial W^i}{\partial x^j} - W^j \frac{\partial V^i}{\partial x^j} \right) \frac{\partial}{\partial x^i}. \qquad (4.6.7)$$

The proof is easy: The components $[\mathbf{V}, \mathbf{W}]^i$ of $[\mathbf{V}, \mathbf{W}]$ relative to (U, φ) are given by $[\mathbf{V}, \mathbf{W}]^i = [\mathbf{V}, \mathbf{W}](x^i) = \mathbf{V}(\mathbf{W}x^i) - \mathbf{W}(\mathbf{V}x^i) = \mathbf{V}(W^i) - \mathbf{W}(V^i) = (V^j \frac{\partial}{\partial x^j})(W^i) - (W^j \frac{\partial}{\partial x^j})(V^i) = V^j \frac{\partial W^i}{\partial x^j} - W^j \frac{\partial V^i}{\partial x^j}$ which gives (4.6.7).

A **1-form** on a smooth manifold X is a map $\boldsymbol{\Theta}$ that assigns to each $p \in X$ a covector $\boldsymbol{\Theta}(p)$, also written $\boldsymbol{\Theta}_p$, in $T_p^*(X)$. If (U, φ) is a chart on X with coordinate functions x^i, $i = 1, \ldots, n$, and $p \in U$, then, by Exercise 4.4.14, $\boldsymbol{\Theta}(p)$ can be written $\boldsymbol{\Theta}(p) = \boldsymbol{\Theta}_p(\frac{\partial}{\partial x^i}|_p)\, dx^i_{\ p}$. The real-valued functions $\Theta_i : U \to \mathbb{R}$ defined by $\Theta_i(p) = \boldsymbol{\Theta}_p(\frac{\partial}{\partial x^i}|_p)$ are called the **components of $\boldsymbol{\Theta}$ relative to** (U, φ). If these component functions Θ_i are continuous, or C^∞ for all charts in (some atlas for) the differentiable structure of X, then we say that the 1-form itself is **continuous**, or $\mathbf{C}^\infty$ (**smooth**). For example, for any $f \in C^\infty(X)$, the map that sends each $p \in X$ to $df_p \in T_p^*(X)$ is a smooth 1-form on X because its component functions in any chart are just the partial derivatives $\frac{\partial f}{\partial x^i}$. We denote this 1-form df.

The collection of all C^∞ 1-forms on X is denoted $\boldsymbol{\mathcal{X}}^*(X)$ and has the obvious structure of a real vector space and a module over $C^\infty(X)$: For $\boldsymbol{\Theta}, \omega \in \boldsymbol{\mathcal{X}}^*(X)$, $a \in \mathbb{R}$ and $f \in C^\infty(X)$ one defines $\boldsymbol{\Theta} + \omega$, $a\boldsymbol{\Theta}$ and $f\boldsymbol{\Theta}$ by $(\boldsymbol{\Theta} + \omega)(p) = \boldsymbol{\Theta}(p) + \omega(p)$, $(a\boldsymbol{\Theta})(p) = a\boldsymbol{\Theta}(p)$ and $(f\boldsymbol{\Theta})(p) = f(p)\boldsymbol{\Theta}(p)$. As for vector fields (Exercise 4.6.2) any 1-form defined only on an open subset of X may, by judicuous use of a bump function, be regarded as an element of $\boldsymbol{\mathcal{X}}^*(X)$.

Exercise 4.6.7 Show that the operator $d : C^\infty(X) \to \boldsymbol{\mathcal{X}}^*(X)$, called **exterior differentiation on** $C^\infty(X)$, that takes f to df has all of the following properties:

1. $d(af + bg) = a\,df + b\,dg$ for all $a, b \in \mathbb{R}$ and $f, g \in C^\infty(X)$.

2. $d(fg) = fdg + gdf$ for all $f, g \in C^\infty(X)$.

3. For any $f \in C^\infty(X)$ and $h \in C^\infty(\mathbb{R})$, $d(h \circ f) = (h' \circ f)df$, where h' is the ordinary derivative of $h : \mathbb{R} \to \mathbb{R}$.

Now let $\boldsymbol{\Theta} \in \boldsymbol{\mathcal{X}}^*(X)$ and $\mathbf{V} \in \boldsymbol{\mathcal{X}}(X)$. Define a real-valued function $\boldsymbol{\Theta}(\mathbf{V})$, also written $\boldsymbol{\Theta}\mathbf{V}$, on X by $(\boldsymbol{\Theta}(\mathbf{V}))(p) = (\boldsymbol{\Theta}\mathbf{V})(p) = \boldsymbol{\Theta}_p(\mathbf{V}_p)$ for each $p \in X$. If (U, φ) is any chart for X and we write, for each $p \in U$, $\boldsymbol{\Theta}_p = \Theta_i(p)dx^i{}_p$ and $\mathbf{V}_p = V^j(p)\frac{\partial}{\partial x^j}|_p$, then

$$\begin{aligned}(\boldsymbol{\Theta}\mathbf{V})(p) &= \left(\Theta_i(p)dx^i{}_p\right)\left(V^j(p)\left.\frac{\partial}{\partial x^j}\right|_p\right) = \Theta_i(p)V^j(p)dx^i{}_p\left(\left.\frac{\partial}{\partial x^j}\right|_p\right)\\ &= \Theta_i(p)V^j(p)\delta^i_j = \Theta_i(p)V^i(p)\end{aligned}$$

which is C^∞ on U. Thus, $\boldsymbol{\Theta}\mathbf{V} \in C^\infty(X)$. Note, in particular, that $\boldsymbol{\Theta}(\frac{\partial}{\partial x^i}) = \Theta_i$ for $i = 1, \ldots, n$.

Thus, a 1-form $\boldsymbol{\Theta}$ determines a map from $\boldsymbol{\mathcal{X}}(X)$ to $C^\infty(X)$ that carries $\mathbf{V}$ to $\boldsymbol{\Theta}\mathbf{V}$. This map is not only linear, but is, in fact, a $C^\infty(X)$-module homomorphism, i.e.,

$$\boldsymbol{\Theta}(\mathbf{V} + \mathbf{W}) = \boldsymbol{\Theta}\mathbf{V} + \boldsymbol{\Theta}\mathbf{W} \quad \text{for all} \quad \mathbf{V}, \mathbf{W} \in \boldsymbol{\mathcal{X}}(X), \quad \text{and} \tag{4.6.8}$$

$$\boldsymbol{\Theta}(f\mathbf{V}) = f(\boldsymbol{\Theta}\mathbf{V}) \quad \text{for all} \quad \mathbf{V} \in \boldsymbol{\mathcal{X}}(X) \quad \text{and} \quad f \in C^\infty(X). \tag{4.6.9}$$

Exercise 4.6.8 Prove (4.6.8) and (4.6.9) and observe that $\boldsymbol{\Theta}(a\mathbf{V}) = a(\boldsymbol{\Theta}\mathbf{V})$ for $\mathbf{V} \in \boldsymbol{\mathcal{X}}(X)$ and $a \in \mathbb{R}$ follows from (4.6.9).

We wish to show that, conversely, any $C^\infty(X)$-module homomorphism $A : \boldsymbol{\mathcal{X}}(X) \to C^\infty(X)$ determines a unique 1-form $\boldsymbol{\Theta}$ on X with $\boldsymbol{\Theta}\mathbf{V} = A(\mathbf{V})$ for every $\mathbf{V} \in \boldsymbol{\mathcal{X}}(X)$.

Lemma 4.6.1 *Let X be a smooth manifold and $A : \boldsymbol{\mathcal{X}}(X) \to C^\infty(X)$ a map satisfying $A(\mathbf{V} + \mathbf{W}) = A(\mathbf{V}) + A(\mathbf{W})$ and $A(f\mathbf{V}) = fA(\mathbf{V})$ for all $\mathbf{V}, \mathbf{W} \in \boldsymbol{\mathcal{X}}(X)$ and all $f \in C^\infty(X)$. If $\mathbf{V} \in \boldsymbol{\mathcal{X}}(X)$ and p is a point in X at which $\mathbf{V}_p = 0 \in T_p(X)$, then $A(\mathbf{V})(p) = 0$.*

Proof: Let (U, φ) be a chart at p with coordinate functions x^i, $i = 1, \ldots, n$. Then $\mathbf{V} = V^i\frac{\partial}{\partial x^i}$ on U, where $V^i(p) = 0$ for each $i = 1, \ldots, n$. Now let g be a bump function at p in U. Then gV^i agrees with V^i on some nbd of p and, on that same nbd, $g\frac{\partial}{\partial x^i}$ agrees with $\frac{\partial}{\partial x^i}$. Define $\tilde{V}^i \in C^\infty(X)$ and $\widetilde{\frac{\partial}{\partial x^i}} \in \boldsymbol{\mathcal{X}}(X)$ by $\tilde{V}^i(x) = \begin{cases}(gV^i)(x), & x \in U\\ 0, & x \notin U\end{cases}$ and $\widetilde{\frac{\partial}{\partial x^i}}(x) = \begin{cases}(g\frac{\partial}{\partial x^i})(x), & x \in U\\ 0, & x \notin U\end{cases}$.

Exercise 4.6.9 Show that $g^2\mathbf{V} = \tilde{V}^i\widetilde{\frac{\partial}{\partial x^i}}$ on all of X, where $g^2(x) = (g(x))^2$.

Thus, $g^2A(\mathbf{V}) = A(g^2\mathbf{V}) = A(\tilde{V}^i \widetilde{\frac{\partial}{\partial x^i}}) = \tilde{V}^i A(\widetilde{\frac{\partial}{\partial x^i}})$ everywhere on X. At p, $g^2(p) = (g(p))^2 = 1$ and $\tilde{V}^i(p) = V^i(p) = 0$ so $A(\mathbf{V})(p) = 0$ as required. ■

Now, suppose A is as in Lemma 4.6.1 and $\mathbf{V}$ and $\mathbf{W}$ are two smooth vector fields on X that take the same value at p ($\mathbf{V}(p) = \mathbf{W}(p)$). Then $\mathbf{V} - \mathbf{W} \in \boldsymbol{\mathcal{X}}(X)$ vanishes at p so $A(\mathbf{V} - \mathbf{W})(p) = 0$, i.e., $A(\mathbf{V})(p) = A(\mathbf{W})(p)$. We conclude that the value of $A(\mathbf{V})$ at any p is completely determined by the value of $\mathbf{V}$ *at* p. Consequently, we may define an element A_p of $T_p^*(X)$ by

$$A_p(\mathbf{v}) = A(\mathbf{V})(p), \tag{4.6.10}$$

where $\mathbf{V}$ is *any* element of $\boldsymbol{\mathcal{X}}(X)$ with $\mathbf{V}(p) = \mathbf{v}$.

Exercise 4.6.10 Show that such $\mathbf{V}$'s exist, i.e., that, for any $p \in X$ and any $\mathbf{v} \in T_p(X)$, there exists a $\mathbf{V} \in \boldsymbol{\mathcal{X}}(X)$ with $\mathbf{V}(p) = \mathbf{v}$.

Thus, any $C^\infty(X)$-module homomorphism $A : \boldsymbol{\mathcal{X}}(X) \to C^\infty(X)$ determines a map that assigns to each $p \in X$ a covector $A_p \in T_p^*(X)$ defined by (4.6.10). This 1-form $p \to A_p$, which we denote $\boldsymbol{\Theta}$, is C^∞ because, for any $\mathbf{V} \in \boldsymbol{\mathcal{X}}(X)$, $\boldsymbol{\Theta}\mathbf{V} : X \to \mathbb{R}$ is given by $(\boldsymbol{\Theta}\mathbf{V})(p) = \boldsymbol{\Theta}(p)(\mathbf{V}(p)) = A_p(\mathbf{V}_p) = A(\mathbf{V})(p)$ so $\boldsymbol{\Theta}\mathbf{V} = A(\mathbf{V})$ and this is in $C^\infty(X)$ by assumption. Since $\boldsymbol{\Theta}$ is clearly uniquely determined by the requirement that $\boldsymbol{\Theta}\mathbf{V} = A(\mathbf{V})$ for each $\mathbf{V} \in \boldsymbol{\mathcal{X}}(X)$ (check the components in any chart) we have proved our claim and thereby established a one-to-one correspondence between 1-forms on X and $C^\infty(X)$-module homomorphisms of $\boldsymbol{\mathcal{X}}(X)$ to $C^\infty(X)$.

Exercise 4.6.11 The collection of all $C^\infty(X)$-homomorphisms of $\boldsymbol{\mathcal{X}}(X)$ to $C^\infty(X)$ has a natural $C^\infty(X)$-module structure ($(A_1 + A_2)(\mathbf{V}) = A_1(\mathbf{V}) + A_2(\mathbf{V})$ and $(fA)(\mathbf{V}) = fA(\mathbf{V})$). Show that our one-to-one correspondence is an isomorphism.

Henceforth we will therefore not distinguish, either conceptually or notationally, between these two interpretations of a 1-form.

The derivative of a smooth map $f : X \to Y$ carries tangent vectors to X onto tangent vectors to Y, but it cannot, in general, carry vector fields on X to vector fields on Y (e.g., if f is not one-to-one, f_* could assign two different vectors to the same point $f(p_1) = f(p_2)$ in Y).

Exercise 4.6.12 Suppose, however, that f is a diffeomorphism of X onto Y and $\mathbf{V} \in \boldsymbol{\mathcal{X}}(X)$. Define a vector field $f_*\mathbf{V}$ on Y by $(f_*\mathbf{V})_q = f_{*p}(\mathbf{V}_p)$, where $p = f^{-1}(q)$, for each $q \in Y$. Show that $f_*\mathbf{V} \in \boldsymbol{\mathcal{X}}(Y)$.

Exercise 4.6.13 If $f : X \to Y$ is any C^∞ map, $\mathbf{V} \in \boldsymbol{\mathcal{X}}(X)$ and $\mathbf{V}' \in \boldsymbol{\mathcal{X}}(Y)$, then we say that $\mathbf{V}$ and $\mathbf{V}'$ are **f-related** if $f_{*p}(\mathbf{V}_p) = \mathbf{V}'_{f(p)}$

for each $p \in X$. Show that if $\mathbf{V}$ is f-related to $\mathbf{V}'$ and $\mathbf{W}$ is f-related to $\mathbf{W}'$, then $[\mathbf{V}, \mathbf{W}]$ is f-related to $[\mathbf{V}', \mathbf{W}']$.

However, even though f generally cannot push vector fields forward from X to Y, we show now that it can pull 1-forms back from Y to X.

Let $f : X \to Y$ be an arbitrary smooth map and let $\boldsymbol{\Theta}$ be a 1-form on Y. We define a 1-form $f^*\boldsymbol{\Theta}$ on X, called the **pullback** of $\boldsymbol{\Theta}$ to X by f, as follows: For each $p \in X$ and each $\mathbf{v} \in T_p(X)$ we define $(f^*\boldsymbol{\Theta})_p(\mathbf{v}) = \boldsymbol{\Theta}_{f(p)}(f_{*p}(\mathbf{v}))$ ($f^*\boldsymbol{\Theta}$ acts on $\mathbf{v}$ by pushing $\mathbf{v}$ forward by f_* and letting $\boldsymbol{\Theta}$ act on this). Since $(f^*\boldsymbol{\Theta})_p$ is clearly in $T_p^*(X)$ we have defined a 1-form which we now proceed to show is smooth if $\boldsymbol{\Theta}$ is smooth. Select charts (U, φ) on X and (V, ψ) on Y with coordinate functions $x^1, \ldots, x^n$ and $y^1, \ldots, y^m$, respectively, and assume, without loss of generality, that $\varphi(U) \subseteq V$. Let $\psi \circ f \circ \varphi^{-1}$ have coordinate functions $f^1, \ldots, f^m$ and write $\boldsymbol{\Theta} = \Theta_i dy^i$. Then the components of $f^*\boldsymbol{\Theta}$ relative to (U, φ) are given by $(f^*\boldsymbol{\Theta})_j = (f^*\boldsymbol{\Theta})(\frac{\partial}{\partial x^j})$. Thus, at each $p \in U$, $(f^*\boldsymbol{\Theta})_j(p) = (f^*\boldsymbol{\Theta})_p(\frac{\partial}{\partial x^j}|_p) = \boldsymbol{\Theta}_{f(p)}(f_{*p}(\frac{\partial}{\partial x^j}|_p)) = \boldsymbol{\Theta}_{f(p)}(\frac{\partial f^i}{\partial x^j}(p)\, \frac{\partial}{\partial y^i}|_{f(p)}) = \frac{\partial f^i}{\partial x^j}(p)\boldsymbol{\Theta}_{f(p)}(\frac{\partial}{\partial y^i}|_{f(p)}) = \frac{\partial f^i}{\partial x^j}(p)\Theta_i(f(p))$. If $\boldsymbol{\Theta}$ is C^∞, then each Θ_i is C^∞ on V and consequently $(f^*\boldsymbol{\Theta})_j$ is C^∞ on U as required. It will be useful to record for future reference the coordinate formula for the pullback that we have just proved.

$$\boldsymbol{\Theta} = \Theta_i dy^i \Longrightarrow f^*\boldsymbol{\Theta} = \frac{\partial f^i}{\partial x^j}\,(\Theta_i \circ f)\, dx^j\,. \tag{4.6.11}$$

Notice that this is particularly easy to remember since the expression for $f^*\boldsymbol{\Theta}$ is just what one would obtain by formally substituting $y^i = f^i(x^1, \ldots, x^n)$ into $\Theta_i(y^1, \ldots, y^m)dy^i$ and computing differentials:

$$\begin{aligned}(f^*\boldsymbol{\Theta})\left(x^1, \ldots, x^n\right) = \Theta_i\,\big(f^1(x^1, \ldots, x^n), \ldots,&\\ f^m(x^1, \ldots, x^n)\big)\, d\left(f^i(x^1, \ldots, x^n)\right)&\end{aligned} \tag{4.6.12}$$

(we do an explicit calculation shortly).

Exercise 4.6.14 Let $f : X \to Y$ be smooth and $g \in C^\infty(Y)$. Then $dg \in \boldsymbol{\mathcal{X}}^*(Y)$. Show that

$$f^*(dg) = d(g \circ f)\,. \tag{4.6.13}$$

Because of (4.6.13) one often defines the pullback of an element $g \in C^\infty(Y)$ by f to be the element $f^*(g) = g \circ f$ of $C^\infty(X)$. Then (4.6.13) reads

$$f^*(dg) = d(f^*g) \tag{4.6.14}$$

which one interprets as saying that pullback commutes with the differential. Symbolically, $f^* \circ d = d \circ f^*$.

Exercise 4.6.15 Let $f : X \to Y$ and $g : Y \to Z$ be smooth. Show that

$$(g \circ f)^* = f^* \circ g^* \,, \tag{4.6.15}$$

i.e., that $(g \circ f)^* \Theta = f^*(g^* \Theta)$ for every $\Theta \in \mathcal{X}^*(Z)$.

A useful example of a pullback is obtained as follows. Suppose Y' is a submanifold of Y. Then the inclusion map $\iota : Y' \hookrightarrow Y$ is smooth (in fact, an imbedding) so, for every $\Theta \in \mathcal{X}^*(Y)$, $\iota^* \Theta$ is in $\mathcal{X}(Y')$. Now, for each $p \in Y'$, ι_{*p} identifies $T_p(Y')$ with a subspace of $T_p(Y)$ and $(\iota^* \Theta)_p(\mathbf{v}) = \Theta_p(\iota_{*p}(\mathbf{v}))$ so one may write, somewhat loosely, $(\iota^* \Theta)_p(\mathbf{v}) = \Theta_p(\mathbf{v})$. For this reason, $\iota^* \Theta$ is called the **restriction of Θ to Y'**. Now suppose $f : X \to Y'$ is a smooth map and $\Theta \in \mathcal{X}^*(Y)$. Then $\iota^* \Theta \in \mathcal{X}^*(Y')$ so $f^*(\iota^* \Theta) \in \mathcal{X}(X)$. Furthermore, (4.6.15) gives

$$f^* \left(\iota^* \Theta \right) = \left(\iota \circ f \right)^* \Theta \,. \tag{4.6.16}$$

Intuitively, one computes the pullback by f of the restriction $\iota^* \Theta$ by regarding f as a map into Y and pulling back Θ.

To illustrate these ideas we now perform a concrete calculation, the result of which will eventually be of great significance to us. We wish to consider the Hopf bundle $(S^3, \mathbb{CP}^1, \mathcal{P}, U(1))$, a 1-form on S^3 (which will be the restriction to S^3 of a 1-form on $\mathbb{R}^4$) and the pullback of this form by the canonical local cross-sections of the bundle (Section 3.3). In order to maximize the resemblance of the result to our discussion of monopoles in Chapter 0, however, we prefer to identify $\mathbb{CP}^1$ with S^2 (see (1.2.8)).

Exercise 4.6.16 Show that the diffeomorphism $[z^1, z^2] \to (2z^1 \bar{z}^2, |z^1|^2 - |z^2|^2)$ of $\mathbb{CP}^1$ onto $S^2 \subseteq \mathbb{C} \times \mathbb{R}$ (Exercise 4.3.10) can be written

$$\left[z^1, z^2 \right] \longrightarrow \left(z^1 \bar{z}^2 + \bar{z}^1 z^2, -\mathbf{i}\, z^1 \bar{z}^2 + \mathbf{i}\, \bar{z}^1 z^2, |z^1|^2 - |z^2|^2 \right)$$

(cf., (0.3.2)) and carries U_1 onto U_N and U_2 onto U_S.

For convenience, we will suppress the diffeomorphism of Exercise 4.6.16, identify $\mathbb{CP}^1$ with S^2 and thereby regard the Hopf projection $\mathcal{P}$ as a map of S^3 onto S^2 given by

$$\mathcal{P}\left(z^1, z^2 \right) = \left(z^1 \bar{z}^2 + \bar{z}^1 z^2, -\mathbf{i}\, z^1 \bar{z}^2 + \mathbf{i}\, \bar{z}^1 z^2, |z^1|^2 - |z^2|^2 \right) .$$

The trivializing nbds are then U_N and U_S so that the associated canonical cross-sections are $s_N : U_N \to \mathcal{P}^{-1}(U_N)$ and $s_S : U_S \to \mathcal{P}^{-1}(U_S)$. We will write these out explicitly in terms of standard spherical coordinates ϕ and θ on S^2 (see the comments following Exercise 4.5.4).

Remark: It is a time-honored tradition to be a bit sloppy when dealing with spherical coordinates. Strictly speaking, one needs four separate spherical coordinate charts to cover all of S^2 (see Exercise 4.2.3 and the remarks that follow). These, however, differ rather trivially from each other (e.g., the range of values assumed by θ) so all calculations are the same in any one of the four. The sensible thing to do then is to refer to (ϕ, θ) as "standard spherical coordinates on S^2" with the understanding that it doesn't matter which of the four charts we happen to have in mind at the moment.

We have already observed (Section 0.3) that every $(z^1, z^2) \in S^3$ can be written as $(z^1, z^2) = (\cos\frac{\phi}{2}\, e^{\mathbf{i}\xi_1}, \sin\frac{\phi}{2}\, e^{\mathbf{i}\xi_2})$ for some $0 \le \phi \le \pi$ and $\xi_1, \xi_2 \in \mathbb{R}$ and that the image of such a (z^1, z^2) under $\mathcal{P}$ is $(\sin\phi\cos\theta, \sin\phi\sin\theta, \cos\phi)$, where $\theta = \xi_1 - \xi_2$. On the other hand, given a point $(\sin\phi\cos\theta, \sin\phi\sin\theta, \cos\phi)$ in S^2, *any* $(z^1, z^2) = (\cos\frac{\phi}{2}\, e^{\mathbf{i}\xi_1}, \sin\frac{\phi}{2}\, e^{\mathbf{i}\xi_2}) \in S^3$ with $\xi_1 - \xi_2 = \theta$ satisfies

$$\left(|z^1|, \frac{z^2\,|z^1|}{z^1}\right) = \left(\cos\frac{\phi}{2}, \sin\frac{\phi}{2}\, e^{-\mathbf{i}\theta}\right)$$

and

$$\left(\frac{z^1\,|z^2|}{z^2}, |z^2|\right) = \left(\cos\frac{\phi}{2}\, e^{\mathbf{i}\theta}, \sin\frac{\phi}{2}\right).$$

Consequently (see Section 3.3),

$$s_N(\sin\phi\cos\theta, \sin\phi\sin\theta, \cos\phi) = \left(\cos\frac{\phi}{2}, \sin\frac{\phi}{2}\, e^{-\mathbf{i}\theta}\right)$$

and

$$s_S(\sin\phi\cos\theta, \sin\phi\sin\theta, \cos\phi) = \left(\cos\frac{\phi}{2}\, e^{\mathbf{i}\theta}, \sin\frac{\phi}{2}\right).$$

Identifying $\mathbb{C}^2$ with $\mathbb{R}^4$ as usual and letting $\iota : S^3 \hookrightarrow \mathbb{R}^4$ be the inclusion map, we therefore have

$$\iota \circ s_N(\sin\phi\cos\theta, \sin\phi\sin\theta, \cos\phi) = \left(\cos\frac{\phi}{2}, 0, \sin\frac{\phi}{2}\cos\theta, -\sin\frac{\phi}{2}\sin\theta\right) \tag{4.6.17}$$

$$\iota \circ s_S(\sin\phi\cos\theta, \sin\phi\sin\theta, \cos\phi) = \left(\cos\frac{\phi}{2}\cos\theta, \cos\frac{\phi}{2}\sin\theta, \sin\frac{\phi}{2}, 0\right). \tag{4.6.18}$$

Now let us consider a 1-form $\tilde{\Theta}$ on $\mathbb{R}^4$ given, relative to the standard, global chart on $\mathbb{R}^4$ by

$$\tilde{\Theta} = -x^2dx^1 + x^1dx^2 - x^4dx^3 + x^3dx^4$$

and its restriction

$$\boldsymbol{\Theta} = \iota^* \widetilde{\boldsymbol{\Theta}}$$

to S^3 (except for a factor of **i**, $\boldsymbol{\Theta}$ will eventually emerge as the "natural connection form" on the Hopf bundle $S^1 \to S^3 \to S^2$). Now we pull $\boldsymbol{\Theta}$ back to $U_N \subseteq S^2$ by the cross-section s_N. First observe that (4.6.16) gives $s_N^* \boldsymbol{\Theta} = (\iota \circ s_N)^* \widetilde{\boldsymbol{\Theta}}$ and this we compute using (4.6.12) and (4.6.17).

$$\begin{aligned}
(s_N^* \boldsymbol{\Theta})(\phi, \theta) &= \left((\iota \circ s_N)^* \widetilde{\boldsymbol{\Theta}} \right)(\phi, \theta) \\
&= -0 \cdot d\left(\cos\frac{\phi}{2}\right) + \cos\frac{\phi}{2}\, d(0) + \sin\frac{\phi}{2}\sin\theta\, d\left(\sin\frac{\phi}{2}\cos\theta\right) \\
&\quad + \sin\frac{\phi}{2}\cos\theta d\left(-\sin\frac{\phi}{2}\sin\theta\right) \\
&= \sin\frac{\phi}{2}\sin\theta\left(\frac{1}{2}\cos\frac{\phi}{2}\cos\theta d\phi - \sin\frac{\phi}{2}\sin\theta d\theta\right) \\
&\quad - \sin\frac{\phi}{2}\cos\theta\left(\frac{1}{2}\cos\frac{\phi}{2}\sin\theta d\phi + \sin\frac{\phi}{2}\cos\theta d\theta\right) \\
&= -\sin^2\frac{\phi}{2} d\theta \\
&= -\frac{1}{2}(1 - \cos\phi) d\theta\,.
\end{aligned}$$

Exercise 4.6.17 Show that, on U_S, $(s_S^* \boldsymbol{\Theta})(\phi, \theta) = \frac{1}{2}(1 + \cos\phi) d\theta$.

We call the readers attention to the potential 1-forms $\mathbf{A}_N$ and $\mathbf{A}_S$ for the Dirac monopole in (0.4.2) and (0.4.3). Their complex versions, introduced in (0.4.5), are $\boldsymbol{\mathcal{A}}_N = -\frac{1}{2}\mathbf{i}(1 - \cos\phi) d\theta$ and $\boldsymbol{\mathcal{A}}_S = \frac{1}{2}\mathbf{i}(1 + \cos\phi) d\theta$ and these are (except for the factor of **i**) just the pullbacks of $\boldsymbol{\Theta}$ to S^2 by the canonical cross-sections of the Hopf bundle $S^1 \to S^3 \to S^2$.

The final item on the agenda for this section is the generalization to manifolds of a notion familiar from elementary differential equations. Let **V** be a smooth vector field on the manifold X. A smooth curve $\alpha : \mathcal{I} \to X$ in X is an **integral curve for V** if its velocity vector at each point coincides with the vector assigned to that point by **V**, i.e., if

$$\alpha'(t) = \mathbf{V}\,(\alpha(t)) \qquad (4.6.19)$$

for each $t \in \mathcal{I}$. We write out (4.6.19) in local coordinates as follows: Let (U, φ) be a chart for X with coordinate functions x^i, $i = 1, \ldots, n$, such that U intersects the image of α. Then, whenever $\alpha(t) \in U$ we have

$$\alpha'(t) = \alpha'(t)\,(x^i)\, \frac{\partial}{\partial x^i}\bigg|_{\alpha(t)} = \frac{d(x^i \circ \alpha)}{dt}(t)\, \frac{\partial}{\partial x^i}\bigg|_{\alpha(t)}.$$

We also write $\mathbf{V} = \mathbf{V}(x^i)\frac{\partial}{\partial x^i}$ on U. Since $\mathbf{V}(x^i) \in C^\infty(U)$ we can write its coordinate expression as $\mathbf{V}(x^i) \circ \varphi^{-1}(x^1, \ldots, x^n) = F^i(x^1, \ldots, x^n)$ for C^∞ functions F^i, $i = 1, \ldots, n$, on U. Then

$$\mathbf{V}\left(\alpha(t)\right) = F^i\left(x^1\left(\alpha(t)\right), \ldots, x^n\left(\alpha(t)\right)\right) \left.\frac{\partial}{\partial x^i}\right|_{\alpha(t)} .$$

Thus, α is an integral curve for $\mathbf{V}$ (on U) iff

$$\frac{d(x^i \circ \alpha)}{dt}(t) = F^i\left((x^1 \circ \alpha)(t), \ldots, (x^n \circ \alpha)(t)\right), \quad i = 1, \ldots, n. \tag{4.6.20}$$

Now, (4.6.20) is a system of ordinary differential equations for the unknown functions $x^i \circ \alpha(t)$, $i = 1, \ldots, n$ (the coordinate functions of α). We shall now appeal to basic results from the theory of such systems (taken from [**Hure**]) to obtain the facts we need.

Theorem 4.6.2 *Let* $\mathbf{V}$ *be a smooth vector field on the differentiable manifold* X *and* p *a point in* X*. Then there exists an interval* $(a(p), b(p))$ *in* $\mathbb{R}$ *and a smooth curve* $\alpha_p : (a(p), b(p)) \to X$ *such that*

1. $0 \in (a(p), b(p))$ *and* $\alpha_p(0) = p$.
2. α_p *is an integral curve for* $\mathbf{V}$.
3. *If* (c, d) *is an interval containing* 0 *and* $\beta : (c, d) \to X$ *is an integral curve for* $\mathbf{V}$ *with* $\beta(0) = p$*, then* $(c, d) \subseteq (a(p), b(p))$ *and* $\beta = \alpha_p|(c, d)$ *(thus,* α_p *is called the* **maximal** *integral curve of* $\mathbf{V}$ *through* p *at* $t = 0$*).*

Proof: The fundamental existence theorem for the system (4.6.20) (Theorem 4, page 28, of [**Hure**]) implies that there exists an integral curve α of $\mathbf{V}$ defined on some interval about 0 with $\alpha(0) = p$. Let $(a(p), b(p))$ be the union of all open intervals about 0 that are domains of such integral curves. We show that any two such curves must agree on the intersection of their domains so that the Glueing Lemma 1.2.3 yields a continuous curve $\alpha_p : (a(p), b(p)) \to X$ that is clearly smooth and the required integral curve for $\mathbf{V}$.

Thus, let $\alpha_1 : (a_1, b_1) \to X$ and $\alpha_2 : (a_2, b_2) \to X$ be two integral curves for $\mathbf{V}$ with 0 in (a_1, b_1) and (a_2, b_2) and $\alpha_1(0) = \alpha_2(0) = p$. The intersection $(a_1, b_1) \cap (a_2, b_2)$ is an open interval (a, b) about 0. Let $S = \{t \in (a, b) : \alpha_1(t) = \alpha_2(t)\}$. Then $S \neq \emptyset$ since $0 \in S$. S is closed by continuity of α_1 and α_2. Now suppose $t_0 \in S$. Then $\alpha_1(t_0) = \alpha_2(t_0)$ so the fundamental uniqueness theorem for (4.6.20) (Theorem 3, page 28, of [**Hure**]) implies that α_1 and α_2 agree on some interval $|t - t_0| < h$ about t_0. This entire interval is contained in S so S is open. But (a, b) is connected so $S = (a, b)$ by Exercise 1.5.1. ∎

Now, for each $t \in \mathbb{R}$ we denote by $\mathcal{D}_t$ the set of all $p \in X$ for which the maximal integral curve α_p is defined at t, i.e., $\mathcal{D}_t = \{p \in X : t \in (a(p), b(p))\}$. For sufficiently large t, $\mathcal{D}_t$ might well be empty, but Theorem 4.6.2 implies that $X = \bigcup_{t>0} \mathcal{D}_t = \bigcup_{t<0} \mathcal{D}_t$. Of course, $\mathcal{D}_0 = X$. Now, for each $t \in \mathbb{R}$, define a map $\mathbf{V}_t : \mathcal{D}_t \to X$ by

$$\mathbf{V}_t(p) = \alpha_p(t), \quad p \in \mathcal{D}_t\,.$$

Thus, $\mathbf{V}_t$ pushes p t units along the maximal integral curve of $\mathbf{V}$ that starts at p. According to Theorem 7, page 29, of [**Hure**] we may select, for each x in X, an open nbd U of x and an $\varepsilon > 0$ (both depending on x, in general) such that the map

$$(t, p) \longrightarrow \mathbf{V}_t(p) \tag{4.6.21}$$

is defined on all of $(-\varepsilon, \varepsilon) \times U$. In fact, Theorem 9, page 29, of [**Hure**] on the differentiability of solutions to (4.6.20) with respect to their initial values, implies that the map (4.6.21) is C^∞ on $(-\varepsilon, \varepsilon) \times U$.

Next we observe the following. Suppose $p \in \mathcal{D}_t$, i.e., $t \in (a(p), b(p))$. Consider the curve $\beta : (a(p) - t, b(p) - t) \to X$ defined by $\beta(s) = \alpha_p(s + t)$ (if either $a(p) = -\infty$ or $b(p) = \infty$ we interpret $a(p) - t$ and $b(p) - t$ to be $-\infty$ and ∞, respectively). Then $\beta(0) = \alpha_p(t)$ and, by Exercise 4.4.13, β is an integral curve of $\mathbf{V}$. Furthermore, the domain of β is maximal for, otherwise, $(a(p), b(p))$ would not be the maximal domain of α_p. Thus, $(a(p) - t, b(p) - t) = (\,a(\alpha_p(t)), b(\alpha_p(t))\,)$ and $\beta = \alpha_{\alpha_p(t)}$, i.e.,

$$\alpha_{\alpha_p(t)}(s) = \alpha_p(s + t)\,. \tag{4.6.22}$$

Lemma 4.6.3 *Let $\mathbf{V} \in \mathcal{X}(X)$ and let s and t be real numbers. Then the domain of $\mathbf{V}_s \circ \mathbf{V}_t$ is contained in $\mathcal{D}_{s+t}$ and, on this domain, $\mathbf{V}_s \circ \mathbf{V}_t = \mathbf{V}_{s+t}$.*

Proof: Let p be in the domain of $\mathbf{V}_s \circ \mathbf{V}_t$. Then $t \in (a(p), b(p))$ and, since $\mathbf{V}_t(p) = \alpha_p(t)$, $s \in (\,a(\alpha_p(t)), b(\alpha_p(t))\,) = (a(p) - t, b(p) - t)$. In particular, $s + t \in (a(p), b(p))$ so $p \in \mathcal{D}_{s+t}$. Moreover, (4.6.22) gives $\mathbf{V}_s \circ \mathbf{V}_t(p) = \mathbf{V}_s(\mathbf{V}_t(p)) = \mathbf{V}_s(\alpha_p(t)) = \alpha_{\alpha_p(t)}(s) = \alpha_p(s + t) = \mathbf{V}_{s+t}(p)$. ∎

Exercise 4.6.18 Show that, if s and t have the same sign, then the domain of $\mathbf{V}_s \circ \mathbf{V}_t$ equals $\mathcal{D}_{s+t}$.

Theorem 4.6.4 *Let $\mathbf{V} \in \mathcal{X}(X)$. Then, for each real number t, $\mathcal{D}_t$ is an open set in X and $\mathbf{V}_t$ is a diffeomorphism of $\mathcal{D}_t$ onto $\mathcal{D}_{-t}$ with inverse $\mathbf{V}_{-t}$.*

Proof: If $t = 0$, then $\mathcal{D}_0 = X$ and $\mathbf{V}_0 = id_X$ so the result is trivial. Suppose then that $t > 0$ (the case in which $t < 0$ is left to the reader in

Exercise 4.6.19). Fix a $p \in \mathcal{D}_t$. Since $[0,t]$ is compact we may select an open set U in X containing $\alpha_p([0,t])$ and an $\varepsilon > 0$ such that the map (4.6.21) is defined and C^∞ on $(-\varepsilon, \varepsilon) \times U$. Now choose a positive integer n large enough that $0 < \frac{t}{n} < \varepsilon$. Let $F_1 = \mathbf{V}_{t/n}|U$ and $U_1 = F_1^{-1}(U)$. For each $i = 2, \ldots, n$ we inductively define $F_i = \mathbf{V}_{t/n}|U_{i-1}$ and $U_i = F_i^{-1}(U_{i-1})$. Each F_i is a C^∞ map on the open set $U_{i-1} \subseteq U$. In particular, U_n is an open set contained in U. We show that $p \in U_n \subseteq \mathcal{D}_t$ and thus conclude that $\mathcal{D}_t$ is open. By Lemma 4.6.3 and Exercise 4.6.18, the domain of $\mathbf{V}_{t/n} \circ \cdots \circ \mathbf{V}_{t/n}$ (n times) equals the domain of $\mathbf{V}_{t/n + \cdots + t/n} = \mathbf{V}_t$ which contains p so $\mathbf{V}_{t/n} \circ \cdots \circ \mathbf{V}_{t/n}(p) = \mathbf{V}_t(p) = \alpha_p(t)$ and this is contained in U. Thus, $p \in U_n$. Furthermore, for any $q \in U_n$, $F_1 \circ \cdots \circ F_n(q)$ is defined and $F_1 \circ \cdots \circ F_n(q) = \mathbf{V}_{t/n} \circ \cdots \circ \mathbf{V}_{t/n}(q) = \mathbf{V}_t(q)$ so $q \in \mathcal{D}_t$. Thus, $U_n \subseteq \mathcal{D}_t$, as required.

Since we have just shown that $\mathbf{V}_t|U_n = F_1 \circ \cdots \circ F_n|U_n$, $\mathbf{V}_t|U_n$ is a composition of C^∞ maps and so is C^∞. Moreover, since $p \in \mathcal{D}_t$ was arbitrary, it follows that $\mathbf{V}_t$ is C^∞ on $\mathcal{D}_t$. $\mathbf{V}_t$ is one-to-one on $\mathcal{D}_t$ by the uniqueness theorem for (4.6.20) because $\mathbf{V}_t(p) = \mathbf{V}_t(q)$ implies $\alpha_p(t) = \alpha_q(t)$. Moreover, each $\mathbf{V}_t(p) = \alpha_p(t)$ is in $\mathcal{D}_{-t}$, i.e., $-t \in (\, a(\alpha_p(t)), b(\alpha_p(t))\,)$, because $(\, a(\alpha_p(t)), b(\alpha_p(t))\,) = (a(p) - t, b(p) - t)$ and $0 \in (a(p), b(p))$. Furthermore, $\mathbf{V}_t$ maps $\mathcal{D}_t$ onto $\mathcal{D}_{-t}$ since $q \in \mathcal{D}_{-t}$ implies $-t \in (a(q), b(q))$, $\alpha_q(-t) \in \mathcal{D}_t$ and $\mathbf{V}_t(\alpha_q(-t)) = \alpha_{\alpha_q(-t)}(t) = \alpha_q(t + (-t)) = \alpha_q(0) = q$. Thus, $\mathbf{V}_t$ is a C^∞ bijection of $\mathcal{D}_t$ onto $\mathcal{D}_{-t}$.

Exercise 4.6.19 Show, similarly, that if $t < 0$, then $\mathcal{D}_t$ is open and $\mathbf{V}_t$ is a C^∞ bijection of $\mathcal{D}_t$ onto $\mathcal{D}_{-t}$.

Returning to the case in which $t > 0$, we conclude from Exercise 4.6.19 that $\mathbf{V}_{-t}$ is a C^∞ bijection of $\mathcal{D}_{-t}$ onto $\mathcal{D}_t$. Thus, the domains of $\mathbf{V}_{-t} \circ \mathbf{V}_t$ and $\mathbf{V}_t \circ \mathbf{V}_{-t}$ are, respectively, $\mathcal{D}_t$ and $\mathcal{D}_{-t}$ and, on these domains, both compositions are the identity by Lemma 4.6.3. Consequently, $\mathbf{V}_t$ and $\mathbf{V}_{-t}$ are inverse diffeomorphisms. ∎

A vector field $\mathbf{V} \in \mathcal{X}(X)$ is said to be **complete** if $\mathcal{D}_t = X$ for every $t \in \mathbb{R}$, i.e., if each integral curve α_p is defined on all of $\mathbb{R}$. In this case, $\{\mathbf{V}_t\}_{t \in \mathbb{R}}$ is a collection of diffeomorphisms of X onto itself and satisfies

1. $\mathbf{V}_{s+t} = \mathbf{V}_s \circ \mathbf{V}_t$ for all $s,\ t \in \mathbb{R}$.
2. $\mathbf{V}_0 = id_X$.

In other words, the map $t \to \mathbf{V}_t$ is a homomorphism of the additive group $\mathbb{R}$ of real numbers into the group Diff(X) of all diffeomorphisms of X onto itself under composition (called the **diffeomorphism group of** X). The collection $\{\mathbf{V}_t\}_{t \in \mathbb{R}}$ is called the **1-parameter group** of diffeomorphisms of X **generated by** $\mathbf{V}$. If $\mathbf{V}$ is not complete, then the domains of the

$\mathbf{V}_t$ vary with t and one refers to the collection of all the diffeomorphisms $\mathbf{V}_t : \mathcal{D}_t \to \mathcal{D}_{-t}$ as the **local 1-parameter group generated by V**.

Exercise 4.6.20 Let $\mathbf{V} \in \boldsymbol{\mathcal{X}}(X)$ and $f \in C^\infty(X)$. Show that $\mathbf{V}f$ is the derivative of f along the integral curves of $\mathbf{V}$. More precisely, show that, for each $p \in X$, $\mathbf{V}f(p) = \lim_{t\to 0} \frac{1}{t}(f(\alpha_p(t)) - f(p))$.

Exercise 4.6.21 Show that if X is compact, then any $\mathbf{V} \in \boldsymbol{\mathcal{X}}(X)$ is complete. **Hint:** Use compactness to show that there exists an $\varepsilon > 0$ such that the map (4.6.21) is defined and C^∞ on $(-\varepsilon, \varepsilon) \times X$.

There is a useful formula for the Lie bracket $[\mathbf{V}, \mathbf{W}]$ of two vector fields analogous to the formula for $\mathbf{V}f$ in Exercise 4.6.20. In effect, it says that $[\mathbf{V}, \mathbf{W}]$ can be computed by differentiating $\mathbf{W}$ along the integral curves of $\mathbf{V}$. More precisely, we fix a point $p \in X$ and consider the local 1-parameter group $\{\mathbf{V}_t\}$ generated by $\mathbf{V}$. Choose $\varepsilon > 0$ sufficiently small that $p \in \mathcal{D}_t$ for all t in $(-\varepsilon, \varepsilon)$; henceforth, we consider only such t's. Now, $\mathbf{V}_{-t} : \mathcal{D}_{-t} \to \mathcal{D}_t$ is a diffeomorphism so, by Exercise 4.6.12, $(\mathbf{V}_{-t})_*\mathbf{W}$ is a smooth vector field on $\mathcal{D}_t$ whose value at p is

$$((\mathbf{V}_{-t})_*\mathbf{W})_p = (\mathbf{V}_{-t})_{*\mathbf{V}_t(p)}\left(\mathbf{W}_{\mathbf{V}_t(p)}\right) = (\mathbf{V}_{-t})_{*\alpha_p(t)}\left(\mathbf{W}_{\alpha_p(t)}\right)$$

(evaluate $\mathbf{W}$ along the integral curve $\alpha_p(t)$ and move these values back to $T_p(X)$ by $(\mathbf{V}_{-t})_*$). Thus, $t \to ((\mathbf{V}_{-t})_*\,\mathbf{W})_p$ gives a smooth curve in $T_p(X)$ whose derivative at 0 we claim is just $[\mathbf{V}, \mathbf{W}]_p$:

$$[\mathbf{V}, \mathbf{W}]_p = \lim_{t\to 0} \frac{(\mathbf{V}_{-t})_{*\alpha_p(t)}\left(\mathbf{W}_{\alpha_p(t)}\right) - \mathbf{W}_p}{t}. \qquad (4.6.23)$$

Here we remind the reader of our decision to identify the tangent spaces to a vector space with the vector space itself via the canonical isomorphism (Exercise 4.4.9). Thus, the limit on the right-hand side of (4.6.23) is the ordinary (componentwise) limit in the natural topology of $T_p(X)$. We prove (4.6.23) by showing that each side has the same value at an arbitrary $f \in C^\infty(X)$. First observe that

$$\begin{aligned}\left(\lim_{t\to 0} \frac{(\mathbf{V}_{-t})_{*\alpha_p(t)}(\mathbf{W}_{\alpha_p(t)}) - \mathbf{W}_p}{t}\right)(f) &= \lim_{t\to 0}\left(\frac{(\mathbf{V}_{-t})_{*\alpha_p(t)}(\mathbf{W}_{\alpha_p(t)}) - \mathbf{W}_p}{t}(f)\right)\\ &= \frac{d}{dt}\left((\mathbf{V}_{-t})_{*\alpha_p(t)}(\mathbf{W}_{\alpha_p(t)})\right)(f)\big|_{t=0}\\ &= \frac{d}{dt}\left(\mathbf{W}_{\alpha_p(t)}\left(f \circ \mathbf{V}_{-t}\right)\right)\big|_{t=0}.\end{aligned}$$

Now define a real-valued function F on a nbd of $(0,0)$ in $\mathbb{R}^2$ by $F(t,u) =$

$f(\mathbf{V}_{-t}(\mathbf{W}_u(\alpha_p(t))))$. We write out the definition of $D_2F(t,0)$:

$$D_2 F(t,0) = \lim_{h\to 0} \frac{F(t,h) - F(t,0)}{h}$$
$$= \lim_{h\to 0} \frac{(f\circ \mathbf{V}_{-t})(\mathbf{W}_h(\alpha_p(t))) - (f\circ \mathbf{V}_{-t})(\alpha_p(t))}{h}.$$

Since $h \to \mathbf{W}_h(\alpha_p(t))$ is the integral curve of $\mathbf{W}$ starting at $\alpha_p(t)$ we conclude from Exercise 4.6.20 that

$$D_2 F(t,0) = \mathbf{W}\,(f\circ \mathbf{V}_{-t})\,(\alpha_p(t)) = \mathbf{W}_{\alpha_p(t)}\,(f\circ \mathbf{V}_{-t}).$$

Thus, we must show that

$$[\mathbf{V},\mathbf{W}]_p\,(f) = D_{2,1}\,F(0,0). \tag{4.6.24}$$

To evaluate the derivative on the right-hand side we consider the real-valued function G defined on a nbd of $(0,0,0)$ in $\mathbb{R}^3$ by $G(t,u,s) = f(\mathbf{V}_s(\mathbf{W}_u(\alpha_p(t))))$. Then $F(t,u) = G(t,u,-t)$ so, by the Chain Rule, $D_{2,1}F(0,0) = D_{2,1}G(0,0,0) - D_{2,3}G(0,0,0)$. But $G(t,u,0) = f(\mathbf{W}_u(\alpha_p(t)))$ so $D_2G(t,0,0) = \mathbf{W}_{\alpha_p(t)}(f)$, by Exercise 4.6.20. Thus, $D_2G(t,0,0) = (\mathbf{W}f)(\alpha_p(t))$ so, again by Exercise 4.6.20, $D_{2,1}G(0,0,0) = \mathbf{V}(\mathbf{W}f)(p)$.

Exercise 4.6.22 Show, similarly, that $D_{2,3}G(0,0,0) = \mathbf{W}(\mathbf{V}f)(p)$, conclude that $D_{2,1}F(0,0) = \mathbf{V}(\mathbf{W}f)(p) - \mathbf{W}(\mathbf{V}f)(p)$ and so complete the proof of (4.6.24) and, consequently, that of (4.6.23).

4.7 Matrix Lie Groups

We begin with some preliminary material on Lie groups in general, but soon restrict our attention to the classical Lie groups of matrices. A Lie group, as defined in Section 4.3, is a differentiable manifold G that is also a group in which the operations of multiplication $(x,y) \to xy$ and inversion $x \to x^{-1}$ are C^∞. It follows, just as in Exercise 1.6.1, that this is the case iff the map $(x,y) \to x^{-1}y$ is C^∞. Somewhat more surprising is the fact that it would suffice to assume that multiplication alone is C^∞.

Lemma 4.7.1 *Let G be a differentiable manifold that is also a group for which the group multiplication $(x,y) \to xy$ is a C^∞ map of $G \times G$ to G. Then G is a Lie group.*

Proof: We ask the reader to get the ball rolling.

Exercise 4.7.1 Show that, for each fixed $g \in G$, the left and right translation maps $L_g, R_g : G \to G$ defined by $L_g(x) = gx$ and $R_g(x) = xg$ are diffeomorphisms of G onto G.

Next we observe that the map $m : G \times G \to G$ defined by $m(x, y) = xy$ is a submersion at (e, e), where e is the identity element of G. To see this we must show that $m_* : T_{(e,e)}(G \times G) \to T_e(G)$ is surjective. We identify $T_{(e,e)}(G \times G)$ with $T_e(G) \times T_e(G)$ (Section 4.7). Any element of $T_e(G)$ is $\alpha'(0)$ for some smooth curve α in G with $\alpha(0) = e$. For each such we define a smooth curve $\tilde{\alpha}$ in $G \times G$ by $\tilde{\alpha}(t) = (\alpha(t), e)$. Then $m_*(\tilde{\alpha}'(0)) = (m \circ \tilde{\alpha})'(0) = (\alpha(t)e)'(0) = \alpha'(0)$ as required.

Now define a map $f : G \times G \to G \times G$ by $f(x, y) = (x, xy)$. Letting $\mathcal{P}_1 : G \times G \to G$ be the projection onto the first factor, we have $f = (\mathcal{P}_1, m)$ so, at (e, e), $f_* : T_e(G) \times T_e(G) \to T_e(G) \times T_e(G)$ is given by $f_* = ((\mathcal{P}_1)_*, m_*)$ (Section 4.4). Since $(\mathcal{P}_1)_*$ is surjective by Exercise 4.4.18 and we have just shown that m_* is surjective, it follows that f_* is surjective. But the domain and range of f_* have the same dimension so, in fact, f_* is an isomorphism. By Corollary 4.4.8, f is a diffeomorphism on some nbd of (e, e) in $G \times G$. But the inverse of f is clearly given (on all of $G \times G$) by $f^{-1}(x, z) = (x, x^{-1}z)$ so this map must be C^∞ on some nbd $U \times U$ of (e, e) in $G \times G$. Consequently, on U, the map

$$x \longrightarrow (x, e) \xrightarrow{f^{-1}} \left(x, x^{-1}e\right) = \left(x, x^{-1}\right) \longrightarrow x^{-1}$$

is a composition of C^∞ maps and is therefore also C^∞. We have shown therefore that inversion $x \to x^{-1}$ is C^∞ on a nbd U of e. To see that it is C^∞ on a nbd of any point in G (and therefore C^∞) fix a $y \in G$. Then yU is an open nbd of y in G (Exercise 1.6.4 (a)) and every $z \in yU$ has a unique representation of the form $z = yx$ for some $x \in U$. Thus, on yU, the map

$$z = yx \xrightarrow{L_{y^{-1}}} x \longrightarrow x^{-1} \xrightarrow{R_{y^{-1}}} x^{-1}y^{-1} = (yx)^{-1} = z^{-1}$$

is C^∞. ∎

All sorts of examples of Lie groups present themselves immediately: Any $\mathbb{R}^n$ with vector addition. The nonzero real numbers, complex numbers and quaternions with their respective multiplications. The circle S^1 with complex multiplication. The 3-sphere S^3 with quaternion multiplication. The general linear groups $GL(n, \mathbb{R})$, $GL(n, \mathbb{C})$ and $GL(n, \mathbb{H})$ with matrix multiplication (in the quaternionic case we lack a determinant and therefore a simple formula for inverses, but the smoothness of inversion follows from Lemma 4.7.1). Any product of Lie groups is a Lie group. Furthermore, a subgroup H of a Lie group G that is also a submanifold of G is itself a Lie group by Exercise 4.5.3. Thus, for example, $O(n)$ and $SO(n)$, being submanifolds of $GL(n, \mathbb{R})$, are Lie groups (see Section 4.5). $SO(n)$ is actually the connected component of $O(n)$ containing the identity. According to Theorem 1.6.4 the connected component containing the identity in any Lie group is a subgroup and also an open submanifold and therefore is itself a Lie group. Shortly we will show that $U(n)$ and $Sp(n)$ are submanifolds of their respective general linear groups and that $SU(n)$ is a submanifold of $U(n)$ so that all of these are also Lie groups.

A vector field $\mathbf{V}$ on a Lie group G is said to be **left invariant** if, for each $g \in G$, $(L_g)_* \circ \mathbf{V} = \mathbf{V} \circ L_g$, where L_g is left translation by g (see Exercise 4.7.1), i.e., iff $(L_g)_{*h}(\mathbf{V}_h) = \mathbf{V}_{gh}$ for all $g, h \in G$. Note that we do not assume that $\mathbf{V}$ is smooth (see Theorem 4.7.2).

Exercise 4.7.2 Show that $\mathbf{V}$ is left invariant iff $(L_g)_{*e}(\mathbf{V}_e) = \mathbf{V}_g$ for each $g \in G$.

Thus, given any $\mathbf{v} \in T_e(G)$ there exists a unique left invariant vector field $\mathbf{V}$ on G whose value at e is $\mathbf{V}_e = \mathbf{v}$.

Theorem 4.7.2 *A left invariant vector field* $\mathbf{V}$ *on a Lie group* G *is* C^∞.

Proof: It will suffice to show that $\mathbf{V}$ is C^∞ on a nbd of e since then the diffeomorphism L_g carries $\mathbf{V}$ onto a C^∞ vector field $(L_g)_*\mathbf{V} = \mathbf{V}$ on a nbd of g in G. Choose a chart (U, φ) at e with coordinate functions x^i, $i = 1, \ldots, n$, and a nbd U' of e such that $a, b \in U'$ implies $ab \in U'U' \subseteq U$. The component functions of $\mathbf{V}$ are $\mathbf{V}x^i$ and we show that these are C^∞ on U' (note that $U' \subseteq U$ since $e \in U'$). For any $a \in U'$, $\mathbf{V}x^i(a) = \mathbf{V}_a(x^i) = ((L_a)_{*e}(\mathbf{V}_e))\,(x^i) = \mathbf{V}_e(x^i \circ L_a)$ (note that L_a carries U' into U so that $x^i \circ L_a$ is defined and C^∞ on U' for each fixed $a \in U'$). Now let $\mathbf{V}_e = \zeta^j \left.\frac{\partial}{\partial x^j}\right|_e$ so that

$$\mathbf{V}x^i(a) = \zeta^j \left.\frac{\partial(x^i \circ L_a)}{\partial x^j}\right|_e . \qquad (4.7.1)$$

Now, the right-hand side of (4.7.1) is clearly C^∞ on U' for each fixed $a \in U'$, but we must show that it is C^∞ in a. To do this we observe that, for any $b \in U'$, $ab = L_a(b) \in U$ so $(x^i \circ L_a)(b) = x^i(ab)$ is defined and the composition

$$(a, b) \longrightarrow ab \longrightarrow \varphi(ab) = \left(x^1(ab), \ldots, x^n(ab)\right)$$

is C^∞ on $U' \times U'$. Using the chart $\varphi \times \varphi : U' \times U' \to \varphi(U') \times \varphi(U')$ on $U' \times U'$ we have $(\varphi \times \varphi)(a, b) = (\varphi(a), \varphi(b)) = (x^1(a), \ldots, x^n(a), x^1(b), \ldots, x^n(b))$ so we may write

$$\left(x^i \circ L_a\right)(b) = x^i(ab) = f^i\left(x^1(a), \ldots, x^n(a), x^1(b), \ldots, x^n(b)\right)$$

for some C^∞ functions f^i, $i = 1, \ldots, n$, on $\varphi(U') \times \varphi(U')$. Thus,

$$\left.\frac{\partial(x^i \circ L_a)}{\partial x^j}\right|_e = D_{n+j} f^i\left(x^1(a), \ldots, x^n(a), x^1(e), \ldots, x^n(e)\right)$$

is C^∞ in a on U' and so, by (4.7.1), $\mathbf{V}x^i$ is C^∞ on U' as required. ∎

Lemma 4.7.3 *The set $\mathcal{G}$ of all left invariant vector fields on a Lie group G is a linear subspace of $\boldsymbol{\mathcal{X}}(G)$ and the map $\mathbf{V} \to \mathbf{V}_e$ from $\mathcal{G}$ to $T_e(G)$ is an isomorphism. In particular,* $\dim \mathcal{G} = \dim T_e(G) = \dim G$.

Exercise 4.7.3 Prove Lemma 4.7.3. ∎

Theorem 4.7.4 *Let G be a Lie group, $\mathcal{G}$ its vector space of left invariant vector fields and $\mathbf{V}, \mathbf{W} \in \mathcal{G}$. Then $[\mathbf{V}, \mathbf{W}] \in \mathcal{G}$.*

Proof: Since $\mathbf{V}$ is left invariant, $(L_g)_{*h}(\mathbf{V}_h) = \mathbf{V}_{L_g(h)}$ for all g and h in G, i.e., $\mathbf{V}$ is L_g-related to itself (see Exercise 4.6.13). Similarly, $\mathbf{W}$ is L_g-related to $\mathbf{W}$. By Exercise 4.6.13, $[\mathbf{V}, \mathbf{W}]$ is L_g-related to $[\mathbf{V}, \mathbf{W}]$, i.e., $[\mathbf{V}, \mathbf{W}]$ is left invariant. ∎

Thus, the collection $\mathcal{G}$ of left invariant vector fields on a Lie group G is closed under the formation of Lie brackets. Recall that the Lie bracket is bilinear ((4.6.3) and (4.6.4)), skew-symmetric ((4.6.2)) and satisfies the Jacobi identity ((4.6.6)). In general, a **Lie algebra** is real vector space $\mathcal{A}$ on which is defined a bilinear operation $[\ ,\ ] : \mathcal{A} \times \mathcal{A} \to \mathcal{A}$, called **bracket**, such that $[y, x] = -[x, y]$ and $[\,[x, y], z] + [\,[z, x], y] + [\,[y, z], x] = 0$ for all x, y and z in $\mathcal{A}$. Thus, $\mathcal{G}$ is a Lie algebra under the Lie bracket operation and is called the **Lie algebra of** G. There are many other familiar examples of Lie algebras. $\mathbb{R}^3$ with its usual cross product $\times$ as the bracket operation is one such. Define, for any two $n \times n$ matrices A and B, their **commutator** $[A, B]$ by $[A, B] = AB - BA$. Then the collection of all $n \times n$ real matrices forms a Lie algebra of dimension n^2 under commutator. The same is true of $n \times n$ complex or quaternionic matrices provided the collections of all such are regarded as *real* vector spaces (of dimension $2n^2$ and $4n^2$, respectively). Real linear subspaces that are closed under the formation of commutators are likewise Lie algebras. Note that $[y, x] = -[x, y]$ implies $[x, x] = 0$ and so the bracket operation on any 1-dimensional Lie algebra is necessarily **trivial**, i.e., satisfies $[x, y] = 0$ for all x and y.

Exercise 4.7.4 $\mathbb{R}^n$ is a Lie group under (vector) addition. Show that if $x^1, \ldots, x^n$ are standard coordinate functions on $\mathbb{R}^n$, then the coordinate vector fields $\frac{\partial}{\partial x^i}$ are left invariant. Conclude from Exercise 4.6.5 that the bracket operation on the Lie algebra of $\mathbb{R}^n$ is trivial.

Exercise 4.7.5 Show that the collection of all $n \times n$ real, skew-symmetric ($A^T = -A$) matrices forms a Lie algebra under commutator.

Exercise 4.7.6 On the set $\operatorname{Im}\mathbb{H}$ of pure imaginary quaternions define $[x, y] = xy - yx = 2\operatorname{Im}(xy)$ (Exercise 1.1.11). Show that, with this as the bracket operation, $\mathrm{Im}\,\mathbb{H}$ is a 3-dimensional Lie algebra.

If $\mathcal{A}_1$ and $\mathcal{A}_2$ are Lie algebras with brackets $[\ ,\ ]_1$ and $[\ ,\ ]_2$, respectively, then a linear isomorphism $T : \mathcal{A}_1 \to \mathcal{A}_2$ that satisfies $T([x,y]_1) = [T(x), T(y)]_2$ for all $x, y \in \mathcal{A}_1$ is called a **Lie algebra isomorphism** and we say that $\mathcal{A}_1$ and $\mathcal{A}_2$ are **isomorphic** as Lie algebras. Lemma 4.7.3 provides a linear isomorphism from the Lie algebra $\mathcal{G}$ of any Lie group G onto the tangent space $T_e(G)$ to G at the identity. Our next objective is to show that, in the cases of interest to us, $T_e(G)$ can be identified with a collection of matrices that is closed under commutator and so forms a Lie algebra and that, with this structure, the linear isomorphism of Lemma 4.7.3 is actually a Lie algebra isomorphism.

We begin with $G = GL(n, \mathbb{R})$. Since $GL(n, \mathbb{R})$ is an open submanifold of $\mathbb{R}^{n^2}$, the tangent space $T_{id}(GL(n, \mathbb{R}))$ is linearly isomorphic to $\mathbb{R}^{n^2}$. We let x^{ij}, $i, j = 1, \ldots, n$, denote the standard coordinate functions on $\mathbb{R}^{n^2}$ and identify any real $n \times n$ matrix $A(A^{ij})$ with the tangent vector $A = A^{ij} \frac{\partial}{\partial x^{ij}}\big|_{id}$. Denote by $\mathbf{A}$ the unique left invariant field on $GL(n, \mathbb{R})$ with $\mathbf{A}(id) = A$. We compute its component functions $\mathbf{A}x^{kl} : GL(n, \mathbb{R}) \to \mathbb{R}$. For each $g \in GL(n, \mathbb{R})$ we have

$$\mathbf{A}x^{kl}(g) = \mathbf{A}_g\left(x^{kl}\right) = \left(L_g\right)_{*id}(A)\left(x^{kl}\right) = A\left(x^{kl} \circ L_g\right). \tag{4.7.2}$$

Now, $x^{kl} \circ L_g : GL(n, \mathbb{R}) \to \mathbb{R}$ is given by $(x^{kl} \circ L_g)(h) = x^{kl}(gh) = kl$-entry of $gh = \sum_{\alpha=1}^{n} g^{k\alpha} h^{\alpha l}$ for each $h \in GL(n, \mathbb{R})$. Thus, $x^{kl} \circ L_g$ is linear (g is fixed here and the h^{mn} are the standard coordinates of h in $GL(n, \mathbb{R})$ so this last sum is the standard coordinate expression for $x^{kl} \circ L_g$). Note also that

$$\frac{\partial}{\partial x^{ij}}\left(x^{kl} \circ L_g\right) = \begin{cases} 0, & \text{if } j \neq l \\ g^{ki}, & \text{if } j = l \end{cases}$$

so that, by (4.7.2),

$$\mathbf{A}x^{kl}(g) = \left(A^{ij} \frac{\partial}{\partial x^{ij}}\bigg|_{id}\right)\left(x^{kl} \circ L_g\right) = \sum_{i=1}^{n} A^{il} g^{ki} = \sum_{i=1}^{n} g^{ki} A^{il}, \tag{4.7.3}$$

which is the kl-entry of the matrix product of gA. Thus, identifying matrices with elements of the various tangent spaces to $GL(n, \mathbb{R})$ we may write

$$\mathbf{A}(g) = \left(L_g\right)_{*id}(A) = gA. \tag{4.7.4}$$

If, in (4.7.3), we regard A as fixed and g^{mn} as the standard coordinates of g in $GL(n, \mathbb{R})$, then the functions $\mathbf{A}x^{kl}$ are seen to be linear and

$$\frac{\partial}{\partial x^{ij}}\left(\mathbf{A}x^{kl}\right) = \begin{cases} 0, & k \neq i \\ A^{jl}, & k = i \end{cases}.$$

Now suppose $B = (B^{ij})$ is another real $n \times n$ matrix and identify B with the tangent vector $B = B^{ij} \frac{\partial}{\partial x^{ij}}\big|_{id}$. Then $B(\mathbf{A}x^{kl}) = B^{ij} \frac{\partial}{\partial x^{ij}}\big|_{id}(\mathbf{A}x^{kl}) = \sum_{j=1}^{n} B^{kj}A^{jl}$, which is the kl-entry of BA. Switching A and B gives $A(\mathbf{B}x^{kl}) = (AB)^{kl}$. Thus,

$$[\mathbf{A}, \mathbf{B}]_{id}(x^{kl}) = A(\mathbf{B}x^{kl}) - B(\mathbf{A}x^{kl}) = (AB)^{kl} - (BA)^{kl} = (AB - BA)^{kl}.$$

The linear isomorphism $\mathbf{A} \to \mathbf{A}_{id} = A$ of the Lie algebra $\mathcal{GL}(n, \mathbb{R})$ of $GL(n, \mathbb{R})$ onto $T_{id}(GL(n, \mathbb{R}))$ therefore sends the Lie bracket $[\mathbf{A}, \mathbf{B}]$ onto the commutator $[A, B]$ and so is a Lie algebra isomorphism. These two views of $\mathcal{GL}(n, \mathbb{R})$ (left invariant vector fields under Lie bracket and $n \times n$ matrices under commutator) are both very convenient and we will make extensive use of each.

In order to obtain the Lie algebras of $O(n)$ and $SO(n)$ we prove a general result about subgroups of Lie groups. Suppose then that G is a Lie group and H is a subgroup of G that is also a submanifold of G. Then the inclusion map $\iota : H \hookrightarrow G$ is an embedding and $\iota_{*h} : T_h(H) \to T_h(G)$ identifies $T_h(H)$ with a subspace of $T_h(G)$ for each $h \in H$. For each such h we have two translation maps $L_h : H \to H$ and $\hat{L}_h : G \to G$ that are related by $\hat{L}_h \circ \iota = \iota \circ L_h$ so that, at each point in H, $(\hat{L}_h)_* \circ \iota_* = \iota_* \circ (L_h)_*$. For any left invariant vector field $\mathbf{V}$ on H we have $\iota_{*h}(\mathbf{V}_h) = \iota_{*h}\left((L_h)_{*e}(\mathbf{V}_e) \right) = (\hat{L}_h)_{*e}\left(\iota_{*e}(\mathbf{V}_e) \right)$. Now, $(\hat{L}_h)_{*e}(\iota_{*e}(\mathbf{V}_e))$ is the value at h of the left invariant vector field $\mathbf{V}'$ on G whose value at e is $\iota_{*e}(\mathbf{V}_e)$. Thus, $\iota_{*h}(\mathbf{V}_h) = \mathbf{V}'_h = \mathbf{V}'_{\iota(h)}$ so $\mathbf{V}$ and $\mathbf{V}'$ are ι-related. If $\mathbf{W}$ is any other left invariant vector field on H and $\mathbf{W}'$ is the left invariant vector field on G whose value at e is $\iota_{*e}(\mathbf{W}_e)$, then $\mathbf{W}$ and $\mathbf{W}'$ are ι-related. By Exercise 4.6.13, $[\mathbf{V}, \mathbf{W}]$ is ι-related to $[\mathbf{V}', \mathbf{W}']$, i.e.,

$$[\mathbf{V}', \mathbf{W}']_h = \iota_{*h}\left([\mathbf{V}, \mathbf{W}]_h\right), \tag{4.7.5}$$

for each $h \in H$. In particular, $[\mathbf{V}', \mathbf{W}']_e = \iota_{*e}([\mathbf{V}, \mathbf{W}]_e)$. If one regards the maps ι_* as inclusions and not worth mentioning then one may say that the left invariant vector fields on H are just the restrictions to H of left invariant vector fields on G and that these have the same Lie brackets in H and G.

Now suppose $G = GL(n, \mathbb{R})$ and $H = O(n)$. We identify the Lie algebra $\mathcal{GL}(n, \mathbb{R})$ with the set of all $n \times n$ real matrices under commutator. Then, as we have just shown, the Lie algebra $\mathcal{O}(n)$ of $O(n)$ is a linear subspace of $\mathcal{GL}(n, \mathbb{R})$ whose bracket is also the commutator so we need only identify it as a *set*. Now, any element of $T_{id}(O(n))$ is $A'(0)$ for some smooth curve $A : (-\varepsilon, \varepsilon) \to O(n)$ with $A(0) = id$. Since $O(n)$ is a submanifold of $\mathbb{R}^{n^2}$ we may regard A as a curve in $\mathbb{R}^{n^2}$ and use standard coordinates to differentiate entrywise. Thus, the components of $A'(0)$ relative to $\frac{\partial}{\partial x^{ij}}\big|_{id}$ are $((A^{ij})'(0))$. Since $A(t) \in O(n)$ for each t, $A(t)(A(t))^T = id$, i.e., $\sum_{k=1}^{n} A^{ik}(t)A^{jk}(t) =$

δ^{ij} for each t. Differentiating at $t = 0$ gives

$$\sum_{k=1}^{n} \big((A^{ik})'(0)\delta^{jk} + \delta^{ik}(A^{jk})'(0) \big) = 0$$

$$(A^{ij})'(0) + (A^{ji})'(0) = 0$$

$$(A^{ji})'(0) = -(A^{ij})'(0)$$

so $A'(0)$ is, as a real $n \times n$ matrix, skew-symmetric. Thus, $T_{id}(O(n))$ is contained in the subspace of $\mathcal{GL}(n, \mathbb{R})$ consisting of skew-symmetric matrices. But this latter subspace has dimension $\frac{n(n-1)}{2}$ and this is precisely the dimension of $O(n)$ (Section 4.5). Thus, the Lie algebra $\mathcal{O}(n)$ of $O(n)$ is precisely the set of real $n \times n$ skew-symmetric matrices under commutator. Moreover, by Exercise 4.5.15, $SO(n)$ is an open submanifold of $O(n)$ so its tangent spaces coincide with those of $O(n)$ and, in particular, its Lie algebra $\mathcal{SO}(n)$ coincides with $\mathcal{O}(n)$.

The complex general linear group $GL(n, \mathbb{C})$ is handled in precisely the same way as $GL(n, \mathbb{R})$, but everything is twice as long. $GL(n, \mathbb{C})$ is an open submanifold of $\mathbb{R}^{2n^2}$. We denote by $\{x^{11}, y^{11}, \ldots, x^{nn}, y^{nn}\}$ the standard coordinate functions on $\mathbb{R}^{2n^2}$. Then, for each $g \in GL(n, \mathbb{C})$, $T_g(GL(n, \mathbb{C}))$ consists of all $a^{11}\frac{\partial}{\partial x^{11}} + b^{11}\frac{\partial}{\partial y^{11}} + \cdots + a^{nn}\frac{\partial}{\partial x^{nn}} + b^{nn}\frac{\partial}{\partial y^{nn}}$ evaluated at $(x^{11}(g), y^{11}(g), \ldots, x^{nn}(g), y^{nn}(g))$. We denote this

$$a^{ij} \left.\frac{\partial}{\partial x^{ij}}\right|_g + b^{ij} \left.\frac{\partial}{\partial y^{ij}}\right|_g . \tag{4.7.6}$$

There is an obvious isomorphism from $T_g(GL(n, \mathbb{C}))$ to the set of all $n \times n$ complex matrices that carries (4.7.6) onto $(z^{ij}) = (a^{ij} + b^{ij}\,\mathbf{i})$. At $g = id \in GL(n, \mathbb{C})$, Lemma 4.7.3 identifies the tangent space $T_{id}(GL(n, \mathbb{C}))$ with the Lie algebra $\mathcal{GL}(n, \mathbb{C})$ of $GL(n, \mathbb{C})$ and we wish to show that this isomorphism carries the Lie bracket in $\mathcal{GL}(n, \mathbb{C})$ onto the commutator of the corresponding complex matrices.

Exercise 4.7.7 Fill in the details of the following argument to establish this and conclude that $\mathcal{GL}(n, \mathbb{C})$ can be identified with the set of all $n \times n$ complex matrices under commutator.

Denote by $\mathbf{A}$ the unique left invariant vector field on $GL(n, \mathbb{C})$ whose value at $id \in GL(n, \mathbb{C})$ is $A = a^{ij}\frac{\partial}{\partial x^{ij}}|_{id} + b^{ij}\frac{\partial}{\partial y^{ij}}|_{id}$. Its component functions are $\mathbf{A}x^{kl}$ and $\mathbf{A}y^{kl}$, $k, l = 1, \ldots, n$. For each $g \in GL(n, \mathbb{C})$, $\mathbf{A}x^{kl}(g) = A(x^{kl} \circ L_g)$ and $\mathbf{A}y^{kl}(g) = A(y^{kl} \circ L_g)$. But, for each $h \in GL(n, \mathbb{C})$, $(x^{kl} \circ L_g)(h) = \mathrm{Re}\,((gh)^{kl})$ and $(y^{kl} \circ L_g)(h) = \mathrm{Im}\,((gh)^{kl})$. Regarding $g = (g^{ij}) \in GL(n, \mathbb{C})$ as fixed and $h = (h^{ij})$ as a variable point in $GL(n,\mathbb{C})$, $x^{kl}\circ L_g$ and $y^{kl}\circ L_g$ are therefore linear and

$$\frac{\partial}{\partial x^{ij}}\left(x^{kl}\circ L_g\right)=\begin{cases}0, & j\neq l\\ \operatorname{Re} g^{ki}, & j=l\end{cases},\quad \frac{\partial}{\partial y^{ij}}\left(x^{kl}\circ L_g\right)=\begin{cases}0, & j\neq l\\ -\operatorname{Im} g^{ki}, & j=l\end{cases}$$

$$\frac{\partial}{\partial x^{ij}}\left(y^{kl}\circ L_g\right)=\begin{cases}0, & j\neq l\\ \operatorname{Im} g^{ki}, & j=l\end{cases},\quad \frac{\partial}{\partial y^{ij}}\left(y^{kl}\circ L_g\right)=\begin{cases}0, & j\neq l\\ \operatorname{Re} g^{ki}, & j=l\end{cases}.$$

Thus, $\mathbf{A}x^{kl}(g)=\sum_{i=1}^n\left((\operatorname{Re} g^{ki})\,a^{il}-(\operatorname{Im} g^{ki})\,b^{il}\right)$ and $\mathbf{A}y^{kl}(g)=\sum_{i=1}^n\left((\operatorname{Im} g^{ki})\,a^{il}+(\operatorname{Re} g^{ki})\,b^{il}\right)$ so that

$$\mathbf{A}_g=\left[\sum_{i=1}^n\left((\operatorname{Re} g^{ki})\,a^{il}-(\operatorname{Im} g^{ki})\,b^{il}\right)\right]\frac{\partial}{\partial x^{kl}}\bigg|_g+\left[\sum_{i=1}^n\left((\operatorname{Im} g^{ki})\,a^{il}+(\operatorname{Re} g^{ki})\,b^{il}\right)\right]\frac{\partial}{\partial y^{kl}}\bigg|_g.$$

In our expressions for $\mathbf{A}x^{kl}(g)$ and $\mathbf{A}y^{kl}(g)$ we now regard the a^{il} and b^{il} as fixed and g as the variable in $GL(n,\mathbb{C})$ and find that

$$\frac{\partial}{\partial x^{ij}}\left(\mathbf{A}x^{kl}\right)=\begin{cases}0, & i\neq k\\ a^{jl}, & i=k\end{cases},\quad \frac{\partial}{\partial y^{ij}}\left(\mathbf{A}x^{kl}\right)=\begin{cases}0, & i\neq k\\ -b^{jl}, & i=k\end{cases}$$

$$\frac{\partial}{\partial x^{ij}}\left(\mathbf{A}y^{kl}\right)=\begin{cases}0, & i\neq k\\ b^{jl}, & i=k\end{cases},\quad \frac{\partial}{\partial y^{ij}}\left(\mathbf{A}y^{kl}\right)=\begin{cases}0, & i\neq k\\ a^{jl}, & i=k\end{cases}.$$

Now, let $B=c^{ij}\frac{\partial}{\partial x^{ij}}|_{id}+d^{ij}\frac{\partial}{\partial y^{ij}}|_{id}$ be another element of $T_{id}(GL(n,\mathbb{C}))$ corresponding to $\mathbf{B}\in\mathcal{GL}(n,\mathbb{C})$ and compute $B(\mathbf{A}x^{kl})$ and $B(\mathbf{A}y^{kl})$. Switch the roles of B and A and compute $[\mathbf{A},\mathbf{B}]_{id}(x^{kl})$ and $[\mathbf{A},\mathbf{B}]_{id}(y^{kl})$, then compare with the commutator of the matrices $(a^{ij}+b^{ij}\,\mathbf{i})$ and $(c^{ij}+d^{ij}\,\mathbf{i})$.

Any subgroup of some $GL(n,\mathbb{C})$ that is also a submanifold is called a **matrix Lie group** and we will henceforth restrict our attention to these.

Exercise 4.7.8 Show that $GL(n,\mathbb{R})$ is a subgroup of $GL(n,\mathbb{C})$ that is also a submanifold. **Hint:** Exercise 4.5.4.

We have seen that any matrix Lie group G has a Lie algebra $\mathcal{G}$ that can be identified with a set of $n\times n$ complex matrices under commutator so finding $\mathcal{G}$ simply amounts to identifying it as a subset of $\mathcal{GL}(n,\mathbb{C})$. We would like to apply this procedure to $U(n)$ and $SU(n)$, but, unfortunately, we have not yet even shown that these are manifolds. To remedy this situation, and for a great many other purposes as well, we introduce a new tool.

Remark: One can actually show that any closed subgroup G of $GL(n, \mathbb{C})$ is necessarily a submanifold and therefore a matrix Lie group. There is a nice, elementary proof of this in [**Howe**]. For even more general, and more difficult, results consult [**Warn**].

For each $n \times n$ complex matrix $A \in \mathcal{GL}(n, \mathbb{C})$ we define $\exp(A) = e^A = \sum_{k=0}^{\infty} \frac{1}{k!} A^k$; that is, the ij-entry in $\exp(A)$ is $\sum_{k=0}^{\infty} \frac{1}{k!}(A^{ij})^k$. The series not only converges entrywise for each fixed A, but does so uniformly on every bounded region in $\mathcal{GL}(n, \mathbb{C})$ $(= \mathbb{C}^{n^2})$. Indeed, on such a region one can choose a constant m such that $|A^{ij}| \leq m$ for every A. By induction, $|(A^k)^{ij}| \leq n^{k-1} m^k$ for each $k = 1, 2, \ldots$ so the result follows from the Weierstrass M-test (Theorem 9.6 of [**Apos**]). We collect together a few elementary properties of matrix exponentiation that will be of use.

Lemma 4.7.5 *The series* $\exp(A) = e^A = \sum_{k=0}^{\infty} \frac{1}{k!} A^k$ *converges absolutely and uniformly on any bounded region in* $\mathcal{GL}(n, \mathbb{C})$. *Moreover,*

1. $e^A e^B = e^{A+B}$ *if* $A, B \in \mathcal{GL}(n, \mathbb{C})$ *and* $AB = BA$.
2. $\det(e^A) = e^{\operatorname{trace}(A)}$ *(so, in particular,* $e^A \in GL(n, \mathbb{C})$*).*
3. *The map* $\exp : \mathcal{GL}(n, \mathbb{C}) \to GL(n, \mathbb{C})$ *defined by* $A \to e^A$ *is* C^∞.
4. *For any* $A \in \mathcal{GL}(n, \mathbb{C})$*, the curve* $t \to e^{tA}$ *is a smooth homomorphism of the additive group* $\mathbb{R}$ *into* $GL(n, \mathbb{C})$ *whose velocity vector at* $t = 0$ *is* A.
5. $\exp_{*0} : T_0(\mathcal{GL}(n, \mathbb{C})) \to \mathcal{GL}(n, \mathbb{C})$ *is the identity map (here* 0 *is the* $n \times n$ *zero matrix and* $T_0(\mathcal{GL}(n, \mathbb{C}))$ *is identified with* $\mathcal{GL}(n, \mathbb{C})$ *via the canonical isomorphism of Exercise 4.4.9).*

Proof: The first statement has already been established and, together with the commutativity of A and B, justifies the rearrangements in the following calculation.

$$
\begin{aligned}
e^A e^B &= \left(\sum_{k=0}^{\infty} \frac{1}{k!} A^k \right) \left(\sum_{l=0}^{\infty} \frac{1}{l!} B^l \right) = \sum_{k,l=0}^{\infty} \frac{1}{k!l!} A^k B^l \\
&= \sum_{N=0}^{\infty} \sum_{k=0}^{N} \frac{1}{k!(N-k)!} A^k B^{N-k} = \sum_{N=0}^{\infty} \frac{1}{N!} \sum_{k=0}^{N} \binom{N}{k} A^k B^{N-k} \\
&= \sum_{N=0}^{\infty} \frac{1}{N!} (A+B)^N = e^{A+B}.
\end{aligned}
$$

Exercise 4.7.9 Show that, for any $A \in \mathcal{GL}(n, \mathbb{C})$ and any $g \in GL(n, \mathbb{C})$, $g e^A g^{-1} = e^{gAg^{-1}}$. **Hint:** The map $C \to gCg^{-1}$ of $\mathcal{GL}(n, \mathbb{C})$ to itself is continuous.

From Exercise 4.7.9 we find that $\det(e^{gAg^{-1}}) = \det(ge^A g^{-1}) = \det(e^A)$. Furthermore, $\text{trace}\,(gAg^{-1}) = \text{trace}\,(A)$. Now, for any $A \in \mathcal{GL}(n, \mathbb{C})$ there exists a $g \in GL(n, \mathbb{C})$ such that gAg^{-1} is upper triangular (Corollary 2, Section 1, Chapter $\underline{\text{X}}$ of [**Lang**]) so we need only prove (2) for upper triangular matrices. But if A is upper triangular with diagonal entries $\lambda_1, \ldots, \lambda_n$, then e^A is upper triangular with diagonal entries $e^{\lambda_1}, \ldots, e^{\lambda_n}$. Thus, $\det(e^A) = e^{\lambda_1} \cdots e^{\lambda_n} = e^{\lambda_1 + \cdots + \lambda_n} = e^{\text{trace}\,(A)}$ as required.

Property (3) is clear since the entries in e^A are convergent power series in the entries of A and so their real and imaginary parts (i.e., the coordinate functions of exp) are C^∞. In particular, for each fixed A, the curve $\alpha(t) = e^{tA}$, $t \in \mathbb{R}$, is smooth. Since $t_1 A$ and $t_2 A$ commute, $\alpha(t_1 + t_2) = e^{(t_1+t_2)A} = e^{t_1 A + t_2 A} = e^{t_1 A} e^{t_2 A} = \alpha(t_1)\alpha(t_2)$ so α is also a homomorphism. Differentiating the entry power series term-by-term with respect to t gives $\alpha'(0) = A$.

Finally, for any $A \in \mathcal{GL}(n, \mathbb{C})$, $\beta(t) = 0 + tA = tA$ is a smooth curve in $\mathcal{GL}(n, \mathbb{C})$ whose velocity vector at $t = 0$ is identified with A under the canonical isomorphism. Thus, $\exp_{0*}(A)$ is the velocity vector at $t = 0$ of the curve $\exp \circ \beta(t) = e^{tA} = \alpha(t)$ and we have just shown that $\alpha'(0) = A$. Thus, $\exp_{0*}(A) = A$ for each $A \in \mathcal{GL}(n, \mathbb{C})$. ∎

From Lemma 4.7.5 (5) and Corollary 4.4.8 it follows that $\exp : \mathcal{GL}(n, \mathbb{C}) \to GL(n, \mathbb{C})$ is a diffeomorphism of some open nbd V of $0 \in \mathcal{GL}(n, \mathbb{C})$ onto an open neighborhood U of $id \in GL(n, \mathbb{C})$, i.e., $(U, (\exp|V)^{-1})$ is a chart at id in $GL(n, \mathbb{C})$. By restricting exp to various linear subspaces of $\mathcal{GL}(n, \mathbb{C})$ we will similarly obtain charts at the identity for $U(n)$, $SU(n)$ and $Sp(n)$.

Theorem 4.7.6 *Let A be in $\mathcal{GL}(n, \mathbb{C})$.*

1. *If A is skew-Hermitian ($\bar{A}^T = -A$), then $e^A \in U(n)$.*
2. *A is skew-Hermitian and has* trace $(A) = 0$, *then $e^A \in SU(n)$.*
3. *If $n = 2m$ and A is skew-Hermitian and satisfies $JA + A^T J = 0$, where $J = \begin{pmatrix} 0 & id \\ -id & 0 \end{pmatrix}$ and id is the $m \times m$ identity matrix, then $e^A \in Sp(m) = \{M \in U(2m) : M^T J M = J\}$.*

Furthermore, there exist open nbds V of 0 in $\mathcal{GL}(n, \mathbb{C})$ and U of id in $GL(n, \mathbb{C})$ such that $\exp : V \to U$ *is a diffeomorphism and, on V, the converse of each implication in (1), (2) and (3) is true.*

Proof: First note that $\bar{A}^T = -A$ implies $\overline{(e^A)}^T = e^{\bar{A}^T} = e^{-A} = (e^A)^{-1}$ so $e^A \in U(n)$. If, in addition, $\text{trace}\,(A) = 0$, then, by Lemma 4.7.5 (2),

$\det(e^A) = e^0 = 1$ so $e^A \in SU(n)$. Next assume that $n = 2m$, $\bar{A}^T = -A$ and $JA + A^T J = 0$. Then $JAJ^{-1} = -A^T$ so $e^{JAJ^{-1}} = e^{-A^T} = (e^{-A})^T = ((e^A)^{-1})^T = ((e^A)^T)^{-1}$. But then, by Exercise 4.7.9, $Je^AJ^{-1} = ((e^A)^T)^{-1}$ so $(e^A)^T Je^A = J$ as required.

To show that the converses are all locally true we begin with an open nbd W of $0 \in \mathcal{GL}(n, \mathbb{C})$ on which exp is a diffeomorphism onto some open nbd of $id \in GL(n, \mathbb{C})$. By shrinking W if necessary we may assume that $|\text{trace}\,(A)| < 2\pi$ for every $A \in W$. Let $\bar{W}$ be the set of all $\bar{A}$ for $A \in W$, W^T the set of all A^T for $A \in W$, $-W$ the set of all $-A$ for $A \in W$ and JWJ^{-1} the set of all JAJ^{-1} for $A \in W$.

Exercise 4.7.10 Show that $V = W \cap \bar{W} \cap W^T \cap (-W) \cap (JWJ^{-1})$ is an open nbd of $0 \in \mathcal{GL}(n, \mathbb{C})$ that is closed under complex conjugation, transposition, negation and conjugation by J.

Of course, $|\text{trace}\,(A)| < 2\pi$ for each $A \in V$ and exp carries V diffeomorphically onto an open nbd U of $id \in GL(n, \mathbb{C})$. Now suppose $A \in V$ and $e^A \in U(n) \cap U$. Then $e^{-A} = (e^A)^{-1} = (\overline{e^A})^T = e^{\bar{A}^T}$. But exp is one-to-one on V so $-A = \bar{A}^T$, i.e., A is skew-Hermitian. Next suppose that $A \in V$ and $e^A \in SU(n) \cap U$. Then $\det(e^A) = 1$ so, by Lemma 4.7.5 (2), $\text{trace}\,(A) = 2k\pi\,\mathbf{i}$ for some integer k. But $A \in V$ implies $|\text{trace}\,(A)| < 2\pi$ so $\text{trace}\,(A) = 0$ as required.

Exercise 4.7.11 Assume $n = 2m$, $A \in V$ and $e^A \in Sp(m)$ and show that $JA + A^T J = 0$. ∎

Let us define, for each positive integer n,

$$\mathcal{U}(n) = \left\{A \in \mathcal{GL}(n, \mathbb{C}) : \bar{A}^T = -A\right\},$$

$$\mathcal{SU}(n) = \left\{A \in \mathcal{GL}(n, \mathbb{C}) : \bar{A}^T = -A \text{ and } \text{trace}\,(A) = 0\right\},$$

$$\mathcal{SP}(n) = \left\{A \in \mathcal{GL}(2n, \mathbb{C}) : \bar{A}^T = -A \text{ and } JA + A^T J = 0\right\}.$$

These (real) vector subspaces are of dimension n^2, $n^2 - 1$ and $2n^2 + n$, respectively, and are carried by the exponential map to $U(n)$, $SU(n)$ and $Sp(n)$. Moreover, if U and V are as described in Theorem 4.7.6, then the restriction of the chart $(U, (\exp|V)^{-1})$ to one of these subgroups gives a coordinate slice (submanifold chart) at the identity.

Exercise 4.7.12 Show that, by composing exp with left translation, one can similarly obtain submanifold charts at any point of $U(n)$, $SU(n)$, or $Sp(n)$. Conclude that $U(n)$ and $SU(n)$ are submanifolds of $GL(n, \mathbb{C})$ and $Sp(n)$ is a submanifold of $GL(2n, \mathbb{C})$ and so all of these are matrix Lie groups. Show, furthermore, that $Sp(1)$ and $SU(2)$ are **isomorphic** as Lie groups, i.e., that there is a group isomorphism of one onto the other that is also a diffeomorphism.

Exercise 4.7.13 Show that the Lie algebras of $U(n)$, $SU(n)$ and $Sp(n)$ are $\mathcal{U}(n)$, $\mathcal{SU}(n)$ and $\mathcal{SP}(n)$, respectively. **Hint:** Lemma 4.7.5 (4).

Exercise 4.7.14 Show that the Lie algebra $\mathcal{U}(1)$ of $U(1)$ is isomorphic to the pure imaginary complex numbers $\operatorname{Im}\mathbb{C}$ with trivial bracket and that the Lie algebra $\mathcal{SU}(2)$ of $SU(2)$ is isomorphic to the pure imaginary quaternions $\operatorname{Im}\mathbb{H}$ with bracket $[x, y] = xy - yx = 2\operatorname{Im}(xy)$ (Exercise 4.7.5).

If G is a matrix Lie group we will denote elements of its Lie algebra $\mathcal{G}$ by $A, B, \ldots$ when they are to be thought of as complex matrices (or when it doesn't matter how you think of them) and we will write $\mathbf{A}, \mathbf{B}, \ldots$ for the corresponding left invariant vector fields ($\mathbf{A}(id) = A$, etc.). Any $A \in \mathcal{G}$ can be thought of as $\alpha'(0)$, where $\alpha(t) = \exp(tA)$. Notice that, since α is a homomorphism of the additive group $\mathbb{R}$ into G, $\alpha \circ L_{t_0} = L_{\alpha(t_0)} \circ \alpha$ for any t_0 so $\alpha_{*t_0} \circ (L_{t_0})_{*0} = (L_{\alpha(t_0)})_{*id} \circ \alpha_{*0}$.

Exercise 4.7.15 Show that $\mathbf{A}(\alpha(t_0)) = \alpha'(t_0)$. **Hint:** Use (4.4.4) and Exercise 4.7.2.

Now, fix a $g \in G$ and define $\alpha_g : \mathbb{R} \to G$ by $\alpha_g(t) = (L_g \circ \alpha)(t) = g\exp(tA)$. We claim that α_g is the unique maximal integral curve of $\mathbf{A}$ that starts at g, i.e., that $\alpha_g(0) = g$ and

$$\mathbf{A}\left(\alpha_g(t)\right) = \alpha_g'(t) \tag{4.7.7}$$

for each t in $\mathbb{R}$ (so, in particular, $\mathbf{A}$ is complete). To see this, fix a t_0 in $\mathbb{R}$ and compute

$$\begin{aligned}\mathbf{A}\left(\alpha_g(t_0)\right) &= \mathbf{A}\left(L_g(\alpha(t_0))\right) = (L_g)_{*\alpha(t_0)}\left(\mathbf{A}(\alpha(t_0))\right)\\ &= (L_g)_{*\alpha(t_0)}\left(\alpha'(t_0)\right) = (L_g)_{*\alpha(t_0)} \circ \alpha_{*t_0}\left(\left.\frac{d}{dt}\right|_{t=t_0}\right)\\ &= (L_g \circ \alpha)_{*t_0}\left(\left.\frac{d}{dt}\right|_{t=t_0}\right) = \alpha_g'(t_0)\,.\end{aligned}$$

We now have an ample supply of matrix Lie groups and associated Lie algebras, but still require a few more tools before we can effectively exploit them. First we extend the notion of left invariance to 1-forms. A 1-form on a Lie group G is a map $\boldsymbol{\Theta}$ that assigns to each $g \in G$ an element $\boldsymbol{\Theta}(g) = \boldsymbol{\Theta}_g$ of $T_g^*(G)$. $\boldsymbol{\Theta}$ is said to be **left invariant** if, for each $g \in G$, $(L_g)^*\boldsymbol{\Theta} = \boldsymbol{\Theta}$. More explicitly this means that

$$\boldsymbol{\Theta}(h) = \left(L_g\right)^*\left(\boldsymbol{\Theta}(gh)\right),$$

or, equivalently,

$$\Theta(gh) = \left(L_{g^{-1}}\right)^*(\Theta(h))$$

for all $g, h \in G$.

Exercise 4.7.16 Show that Θ is left invariant if and only if $\Theta(g) = (L_{g^{-1}})^*(\Theta(id))$ for all $g \in G$.

Thus, a left invariant 1-form is completely determined by its value at the identity id in G and, moreover, given any covector Θ_{id} at id there exists a unique left invariant 1-form Θ on G whose value at id is $\Theta(id) = \Theta_{id}$. Just as for vector fields, left invariance assures smoothness for a 1-form Θ. To prove this it will suffice to show that $\Theta\mathbf{V}$ is in $C^\infty(G)$ for every $\mathbf{V} \in \mathcal{X}(G)$. Note first that if $\mathbf{V} \in \mathcal{G} \subseteq \mathcal{X}(G)$, then $\Theta\mathbf{V}$ is actually constant on G since $(\Theta\mathbf{V})(g) = \Theta_g(\mathbf{V}(g)) = ((L_{g^{-1}})^*(\Theta_{id}))(\mathbf{V}_g) = \Theta_{id}((L_{g^{-1}})_*(\mathbf{V}_g)) = \Theta_{id}((L_{g^{-1}})_*((L_g)_*(\mathbf{V}_{id}))) = \Theta_{id}((L_{g^{-1}} \circ L_g)_*(\mathbf{V}_{id})) = \Theta_{id}(\mathbf{V}_{id}) = (\Theta\mathbf{V})(id)$. Now suppose $\mathbf{V} \in \mathcal{X}(G)$ is arbitrary. Let $\{\mathbf{V}_1, \ldots, \mathbf{V}_n\}$ be a family of left invariant vector fields on G for which $\{\mathbf{V}_1(id), \ldots, \mathbf{V}_n(id)\}$ is a basis for $T_{id}(G)$. Then, since $\mathbf{V}_i(g) = (L_g)_*(\mathbf{V}_i(id))$ for $i = 1, \ldots, n$, $\{\mathbf{V}_1(g), \ldots, \mathbf{V}_n(g)\}$ is a basis for $T_g(G)$ for each $g \in G$. Thus, we may write $\mathbf{V}(g) = f^i(g)\mathbf{V}_i(g)$ for some real-valued functions $f^1, \ldots, f^n$ on G. Now, $\mathbf{V}$ is C^∞ and each $\mathbf{V}_i$, being left invariant, is C^∞ so, writing $\mathbf{V} = f^i\mathbf{V}_i$ out in local coordinates, we find that $f^i \in C^\infty(G)$ for each $i = 1, \ldots, n$. Thus, $\Theta\mathbf{V} = \Theta(f^i\mathbf{V}_i) = f^i(\Theta\mathbf{V}_i)$ is C^∞ because each $\Theta\mathbf{V}_i$ is a constant.

Exercise 4.7.17 An n-dimensional differentiable manifold X is said to be **parallelizable** if there exist $\mathbf{V}_1, \ldots, \mathbf{V}_n \in \mathcal{X}(X)$ such that $\{\mathbf{V}_1(p), \ldots, \mathbf{V}_n(p)\}$ is a basis for $T_p(X)$ for each $p \in X$. Show that any Lie group is parallelizable. Conclude that S^1 and S^3 are parallelizable. Any thoughts on S^2?

Let G be a matrix Lie group and g some fixed element of G. Like any other group, G is closed under conjugation so we may define a map $Ad_g : G \to G$ by $Ad_g(h) = ghg^{-1}$ for every $h \in G$. Ad_g is clearly a diffeomorphism. Indeed, $Ad_g = L_g \circ R_{g^{-1}} = R_{g^{-1}} \circ L_g$. Furthermore, $Ad_g(id) = id$ so the derivative of Ad_g at the identity carries $\mathcal{G}$ isomorphically onto $\mathcal{G}$. We denote this map

$$ad_g : \mathcal{G} \longrightarrow \mathcal{G}$$

and call the assignment $g \to ad_g$ the **adjoint representation** of G. Thus, $ad_g = (Ad_g)_{*id} = (L_g)_{*g^{-1}} \circ (R_{g^{-1}})_{*id} = (R_{g^{-1}})_{*g} \circ (L_g)_{*id}$. It follows that $ad_{gh} = ad_g \circ ad_h$ for all g and h in G so that the map $g \to ad_g$ is a homomorphism from G into the group of nonsingular linear transformations on the Lie algebra $\mathcal{G}$. In general, if G is a Lie group and $\mathcal{V}$ is a finite dimensional vector space, then a homomorphism of G into the group of nonsingular linear transformations on $\mathcal{V}$ is called a **representation** of G on $\mathcal{V}$. Choosing a basis for $\mathcal{V}$ one can regard a representation of G as a homomorphism into some general linear group. The representation is said to be **continuous** (**smooth**) if this corresponding matrix-valued map is continuous (smooth). Observe that this definition clearly does not depend on the choice of basis for $\mathcal{V}$. Although we will not require this result we point out that one can actually show that a continuous representation of a Lie group is necessarily smooth (see Theorem 3.39 of [**Warn**]).

Now, any element of $\mathcal{G}$ is $\alpha'(0)$ for some smooth curve α in G with $\alpha(0) = id$ and $ad_g(\alpha'(0)) = (Ad_g)_{*id}(\alpha'(0)) = (Ad_g \circ \alpha)'(0)$. But $(Ad_g \circ \alpha)(t) = Ad_g(\alpha(t)) = g\,\alpha(t)\,g^{-1}$. Differentiating entrywise gives $ad_g(\alpha'(0)) = g\,\alpha'(0)\,g^{-1}$. In particular, $\mathcal{G}$ (as a set of matrices) is closed under conjugation by elements of G and we have proved the following Lemma.

Lemma 4.7.7 *Let G be a matrix Lie group, $\mathcal{G}$ its Lie algebra and $g \in G$. Then, for each $A \in \mathcal{G}$, $gAg^{-1} \in \mathcal{G}$ and the isomorphism $ad_g : \mathcal{G} \to \mathcal{G}$ is given by $ad_g(A) = gAg^{-1}$.*

Let G be a Lie group with Lie algebra $\mathcal{G}$. There is a general procedure, utilizing the adjoint representation, for defining on $\mathcal{G}$ a natural symmetric, bilinear form $K : \mathcal{G} \times \mathcal{G} \to \mathbb{R}$ called the *Killing form* of $\mathcal{G}$. For certain Lie algebras (the *semisimple* ones) this bilinear form is nondegenerate. If, in addition, G is compact, K is negative definite so that, with an extra minus sign, it gives rise to an inner product on $\mathcal{G}$. We will not require this general construction (see Chapter II, Section 6, of [**Helg**] if you're interested), but only its result when applied to $G = SU(2)$ (which we know to be compact and also happens to be semisimple).

Exercise 4.7.18 Let $G = SU(2)$ and $\mathcal{G} = \mathcal{SU}(2) = \{A \in \mathcal{GL}(n, \mathbb{C}) : \bar{A}^T = -A \text{ and } \operatorname{trace}(A) = 0\}$. For $A, B \in \mathcal{SU}(2)$ define $< A, B >= -\operatorname{trace}(AB)$.

(a) Show that $<\ ,\ >$ is a nondegenerate, symmetric, real-valued bilinear form on $\mathcal{SU}(2)$. **Hint:** Note that every $A \in \mathcal{SU}(2)$ can be written in the form

$$A = \begin{pmatrix} a_1\,\mathbf{i} & a_2 + a_3\,\mathbf{i} \\ -a_2 + a_3\,\mathbf{i} & -a_1\,\mathbf{i} \end{pmatrix}.$$

(b) Show that, under the natural identification $A \to a_1\,\mathbf{i} + (a_2 + a_3\,\mathbf{i})\,\mathbf{j} = a_1\,\mathbf{i} + a_2\,\mathbf{j} + a_3\,\mathbf{k}$ of $\mathcal{SU}(2)$ and $\operatorname{Im}\mathbb{H}$, $<\ ,\ >$ is just twice the usual inner product, i.e., $<A, B> = 2(a_1 b_1 + a_2 b_2 + a_3 b_3)$.

We close this section by utilizing much of the machinery we have developed to build an object associated with any smooth principal bundle (Section 4.3) that will be crucial in our discussion of connections. The idea is that, in such a bundle $\mathcal{P} : P \to X$, each fiber is a submanifold of P (Corollary 4.5.7) diffeomorphic to the Lie group G acting on P. Since each $p \in P$ is contained in such a fiber, each $T_p(P)$ contains a subspace isomorphic to $T_p(G)$ which is, in turn, isomorphic to $\mathcal{G}$. The action of G on P provides a natural way of identifying these copies of $T_p(G)$ with the Lie algebra $\mathcal{G}$.

Begin by considering a matrix Lie group G, a manifold P and a smooth right action $\sigma : P \times G \to P$, $\sigma(p,g) = p \cdot g$, of G on P. For each $p \in P$ define $\sigma_p : G \to P$ by $\sigma_p(g) = p \cdot g$. We define a map, also denoted σ, from $\mathcal{G}$ to $\boldsymbol{\mathcal{X}}(P)$ as follows: For each $A \in \mathcal{G}$ (identified with a set of matrices) we define $\sigma(A) \in \boldsymbol{\mathcal{X}}(P)$ by $\sigma(A)(p) = (\sigma_p)_{*id}(A)$. To see that $\sigma(A)$ is, indeed, C^∞ we write the definition out more explicitly. Define $\alpha : \mathbb{R} \to G$ by $\alpha(t) = \exp(tA) = e^{tA}$. Then $\alpha'(0) = A$ and $\sigma(A)(p) = (\sigma_p)_{*id}(A) = (\sigma_p)_{*id}(\alpha'(0)) = (\sigma_p \circ \alpha)'(0)$. But $(\sigma_p \circ \alpha)(t) = \sigma_p(\alpha(t)) = \sigma_p(e^{tA}) = \sigma(p, \exp(tA))$. Since the map of $P \times \mathcal{G}$ into P given by $(p, B) \to (p, \exp(B)) \to \sigma(p, \exp(B))$ is C^∞, it has smooth local coordinate expressions and therefore so does $\sigma(A)(p)$. Furthermore, we may write

$$\sigma(A)(p) = \frac{d}{dt}(p \cdot \exp(tA))\bigg|_{t=0}. \tag{4.7.8}$$

Exercise 4.7.19 Show that $\alpha_p : \mathbb{R} \to P$ given by $\alpha_p(t) = p \cdot \exp(tA)$ is the maximal integral curve of $\sigma(A)$ through p at $t = 0$ so that $\sigma(A)$ is complete and its 1-parameter group of diffeomorphisms $\{(\sigma(A))_t\}_{t \in \mathbb{R}}$ is given by $(\sigma(A))_t(p) = p \cdot \exp(tA)$.

Theorem 4.7.8 *Let P be a differentiable manifold, G a matrix Lie group and $\sigma : P \times G \to P$, $\sigma(p,g) = p \cdot g$, a smooth right action of G on P. Then, for each $A \in \mathcal{G}$, $\sigma(A)$ is in $\boldsymbol{\mathcal{X}}(P)$ and the map $A \to \sigma(A)$ of $\mathcal{G}$ to $\boldsymbol{\mathcal{X}}(P)$ is linear and satisfies $\sigma([A,B]) = [\sigma(A), \sigma(B)]$ for all $A, B \in \mathcal{G}$. If the action σ is effective and $A \neq 0$, then $\sigma(A)$ is not the zero vector field on P. If the action σ is free and $A \neq 0$, then $\sigma(A)$ is never zero.*

Proof: We have already shown that $\sigma(A) \in \boldsymbol{\mathcal{X}}(P)$. Linearity is clear since $\sigma(A)(p) = (\sigma_p)_{*id}(A)$ and $(\sigma_p)_{*id}$ is linear. Next we let A and B be in $\mathcal{G}$ and $\mathbf{A}$ and $\mathbf{B}$ the corresponding left invariant vector fields on G. Then $[A,B] = [\mathbf{A}, \mathbf{B}]_{id}$, which we compute using (4.6.23). By (4.7.7), the 1-parameter group $\{\mathbf{A}_t\}_{t \in \mathbb{R}}$ of $\mathbf{A}$ is given by $\mathbf{A}_t(g) = g \exp(tA)$ so

$$[A,B] = [\mathbf{A}, \mathbf{B}]_{id} = \lim_{t \to 0} \frac{1}{t}\left[(\mathbf{A}_{-t})_{*\exp(tA)}\left(\mathbf{B}_{\exp(tA)}\right) - \mathbf{B}_{id}\right]$$

$$= \lim_{t\to 0} \frac{1}{t} \left[(R_{\exp(-tA)})_{*\exp(tA)} \left(\mathbf{B}(\exp(tA)) \right) - B \right]$$

$$= \lim_{t\to 0} \frac{1}{t} \left[(R_{\exp(-tA)})_{*\exp(tA)} \circ (L_{\exp(tA)})_{*id} \left(\mathbf{B}(id) \right) - B \right]$$

$$= \lim_{t\to 0} \frac{1}{t} \left[ad_{\exp(tA)}(B) - B \right].$$

Next we compute $[\sigma(A), \sigma(B)]_p$. For convenience, we denote the 1-parameter group of $\sigma(A)$ by $\{\xi_t\}_{t\in\mathbb{R}}$ so that, by Exercise 4.7.19, $\xi_t(p) = p\cdot\exp(tA)$. Thus,

$$[\sigma(A), \sigma(B)]_p = \lim_{t\to 0} \frac{1}{t} \left[(\xi_{-t})_{*\xi_t(p)} \left(\sigma(B)_{\xi_t(p)} \right) - \sigma(B)_p \right]$$

$$= \lim_{t\to 0} \frac{1}{t} \left[(\xi_{-t})_{*p\cdot\exp(tA)} \left((\sigma_{p\cdot\exp(tA)})_{*id}(B) \right) - (\sigma_p)_{*id}(B) \right]$$

$$= \lim_{t\to 0} \frac{1}{t} \left[(\xi_{-t} \circ \sigma_{p\cdot\exp(tA)})_{*id}(B) - (\sigma_p)_{*id}(B) \right].$$

Exercise 4.7.20 Show that $\xi_{-t} \circ \sigma_{p\cdot\exp(tA)} = \sigma_p \circ Ad_{\exp(tA)}$.

Thus, $(\xi_{-t} \circ \sigma_{p\cdot\exp(tA)})_{*id} = (\sigma_p)_{*id} \circ ad_{\exp(tA)}$ so

$$[\sigma(A), \sigma(B)]_p = \lim_{t\to 0} \frac{1}{t} \left[(\sigma_p)_{*id} \left(ad_{\exp(tA)}(B) \right) - (\sigma_p)_{*id}(B) \right]$$

$$= (\sigma_p)_{*id} \left(\lim_{t\to 0} \frac{1}{t} \left[ad_{\exp(tA)}(B) - B \right] \right)$$

$$= (\sigma_p)_{*id}([A,B]) = \sigma([A,B])(p)$$

so $[\sigma(A), \sigma(B)] = \sigma([A,B])$.

Now suppose that the action σ is effective (Section 1.7) and $A \neq 0$. To show that $\sigma(A)$ is not the zero vector field we assume to the contrary that $\sigma(A)(p) = 0$ for every $p \in P$. Then every integral curve of $\sigma(A)$ is constant so $\xi_t(p) = p$ for each t, i.e., $p\cdot\exp(tA) = p$ for each t. Since σ is effective, $\exp(tA) = id$ for each t. But exp is a local diffeomorphism near 0 in $\mathcal{G}$ so this contradicts $A \neq 0$.

Exercise 4.7.21 Show that if the action σ is free and $A \neq 0$, then $\sigma(A)$ is never zero. ∎

We apply Theorem 4.7.8 to the case of a smooth principal G-bundle $\mathcal{B} = (P, \mathcal{P}, \sigma)$ over X, where G is assumed to be a matrix Lie group. For each $A \in \mathcal{G}$ we denote the vector field $\sigma(A) \in \boldsymbol{\mathcal{X}}(P)$ by $A^{\#}$ and call it the **fundamental vector field on P determined by** A. Recall that the fibers $\mathcal{P}^{-1}(x)$ of $\mathcal{P}$ are all submanifolds of P diffeomorphic to G. Thus, for each $p \in P$, the tangent space $T_p(P)$ contains a subspace $\mathrm{Vert}_p(P)$ (consisting of tangent vectors at p to smooth curves in the fiber containing p) that is isomorphic to $T_p(G)$ (and therefore to $\mathcal{G}$). We call $\mathrm{Vert}_p(P)$ the **vertical subspace** of $T_p(P)$ and refer to its elements as **vertical vectors** at p. Moreover, since the fibers of $\mathcal{P}$ are invariant under the action σ and, for each $A \in \mathcal{G}$, $A^{\#}(p)$ is the velocity vector of $t \to p \cdot \exp(tA)$ at $t = 0$, $A^{\#}(p)$ is a vertical vector at p for each $A \in \mathcal{G}$. Thus, if we fix $p \in P$, the assignment $A \to A^{\#}(p)$ is a linear mapping of $\mathcal{G}$ to $\mathrm{Vert}_p(P)$. Since the action σ of G on P is free (Exercise 3.1.1), $A \neq 0$ implies $A^{\#}(p) \neq 0$, i.e., $A \to A^{\#}(p)$ is one-to-one. Finally, since $\dim \mathcal{G} = \dim \mathrm{Vert}_p(P)$, this map must, in fact, be an isomorphism and we have completed the proof of the following.

Corollary 4.7.9 *Let $\mathcal{B} = (P, \mathcal{P}, \sigma)$ be a smooth principal G-bundle over X. For each $p \in P$ the mapping $A \to A^{\#}(p)$ is an isomorphism of the Lie algebra $\mathcal{G}$ of G onto the vertical subspace* $\mathrm{Vert}_p(P)$ *of $T_p(P)$.*

For each $g \in G$ the map $\sigma_g : P \to P$ given by $\sigma_g(p) = p \cdot g$ is a diffeomorphism of P onto itself. Thus, for any $A \in \mathcal{G}$, $(\sigma_g)_*(A^{\#})$ is a smooth vector field on P (Exercise 4.6.12). We claim that it is, in fact, the fundamental vector field determined by $ad_{g^{-1}}(A) \in \mathcal{G}$, i.e.,

$$(\sigma_g)_*(A^{\#}) = \left(ad_{g^{-1}}(A)\right)^{\#}. \tag{4.7.9}$$

Since $\sigma_g(p \cdot g^{-1}) = p$, we must prove that $(\sigma_g)_{*p\cdot g^{-1}}(A^{\#}(p \cdot g^{-1})) = (ad_{g^{-1}}(A))^{\#}(p)$. Now, $A^{\#}(p \cdot g^{-1}) = \beta'(0)$, where $\beta(t) = (p \cdot g^{-1}) \cdot \exp(tA) = p \cdot (g^{-1}\exp(tA))$ so $(\sigma_g)_{*p\cdot g^{-1}}(A^{\#}(p \cdot g^{-1})) = (\sigma_g)_{*p\cdot g^{-1}}(\beta'(0)) = (\sigma_g \circ \beta)'(0)$. But $(\sigma_g \circ \beta)(t) = \sigma_g(\beta(t)) = \beta(t) \cdot g = (p \cdot (g^{-1}\exp(tA))) \cdot g = p \cdot (g^{-1}\exp(tA)g) = p \cdot (\exp(g^{-1}(tA)g)) = p \cdot (\exp(t(g^{-1}Ag))) = p \cdot (\exp(t\, ad_{g^{-1}}(A)))$ so $(\sigma_g \circ \beta)'(0)$ is, by definition, $(ad_{g^{-1}}(A))^{\#}(p)$.

4.8 Vector-Valued 1-Forms

In Sections 0.4 and 0.5 we suggested that the 1-forms of real interest in gauge theory are not the real-valued variety that we have considered thus far, but those that take values in the Lie algebra $\mathcal{G}$ of some Lie group G. Here we define, more generally, 1-forms with values in an arbitrary vector space and compute some important examples.

Let $\mathcal{V}$ be a d-dimensional real vector space and $\mathcal{V}^*$ its dual space. If X is a differentiable manifold, then a **$\mathcal{V}$-valued 1-form on X** is a map ω on X that assigns to every $p \in X$ a linear transformation $\omega(p) = \omega_p$ from $T_p(X)$ to $\mathcal{V}$. Thus, an ordinary 1-form on X as defined in Section 4.6 is just an $\mathbb{R}$-valued 1-form on X. If $\{e_1, \ldots, e_d\}$ is a basis for $\mathcal{V}$, then, for any $\mathbf{v} \in T_p(X)$, we may write $\omega(p)(\mathbf{v}) = \omega_p(\mathbf{v}) = \omega_p^1(\mathbf{v})\, e_1 + \cdots + \omega_p^d(\mathbf{v})\, e_d = \omega_p^i(\mathbf{v})\, e_i$. The ω_p^i are real-valued linear maps on $T_p(X)$ so, defining ω^i on X by $\omega^i(p) = \omega_p^i$ we find that each ω^i, $i = 1, \ldots, n$, is an $\mathbb{R}$-valued 1-form on X. These ω^i are called the **components** of the $\mathcal{V}$-valued 1-form ω with respect to the basis $\{e_1, \ldots, e_d\}$ for $\mathcal{V}$. We will say that ω is **smooth** if each ω^i is smooth.

Exercise 4.8.1 Show that this definition does not depend on the choice of basis in $\mathcal{V}$ and, in fact, is equivalent to the requirement that, for every $\lambda \in \mathcal{V}^*$, the $\mathbb{R}$-valued 1-form $\lambda\omega$ defined on X by $(\lambda\omega)(p) = \lambda \circ \omega_p$ is smooth.

Conversely, given a family $\omega^1, \ldots, \omega^d$ of smooth $\mathbb{R}$-valued 1-forms on X and a basis $\{e_1, \ldots, e_d\}$ for $\mathcal{V}$ one can build a smooth $\mathcal{V}$-valued 1-form $\omega = \omega^1 e_1 + \cdots + \omega^d e_d = \omega^i e_i$ on X defined by $\omega(p)(\mathbf{v}) = \omega^1(p)(\mathbf{v})e_1 + \cdots + \omega^d(p)(\mathbf{v})e_d$ for each $p \in X$ and $\mathbf{v} \in T_p(X)$.

Exercise 4.8.2 If $f : X \to Y$ is a smooth map and ω is a $\mathcal{V}$-valued 1-form on Y, the **pullback of ω by f** is the $\mathcal{V}$-valued 1-form $f^*\omega$ on X defined by $(f^*\omega)_p(\mathbf{v}) = \omega_{f(p)}(f_{*p}(\mathbf{v}))$ for all $p \in X$ and $\mathbf{v} \in T_p(X)$. Show that if $\omega = \omega^i e_i$, then $f^*\omega = (f^*\omega^i)\, e_i$ so that $f^*\omega$ is smooth if ω is smooth.

Common choices for $\mathcal{V}$ in our work are $\mathbb{C}$, $\operatorname{Im}\mathbb{C}$, $\mathbb{H}$ and $\operatorname{Im}\mathbb{H}$ and, for each of these, we will invariably use the natural basis, i.e., $\{1, \mathbf{i}\}$, $\{\mathbf{i}\}$, $\{1, \mathbf{i}, \mathbf{j}, \mathbf{k}\}$ and $\{\mathbf{i}, \mathbf{j}, \mathbf{k}\}$, respectively. For example, if $\mathcal{V} = \mathbb{C}$ and $X = \mathbb{R}^2 = \mathbb{C}$ (with standard coordinate functions x and y), we define two $\mathbb{C}$-valued 1-forms dz and $d\bar{z}$ on $\mathbb{R}^2 = \mathbb{C}$ by $dz = dx + dy\,\mathbf{i}$ and $d\bar{z} = dx - dy\,\mathbf{i}$. At each $p \in \mathbb{C}$ the canonical isomorphism (Exercise 4.4.9) identifies $v \in \mathbb{C}$ with $\mathbf{v}_p = \frac{d}{dt}(p + tv)|_{t=0} \in T_p(\mathbb{C})$ so $dz_p(\mathbf{v}_p) = v$ and $d\bar{z}_p(\mathbf{v}_p) = \bar{v}$. Similarly, one defines $\mathbb{H}$-valued 1-forms dq and $d\bar{q}$ on $\mathbb{R}^4 = \mathbb{H}$ by $dq = dx + dy\,\mathbf{i} + du\,\mathbf{j} + dv\,\mathbf{k}$ and $d\bar{q} = dx - dy\,\mathbf{i} - du\,\mathbf{j} - dv\,\mathbf{k}$ so that $dq_p(\mathbf{v}_p) = v$ and $d\bar{q}_p(\mathbf{v}_p) = \bar{v}$ for all $v \in \mathbb{H}$.

The exterior derivative df of a 0-form (smooth, real-valued function) f on X is a 1-form on X. Similarly, we define a **$\mathcal{V}$-valued 0-form** on X to be a smooth map ϕ from X into $\mathcal{V}$. The **exterior derivative** $d\phi$ of ϕ, computed componentwise relative to any basis for $\mathcal{V}$, is then a well-defined $\mathcal{V}$-valued 1-form on X. We will find, in Section 5.8, that when $\mathcal{V} = \mathbb{C}^k$ (regarded as a

$2k$-dimensional real vector space) such vector-valued 0-forms represent the "matter fields" that respond to (i.e., are coupled to) external gauge fields.

The case of most interest to us arises in the following way. We let $X = G$ be a matrix Lie group and $\mathcal{V} = \mathcal{G}$ its Lie algebra (which we now regard as $T_{id}(G)$). The **Cartan (canonical) 1-form** on G is the $\mathcal{G}$-valued 1-form Θ on G defined as follows: For each $g \in G$, $\Theta(g) = \Theta_g : T_g(G) \to \mathcal{G}$ is given by

$$\Theta(g)(\mathbf{v}) = \Theta_g(\mathbf{v}) = (L_{g^{-1}})_{*g}(\mathbf{v}) \qquad (4.8.1)$$

(push $\mathbf{v}$ back to $id \in G$ by left translation). Equivalently,

$$\Theta(g)(\mathbf{A}(g)) = \mathbf{A}(id) \qquad (4.8.2)$$

for every left invariant vector field $\mathbf{A}$ on G.

Lemma 4.8.1 *Suppose G is a matrix Lie group and $\mathcal{G}$ is its Lie algebra. Let $\{e_1, \ldots, e_n\}$ be a basis for $\mathcal{G}$ and let $\{\Theta^1, \ldots, \Theta^n\}$ be the unique left invariant $\mathbb{R}$-valued 1-forms on G for which $\{\Theta^1(id), \ldots, \Theta^n(id)\}$ is the dual basis to $\{e_1, \ldots, e_n\}$ (i.e., $\Theta^i(id)(e_j) = \delta^i_j$ for $i, j = 1, \ldots, n$). Then the Cartan 1-form Θ on G is given by $\Theta = \Theta^i e_i = \Theta^1 e_1 + \cdots + \Theta^n e_n$.*

Proof: For any $g \in G$ and any $\mathbf{v} \in T_g(G)$, $((\Theta^i e_i)(g))(\mathbf{v}) = (\Theta^i(g)(\mathbf{v}))e_i = (((L_{g^{-1}})^*(\Theta^i(id)))(\mathbf{v}))e_i = \Theta^i(id)((L_{g^{-1}})_{*g}(\mathbf{v}))e_i = (L_{g^{-1}})_{*g}(\mathbf{v}) = \Theta_g(\mathbf{v})$. ∎

Exercise 4.8.3 Show that the Cartan 1-form Θ on G is left invariant, i.e., that $(L_g)^* \Theta = \Theta$ for each $g \in G$.

A $\mathcal{G}$-valued 1-form ω on G is said to be **right equivariant** if $(R_g)^* \omega = ad_{g^{-1}} \circ \omega$ for each $g \in G$. In more detail, right equivariance requires that for all $g, h \in G$ and all $\mathbf{v} \in T_{hg^{-1}}(G)$,

$$\omega_h\left((R_g)_{*hg^{-1}}(\mathbf{v})\right) = ad_{g^{-1}}\left(\omega_{hg^{-1}}(\mathbf{v})\right), \qquad (4.8.3)$$

where $ad_{g^{-1}} = (L_{g^{-1}})_{*g} \circ (R_g)_{*id}$ (Section 4.7). We show that the Cartan 1-form Θ is right equivariant as follows:

$$\begin{aligned}
\Theta_h\left((R_g)_{*hg^{-1}}(\mathbf{v})\right) &= (L_{h^{-1}})_{*h}\left((R_g)_{*hg^{-1}}(\mathbf{v})\right) \\
&= (L_{g^{-1}} \circ L_{gh^{-1}})_{*h}\left((R_g)_{*hg^{-1}}(\mathbf{v})\right) \\
&= (L_{g^{-1}})_{*g} \circ (L_{gh^{-1}})_{*h} \circ (R_g)_{*hg^{-1}}(\mathbf{v}) \\
&= (L_{g^{-1}})_{*g}\left((L_{gh^{-1}} \circ R_g)_{*hg^{-1}}(\mathbf{v})\right) \\
&= (L_{g^{-1}})_{*g}\left((R_g \circ L_{gh^{-1}})_{*hg^{-1}}(\mathbf{v})\right) \\
&= \left((L_{g^{-1}})_{*g} \circ (R_g)_{*id}\right)\left((L_{gh^{-1}})_{*hg^{-1}}(\mathbf{v})\right) \\
&= ad_{g^{-1}}\left(\Theta_{hg^{-1}}(\mathbf{v})\right).
\end{aligned}$$

The notion of right equivariance extends to $\mathcal{G}$-valued 1-forms on any manifold on which G acts. Specifically, if σ is a smooth right action of G on P, $\sigma(p,g) = p \cdot g$, and $\boldsymbol{\omega}$ is a $\mathcal{G}$-valued 1-form on P, then $\boldsymbol{\omega}$ is said to be **right equivariant under** σ if $(\sigma_g)^*\boldsymbol{\omega} = ad_{g^{-1}} \circ \boldsymbol{\omega}$ for each g in G.

Next we explicitly compute the Cartan 1-form $\boldsymbol{\Theta}$ for the real general linear group. $GL(n,\mathbb{R})$ is an open submanifold of $\mathbb{R}^{n^2}$ and we denote by x^{ij}, $i,j = 1,\ldots,n$, the standard coordinate (entry) functions on $\mathbb{R}^{n^2}$. Thus, for each $g \in GL(n,\mathbb{R})$, $x^{ij}(g) = g^{ij}$ is the ij-entry in g. Any $A \in \mathcal{GL}(n,\mathbb{R})$ is an $n \times n$ real matrix (A^{ij}) and gives rise to a unique left invariant vector field $\mathbf{A}$ on $GL(n,\mathbb{R})$ satisfying $\mathbf{A}(id) = A^{ij} \frac{\partial}{\partial x^{ij}}|_{id}$. By (4.7.3),

$$\mathbf{A}(g) = \left(\sum_{k=1}^{n} g^{ik} A^{kj} \right) \frac{\partial}{\partial x^{ij}}\bigg|_g = \left(\sum_{k=1}^{n} x^{ik}(g) A^{kj} \right) \frac{\partial}{\partial x^{ij}}\bigg|_g ,$$

which one often abbreviates as $\mathbf{A}(g) = gA$ (see (4.7.4)).

Now we construct $\boldsymbol{\Theta}$ using Lemma 4.8.1 and the basis $\{\frac{\partial}{\partial x^{ij}}|_{id}\}_{i,j=1,\ldots,n}$ for $\mathcal{GL}(n,\mathbb{R})$. The corresponding dual basis is $\{dx^{ij}(id)\}_{i,j=1,\ldots,n}$ so we must find left invariant $\mathbb{R}$-valued 1-forms $\{\boldsymbol{\Theta}^{ij}\}_{i,j=1,\ldots,n}$ such that $\boldsymbol{\Theta}^{ij}(id) = dx^{ij}(id)$ for all i and j (unfortunately, the dx^{ij} themselves are not left invariant, as we shall see). Left invariance requires that $\boldsymbol{\Theta}^{ij}(g) = (L_{g^{-1}})^*(\boldsymbol{\Theta}^{ij}(id))$ so, for each $\mathbf{v} \in T_g(GL(n,\mathbb{R}))$, $\boldsymbol{\Theta}^{ij}(g)(\mathbf{v}) = \boldsymbol{\Theta}^{ij}(id)((L_{g^{-1}})_{*g}(\mathbf{v})) = dx^{ij}(id)((L_{g^{-1}})_{*g}(\mathbf{v}))$. Let $\alpha: (-\varepsilon,\varepsilon) \to GL(n,\mathbb{R})$, $\alpha(t) = (\alpha^{ij}(t))$, be a smooth curve in $GL(n,\mathbb{R})$ with $\alpha'(0) = \mathbf{v}$. Then $\boldsymbol{\Theta}^{ij}(g)(\mathbf{v}) = dx^{ij}(id)((L_{g^{-1}})_{*g}(\alpha'(0))) = dx^{ij}(id)((L_{g^{-1}} \circ \alpha)'(0)) = dx^{ij}(id)(\frac{d}{dt}(g^{-1}\alpha(t))|_{t=0})$. Now, $g^{-1}\alpha(t) = (\sum_{k=1}^{n}(g^{-1})^{ik}\alpha^{kj}(t))$ so

$$\frac{d}{dt}(g^{-1}\alpha(t))\big|_{t=0} = \left(\frac{d}{dt} \left(\sum_{k=1}^{n} (g^{-1})^{ik} \alpha^{kj}(t) \right)\bigg|_{t=0} \right) \frac{\partial}{\partial x^{ij}}\bigg|_{id}$$

$$= \left(\sum_{k=1}^{n} (g^{-1})^{ik} v^{kj} \right) \frac{\partial}{\partial x^{ij}}\bigg|_{id} ,$$

where $v = v^{kj} \frac{\partial}{\partial x^{kj}}|_g$. Thus,

$$\boldsymbol{\Theta}^{ij}(g)(\mathbf{v}) = \sum_{k=1}^{n} (g^{-1})^{ik} v^{kj} = \sum_{k=1}^{n} x^{ik}(g^{-1}) \left(dx^{kj}(g) \right)(\mathbf{v})$$

and so

$$\boldsymbol{\Theta}^{ij}(g) = \sum_{k=1}^{n} x^{ik}(g^{-1})\, dx^{kj}(g) .$$

Lemma 4.8.1 thus gives

$$\Theta(g) = \Theta^{ij}(g)\,\frac{\partial}{\partial x^{ij}}\bigg|_{id} = \left(\sum_{k=1}^{n} x^{ik}\,(g^{-1})\,dx^{kj}\,(g)\right)\frac{\partial}{\partial x^{ij}}\bigg|_{id}. \tag{4.8.4}$$

Remark: The matrix of coefficients in (4.8.4) is a matrix of $\mathbb{R}$-valued 1-forms on $GL\,(n,\mathbb{R})$ that is often conveniently identified with $\Theta(g)$. Note that it is the formal matrix product of g^{-1} and the matrix of global coordinate differentials $dx^{ij}(g)$. For this reason one might abbreviate (4.8.4) as

$$\Theta(g) = g^{-1}\,dx\,(g)\,, \tag{4.8.5}$$

where $dx(g)$ is the $n \times n$ matrix $(\,dx^{ij}(g)\,)$. As a practical matter, one computes $\Theta(g)$ as the matrix product of g^{-1} and

$$\begin{pmatrix} dx^{11} & \cdots & dx^{1n} \\ \vdots & & \vdots \\ dx^{n1} & \cdots & dx^{nn} \end{pmatrix}.$$

The result is a matrix of $\mathbb{R}$-valued 1-forms, each of which is evaluated at any given $\mathbf{v} \in T_g\,(\,GL\,(n,\mathbb{R}\,)\,)$ to yield a matrix in $\mathcal{GL}\,(n,\mathbb{R}\,)$.

In order to obtain the Cartan 1-forms for $O(n)$ and $SO(n)$ we formulate a general result on subgroups. Thus, suppose G is a matrix Lie group with Lie algebra $\mathcal{G}$, H is a subgroup of G that is also a submanifold and $\iota : H \hookrightarrow G$ is the inclusion. Then ι is an embedding and we identify the Lie algebra $\mathcal{H}$ of H with the subspace $\iota_{*id}(\,T_{id}\,(H)\,)$ of $\mathcal{G}$. Let Θ_H and Θ_G be the Cartan 1-forms on H and G, respectively. Then we claim that

$$\Theta_H \;=\; \iota^*\,\Theta_G\,. \tag{4.8.6}$$

To see this we fix an $h \in H$. Then, for every $\mathbf{v} \in T_h\,(H)$, $(\,(\iota^*\Theta_G)\,(h)\,)\,(\mathbf{v}) = (\,\Theta_G\,(\,\iota\,(h)\,)\,)\,(\,\iota_{*h}\,(\mathbf{v})\,) = (\,\Theta_G\,(h)\,)\,(\,\iota_{*h}\,(\mathbf{v})\,)$ so we must show that this is equal to $(\,\Theta_H\,(h)\,)\,(\mathbf{v})$. Let $L_{h^{-1}} : H \rightarrow H$ and $\hat{L}_{h^{-1}} : G \rightarrow G$ be the left translation maps on H and G. Then $L_{h^{-1}} = \hat{L}_{h^{-1}} \circ \iota$ so $(\,\Theta_H\,(h)\,)\,(\mathbf{v}) = (L_{h^{-1}})_{*h}\,(\mathbf{v}) = (\hat{L}_{h^{-1}} \circ \iota)_{*h}\,(\mathbf{v}) = (\hat{L}_{h^{-1}})_{*\iota(h)}(\,\iota_{*h}\,(\mathbf{v})\,) = (\hat{L}_{h^{-1}})_{*h}\,(\,\iota_{*h}\,(\mathbf{v})\,) = (\,\Theta_G\,(h)\,)\,(\,\iota_{*h}\,(\mathbf{v})\,)$ as required.

Suppose, for example, that $G = GL(2,\mathbb{R}\,)$, $H = SO(2)$ and $\iota : SO(2) \hookrightarrow GL(2,\mathbb{R}\,)$ is the inclusion. For each $g = \begin{pmatrix} a & b \\ c & d \end{pmatrix} \in GL\,(2,\mathbb{R}\,)$, $\Theta_G(g)$ is given by

$$\boldsymbol{\Theta}_G(g) = g^{-1}\,dx = \frac{1}{ad-bc}\begin{pmatrix} d & -b \\ -c & a \end{pmatrix}\begin{pmatrix} dx^{11} & dx^{12} \\ dx^{21} & dx^{22} \end{pmatrix}$$

$$= \frac{1}{ad-bc}\begin{pmatrix} d\,dx^{11} - b\,dx^{21} & d\,dx^{12} - b\,dx^{22} \\ -c\,dx^{11} + a\,dx^{21} & -c\,dx^{12} + a\,dx^{22} \end{pmatrix}.$$

Now, if $g \in SO(2)$, then $d = a$, $c = -b$ and $a^2 + b^2 = 1$ so this reduces to

$$\begin{pmatrix} a\,dx^{11} - b\,dx^{21} & a\,dx^{12} - b\,dx^{22} \\ b\,dx^{11} + a\,dx^{21} & b\,dx^{12} + a\,dx^{22} \end{pmatrix}.$$

Furthermore, (4.8.6) gives $\boldsymbol{\Theta}_H(g) = \boldsymbol{\Theta}_G(g) \circ \iota_{*g}$ and ι_{*g} is just the inclusion of $T_g(SO(2))$ in $T_g(GL(2,\mathbb{R}))$. There is more to be said, however, because some of the 1-forms in this last matrix simplify considerably when restricted to $T_g(SO(2))$. To see this note first that any element of $T_g(SO(2))$ can be written as $\alpha'(0)$, where α is a smooth curve in $SO(2)$ with $\alpha(0) = g$. Now, $\iota_{*g}(\alpha'(0)) = (\iota \circ \alpha)'(0)$ and $\iota \circ \alpha$ is a curve in $GL(2,\mathbb{R})$ which we may write in terms of the coordinates x^{ij} as

$$(\iota \circ \alpha)(t) = \begin{pmatrix} x^{11}((\iota\circ\alpha)(t)) & x^{12}((\iota\circ\alpha)(t)) \\ x^{21}((\iota\circ\alpha)(t)) & x^{22}((\iota\circ\alpha)(t)) \end{pmatrix} = \begin{pmatrix} a(t) & b(t) \\ -b(t) & a(t) \end{pmatrix}.$$

But $(a(t))^2 + (b(t))^2 = 1$ for all t so differentiation at $t = 0$ gives $a(0)\,a'(0) + b(0)\,b'(0) = 0$, i.e., $aa'(0) + bb'(0) = 0$. Observe that $a'(0) = dx^{11}(\iota_{*g}(\alpha'(0))) = dx^{22}(\iota_{*g}(\alpha'(0)))$ and $b'(0) = dx^{12}(\iota_{*g}(\alpha'(0))) = -\,dx^{21}(\iota_{*g}(\alpha'(0)))$. Thus, $a\,dx^{11} - b\,dx^{21}$ and $b\,dx^{12} + a\,dx^{22}$ both vanish at $\iota_{*g}(\alpha'(0))$ and therefore on all of $T_g(SO(2))$. Furthermore, $b\,dx^{11}(\iota_{*g}(\alpha'(0))) + a\,dx^{21}(\iota_{*g}(\alpha'(0))) = b\,dx^{22}(\iota_{*g}(\alpha'(0))) - a\,dx^{12}(\iota_{*g}(\alpha'(0)))$. We conclude that the Cartan 1-form for $SO(2)$ is, at each $g = \begin{pmatrix} a & b \\ -b & a \end{pmatrix} \in SO(2)$, the restriction to $T_g(SO(2))$ of

$$\begin{pmatrix} 0 & a\,dx^{12} - b\,dx^{22} \\ -a\,dx^{12} + b\,dx^{22} & 0 \end{pmatrix}.$$

Finally, noting that $a = x^{22}(g)$ and $b = x^{12}(g)$ we arrive at the matrix of $\mathbb{R}$-valued 1-forms representing the Cartan 1-form for $SO(2)$:

$$\begin{pmatrix} 0 & x^{22}\,dx^{12} - x^{12}\,dx^{22} \\ -\,x^{22}\,dx^{12} + x^{12}\,dx^{22} & 0 \end{pmatrix}.$$

The calculations for $GL(n,\mathbb{C})$ are virtually identical to, albeit twice as long as, those for $GL(n,\mathbb{R})$ and we feel comfortable leaving them in the hands of the reader.

Exercise 4.8.4 Let $\{x^{11}, y^{11}, \ldots, x^{nn}, y^{nn}\}$ be the standard coordinate functions on $GL(n,\mathbb{C}) \subseteq \mathbb{R}^{2n^2}$ and take $\{\frac{\partial}{\partial x^{11}}|_{id}, \frac{\partial}{\partial y^{11}}|_{id}, \ldots, \frac{\partial}{\partial x^{nn}}|_{id}, \frac{\partial}{\partial y^{nn}}|_{id}\}$ as a basis for $\mathcal{GL}(n,\mathbb{C})$. Show that the $\mathbb{R}$-valued 1-forms $\{\boldsymbol{\omega}^{11}, \boldsymbol{\eta}^{11}, \ldots, \boldsymbol{\omega}^{nn}, \boldsymbol{\eta}^{nn}\}$ given by

$$\boldsymbol{\omega}^{ij}(g) = \sum_{k=1}^{n} \left(x^{ik}(g^{-1})\, dx^{kj}(g) - y^{ik}(g^{-1})\, dy^{kj}(g) \right)$$

and

$$\boldsymbol{\eta}^{ij}(g) = \sum_{k=1}^{n} \left(x^{ik}(g^{-1})\, dy^{kj}(g) - y^{ik}(g^{-1})\, dx^{kj}(g) \right)$$

for each $g \in GL(n,\mathbb{C})$ are left invariant and satisfy $\boldsymbol{\omega}^{ij}(id) = dx^{ij}(id)$ and $\boldsymbol{\eta}^{ij}(id) = dy^{ij}(id)$. Conclude that the Cartan 1-form $\boldsymbol{\Theta}$ for $GL(n,\mathbb{C})$ is given by

$$\boldsymbol{\Theta}(g) = \boldsymbol{\omega}^{ij}(g)\, \frac{\partial}{\partial x^{ij}}\bigg|_{id} + \boldsymbol{\eta}^{ij}(g)\, \frac{\partial}{\partial y^{ij}}\bigg|_{id}.$$

Identifying $\mathcal{GL}(n,\mathbb{C})$ with the $n \times n$ complex matrices as in Exercise 4.7.7 one then identifies $\boldsymbol{\Theta}(g)$ with the matrix of complex-valued 1-forms $(\boldsymbol{\omega}^{ij}(g) + \boldsymbol{\eta}^{ij}(g)\,\mathbf{i}\,)_{i,j=1,\ldots,n}$. Show that $\boldsymbol{\omega}^{ij}(g)$ and $\boldsymbol{\eta}^{ij}(g)$ are, respectively, the real and imaginary parts of the ij-entry in the formal matrix product of g^{-1} and $(dx^{ij}(g) + dy^{ij}(g)\,\mathbf{i}\,)_{i,j=1,\ldots,n}$. Conclude that, just as in the real case, one can write $\boldsymbol{\Theta}(g) = g^{-1}dz(g)$, where

$$dz = \begin{pmatrix} dz^{11} & \cdots & dz^{1n} \\ \vdots & & \vdots \\ dz^{n1} & \cdots & dz^{nn} \end{pmatrix},$$

with $dz^{ij} = dx^{ij} + dy^{ij}\,\mathbf{i}$.

Subgroups of $GL(n,\mathbb{C})$ such as $U(n)$ and $SU(n)$ are again handled by (4.8.6), i.e., by simple restriction. We ask the reader to mimic our earlier discussion of $SO(2)$ to arrive at the Cartan 1-form for $SU(2)$.

Exercise 4.8.5 Show that the Cartan 1-form for $SU(2)$ is, for each $g = \begin{pmatrix} \alpha & \beta \\ -\bar{\beta} & \bar{\alpha} \end{pmatrix} \in SU(2)$, the restriction of

$$\begin{pmatrix} \bar{\alpha}\,dz^{11} - \beta\,dz^{21} & \bar{\alpha}\,dz^{12} - \beta\,dz^{22} \\ \bar{\beta}\,dz^{11} + \alpha\,dz^{21} & \bar{\beta}\,dz^{12} + \alpha\,dz^{22} \end{pmatrix}$$

to $T_g\,(\,SU(2)\,)$. Show directly that, at any vector in $T_g\,(\,SU(2)\,)$, the 11-entry $\bar{\alpha}\,dz^{11} - \beta\,dz^{21}$ is pure imaginary.

Before writing out a few more examples we pause momentarily to have a closer look at the 11-entry in the Cartan 1-form for $SU(2)$ (Exercise 4.8.5). Observe that, since $z^{11}(g) = \alpha \;\; z^{21}(g) = -\bar{\beta}$ we may write this complex-valued 1-form as

$$\begin{aligned} \bar{z}^{11}\,dz^{11} + \bar{z}^{21}\,dz^{21} &= (\,x^{11} - y^{11}\,\mathbf{i}\,)\,(\,dx^{11} + dy^{11}\,\mathbf{i}\,) + \\ &\qquad (\,x^{21} - y^{21}\,\mathbf{i}\,)\,(\,dx^{21} + dy^{21}\,\mathbf{i}\,) \\ &= (\,x^{11}\,dx^{11} + y^{11}\,dy^{11} + x^{21}\,dx^{21} + y^{21}\,dy^{21}\,) + \\ &\qquad (\,-y^{11}\,dx^{11} + x^{11}\,dy^{11} - y^{21}\,dx^{21} + x^{21}\,dy^{21}\,)\,\mathbf{i}\,. \end{aligned}$$

The reader has already shown in Exercise 4.8.5 that the real part of this is zero on $SU(2)$ so that

$$\bar{z}^{11}\,dz^{11} + \bar{z}^{21}\,dz^{21} = (-y^{11}\,dx^{11} + x^{11}\,dy^{11} - y^{21}\,dx^{21} + x^{21}\,dy^{21}\,)\,\mathbf{i}$$

(on $SU(2)$). The point here is that we have seen $\mathrm{Im}\,(\bar{z}^{11}\,dz^{11} + \bar{z}^{21}\,dz^{21})$ before. Letting $(x^{11}, y^{11}, x^{21}, y^{21}) = (x^1, x^2, x^3, x^4)$ it is just $-x^2\,dx^1 + x^1\,dx^2 - x^4\,dx^3 + x^3\,dx^4$ and this is precisely the 1-form on $\mathbb{R}^4$ whose restriction to S^3 we found (in Section 4.6) pulled back under the canonical cross-sections of the Hopf bundle $S^1 \to S^3 \to S^2$ to the potential 1-forms for a Dirac monopole. We will have more to say about this apparently miraculous coincidence.

Since we have the symplectic group $Sp\,(n)$ identified with a submanifold of $GL(2n, \mathbb{C}\,)$ we could find its Cartan 1-form by restricting that of the complex general linear group. It will be more convenient, however, to work directly with the quaternions.

Exercise 4.8.6 Show that the Cartan 1-form for $GL(n, \mathbb{H}\,)$ can be identified, at each $g \in GL(n, \mathbb{H}\,)$ with the formal matrix product $g^{-1}dq$ with dq given by

$$dq = \begin{pmatrix} dq^{11} & \cdots & dq^{1n} \\ \vdots & & \vdots \\ dq^{n1} & \cdots & dq^{nn} \end{pmatrix},$$

where $dq^{ij} = dx^{ij} + dy^{ij}\mathbf{i} + du^{ij}\mathbf{j} + dv^{ij}\mathbf{k}$ for $i, j = 1, \ldots, n$.

The Cartan 1-form for $Sp(n)$ is now the restriction of $g^{-1}dq$ to $Sp(n)$. Even the $n=1$ case is of some interest. Here, of course, we identify a 1×1 quaternionic matrix $g=(q)$ with its sole entry q so that $GL(1,\mathbb{H})$ is just the multiplicative group $\mathbb{H}-\{0\}$ and $Sp(1)$ is the group of unit quaternions. Moreover, $g^{-1}dq$ is just the quaternion product $q^{-1}dq=\frac{1}{|q|^2}\,\bar{q}dq$. On $Sp(1)$, $|q|=1$ so this reduces to $\bar{q}dq$. Writing out the product of $\bar{q}=x-y\mathbf{i}-u\mathbf{j}-v\mathbf{k}$ and $dq=dx+dy\mathbf{i}+du\mathbf{j}+dv\mathbf{k}$ one finds that the real part is $x\,dx+y\,dy+u\,du+v\,dv$. But, on $Sp(1)$, $x^2+y^2+u^2+v^2=1$ and this implies, just as for $SO(2)$ and $SU(2)$, that $x\,dx+y\,dy+u\,du+v\,dv=0$ on $Sp(1)$. Thus, the Cartan 1-form for $Sp(1)$ is the restriction of $\mathrm{Im}\,(q^{-1}dq)$ to $Sp(1)$.

Exercise 4.8.7 Calculate the Cartan 1-form for $Sp(2)$ and show that its 11-entry is the restriction to $Sp(2)$ of

$$\mathrm{Im}\,\left(\bar{q}^{11}\,dq^{11}+\bar{q}^{21}\,dq^{21}\right). \tag{4.8.7}$$

Spending one's life hoping for miraculous coincidences is generally a losing proposition. However, just this once, motivated by our discovery of a Dirac monopole lying underneath the 11-entry of the Cartan 1-form for $SU(2)$, we will throw caution to the wind and conduct a search beneath (4.8.7). The omens seem propitious. The 11-entry in the Cartan 1-form for $SU(2)$ is the restriction to $SU(2)$ of $\mathbf{i}$ times $\mathrm{Im}\,(\bar{z}^{11}dz^{11}+\bar{z}^{21}dz^{21})$, while that of $Sp(2)$ is the restriction of $\mathrm{Im}\,(\bar{q}^{11}dq^{11}+\bar{q}^{21}dq^{21})$ to $Sp(2)$. Introducing real coordinates $(x^1,x^2,x^3,x^4)=(x^{11},y^{11},x^{21},y^{21})$ in $\mathbb{C}^2$ gives a 1-form on $\mathbb{R}^4$ that can be restricted to S^3 and so pulled back to S^2 via the canonical cross-sections of the Hopf bundle $S^1\to S^3\to S^2$. This is the monopole. Similarly, introducing real coordinates $(x^1,\ldots,x^8)=(x^{11},\ldots,v^{21})$ in $\mathbb{H}^2$ gives a 1-form on $\mathbb{R}^8$ that can be restricted to S^7. Furthermore, we have at hand a Hopf bundle $S^3\to S^7\to S^4$ which suggests the possibility of getting something interesting by pulling this form back to S^4 by canonical cross-sections. At any rate, it's worth a try.

The program just outlined, however, is much more conveniently carried out entirely in terms of quaternions. Thus, we will identify S^4 with $\mathbb{HP}^1$ (Exercise 4.3.10) and the Hopf map with the projection $S^7\to\mathbb{HP}^1$ (Section 1.3). S^7 is, of course, just the subset $S^7=\{(q^1,q^2)\in\mathbb{H}^2 : |q^1|^2+|q^2|^2=1\}$ of $\mathbb{H}^2$. To see the relationship to $Sp(2)$ and its Cartan 1-form more clearly, however, we recall that S^7 is also homeomorphic to the quotient space $Sp(2)/Sp(1)$ (see (1.6.11)). Indeed, in (1.6.12), we have written out explicitly a homeomorphism $\varphi : Sp(2)/Sp(1)\to S^7$:

$$\varphi(\,[g]\,) = (\alpha, \gamma), \quad \text{where} \quad g = \begin{pmatrix} \alpha & \beta \\ \gamma & \delta \end{pmatrix} \in Sp\,(2)\,.$$

Supplying $Sp\,(2)/Sp\,(1)$ with the differentiable structure that makes φ a diffeomorphism we may identify S^7 and $Sp\,(2)/Sp\,(1)$ as manifolds also.

From our point-of-view the important observation here is that the 11-entry $\mathrm{Im}\,(\bar{q}^{11}dq^{11} + \bar{q}^{21}dq^{21})$ of the Cartan 1-form for $Sp\,(2)$ involves only the 11- and 21-entries in $Sp\,(2)$, i.e., α and γ, and so takes the same value at each point of the coset $[g] = g\,Sp\,(1) = \{\,(\begin{smallmatrix} \alpha & \beta a \\ \gamma & \delta a \end{smallmatrix}) : a \in \mathbb{H}\,,\ |\,a\,| = 1\,\}$ (see Section 1.6). It therefore naturally "descends" to a 1-form on the quotient $Sp\,(2)/Sp\,(1)$ and so, via φ, to S^7. This apparently fortuitous circumstance is actually a particular instance of a general phenomenon first pointed out in [**Trau**] and [**NT**]. The reader may also wish to consult Theorem 11.1, Chapter II, of [**KN 1**] for a still more general perspective. In any case, we now look more closely at the indicated 1-form on S^7 and its relationship to the structure of the quaternionic Hopf bundle.

Before embarking on this, let's get all of the required notation on the table. $S^7 = \{\,(q^1, q^2) \in \mathbb{H}^2 : |\,q^1|^2 + |\,q^2|^2 = 1\,\}$ and $\iota : S^7 \hookrightarrow \mathbb{H}^2$ is the inclusion map. The usual right action of $Sp\,(1)$ on S^7 is $\sigma(\,(q^1, q^2), g\,) = (q^1, q^2) \cdot g = (q^1 g, q^2 g)$. We identify the Lie algebra $\mathcal{SP}\,(1)$ of $Sp\,(1)$ with $\mathrm{Im}\,\mathbb{H}$ and the quaternionic Hopf bundle with $Sp\,(1) \to S^7 \xrightarrow{\mathcal{P}} \mathbb{HP}^1$, where $\mathcal{P}(q^1, q^2) = [q^1, q^2] \in \mathbb{HP}^1 \cong S^4$. The standard trivializations (V_k, Ψ_k), $k = 1, 2$, for this bundle are given as follows: $V_k = \{x = [q^1, q^2] \in \mathbb{HP}^1 : q^k \neq 0\}$ and $\Psi_k : \mathcal{P}^{-1}(V_k) \to V_k \times Sp\,(1)$ is $\Psi_k(p) = (\,\mathcal{P}(p), \psi_k(p)\,)$, where $\psi_k(p) = \psi_k(q^1, q^2) = |\,q^k|^{-1} q^k$. The inverses $\Phi_k = \Psi_k^{-1} : V_k \times Sp\,(1) \to \mathcal{P}^{-1}(V_k)$ are then given by $\Phi_1(\,[q^1, q^2], y\,) = (\,|\,q^1|\,y, q^2(q^1)^{-1}\,|\,q^1|\,y\,)$ and $\Phi_2(\,[q^1, q^2], y\,) = (\,q^1(q^2)^{-1}\,|\,q^2|\,y,\ |\,q^2|\,y\,)$ so that the transition functions $g_{12}, g_{21} : V_1 \cap V_2 \to Sp\,(1)$ are

$$g_{12}(x) = g_{12}(\,[q^1, q^2]\,) = |\,q^1\,|^{-1}\,q^1(q^2)^{-1}\,|\,q^2\,| \ \text{ and } \ g_{21}(x) = (\,g_{12}(x)\,)^{-1}\,.$$

The canonical local cross-sections $s_k : V_k \to S^7$ associated with these trivializations are $s_1(x) = s_1(\,[q^1, q^2]\,) = (\,|\,q^1\,|, q^2(q^1)^{-1}\,|\,q^1\,|\,)$ and $s_2(x) = s_2(\,[q^1, q^2]\,) = (\,q^1(q^2)^{-1}\,|\,q^2\,|,\ |\,q^2\,|\,)$. Of course, V_1 and V_2 are also the standard coordinate nbds on $\mathbb{HP}^1$. The corresponding diffeomorphisms $\varphi_k : V_k \to \mathbb{H} = \mathbb{R}^4$ are $\varphi_1(\,[q^1, q^2]\,) = q^2(q^1)^{-1}$ and $\varphi_2(\,[q^1, q^2]\,) = q^1(q^2)^{-1}$. Their inverses are $\varphi_1^{-1}(q) = [1, q]$ and $\varphi_2^{-1}(q) = [q, 1]$ so that the overlap maps are $\varphi_2 \circ \varphi_1^{-1}(q) = q^{-1} = \varphi_1 \circ \varphi_2^{-1}(q)$ for all $q \in \mathbb{H} - \{0\}$. However, since $(1, q)$ and $(q, 1)$ are generally not in S^7, it will be more convenient for our purposes to use the following equivalent descriptions of φ_1^{-1} and φ_2^{-1}:

$$\begin{aligned}\varphi_1^{-1}(q) &= \left[\left(1+|q|^2\right)^{-\frac{1}{2}}, q\left(1+|q|^2\right)^{-\frac{1}{2}} \right] \\ \varphi_2^{-1}(q) &= \left[q\left(1+|q|^2\right)^{-\frac{1}{2}}, \left(1+|q|^2\right)^{-\frac{1}{2}} \right]\end{aligned} \tag{4.8.8}$$

Exercise 4.8.8 Show that, for all $q \in \mathbb{H} - \{0\}$, $(s_1 \circ \varphi_1^{-1})(q) = \frac{1}{\sqrt{1+|q|^2}}(1,q)$ and $(s_2 \circ \varphi_2^{-1})(q) = \frac{1}{\sqrt{1+|q|^2}}(q,1)$.

Now, the $\mathrm{Im}\,\mathbb{H}$-valued 1-form on S^7 of interest to us is obtained as follows: First define $\tilde{\omega}$ on $\mathbb{H}^2$ by $\tilde{\omega} = \mathrm{Im}\,(\bar{q}^1 dq^1 + \bar{q}^2 dq^2)$ and then let ω be the restriction of $\tilde{\omega}$ to S^7, i.e., $\omega = \iota^* \tilde{\omega}$. Thus, for every $p \in S^7$ and every $\mathbf{v} \in T_p(S^7)$, $\omega_p(\mathbf{v}) = \tilde{\omega}_{\iota(p)}(\iota_{*p}(\mathbf{v})) = \tilde{\omega}_p(\iota_{*p}(\mathbf{v}))$. Suppressing the inclusions we write $p = (p^1, p^2) \in S^7 \subseteq \mathbb{H}^2$ and $\mathbf{v} = (\mathbf{v}^1, \mathbf{v}^2) \in T_p(S^7) \subseteq T_p(\mathbb{H}^2) = T_{p^1}(\mathbb{H}) \times T_{p^2}(\mathbb{H})$ so that $\omega_p(\mathbf{v}) = \mathrm{Im}\,(\bar{q}^1 dq^1 + \bar{q}^2 dq^2)\,(p^1, p^2)\,(\mathbf{v}^1, \mathbf{v}^2) = \mathrm{Im}\,(\bar{p}^1 dq^1(\mathbf{v}^1, \mathbf{v}^2) + \bar{p}^2 dq^2(\mathbf{v}^1, \mathbf{v}^2))$.

Exercise 4.8.9 Show that $dq^i(\mathbf{v}^1, \mathbf{v}^2) = v^i$, where v^i is the element of $\mathbb{H}$ corresponding to $\mathbf{v}^i \in T_{p^i}(\mathbb{H})$ under the canonical isomorphism (Exercise 4.4.9).

Henceforth we will identify a $\mathbf{v} \in T_p(S^7)$ with the pair $(v^1, v^2) \in \mathbb{H}^2$ so that

$$\omega_p(\mathbf{v}) = \omega_{(p^1,p^2)}(v^1, v^2) = \mathrm{Im}\,(\bar{p}^1 v^1 + \bar{p}^2 v^2). \tag{4.8.9}$$

This 1-form has two properties of great significance to us that we wish to establish at once. First, it is right equivariant under the standard action σ of $Sp(1)$ on S^3, i.e., for all $g \in Sp(1)$, $p \in S^7$ and $\mathbf{v} \in T_{p \cdot g^{-1}}(S^7)$,

$$\omega_p\left((\sigma_g)_{*p\cdot g^{-1}}(\mathbf{v})\right) = ad_{g^{-1}}\left(\omega_{p\cdot g^{-1}}(\mathbf{v})\right). \tag{4.8.10}$$

For the proof we simply compute each side.

Exercise 4.8.10 Show that if $\mathbf{v} = (v^1, v^2) \in T_{p\cdot g^{-1}}(S^7)$, then $(\sigma_g)_{*p\cdot g^{-1}}(\mathbf{v}) = (v^1 g, v^2 g)$.

Thus, $\omega_p((\sigma_g)_{*p\cdot g^{-1}}(\mathbf{v})) = \mathrm{Im}\,(\bar{p}^1 v^1 g + \bar{p}^2 v^2 g)$. On the other hand, $\omega_{p\cdot g^{-1}}(\mathbf{v}) = \mathrm{Im}\,((\overline{p^1 g^{-1}})v^1 + (\overline{p^2 g^{-1}})v^2) = \mathrm{Im}\,(\overline{g^{-1}}\,\bar{p}^1 v^1 + \overline{g^{-1}}\,\bar{p}^2 v^2) = \mathrm{Im}\,(g\bar{p}^1 v^1 + g\bar{p}^2 v^2)$ because $g \in Sp(1)$ implies $\overline{g^{-1}} = g$. Thus, $ad_{g^{-1}}(\omega_{p\cdot g^{-1}}(\mathbf{v})) = g^{-1}\omega_{p\cdot g^{-1}}(\mathbf{v})g = g^{-1}\,\mathrm{Im}\,(g\bar{p}^1 v^1 + g\bar{p}^2 v^2)g$.

Exercise 4.8.11 Show that, for any $g \in Sp(1)$ and $h \in \mathbb{H}$, $g^{-1}(\operatorname{Im} h)g = \operatorname{Im}(g^{-1}hg)$.

Thus, $ad_{g^{-1}}(\omega_{p \cdot g^{-1}}(\mathbf{v})) = \operatorname{Im}(g^{-1}(g\bar{p}^1 v^1 + g\bar{p}^2 v^2)g) = \operatorname{Im}(\bar{p}^1 v^1 g + \bar{p}^2 v^2 g)$ $= \omega_p((\sigma_g)_{*p \cdot g^{-1}}(\mathbf{v}))$ as required.

Secondly, we show that ω acts trivially on fundamental vector fields. More precisely, we let A be any element of $\mathcal{SP}(1) = \operatorname{Im}\mathbb{H}$ and $A^\# = \sigma(A)$ the fundamental vector field on S^7 determined by A (and the standard action σ of $Sp(1)$ on S^7). Then $\omega(A^\#)$ is an $\operatorname{Im}\mathbb{H}$-valued function on S^7 defined at each $p \in S^7$ by $\omega(A^\#)(p) = \omega_p(A^\#(p))$. We claim that it is actually a constant function, taking the value A everywhere, i.e.,

$$\omega_p\left(A^\#(p)\right) = A \tag{4.8.11}$$

for each $p \in S^7$. For the proof we recall that $A^\#(p)$ is the velocity vector at $t = 0$ of the curve $\alpha_p(t) = p \cdot \exp(tA) = (p^1 \exp(tA), p^2 \exp(tA))$ (see (4.7.8)) and this is just $(p^1 A, p^2 A)$ by Lemma 4.7.5(4). Thus, $\omega_p(A^\#(p)) = \operatorname{Im}(\bar{p}^1 p^1 A + \bar{p}^2 p^2 A) = \operatorname{Im}((|p^1|^2 + |p^2|^2)A) = \operatorname{Im} A = A$ because $p \in S^7$ and $A \in \operatorname{Im}\mathbb{H}$.

In Section 5.1 we will see that (4.8.10) and (4.8.11) qualify ω as a "connection form" on the Hopf bundle $Sp(1) \to S^7 \to \mathbb{HP}^1$. For the present we are interested in pulling back ω to $\mathbb{HP}^1 \cong S^4$ by the canonical cross-sections of this bundle (in Section 5.1 we will adopt the terminology of the physics literature and refer to these pullbacks as the "gauge potentials" of the connection ω). In order to have some coordinates to write things in, however, we will go one step further and pull these back to $\mathbb{H}$ via the standard charts on $\mathbb{HP}^1$.

Exercise 4.8.12 Show that, for each $q \in \mathbb{H}$, each $\mathbf{v} \in T_q(\mathbb{H})$ and each $k = 1, 2$,

$$\left((s_k \circ \varphi_k^{-1})^* \omega\right)_q (\mathbf{v}) = (s_k^* \omega)_{\varphi_k^{-1}(q)}\left((\varphi_k^{-1})_{*q}(\mathbf{v})\right).$$

The upshot of Exercise 4.8.12 is that if we use the diffeomorphism φ_k^{-1} to identify $\mathbb{H}$ with V_k so that q is identified with $\varphi_k^{-1}(q)$ and $\mathbf{v}$ is identified with $(\varphi_k^{-1})_{*q}(\mathbf{v})$, then $(s_k \circ \varphi_k^{-1})^* \omega$ is identified with $s_k^* \omega$. We will compute $(s_1 \circ \varphi_1^{-1})^* \omega$ and leave $(s_2 \circ \varphi_2^{-1})^* \omega$ for the reader.

To ease the typography a bit we will let $s = s_1 \circ \varphi_1^{-1} : \mathbb{H} \to \mathcal{P}^{-1}(V_1)$ and compute $s^* \omega$. For each $q \in \mathbb{H}$, Exercise 4.8.8 gives $s(q) = (1 + |q|^2)^{-\frac{1}{2}}(1, q)$. Each $\mathbf{v}_q \in T_q(\mathbb{H})$ we identify with $\frac{d}{dt}(q + vt)|_{t=0}$ for the $v \in \mathbb{H}$ corresponding to $\mathbf{v}_q$ under the canonical isomorphism. Thus,

$$s_{*q}(\mathbf{v}_q) = s_{*q}\left(\frac{d}{dt}(q + vt)|_{t=0}\right) = \frac{d}{dt}(s(q + vt))|_{t=0}$$

$$= \frac{d}{dt}\left(\frac{1}{\sqrt{1+|q+vt|^2}}\,(1, q+vt) \right)\Bigg|_{t=0}.$$

Exercise 4.8.13 Show that $|q+vt|^2 = |q|^2 + 2\,\mathrm{Re}\,(v\bar{q})t + |v|^2t^2$ and conclude that

$$\frac{d}{dt}\left(\frac{1}{\sqrt{1+|q+vt|^2}} \right)\Bigg|_{t=0} = -\frac{\mathrm{Re}\,(v\bar{q})}{(1+|q|^2)^{3/2}}.$$

From this show also that

$$\frac{d}{dt}\left(\frac{1}{\sqrt{1+|q+vt|^2}}\,(q+vt) \right)\Bigg|_{t=0} = \frac{1}{\sqrt{1+|q|^2}}\,v - \frac{\mathrm{Re}\,(v\bar{q})}{(1+|q|^2)^{3/2}}\,q.$$

From Exercise 4.8.13 we conclude that

$$s_{*q}\,(\mathbf{v}_q) = \left(-\frac{\mathrm{Re}\,(v\bar{q})}{(1+|q|^2\,)^{3/2}}\,,\, \frac{1}{\sqrt{1+|q|^2}}\,v - \frac{\mathrm{Re}\,(v\bar{q})}{(1+|q|^2)^{3/2}}\,q \right). \quad (4.8.12)$$

Now, $(s^*\boldsymbol{\omega})_q(\mathbf{v}_q) = \boldsymbol{\omega}_{s(q)}(\,s_{*q}(\mathbf{v}_q)\,)$ which we compute as follows:

$$\begin{aligned}(\bar{q}^2\,dq^2)_{s(q)}\,(s_{*q}\,(\mathbf{v}_q)\,) &= \bar{q}^2\,(\,s(q)\,)\,dq^2\,(\,s_{*q}\,(\mathbf{v}_q)\,) \\ &= \left(\frac{1}{\sqrt{1+|q|^2}}\,\bar{q} \right)\left(\frac{1}{\sqrt{1+|q|^2}}\,v - \frac{\mathrm{Re}\,(v\bar{q})}{(1+|q|^2)^{3/2}}\,q \right) \\ &= \frac{1}{1+|q|^2}\,\bar{q}v - \frac{\mathrm{Re}\,(v\bar{q})\,|q|^2}{(1+|q|^2)^2}\end{aligned}$$

(notice that the second term is real).

Exercise 4.8.14 Show that $(\bar{q}^1\,dq^1)_{s(q)}\,(\,s_{*q}\,(\mathbf{v}_q)\,) = -\frac{\mathrm{Re}\,(v\bar{q})}{(1+|q|^2)^2}$ (which is real).

Thus, $\boldsymbol{\omega}_{s(q)}(\,s_{*q}(\mathbf{v}_q)\,) = (\,\mathrm{Im}\,(\bar{q}^1\,dq^1 + \bar{q}^2\,dq^2)\,)_{s(q)}(\,s_{*q}(\mathbf{v}_q)\,) = \frac{\mathrm{Im}\,(\bar{q}\,v)}{1+|q|^2}$ so

$$\left((s_1 \circ \varphi_1^{-1})^*\,\boldsymbol{\omega} \right)_q (\mathbf{v}_q) = \frac{\mathrm{Im}\,(\bar{q}\,v)}{1+|q|^2} = \mathrm{Im}\left(\frac{\bar{q}}{1+|q|^2}\,v \right) \quad (4.8.13)$$

In (4.8.13), q refers to any fixed point in $\mathbb{H}$ and v is the quaternion corresponding to some fixed $\mathbf{v}_q \in T_q(\mathbb{H})$. If we now regard q as the standard quaternionic coordinate function ($q = x + y\mathbf{i} + u\mathbf{j} + v\mathbf{k}$) on $\mathbb{H}$, then one can also regard $\bar{q}$ and $|q|^2$ as functions on $\mathbb{H}$. Letting $dq = dx + dy\mathbf{i} + du\mathbf{j} + dv\mathbf{k}$ as usual we have $dq(\mathbf{v}_q) = v$ so (4.8.13) can be written

$$\left(s_1 \circ \varphi_1^{-1}\right)^* \boldsymbol{\omega} = \operatorname{Im}\left(\frac{\bar{q}}{1+|q|^2}\,dq\right). \tag{4.8.14}$$

Exercise 4.8.15 Show that $(s_2 \circ \varphi_2^{-1})^* \boldsymbol{\omega} = \operatorname{Im}\left(\frac{\bar{q}}{1+|q|^2}\,dq\right)$.

Remark: These last two expressions for $(s_1 \circ \varphi_1^{-1})^* \boldsymbol{\omega}$ and $(s_2 \circ \varphi_2^{-1})^* \boldsymbol{\omega}$ are perhaps just a bit too concise and elegant. The q's, after all, refer to different coordinate functions on S^4 in (4.8.14) and Exercise 4.8.15. In order to dispel any possible confusion we write out in detail what each "really" means. For each $p \in V_1$ and each $\mathbf{X} \in T_p(S^4)$, (4.8.14) and Exercise 4.8.12 imply

$$(s_1^* \boldsymbol{\omega})_p \mathbf{X} = \left((s_1 \circ \varphi_1^{-1})^* \boldsymbol{\omega}\right)_{\varphi_1(p)}\left((\varphi_1)_{*p}\mathbf{X}\right) = \operatorname{Im}\left(\frac{\overline{\varphi_1(p)}\,v}{1+|\varphi_1(p)|^2}\right),$$

where $v = dq\left((\varphi_1)_{*p}\mathbf{X}\right)$. Similarly, for $p \in V_2$ and $\mathbf{X} \in T_p(S^4)$,

$$(s_2^* \boldsymbol{\omega})_p \mathbf{X} = \left((s_2 \circ \varphi_2^{-1})^* \boldsymbol{\omega}\right)_{\varphi_2(p)}\left((\varphi_2)_{*p}\mathbf{X}\right) = \operatorname{Im}\left(\frac{\overline{\varphi_2(p)}\,w}{1+|\varphi_2(p)|^2}\right),$$

where $w = dq\left((\varphi_2)_{*p}\mathbf{X}\right)$.

Exercise 4.8.16 Show that, on $\mathbb{H} - \{0\}$,

$$\operatorname{Im}\left(\frac{\bar{q}}{1+|q|^2}\,dq\right) = \frac{|q|^2}{1+|q|^2}\operatorname{Im}\left(q^{-1}dq\right). \tag{4.8.15}$$

Exercise 4.8.17 Identify $GL(1,\mathbb{H})$ with $\mathbb{H} - \{0\}$ and $Sp(1)$ with the unit quaternions in $\mathbb{H} - \{0\}$. Define $g : \mathbb{H} - \{0\} \to Sp(1)$ by $g(q) = \frac{q}{|q|}$ and let $\iota : Sp(1) \hookrightarrow \mathbb{H} - \{0\}$ be the inclusion so that $\boldsymbol{\Theta} = \iota^*(\operatorname{Im}(q^{-1}dq))$ is the Cartan 1-form for $Sp(1)$. Show that $g^*\boldsymbol{\Theta} = \operatorname{Im}(q^{-1}dq)$.

What these calculations (and Exercise 4.8.12) reveal is that $s_1^*\boldsymbol{\omega}$ and $s_2^*\boldsymbol{\omega}$ are formally identical when $s_1^*\boldsymbol{\omega}$ is expressed in (V_1, φ_1) coordinates and $s_2^*\boldsymbol{\omega}$ is expressed in (V_2, φ_2) coordinates. This is rather like comparing apples and oranges, however. What we really need is a comparison of the two pullbacks in the same coordinate system. Thus, let us fix a $p \in V_1 \cap V_2$ and an $\mathbf{X} \in T_p(S^4)$. Suppose $p = \varphi_1^{-1}(q)$ for $q \in \varphi_1(V_1 \cap V_2) = \mathbb{H} - \{0\}$. Then, since $\varphi_2 \circ \varphi_1^{-1}(q) = q^{-1}$, $p = \varphi_2^{-1}(q^{-1})$. Next suppose that $\mathbf{X} =$

$(\varphi_1^{-1})_{*q}(\mathbf{v}_q)$. Since $(\varphi_2 \circ \varphi_1^{-1})_{*q} = (\varphi_2)_{*p} \circ (\varphi_1^{-1})_{*q}$, $\mathbf{X} = (\varphi_2^{-1})_{*q^{-1}}((\varphi_2 \circ \varphi_1^{-1})_{*q}(\mathbf{v}_q))$. Thus,

$$(s_1^* \boldsymbol{\omega})_p \, \mathbf{X} = \left((s_1 \circ \varphi_1^{-1})^* \boldsymbol{\omega} \right)_q (\mathbf{v}_q) = \operatorname{Im}\left(\frac{\bar{q}\, v}{1 + |q|^2} \right), \tag{4.8.16}$$

where $dq\,(\mathbf{v}_q) = v$ and

$$\begin{aligned} (s_2^* \boldsymbol{\omega})_p \, \mathbf{X} &= \left((s_2 \circ \varphi_2^{-1})^* \boldsymbol{\omega} \right)_{q^{-1}} \left((\varphi_2 \circ \varphi_1^{-1})_{*q}(\mathbf{v}_q) \right) \\ &= \operatorname{Im}\left(\frac{\overline{q^{-1}}\, w}{1 + |q^{-1}|^2} \right), \end{aligned} \tag{4.8.17}$$

where $dq(\,(\varphi_2 \circ \varphi_1^{-1})_{*q}(\mathbf{v}_q)\,) = w$. We compute this last expression by first noting that

$$\frac{\overline{q^{-1}}}{1 + |q^{-1}|^2} = \frac{q}{1 + |q|^2}.$$

Next, $\mathbf{v}_q = \frac{d}{dt}(q + vt)|_{t=0}$ implies

$$\begin{aligned} (\varphi_2 \circ \varphi_1^{-1})_{*q}(\mathbf{v}_q) &= \frac{d}{dt} \left((\varphi_2 \circ \varphi_1^{-1})(q + vt) \right)\Big|_{t=0} \\ &= \frac{d}{dt} (q + vt)^{-1}\Big|_{t=0} = \frac{d}{dt} \left(\frac{\bar{q} + \bar{v}t}{|q + vt|^2} \right)\Bigg|_{t=0}. \end{aligned}$$

Exercise 4.8.18 Use the expression for $|q + vt|^2$ in Exercise 4.8.13 to compute this derivative and show that

$$w = \frac{\bar{v}}{|q|^2} - \frac{2\,\operatorname{Re}(v\bar{q})\,\bar{q}}{|q|^4}.$$

Thus, (4.8.17) gives

$$(s_2^* \boldsymbol{\omega})_p \, \mathbf{X} = \operatorname{Im}\left(\frac{q\bar{v}}{|q|^2\,(1 + |q|^2\,)} \right), \tag{4.8.18}$$

where $p = \varphi_1^{-1}(q)$, $\mathbf{X} = (\varphi_1^{-1})_{*q}(\mathbf{v}_q)$ and $dq\,(\mathbf{v}_q) = v$.

Thus, (4.8.14) and (4.8.18) express both $s_1^* \boldsymbol{\omega}$ and $s_2^* \boldsymbol{\omega}$ in terms of the coordinates on $V_1 \cap V_2$ supplied by $\varphi_1 : V_1 \cap V_2 \to \mathbb{H} - \{0\}$. As above we may write these results as

$$(s_1^* \boldsymbol{\omega})_p \mathbf{X} = \operatorname{Im}\left(\frac{\bar{q}}{1 + |q|^2}\, dq\,(\mathbf{v}_q) \right) \tag{4.8.19}$$

$$(s_2^* \boldsymbol{\omega})_p \mathbf{X} = \operatorname{Im}\left(\frac{q}{|q|^2\,(1 + |q|^2\,)} d\bar{q}\,(\mathbf{v}_q) \right), \tag{4.8.20}$$

where $p = \varphi_1^{-1}(q)$ and $\mathbf{X} = (\varphi_1^{-1})_{*q}(\mathbf{v}_p)$. In somewhat more detail we have, for each $p \in V_1 \cap V_2$ and each $\mathbf{X} \in T_p(S^4)$,

$$\begin{aligned}(s_2^*\omega)_p\mathbf{X} &= \mathrm{Im}\left(\frac{\overline{\varphi_2(p)}\,w}{1+|\varphi_2(p)|^2}\right)\\ &= \mathrm{Im}\left(\frac{\varphi_1(p)\,\bar{v}}{|\varphi_1(p)|^2\,(1+|\varphi_1(p)|^2)}\right),\end{aligned} \tag{4.8.21}$$

where $dq((\varphi_1)_{*p}\mathbf{X}) = v$ and $dq((\varphi_2)_{*p}\mathbf{X}) = w$.

Exercise 4.8.19 Show that, for $p \in V_1 \cap V_2$ and $\mathbf{X} \in T_p(S^4)$,

$$\begin{aligned}(s_1^*\omega)_p\mathbf{X} &= \mathrm{Im}\left(\frac{\overline{\varphi_1(p)}\,v}{1+|\varphi_1(p)|^2}\right)\\ &= \mathrm{Im}\left(\frac{\varphi_2(p)\,\bar{w}}{|\varphi_2(p)|^2\,(1+|\varphi_2(p)|^2)}\right),\end{aligned} \tag{4.8.22}$$

where $dq((\varphi_1)_{*p}\mathbf{X}) = v$ and $dq((\varphi_2)_{*p}\mathbf{X}) = w$.

We have therefore accomplished our stated purpose of pulling back to S^4 via canonical cross-sections of the Hopf bundle the 11-entry in the Cartan 1-form for $Sp(2)$. But, has it been worth the effort? Have we found anything interesting? The answer is every bit as startling as finding a Dirac monopole lurking beneath the complex Hopf bundle. What we have found here is a "BPST instanton". These objects arose first in the physics literature and we will have something to say about their origin and significance in Chapter 5. To conclude this section, though, we wish to show that the two 1-forms $s_1^*\omega$ and $s_2^*\omega$ are related on $V_1 \cap V_2$ by $s_2^*\omega = ad_{g_{12}^{-1}} \circ s_1^*\omega + g_{12}^*\Theta$, where g_{12} is the transition function for the Hopf bundle and Θ is the Cartan 1-form for $Sp(1)$. Since this fact depends only on (4.8.10) and (4.8.11), we will actually prove a much more general result that will be of considerable significance to us in Chapter 5.

Lemma 4.8.2 *Let G be a matrix Lie group with Lie algebra $\mathcal{G}$, $(P, \mathcal{P}, \sigma)$ a smooth principal G-bundle over X and ω a $\mathcal{G}$-valued 1-form on P that satisfies*

1. $(\sigma_g)^*\omega = ad_{g^{-1}} \circ \omega$ *for all* $g \in G$, *and*

2. $\omega(A^\#) = A$ *for all* $A \in \mathcal{G}$.

Let (V_1, Ψ_1) and (V_2, Ψ_2) be trivializations of the bundle with $V_2 \cap V_1 \neq \emptyset$, $g_{12} : V_2 \cap V_1 \to G$ the transition function, $s_k : V_k \to P$, $k = 1, 2,$

the associated canonical cross-sections and $\boldsymbol{\Theta}$ *the Cartan 1-form for G. Letting* $\boldsymbol{\mathcal{A}}_k = s_k^* \boldsymbol{\omega}$ *and* $\boldsymbol{\Theta}_{12} = g_{12}^* \boldsymbol{\Theta}$ *we have*

$$\boldsymbol{\mathcal{A}}_2 = ad_{g_{12}^{-1}} \circ \boldsymbol{\mathcal{A}}_1 + \boldsymbol{\Theta}_{12} \tag{4.8.23}$$

(i.e., for each $x \in V_2 \cap V_1$ *and each* $\mathbf{v} \in T_x(X)$, $\boldsymbol{\mathcal{A}}_2(x)(\mathbf{v}) = ad_{g_{12}(x)^{-1}}(\boldsymbol{\mathcal{A}}_1(x)(\mathbf{v})) + \boldsymbol{\Theta}_{12}(x)(\mathbf{v}) = g_{12}(x)^{-1}(\boldsymbol{\mathcal{A}}_1(x)(\mathbf{v}))\, g_{12}(x) + \boldsymbol{\Theta}_{g_{12}(x)}((g_{12})_{*x}(\mathbf{v}))$*).*

Proof: Fix $x \in V_2 \cap V_1$ and $\mathbf{v} \in T_x(X)$. Exercise 3.3.5 gives $s_2(x) = s_1(x) \cdot g_{12}(x)$. We identify $T_{(p_0, g_0)}(P \times G)$ with $T_{p_0}(P) \times T_{g_0}(G)$ and use Exercise 4.4.17 to write $\sigma_{*(p_0,g_0)}(\mathbf{w}) = (\sigma_1)_{*p_0}(\mathbf{w}_P) + (\sigma_2)_{*g_0}(\mathbf{w}_G)$, where $\mathbf{w} = (\mathbf{w}_P, \mathbf{w}_G) \in T_{(p_0,g_0)}(P \times G)$ and $\sigma_1 : P \to P$ and $\sigma_2 : G \to P$ are defined by $\sigma_1(p) = p \cdot g_0$ and $\sigma_2(g) = p_0 \cdot g$. Now, for any $y \in V_2 \cap V_1$, $s_2(y) = s_1(y) \cdot g_{12}(y) = \sigma(s_1(y), g_{12}(y)) = \sigma \circ (s_1, g_{12})(y)$ so $(s_2)_{*x} = (\sigma \circ (s_1, g_{12}))_{*x} = \sigma_{*(s_1(x), g_{12}(x))} \circ ((s_1)_{*x}, (g_{12})_{*x})$. Thus,

$$\begin{aligned} (s_2)_{*x}(\mathbf{v}) &= (\sigma_1)_{*s_1(x)}((s_1)_{*x}(\mathbf{v})) + (\sigma_2)_{*g_{12}(x)}((g_{12})_{*x}(\mathbf{v})) \\ &= (\sigma_1 \circ s_1)_{*x}(\mathbf{v}) + (\sigma_2 \circ g_{12})_{*x}(\mathbf{v}). \end{aligned}$$

Note that applying $\boldsymbol{\omega}_{s_2(x)}$ to the left-hand side gives $\boldsymbol{\omega}_{s_2(x)}((s_2)_{*x}(\mathbf{v})) = (s_2^* \boldsymbol{\omega})_x(\mathbf{v}) = \boldsymbol{\mathcal{A}}_2(x)(\mathbf{v})$ so we can complete the proof by showing

$$\boldsymbol{\omega}_{s_2(x)}((\sigma_1 \circ s_1)_{*x}(\mathbf{v})) = ad_{g_{12}(x)^{-1}}(\boldsymbol{\mathcal{A}}_1(x)(\mathbf{v})) \tag{4.8.24}$$

$$\boldsymbol{\omega}_{s_2(x)}((\sigma_2 \circ s_2)_{*x}(\mathbf{v})) = \boldsymbol{\Theta}_{12}(x)(\mathbf{v}). \tag{4.8.25}$$

To prove (4.8.24) we write $\sigma_1 \circ s_1(y) = \sigma(s_1(y), g_{12}(x)) = s_1(y) \cdot g_{12}(x) = \sigma_{g_{12}(x)}(s_1(y)) = \sigma_{g_{12}(x)} \circ s_1(y)$ so that $(\sigma_1 \circ s_1)_{*x}(\mathbf{v}) = (\sigma_{g_{12}(x)})_{*s_1(x)}((s_1)_{*x}(\mathbf{v}))$. Thus,

$$\begin{aligned} \boldsymbol{\omega}_{s_2(x)}((\sigma_1 \circ s_1)_{*x}(\mathbf{v})) &= \boldsymbol{\omega}_{s_2(x)}\big((\sigma_{g_{12}(x)})_{*s_1(x)}((s_1)_{*x}(\mathbf{v}))\big) \\ &= \boldsymbol{\omega}_{s_2(x)}\big((\sigma_{g_{12}(x)})_{*s_2(x)\cdot g_{12}(x)^{-1}}((s_1)_{*x}(\mathbf{v}))\big) \\ &= ad_{g_{12}(x)^{-1}}\big(\boldsymbol{\omega}_{s_2(x)\cdot g_{12}(x)^{-1}}((s_1)_{*x}(\mathbf{v}))\big) \\ &= ad_{g_{12}(x)^{-1}}\big(\boldsymbol{\omega}_{s_1(x)}((s_1)_{*x}(\mathbf{v}))\big) \\ &= ad_{g_{12}(x)^{-1}}((s_1^* \boldsymbol{\omega})_x(\mathbf{v})) \\ &= ad_{g_{12}(x)^{-1}}(\boldsymbol{\mathcal{A}}_1(x)(\mathbf{v})) \end{aligned}$$

as required.

To prove (4.8.25) we let $A \in \mathcal{G}$ denote the unique element of the Lie algebra (thought of as $T_e(G)$) whose left-invariant vector field

$\mathbf{A}$ satisfies $\mathbf{A}(g_{12}(x)) = (g_{12})_{*x}(\mathbf{v})$. Then $\boldsymbol{\Theta}_{12}(x)(\mathbf{v}) = (g_{12}^*\boldsymbol{\Theta})_x(\mathbf{v}) = \boldsymbol{\Theta}_{g_{12}(x)}((g_{12})_{*x}(\mathbf{v})) = \boldsymbol{\Theta}_{g_{12}(x)}(\mathbf{A}(g_{12}(x)) = A$ by (4.8.2). Thus, we need only show that $\boldsymbol{\omega}_{s_2(x)}((\sigma_2 \circ s_2)_{*x}(\mathbf{v})) = A$. Let $A^\#$ be the fundamental vector field on P determined by A. By definition, $A^\#(p) = (\sigma_p)_{*e}(A)$ so $A^\#(s_1(x) \cdot g_{12}(x)) = (\sigma_{s_1(x) \cdot g_{12}(x)})_{*e}(A)$. But

$$\begin{aligned}\sigma_{s_1(x)\cdot g_{12}(x)}(g) &= (s_1(x) \cdot g_{12}(x)) \cdot g = s_1(x) \cdot (g_{12}(x)\, g) \\ &= \sigma_2(g_{12}(x)\, g) = \sigma_2(L_{g_{12}(x)}(g)) \\ &= \sigma_2 \circ L_{g_{12}(x)}(g)\end{aligned}$$

so

$$\begin{aligned}\left(\sigma_{s_1(x)\cdot g_{12}(x)}\right)_{*e}(A) &= (\sigma_2)_{*g_{12}(x)}\left((L_{g_{12}(x)})_{*e}(A)\right) \\ &= (\sigma_2)_{*g_{12}(x)}\left(\mathbf{A}(g_{12}(x))\right) \\ &= (\sigma_2)_{*g_{12}(x)}\left((g_{12})_{*x}(\mathbf{v})\right) \\ &= (\sigma_2 \circ g_{12})_{*x}(\mathbf{v}).\end{aligned}$$

Thus, $A^\#(s_1(x) \cdot g_{12}(x)) = (\sigma_2 \circ g_{12})_{*x}(\mathbf{v})$ so

$$\begin{aligned}\boldsymbol{\omega}_{s_2(x)}\left((\sigma_2 \circ g_{12})_{*x}(\mathbf{v})\right) &= \boldsymbol{\omega}_{s_2(x)}\left(A^\#(s_1(x) \cdot g_{12}(x))\right) \\ &= \boldsymbol{\omega}_{s_2(x)}\left(A^\#(s_2(x))\right) \\ &= A\end{aligned}$$

as required. ∎

1-forms of the type described in Lemma 4.8.2 and their pullbacks to the base manifold by canonical cross-sections of the bundle are, as we shall see in Chapter 5, of profound significance to both differential geometry and mathematical physics. The transformation law (4.8.23) is particularly prominent in the physics literature, although it is generally expressed somewhat differently. Letting α be a curve in X with $\alpha'(0) = \mathbf{v}$, let us compute

$$\begin{aligned}\boldsymbol{\Theta}_{g_{12}(x)}((g_{12})_{*x}(\mathbf{v})) = \boldsymbol{\Theta}_{g_{12}(x)}((g_{12})_{*x}(\alpha'(0))) &= \boldsymbol{\Theta}_{g_{12}(x)}((g_{12} \circ \alpha)'(0)) \\ &= \left(L_{(g_{12}(x))^{-1}}\right)_{*g_{12}(x)}((g_{12} \circ \alpha)'(0)) \\ &= \left(L_{(g_{12}(x))^{-1}} \circ g_{12} \circ \alpha\right)'(0) \\ &= \frac{d}{dt}\left[(g_{12}(x))^{-1}(g_{12} \circ \alpha)(t)\right]\Big|_{t=0} \\ &= (g_{12}(x))^{-1}(g_{12} \circ \alpha)'(0).\end{aligned}$$

Now, using standard coordinates (i.e., entries) in the matrix group G this becomes $\Theta_{g_{12}(x)}((g_{12})_{*x}(\mathbf{v})) = (g_{12}(x))^{-1}\, dg_{12}(x)(\mathbf{v})$, where dg_{12} is the entrywise differential of g_{12}. Thus, $\Theta_{12}(x)(\mathbf{v}) = (g_{12}(x))^{-1}\, dg_{12}(x)(\mathbf{v})$, and so $\Theta_{12} = g_{12}^{*}\Theta = g_{12}^{-1} dg_{12}$. Thus, (4.8.23) becomes

$$\mathcal{A}_2 = g_{12}^{-1}\, \mathcal{A}_1\, g_{12} + g_{12}^{-1}\, dg_{12} \tag{4.8.26}$$

and this is the form most often encountered in physics. In this context the local sections s_i are called local "gauges", $\mathcal{A}_i = s_i^{*}\,\omega$ is the "gauge potential" in gauge s_i and (4.8.26) describes the effect of the "gauge transformation" $s_1 \to s_2 = s_1 \cdot g_{12}$ (see Exercise 3.3.5).

Exercise 4.8.20 The transformation law (4.8.23) for the gauge potential can be written $s_2^{*}\omega = ad_{g_{21}} \circ s_1^{*}\omega + g_{12}^{*}\Theta$. Show that the same transformation law relates the coordinate expressions $(s_k \circ \varphi^{-1})^{*}\,\omega$, i.e., prove that

$$\left(s_2 \circ \varphi^{-1}\right)^{*} \omega = ad_{g_{21}\circ\varphi^{-1}} \circ \left(s_1 \circ \varphi^{-1}\right)^{*} \omega + \left(g_{12} \circ \varphi^{-1}\right)^{*} \Theta \tag{4.8.27}$$

for any chart (U, φ) with $U \subseteq V_1 \cap V_2$.

Exercise 4.8.21 A $\mathcal{V}$-valued 0-form on X is a smooth map ϕ from X into $\mathcal{V}$ (with its natural differentiable structure). Let $\{e_1, \ldots, e_d\}$ be a basis for $\mathcal{V}$ and write $\phi = \phi^i e_i$, where $\phi^i \in C^\infty(X)$ for $i = 1, \ldots, d$. Define the exterior derivative $d\phi$ of ϕ componentwise, i.e., by $d\phi = (d\phi^i)\, e_i$. Show that this definition does not depend on the choice of basis for $\mathcal{V}$ and that $d\phi$ is a $\mathcal{V}$-valued 1-form on X. **Hint:** If $\{\hat{e}_1, \ldots, \hat{e}_d\}$ is another basis for $\mathcal{V}$, then $e_i = A^j{}_i\, \hat{e}_j$ for some *constants* $A^j{}_i$, $i, j = 1, \ldots, d$.

4.9 Orientability

Consider an n-dimensional real vector space $\mathcal{V}$ and two *ordered* bases $\{e_1, \ldots, e_n\}$ and $\{\hat{e}_1, \ldots, \hat{e}_n\}$ for $\mathcal{V}$. There exists a unique nonsingular matrix $(A^i{}_j)_{i,j=1,\ldots,n}$ such that $\hat{e}_j = A^i{}_j\, e_i$ for each $j = 1, \ldots, n$. Since $(A^i{}_j)$ is nonsingular, $\det(A^i{}_j) \neq 0$. We define a relation $\sim$ on the set of all ordered bases for $\mathcal{V}$ by $\{\hat{e}_1, \ldots, \hat{e}_n\} \sim \{e_1, \ldots, e_n\}$ iff $\det(A^i{}_j) > 0$.

Exercise 4.9.1 Show that this is an equivalence relation on the set of all ordered bases for $\mathcal{V}$ with precisely two equivalence classes.

The equivalence class containing the ordered basis $\{e_1, \ldots, e_n\}$ is denoted $[e_1, \ldots, e_n]$. Each of the two equivalence classes is called an **orientation** for $\mathcal{V}$. An **oriented vector space** is a vector space together with some specific choice of one of its two orientations. The **standard orientation** for $\mathbb{R}^n$ is $[e_1, \ldots, e_n]$, where $e_1 = (1, 0, \ldots, 0, 0), \ldots, \ e_n = (0, 0, \ldots, 0, 1)$.

Now let X be an n-dimensional smooth manifold. We would like to select orientations for each tangent space $T_p(X)$ that "vary smoothly with p" in some sense. As we shall see, however, this is not always possible. If U is an open subset of X, an **orientation on** U is a function μ that assigns to each $p \in U$ an orientation μ_p for $T_p(X)$ and satisfies the following smoothness condition: For each $p_0 \in U$ there is an open nbd W of p_0 in X with $W \subseteq U$ and smooth vector fields $\mathbf{V}_1, \ldots, \mathbf{V}_n$ on W with $\{\mathbf{V}_1(p), \ldots, \mathbf{V}_n(p)\} \in \mu_p$ for each $p \in W$. For example, if (U, φ) is a chart for X with coordinate functions $x^i = \mathcal{P}^i \circ \varphi$, $i = 1, \ldots, n$, then $p \to [\frac{\partial}{\partial x^1}|_p, \ldots, \frac{\partial}{\partial x^n}|_p]$ is an orientation on U. A manifold for which an orientation exists on all of X is said to be **orientable** and X is said to be **oriented** by any specific choice of an orientation μ on X.

Exercise 4.9.2 Show that any real vector space with its natural differentiable structure is orientable.

Exercise 4.9.3 Show that any Lie group is orientable. **Hint:** Consider a basis for the Lie algebra.

Suppose (U, φ) and (V, ψ) are two charts on X with coordinate functions $x^i = \mathcal{P}^i \circ \varphi$ and $y^i = \mathcal{P}^i \circ \psi$, $i = 1, \ldots, n$, for which $U \cap V \neq \emptyset$. On $\varphi(U \cap V)$ we write $(y^1, \ldots, y^n) = \psi \circ \varphi^{-1}(x^1, \ldots, x^n)$ and denote the Jacobian of this map by $(\frac{\partial y^j}{\partial x^i})_{i,j=1,\ldots,n}$. By Corollary 4.4.5, $\frac{\partial}{\partial x^i}|_p = \frac{\partial y^j}{\partial x^i}(p) \frac{\partial}{\partial y^j}|_p$ for each $p \in U \cap V$ so the orientations on U and V determined by φ and ψ, respectively, will agree on $U \cap V$ iff $\det(\frac{\partial y^j}{\partial x^i}) > 0$ on $U \cap V$. If this is the case, then they together determine an orientation on $U \cap V$. Consequently, an orientation on X will be determined by any atlas $\{(U_\alpha, \varphi_\alpha)\}_{\alpha \in \mathcal{A}}$ for X with the property that the Jacobians of the overlap functions $\varphi_\beta \circ \varphi_\alpha^{-1}$ have positive determinant on their domains for all α and β in $\mathcal{A}$. Such an atlas is called an **oriented atlas** for X. Thus, a manifold X that admits an oriented atlas is orientable and this is often the most convenient way of proving orientability. We illustrate with the spheres S^n.

Consider the circle S^1 and its stereographic projection charts (U_S, φ_S) and (U_N, φ_N). On $\varphi_S(U_N \cap U_S) = \mathbb{R} - \{0\}$, $\varphi_N \circ \varphi_S^{-1}(x) = \frac{1}{x}$ so the Jacobian is the 1×1 matrix $(\frac{d}{dx}(x^{-1})) = (-\frac{1}{x^2})$. The determinant is $-\frac{1}{x^2}$ which is negative everywhere on $\mathbb{R} - \{0\}$. Thus, $\{(U_S, \varphi_S), (U_N, \varphi_N)\}$ is *not* an oriented atlas for S^1. To remedy this we change the sign of φ_N, i.e., we define $\tilde{\varphi}_N : U_N \to \mathbb{R}$ by $\tilde{\varphi}_N(x^1, x^2) = -\varphi_N(x^1, x^2) = -\frac{x^1}{1+x^2}$.

Then $(U_N, \tilde{\varphi}_N)$ is a chart on S^1 and $\tilde{\varphi}_N \circ \varphi_S^{-1}(x) = -\frac{1}{x}$.

Exercise 4.9.4 Show that $\{(U_S, \varphi_S), (U_N, \tilde{\varphi}_N)\}$ is an oriented atlas for S^1.

Now we turn to S^n for $n \geq 2$. Again we consider the stereographic projection charts (U_S, φ_S) and (U_N, φ_N). On $\varphi_S(U_N \cap U_S) = \mathbb{R}^n - \{(0, \ldots, 0)\}$, $\varphi_N \circ \varphi_S^{-1}(x^1, \ldots, x^n) = ((x^1)^2 + \cdots + (x^n)^2)^{-1}(x^1, \ldots, x^n)$ so

$$\frac{\partial y^j}{\partial x^i} = \frac{\partial}{\partial x^i}\left(\frac{x^j}{(x^1)^2 + \cdots + (x^n)^2}\right) = \frac{\| x \|^2 \, \delta_i^j - 2x^i x^j}{\| x \|^4}.$$

In particular, $\frac{\partial y^j}{\partial x^i}(1, 0, \ldots, 0) = \delta_i^j - 2x^i x^j$ so the Jacobian of $\varphi_N \circ \varphi_S^{-1}$ at $(1, 0, \ldots, 0)$ is

$$\begin{pmatrix} -1 & 0 & 0 & \cdots & 0 \\ 0 & 1 & 0 & \cdots & 0 \\ \vdots & \vdots & \vdots & & \vdots \\ 0 & 0 & 0 & \cdots & 1 \end{pmatrix}.$$

The determinant of the Jacobian at $(1, 0, \ldots, 0)$ is therefore -1. But, for $n \geq 2$, $\mathbb{R}^n - \{(0, \ldots, 0)\}$ is connected and the determinant of the Jacobian of $\varphi_N \circ \varphi_S^{-1}$ is continuous and never zero on $\mathbb{R}^n - \{(0, \ldots, 0)\}$. Being negative at $(1, 0, \ldots, 0)$, it must therefore be negative everywhere. Again, $\{(U_S, \varphi_S), (U_N, \varphi_N)\}$ is *not* an oriented atlas for S^n. Define $\tilde{\varphi}_N : U_N \to \mathbb{R}^n$ by changing the sign of the first coordinate function of φ_N:

$$\tilde{\varphi}_N(x^1, \ldots, x^n, x^{n+1}) = \left(-\frac{x^1}{1 + x^{n+1}}, \frac{x^2}{1 + x^{n+1}}, \ldots, \frac{x^n}{1 + x^{n+1}}\right).$$

Exercise 4.9.5 Show that $\{(U_S, \varphi_S), (U_N, \tilde{\varphi}_N)\}$ is an oriented atlas for S^n.

The orientation defined on S^n by the atlas $\{(U_S, \varphi_S), (U_N, \tilde{\varphi}_N)\}$ is called its **standard orientation**.

A real vector space admits precisely two orientations and if η denotes one of them we will denote the other by $-\eta$. If μ is an orientation on a manifold X, then we obtain another orientation $-\mu$ on X by the assignment $p \to -\mu_p$ for each p in X.

Lemma 4.9.1 *Let X be a smooth orientable manifold with orientation μ, U a connected open subset of X and μ' any orientation on U. Then μ' is the restriction to U of either μ or $-\mu$.*

Proof: Let $H = \{p \in U : \mu'_p = \mu_p\}$. Then $U - H = \{p \in U : \mu'_p = -\mu_p\}$. Since U is connected and open in X it will suffice to show that both of these are open in X (and therefore in U) for then one must be empty and the other must be U. Suppose $p_0 \in H$. Select open nbds W and W' of p_0 in X contained in U and vector fields $\mathbf{V}_1, \ldots, \mathbf{V}_n$ and $\mathbf{V}'_1, \ldots, \mathbf{V}'_n$ on W and W', respectively, such that $\{\mathbf{V}_1(p), \ldots, \mathbf{V}_n(p)\} \in \mu_p$ and $\{\mathbf{V}'_1(p'), \ldots, \mathbf{V}'_n(p')\} \in \mu_{p'}$ for all $p \in W$ and all $p' \in W'$. For each $p \in W \cap W'$, there exists a nonsingular matrix $(A^i{}_j(p))$ such that $\mathbf{V}'_j(p) = A^i{}_j(p)\, \mathbf{V}_i(p)$ for $j = 1, \ldots, n$. The functions $A^i{}_j$ are C^∞ on $W \cap W'$ and we intend to leave the proof of this to the reader. However, the most efficient proof uses a notion from Section 4.10 so you will not actually do this until Exercise 4.10.17. Granting this for the moment it follows that the map $p \to \det(A^i{}_j(p))$ is a smooth real-valued function on $W \cap W'$. Since $\det(A^i{}_j(p_0)) > 0$, $\det(A^i{}_j(p)) > 0$ for all p in some open nbd of p_0 in X. Thus, $\mu'_p = \mu_p$ on this nbd and it follows that H is open X and therefore in U.

Exercise 4.9.6 Conclude the proof by showing that $U - H$ is open in X. ■

Taking $U = X$ in Lemma 4.9.1 yields the following result.

Theorem 4.9.2 *A connected orientable manifold admits just two orientations.*

Exercise 4.9.7 Show that if (U, φ) and (V, ψ) are two charts on an orientable manifold with U and V connected, then the determinant of the Jacobian of $\psi \circ \varphi^{-1}$ cannot change sign on its domain. **Hint:** Lemma 4.9.1.

We use Exercise 4.9.7 to show that the real projective plane $\mathbb{RP}^2$ is *not* orientable. Recall that $\mathbb{RP}^2$ has an atlas $\{(U_i, \varphi_i)\}_{i=1,2,3}$, where $U_i = \{[x^1, x^2, x^3] \in \mathbb{RP}^2 : x^i \neq 0\}$ and $\varphi_1([x^1, x^2, x^3]) = (\frac{x^2}{x^1}, \frac{x^3}{x^1})$, $\varphi_2([x^1, x^2, x^3]) = (\frac{x^1}{x^2}, \frac{x^3}{x^2})$, and $\varphi_3([x^1, x^2, x^3]) = (\frac{x^1}{x^3}, \frac{x^2}{x^3})$. Thus, for example, $\varphi_2 \circ \varphi_1^{-1}(x, y) = \varphi_2([1, x, y]) = (\frac{1}{x}, \frac{y}{x})$ on the set of all $(x, y) \in \mathbb{R}^2$ with $x \neq 0$. Consequently, the Jacobian of $\varphi_2 \circ \varphi_1^{-1}$ is

$$\begin{pmatrix} -\dfrac{1}{x^2} & 0 \\ -\dfrac{y}{x^2} & \dfrac{1}{x} \end{pmatrix}$$

and its determinant is $-\frac{1}{x^3}$. This determinant is positive when $x < 0$ and negative when $x > 0$ and therefore changes sign on its domain. Since U_1

and U_2 are both connected (being images of connected sets in S^2 under the projection), Exercise 4.9.7 implies that $\mathbb{RP}^2$ cannot be orientable.

Exercise 4.9.8 Show that $\mathbb{RP}^{n-1}$ is orientable when n is even and nonorientable when n is odd.

Exercise 4.9.9 Let X be an orientable manifold with orientation μ and (U, φ) a chart on X. Then (U, φ) is **consistent with** μ if the orientation determined on U by φ is the restriction of μ to U. Show that X has an atlas of charts that are consistent with μ. Conclude that a manifold is orientable iff it admits an oriented atlas.

Exercise 4.9.10 Show that all of the following are orientable: (a) any open submanifold of $\mathbb{R}^n$, (b) any product of orientable manifolds, and (c) any 1-dimensional manifold.

Suppose $\mathcal{V}$ and $\mathcal{W}$ are oriented, n-dimensional, real vector spaces and $L : \mathcal{V} \to \mathcal{W}$ is an isomorphism. Then L carries any basis for $\mathcal{V}$ onto a basis for $\mathcal{W}$. We say that L is **orientation preserving** if, for every $\{e_1, \ldots, e_n\}$ in the orientation for $\mathcal{V}$, $\{L(e_1), \ldots, L(e_n)\}$ is in the orientation for $\mathcal{W}$.

Exercise 4.9.11 Show that L is orientation preserving iff $\{L(e_1), \ldots, L(e_n)\}$ is in the orientation for $\mathcal{W}$ for *some* $\{e_1, \ldots, e_n\}$ in the orientation for $\mathcal{V}$.

If L is not orientation preserving it is said to be **orientation reversing**. Choosing bases for $\mathcal{V}$ and $\mathcal{W}$ we may identify the isomorphisms of $\mathcal{V}$ onto $\mathcal{W}$ with elements of $GL(n, \mathbb{R})$ and then the orientation preserving/reversing isomorphisms correspond to its two connected components ($\det > 0$ and $\det < 0$).

Now let X and Y be two oriented manifolds of the same dimension n and $f : X \to Y$ a smooth map. If f is a diffeomorphism, then f_{*p} is an isomorphism for every $p \in X$ and we will say that f is **orientation preserving** (respectively, **reversing**) if f_{*p} is orientation preserving (respectively, reversing) for each $p \in X$.

Exercise 4.9.12 Show that, if X is connected, then any diffeomorphism $f : X \to Y$ is either orientation preserving or orientation reversing. Also, construct an example to show that this need not be the case if X has more than one component.

Exercise 4.9.13 Show that $f_{*p} : T_p(X) \to T_{f(p)}(Y)$ is orientation preserving iff for some (any) chart (U, φ) consistent with the orientation of X

and with $p \in U$ and some (any) chart (V, ψ) consistent with the orientation of Y and with $f(p) \in V$, the Jacobian of $\psi \circ f \circ \varphi^{-1}$ has positive determinant at $\varphi(p)$. **Hint:** Exercise 4.4.10.

Exercise 4.9.14 Suppose $\mathbb{R}^n$ and S^n have their standard orientations. Show that $\varphi_S : U_S \to \mathbb{R}^n$ is orientation preserving and $\varphi_N : U_N \to \mathbb{R}^n$ is orientation reversing.

Suppose now that $f : X \to Y$ is smooth, but not necessarily a diffeomorphism. Let $q \in Y$ be a regular value of f.

Remark: According to Theorem 3-14 of [**Sp 1**], the set of critical values of a smooth map $g : A \to \mathbb{R}^n$, A open in $\mathbb{R}^n$, has measure zero in $\mathbb{R}^n$. In particular, regular values always exist. Applying this to any coordinate expression for $f : X \to Y$ we find that it too must have regular values (lots of them).

Corollary 4.5.7 implies that $f^{-1}(q)$ is either empty or a submanifold of X of dimension $n - n = 0$. A manifold of dimension zero is a discrete space. Now let us assume in addition that X is compact. Then a discrete subspace of X must be finite so $f^{-1}(q)$ is a finite (possibly empty) set. If $p \in f^{-1}(q)$, then $f_{*p} : T_p(X) \to T_q(Y)$ is an isomorphism (being a surjective linear map between vector spaces of the same dimension). The orientations of X and Y orient $T_p(X)$ and $T_q(Y)$ so f_{*p} is either orientation preserving or orientation reversing. Define the **sign of** f **at** p, denoted sign (f, p) to be 1 if f_{*p} is orientation preserving and -1 if f_{*p} is orientation reversing. Then the **degree of** f **over** q, denoted $\deg(f, q)$, is defined by

$$\deg(f,q) = \begin{cases} 0, & \text{if } f^{-1}(q) = \emptyset \\ \displaystyle\sum_{p \in f^{-1}(q)} \operatorname{sign}(f,p), & \text{if } f^{-1}(q) \neq \emptyset \end{cases} .$$

It is not difficult to get a feel for what $\deg(f, g)$ means when $f^{-1}(q) \neq \emptyset$. Using Corollary 4.4.8 and the fact that $f^{-1}(q)$ is finite we may select connected open nbds W of q and V_p of each $p \in f^{-1}(q)$ such that f carries each V_p diffeomorphically onto W. Restricting the orientations of X and Y to each V_p and W, respectively, each $f|V_p$ will either preserve or reverse orientation depending on sign (f, p). If each sign (f, p) is 1, then $\deg(f, q)$ is just the number of times f maps onto q, whereas, if each sign (f, p) is -1, it is minus this number. If some are positive and some are negative, $\deg(f, q)$ is the "net" number of times f maps onto q, where, as it were, p_1 and p_2 in $f^{-1}(q)$ "cancel" each other if $f|V_{p_1}$ and $f|V_{p_2}$ cover W in opposite senses ("directions" in the 1-dimensional case).

A manifold that is both compact and connected is said to be **closed**. If X and Y are closed, oriented, n-dimensional manifolds and $f : X \to Y$ is a smooth map, then $\deg(f, q)$ can be shown to be the same for every regular value q of f.

Remark: This, together with most of the results to which we refer below and a few lovely applications, is proved concisely and elegantly in the beautiful little book [**Miln**] of Milnor. Since we could not hope to improve upon the treatment found there we will leave to the reader the great pleasure of reading one of the masters.

As a result one can define the **degree** of f, denoted $\deg f$, to be $\deg(f, q)$, where q is *any* regular value of f. Note that the degree of any nonsurjective map is necessarily 0 (any $q \in Y - f(X)$ is a regular value) and the degree of the identity map is 1. Furthermore, a diffeomorphism $f : X \to Y$ must have $\deg f = \pm 1$ depending on whether it is orientation preserving or reversing. Less obvious, but nevertheless true, is the fact that degree is a homotopy invariant, i.e., if $f, g : X \to Y$ and $f \simeq g$, then $\deg f = \deg g$.

A particularly important special case is that in which $X = Y = S^n$, where $n \geq 1$ and S^n is equipped with its standard orientation. Thus, any smooth map $f : S^n \to S^n$ has associated with it an integer $\deg f$ and homotopic smooth maps are assigned the same degree. Actually, standard approximation theorems (see, e.g., [**Hir**]) show that any *continuous* map $f : S^n \to S^n$ can be arbitrarily well approximated by a smooth map homotopic to it and with this one can define the degree of any continuous map of the sphere to itself. Homotopy invariance then implies that deg can be regarded as a map of $[S^n, S^n]$ to the integers $\mathbb{Z}$. But $[S^n, S^n]$ can be identified with $\pi_n(S^n)$ so we have $\deg : \pi_n(S^n) \to \mathbb{Z}$. This map is, in fact, a homomorphism and so maps onto $\mathbb{Z}$. The hard part of showing that $\pi_n(S^n) \cong \mathbb{Z}$ is a remarkable theorem of Heinz Hopf which asserts that, for maps of the sphere to itself, the degree completely determines the homotopy class, i.e., if $f, g : S^n \to S^n$, then $f \simeq g$ *if and only if* $\deg f = \deg g$. The culmination of Milnor's book [**Miln**] is the proof, via Pontryagin's notion of framed cobordism, of a more general result that implies Hopf's theorem.

4.10 2-Forms and Riemannian Metrics

Let $\mathcal{V}$ be an n-dimensional real vector space. The dual space is the set of all real-valued linear maps on $\mathcal{V}$ and is denoted $\mathcal{V}^*$, or sometimes $\mathcal{T}^1(\mathcal{V})$. The dual space has a natural (pointwise) vector space structure of its own. We denote by $\mathcal{T}^2(\mathcal{V})$ the set of all real-valued *bilinear* maps on $\mathcal{V} \times \mathcal{V}$. Thus, an element of $\mathcal{T}^2(\mathcal{V})$ is a map $A : \mathcal{V} \times \mathcal{V} \to \mathbb{R}$ satisfying $A(a_1 v_1 + a_2 v_2, w) = a_1 A(v_1, w) + a_2 A(v_2, w)$ and $A(w, a_1 v_1 + a_2 v_2) = a_1 A(w, v_1) + a_2 A(w, v_2)$ for all $a_1, a_2 \in \mathbb{R}$ and $v_1, v_2, w \in \mathcal{V}$. $\mathcal{T}^2(\mathcal{V})$ admits a natural (pointwise) vector space structure: For $A, A_1, A_2 \in \mathcal{T}^2(\mathcal{V})$ and $a \in \mathbb{R}$ one defines $A_1 + A_2$ and aA in $\mathcal{T}^2(\mathcal{V})$ by $(A_1 + A_2)(v, w) = A_1(v, w) + A_2(v, w)$ and $(aA)(v, w) = aA(v, w)$ for all $v, w \in \mathcal{V}$. The elements of $\mathcal{T}^2(\mathcal{V})$ are called **covariant tensors of rank 2 on** $\mathcal{V}$. An element A of $\mathcal{T}^2(\mathcal{V})$ is said to be **symmetric** (respectively, **skew-symmetric**) if $A(w, v) = A(v, w)$ (respectively, $A(w, v) = -A(v, w)$) for all $v, w \in \mathcal{V}$. Thus, an inner product (or scalar product) on $\mathcal{V}$ in the usual sense is seen to be a symmetric, covariant tensor g of rank 2 on $\mathcal{V}$ that is **nondegenerate** ($g(v, w) = 0$ for all $v \in \mathcal{V}$ implies $w = 0$) and **positive definite** ($g(v, v) \geq 0$ for all $v \in \mathcal{V}$ with $g(v, v) = 0$ only when $v = 0$).

If α and β are elements of $\mathcal{V}^*$ we define an element of $\mathcal{T}^2(\mathcal{V})$, called the **tensor product of** α **and** β and denoted $\alpha \otimes \beta$, by $(\alpha \otimes \beta)(v, w) = \alpha(v)\, \beta(w)$. In particular, if $\{e_1, \ldots, e_n\}$ is a basis for $\mathcal{V}$ and $\{e^1, \ldots, e^n\}$ is the dual basis for $\mathcal{V}^*$ (so $e^i(e_j) = \delta^i_j$), then each $e^i \otimes e^j$ is an element of $\mathcal{T}^2(\mathcal{V})$ and we claim that the set of all such form a basis for $\mathcal{T}^2(\mathcal{V})$.

Lemma 4.10.1 *If $\{e_1, \ldots, e_n\}$ is a basis for $\mathcal{V}$ and $\{e^1, \ldots, e^n\}$ is the dual basis for $\mathcal{V}^*$, then $\{e^i \otimes e^j : i, j = 1, \ldots, n\}$ is a basis for $\mathcal{T}^2(\mathcal{V})$ and each $A \in \mathcal{T}^2(\mathcal{V})$ can be written $A = A_{ij}\, e^i \otimes e^j$, where $A_{ij} = A(e_i, e_j)$. In particular,* $\dim \mathcal{T}^2(\mathcal{V}) = n^2$.

Proof: If $A \in \mathcal{T}^2(\mathcal{V})$, then, for any $v = v^i e_i$ and $w = w^j e_j$ in $\mathcal{V}$, $A(v, w) = A(v^i e_i, w^j e_j) = v^i w^j A(e_i, e_j) = v^i w^j A_{ij} = A_{ij} v^i w^j$. But $A_{ij}\, e^i \otimes e^j(v, w) = A_{ij}\, e^i(v)\, e^j(w) = A_{ij}\, e^i(v^k e_k)\, e^j(w^l e_l) = A_{ij}(v^k e^i(e_k))(w^l e^j(e_l)) = A_{ij}(v^k \delta^i_k)(w^l \delta^i_l) = A_{ij} v^i w^j = A(v, w)$. Thus, $A = A_{ij}\, e^i \otimes e^j$ and, in particular, $\{e^i \otimes e^j : i, j = 1, \ldots, n\}$ spans $\mathcal{T}^2(\mathcal{V})$. To prove linear independence we suppose $A_{ij}\, e^i \otimes e^j$ is the zero element of $\mathcal{T}^2(\mathcal{V})$. Then, for all k and l, $0 = (A_{ij}\, e^i \otimes e^j)(e_k, e_l) = A_{ij}(e^i(e_k))(e^j(e_l)) = A_{ij}\, \delta^i_k \delta^j_l = A_{kl}$, as required. ■

Exercise 4.10.1 Show that, for all α, β and γ in $\mathcal{V}^*$ and all $a \in \mathbb{R}$, $(\alpha + \beta) \otimes \gamma = \alpha \otimes \gamma + \beta \otimes \gamma$, $\alpha \otimes (\beta + \gamma) = \alpha \otimes \beta + \alpha \otimes \gamma$, $(a\alpha) \otimes \beta = \alpha \otimes (a\beta) = a(\alpha \otimes \beta)$. Show also that $\beta \otimes \alpha$ is generally not equal to $\alpha \otimes \beta$.

We conclude, in particular, from Lemma 4.10.1 that if g is an inner product on $\mathcal{V}$ and $\{e_1, \ldots, e_n\}$ is a basis for $\mathcal{V}$, then $g = g_{ij}\, e^i \otimes e^j$, where the matrix (g_{ij}) of components is symmetric, invertible and positive definite. We will denote the inverse of this matrix by (g^{ij}). A well-known theorem in linear algebra (the Corollary in Section 2, Chapter $\underline{\text{VI}}$ of [**Lang**], or Theorem 4-2 of [**Sp 1**]) asserts that, for any such g, there exists an **orthonormal basis**, i.e., a basis $\{e_1, \ldots, e_n\}$ for $\mathcal{V}$, with $g_{ij} = \delta_{ij}$, for all $i, j = 1, \ldots, n$.

The tensor product $\alpha \otimes \beta$ of two covectors is generally neither symmetric nor skew-symmetric, but it is possible to "symmetrize" and "skew-symmetrize" it by averaging. Specifically, we define the **symmetric product** of α and β to be $\frac{1}{2}(\alpha \otimes \beta + \beta \otimes \alpha)$, while the **skew-symmetric product** is $\frac{1}{2}(\alpha \otimes \beta - \beta \otimes \alpha)$. We will have little occasion to use the symmetric product, but the skew-symmetric product (without the $\frac{1}{2}$) comes up often enough to deserve a symbol of its own. For any two elements α and β of $\mathcal{V}^*$ we define the **wedge product** of α and β, denoted $\alpha \wedge \beta$, by

$$\alpha \wedge \beta = \alpha \otimes \beta - \beta \otimes \alpha .$$

Exercise 4.10.2 Show that $\alpha \wedge \beta$ is skew-symmetric and that $(\alpha + \beta) \wedge \gamma = \alpha \wedge \gamma + \beta \wedge \gamma$, $\alpha \wedge (\beta + \gamma) = \alpha \wedge \beta + \alpha \wedge \gamma$, $(a\alpha) \wedge \beta = \alpha \wedge (a\beta) = a(\alpha \wedge \beta)$ and $\beta \wedge \alpha = -\alpha \wedge \beta$ for all $\alpha, \beta, \gamma \in \mathcal{V}^*$ and all $a \in \mathbb{R}$.

We denote by $\Lambda^2(\mathcal{V})$ the linear subspace of $\mathcal{T}^2(\mathcal{V})$ consisting of all skew-symmetric elements.

Lemma 4.10.2 *If $\{e_1, \ldots, e_n\}$ is a basis for $\mathcal{V}$ and $\{e^1, \ldots, e^n\}$ is the dual basis for $\mathcal{V}^*$, then $\{e^i \wedge e^j : 1 \leq i < j \leq n\}$ is a basis for $\Lambda^2(\mathcal{V})$ and each $\Omega \in \Lambda^2(\mathcal{V})$ can be written $\Omega = \sum_{i<j} \Omega_{ij}\, e^i \wedge e^j = \frac{1}{2}\Omega_{ij}\, e^i \wedge e^j$, where $\Omega_{ij} = \Omega(e_i, e_j)$. In particular, $\dim \Lambda^2(\mathcal{V}) = \binom{n}{2} = \frac{n(n-1)}{2}$.*

Proof: Since $\Lambda^2(\mathcal{V}) \subseteq \mathcal{T}^2(\mathcal{V})$ we can write $\Omega = \Omega_{ij}\, e^i \otimes e^j$, where $\Omega_{ij} = \Omega(e_i, e_j)$. But Ω is skew-symmetric so $\Omega_{ji} = -\Omega_{ij}$ and, in particular, $\Omega_{ii} = 0$ for $i = 1, \ldots, n$. Thus,

$$\begin{aligned}
\Omega &= \sum_{i<j} \Omega_{ij}\, e^i \otimes e^j + \sum_{j<i} \Omega_{ij}\, e^i \otimes e^j = \sum_{i<j} \Omega_{ij}\, e^i \otimes e^j + \sum_{j<i} (-\Omega_{ji}) e^i \otimes e^j \\
&= \sum_{i<j} \Omega_{ij}\, e^i \otimes e^j - \sum_{i<j} \Omega_{ij}\, e^j \otimes e^i = \sum_{i<j} \Omega_{ij} \left(e^i \otimes e^j - e^j \otimes e^i\right) \\
&= \sum_{i<j} \Omega_{ij}\, e^i \wedge e^j
\end{aligned}$$

so $\{e^i \wedge e^j : 1 \leq i < j \leq n\}$ spans $\Lambda^2(\mathcal{V})$.

Exercise 4.10.3 Complete the argument by proving linear independence. ∎

The case of interest to us, of course, is that in which $\mathcal{V}$ is $T_p(X)$ for some point p in a smooth manifold X. A **covariant tensor field of rank 2** on a smooth manifold X is a map $\mathbf{A}$ that assigns to each $p \in X$ an element $\mathbf{A}(p)$, also written $\mathbf{A}_p$, of $\mathcal{T}^2(T_p(X))$. If (U, φ) is a chart on X with coordinate functions x^i, $i = 1, \ldots, n$, and $p \in U$, then, by Lemma 4.10.1, $\mathbf{A}(p)$ can be written $\mathbf{A}(p) = \mathbf{A}_p(\frac{\partial}{\partial x^i}|_p, \frac{\partial}{\partial x^j}|_p)\, dx^i{}_p \otimes dx^j{}_p$. The real-valued functions $A_{ij} : U \to \mathbb{R}$ defined by $A_{ij}(p) = \mathbf{A}_p(\frac{\partial}{\partial x^i}|_p, \frac{\partial}{\partial x^j}|_p)$ are called the **components of A relative to** (U, φ). If these component functions are continuous, or C^∞ for all charts in (some atlas for) the differentiable structure for X, then we say that $\mathbf{A}$ itself is **continuous**, or $\mathbf{C}^\infty$ (**smooth**). We denote by $\mathcal{T}^2(X)$ the set of all C^∞ covariant tensor fields of rank 2 on X and provide it with the obvious (pointwise) structure of a real vector space and a module over $C^\infty(X)$: For $\mathbf{A}, \mathbf{B} \in \mathcal{T}^2(X)$, $a \in \mathbb{R}$ and $f \in C^\infty(X)$ we define $(\mathbf{A} + \mathbf{B})(p) = \mathbf{A}(p) + \mathbf{B}(p)$, $(a\mathbf{A})(p) = a\mathbf{A}(p)$ and $(f\mathbf{A})(p) = f(p)\,\mathbf{A}(p)$. As for vector fields and 1-forms, any $\mathbf{A}$ defined only on an open subset of X may, via a bump function, be regarded as an element of $\mathcal{T}^2(X)$. An element $\mathbf{g}$ of $\mathcal{T}^2(X)$ which, at each $p \in X$, is symmetric, nondegenerate and positive definite is called a **Riemannian metric** on X. Thus, a Riemannian metric on X is a smooth assignment of an inner product to each tangent space of X. An element $\boldsymbol{\Omega}$ of $\mathcal{T}^2(X)$ which, at each $p \in X$, is skew-symmetric is called a **2-form** on X. The collection of all 2-forms on X is denoted $\Lambda^2(X)$ and it inherits all of the algebraic structure of $\mathcal{T}^2(X)$. A pair $(X, \mathbf{g})$ consisting of a smooth manifold X and a Riemannian metric $\mathbf{g}$ on X is called a **Riemannian manifold**.

To produce examples, one may start with 1-forms $\boldsymbol{\Theta}_1$ and $\boldsymbol{\Theta}_2$ on X and define the tensor and wedge products pointwise, i.e., $(\boldsymbol{\Theta}_1 \otimes \boldsymbol{\Theta}_2)(p) = \boldsymbol{\Theta}_1(p) \otimes \boldsymbol{\Theta}_2(p)$ and $(\boldsymbol{\Theta}_1 \wedge \boldsymbol{\Theta}_2)(p) = \boldsymbol{\Theta}_1(p) \wedge \boldsymbol{\Theta}_2(p)$ for each $p \in X$.

Exercise 4.10.4 Show that, if $\boldsymbol{\Theta}_1, \boldsymbol{\Theta}_2 \in \mathcal{X}^*(X)$, then $\boldsymbol{\Theta}_1 \otimes \boldsymbol{\Theta}_2$ and $\boldsymbol{\Theta}_1 \wedge \boldsymbol{\Theta}_2$ are in $\mathcal{T}^2(X)$ and $\boldsymbol{\Theta}_1 \wedge \boldsymbol{\Theta}_2$ is in $\Lambda^2(X)$.

Another useful procedure is the pullback. Suppose $f : X \to Y$ is a smooth map and $\mathbf{A} \in \mathcal{T}^2(Y)$. Then the **pullback** of $\mathbf{A}$ to X by f, denoted $f^*\mathbf{A}$, is defined as follows: For each $p \in X$ and all $\mathbf{v}, \mathbf{w} \in T_p(X)$ we define $(f^*\mathbf{A})_p(\mathbf{v}, \mathbf{w}) = \mathbf{A}_{f(p)}(f_{*p}(\mathbf{v}), f_{*p}(\mathbf{w}))$.

Exercise 4.10.5 Let (U, φ) be a chart on X with coordinate functions $x^1, \ldots, x^n$ and (V, ψ) a chart on Y with coordinate functions $y^1, \ldots, y^m$ and with $f(U) \subseteq V$. Write $\psi \circ f \circ \varphi^{-1} = (f^1, \ldots, f^m)$ and let A_{ij} be the components of $\mathbf{A}$ relative to (V, ψ). Show that the components of $f^*\mathbf{A}$

relative to (U, φ) are given by

$$\frac{\partial f^i}{\partial x^\mu}\frac{\partial f^j}{\partial x^\nu}\left(A_{ij} \circ f\right)$$

for $\mu, \nu = 1, \ldots, n$. Conclude that $f^*\mathbf{A} \in \mathcal{T}^2(X)$.

Exercise 4.10.6 Show that if $\mathbf{g}$ is a Riemannian metric on Y, then $f^*\mathbf{g}$ is a Riemannian metric on X and if $\boldsymbol{\Omega}$ is a 2-form on Y, then $f^*\boldsymbol{\Omega}$ is a 2-form on X.

As usual, if $\iota : X \hookrightarrow Y$ is the inclusion map, then $\iota^*\mathbf{A}$ is called the **restriction of A to** X. For example, the **standard metric on** $\mathbb{R}^{n+1}$ assigns to each $T_p(\mathbb{R}^{n+1}) \cong \mathbb{R}^{n+1}$ the usual Euclidean inner product, i.e., in standard coordinates $x^1, \ldots, x^{n+1}$ it is given by $\mathbf{g} = dx^1 \otimes dx^1 + \cdots + dx^{n+1} \otimes dx^{n+1}$, and the **standard metric on** S^n is its restriction $\iota^*\mathbf{g}$ to S^n, where $\iota : S^n \hookrightarrow \mathbb{R}^{n+1}$.

Exercise 4.10.7 Let $f : X \to Y$ and $g : Y \to Z$ be C^∞. Show that, for any $\mathbf{A} \in \mathcal{T}^2(Z)$, $(g \circ f)^*\mathbf{A} = f^*(g^*\mathbf{A})$, i.e., (4.6.15) remains valid for covariant tensor fields of rank 2.

Exercise 4.10.8 Let $f : X \to Y$ be C^∞ and suppose $\boldsymbol{\Theta}_1$ and $\boldsymbol{\Theta}_2$ are in $\boldsymbol{\mathcal{X}}^*(Y)$. Show that $f^*(\boldsymbol{\Theta}_1 \otimes \boldsymbol{\Theta}_2) = f^*\boldsymbol{\Theta}_1 \otimes f^*\boldsymbol{\Theta}_2$ and $f^*(\boldsymbol{\Theta}_1 \wedge \boldsymbol{\Theta}_2) = f^*\boldsymbol{\Theta}_1 \wedge f^*\boldsymbol{\Theta}_2$.

Suppose X_1 and X_2 are smooth manifolds with Riemannian metrics $\mathbf{g}_1$ and $\mathbf{g}_2$, respectively, and suppose $f : X_1 \to X_2$ is a diffeomorphism of X_1 onto X_2. We say that f is a **conformal diffeomorphism** (and X_1 and X_2 are **conformally equivalent**) if $f^*\mathbf{g}_2 = \lambda\mathbf{g}_1$ for some positive C^∞ function $\lambda : X_1 \to \mathbb{R}$. If the function λ is identically equal to 1, f is called an **isometry** and X_1 and X_2 are said to be **isometric**.

For example, let $\mathbf{g} = dx^1 \otimes dx^1 + \cdots + dx^{n+1} \otimes dx^{n+1}$ be the standard metric on $\mathbb{R}^{n+1}$, $\iota : S^n \hookrightarrow \mathbb{R}^{n+1}$ the inclusion and $\iota^*\mathbf{g}$ the standard metric on S^n. Let $U_S = S^n - \{(0, \ldots, 0, 1)\}$ and $\varphi_S : U_S \to \mathbb{R}^n$ be the stereographic projection from the north pole. We let $f = \varphi_S^{-1} : \mathbb{R}^n \to U_S$ so that $\iota \circ f(y) = (1 + \|y\|^2)^{-1}(2y^1, \ldots, 2y^n, \|y\|^2 - 1)$, where $y^1, \ldots, y^n$ are standard coordinates on $\mathbb{R}^n$. Then $x^i = y^i \circ \varphi_S$, $i = 1, \ldots, n$, are coordinates on U_S. We compute the pullback to $\mathbb{R}^n$ by f of the standard metric $\iota^*\mathbf{g}$ on U_S, i.e., $f^*(\iota^*\mathbf{g}) = (\iota \circ f)^*\mathbf{g}$ (technically, there should be another inclusion $U_S \hookrightarrow S^n$ here, but, for the sake of everyone's sanity, we'll suppress that one). Fix $p \in \mathbb{R}^n$ and $\mathbf{v}_p, \mathbf{w}_p \in T_p(\mathbb{R}^n)$. Then

$((\iota \circ f)^*\mathbf{g})_p(\mathbf{v}_p, \mathbf{w}_p) = g_{(\iota \circ f)(p)}((\iota \circ f)_{*p}(\mathbf{v}_p), (\iota \circ f)_{*p}(\mathbf{w}_p))$. Now,

$$(\iota \circ f)_{*p}(\mathbf{v}_p) = (\iota \circ f)_{*p}\left(\frac{d}{dt}(p+tv)\bigg|_{t=0}\right)$$

$$= \frac{d}{dt}\Big(\big(1+\|p+tv\|^2\big)^{-1}\big(2p^1+2tv^1, \ldots,$$

$$2p^n+2tv^n, \|p+tv\|^2 - 1\big)\Big|_{t=0}.$$

Exercise 4.10.9 Compute these derivatives componentwise and show that

$$(\iota \circ f)_{*p}(\mathbf{v}_p) = (1+\|p\|^2)^{-2}\big(2(1+\|p\|^2)v^1 - 4<p,v>p^1, \ldots,$$

$$2(1+\|p\|^2)v^n - 4<p,v>p^n, 4<p,v>\big),$$

where $<p,v> = p^1v^1 + \cdots + p^nv^n$ is the standard inner product on $\mathbb{R}^n$.

Using this and the analogous expression for $(\iota \circ f)_{*p}(\mathbf{w}_p)$ one computes $\mathbf{g}_{(\iota \circ f)(p)}((\iota \circ f)_{*p}(\mathbf{v}_p), (\iota \circ f)_{*p}(\mathbf{w}_p))$ as the sum of the products of corresponding components (usual inner product in $\mathbb{R}^{n+1}$). The result is $4(1+\|p\|^2)^{-2} <v,w>$ so we obtain

$$(f^*(\iota^* g))_p(\mathbf{v}_p, \mathbf{w}_p) = \frac{4}{(1+\|p\|^2)^2} <v, w>.$$

Denoting by $\mathbf{g}_{S^n}$ the standard metric on S^n (restricted to U_S) and by $\mathbf{g}_{\mathbb{R}^n}$ the standard metric on $\mathbb{R}^n$ and defining $\lambda : \mathbb{R}^n \to \mathbb{R}$ by $\lambda(p) = 4(1+\|p\|^2)^{-2}$ we may write our result as

$$(\varphi_S^{-1})^* \mathbf{g}_{S^n} = \lambda \mathbf{g}_{\mathbb{R}^n}.$$

Thus, φ_S^{-1} is a conformal diffeomorphism of $\mathbb{R}^n$ onto U_S, i.e., $\mathbb{R}^n$ is conformally equivalent to the open submanifold U_S of S^n. Similar calculations give the same result for U_N so we say that S^n is **locally conformally equivalent** to $\mathbb{R}^n$.

Exercise 4.10.10 Show that compositions and inverses of conformal diffeomorphisms are likewise conformal diffeomorphisms.

Exercise 4.10.11 Let $(X_1, \mathbf{g}_1)$ and $(X_2, \mathbf{g}_2)$ be Riemannian manifolds and $f : X_1 \to X_2$ a diffeomorphism. Show that f is a conformal diffeomorphism iff there exists a positive C^∞ function $\lambda : X_1 \to \mathbb{R}$ such that $(f^*\mathbf{g}_2)_p(\mathbf{v}_p, \mathbf{v}_p) = \lambda(p)(\mathbf{g}_1)_p(\mathbf{v}_p, \mathbf{v}_p)$ for all $p \in X_1$ and $\mathbf{v}_p \in T_p(X_1)$.

Hint: Any inner product $< \, , \, >$ satisfies the *polarization identity* $<x,y> = \frac{1}{4}(<x+y,x+y> \; - \; <x-y,x-y>)$.

Let's compute a few more concrete examples. First define $f : \mathbb{H} - \{0\} \to \mathbb{H} - \{0\}$ by $f(q) = q^{-1} = \frac{\bar{q}}{|q|^2}$ (the same calculations will also work for inversion on $\mathbb{C} - \{0\}$). We will denote by $\mathbf{g}$ the usual metric on $\mathbb{H} = \mathbb{R}^4$ (restricted to $\mathbb{H} - \{0\}$) and compute $f^*\mathbf{g}$. Fix a $p \in \mathbb{H} - \{0\}$ and a $\mathbf{v}_p \in T_p(\mathbb{H} - \{0\}) = T_p(\mathbb{H})$. Then, writing $\mathbf{v}_p = \frac{d}{dt}(p+tv)\,|_{t=0}$, $\mathbf{g}_p(\mathbf{v}_p, \mathbf{v}_p) = \|\, v \,\|^2 = v\,\bar{v}$. Now, $(f^*\mathbf{g})_p(\mathbf{v}_p, \mathbf{v}_p) = \mathbf{g}_{f(p)}(f_{*p}(\mathbf{v}_p), f_{*p}(\mathbf{v}_p))$ and

$$f_{*p}(\mathbf{v}_p) = \frac{d}{dt}\,(p+tv)^{-1}\Big|_{t=0} = \frac{d}{dt}\left(\frac{\bar{p}+t\,\bar{v}}{|p+tv|^2}\right)_{t=0}.$$

Exercise 4.10.12 Show that the point in $\mathbb{H}$ corresponding to $f_{*p}(\mathbf{v}_p)$ under the canonical isomorphism is

$$\frac{\bar{v}}{|p|^2} - \frac{2\,\mathrm{Re}\,(v\,\bar{p})\,\bar{p}}{|p|^4}.$$

Hint: Exercise 4.8.18.

Thus, $(f^*\mathbf{g})_p(\mathbf{v}_p, \mathbf{v}_p)$ is the product of this last expression with its conjugate.

Exercise 4.10.13 Show that $(f^*\mathbf{g})_p(\mathbf{v}_p, \mathbf{v}_p) = \frac{1}{|p|^4}\,|v|^2 = \frac{1}{|p|^4}\,\mathbf{g}_p(\mathbf{v}_p, \mathbf{v}_p)$.

Thus, $\mathbf{g}_p(\mathbf{v}_p, \mathbf{v}_p) = |p|^4\,(f^*\mathbf{g})_p(\mathbf{v}_p, \mathbf{v}_p)$ and it follows from Exercise 4.10.11 that f (i.e., inversion) is a conformal diffeomorphism of $\mathbb{H} - \{0\}$ onto itself. Moreover, $f^*\mathbf{g} = \lambda\mathbf{g}$, where $\lambda(p) = |p|^{-4}$ for all $p \in \mathbb{H} - \{0\}$.

Exercise 4.10.14 Let A, B and C be fixed quaternions with A and B nonzero. Show that the linear maps $q \to Aq$, $q \to qB$ and $q \to q+C = C+q$ are all conformal diffeomorphisms of $\mathbb{H}$ onto itself, as are their compositions.

Finally, let us consider the **quaternionic fractional linear transformation** $q \to (aq+b)(cq+d)^{-1}$, where $\begin{pmatrix} a & b \\ c & d \end{pmatrix} \in SL(2, \mathbb{H})$ (see Section 1.2 and note again that the following arguments would apply equally well in the complex case). Since any element of $SL(2, \mathbb{H})$ is invertible, not both of a and c can be zero. If $c = 0$, then $a \neq 0$ and $d \neq 0$ so the map is a linear transformation $q \to aqd^{-1} + bd^{-1}$ and so is a conformal diffeomorphism of $\mathbb{H}$ onto itself by Exercise 4.10.14. Henceforth we assume that

$c \neq 0$. Writing $(aq+b)(cq+d)^{-1} = ac^{-1} + (b - ac^{-1}d)(cq+d)^{-1}$ (obtained by "long division") we find that our map is the composition

$$q \longrightarrow cq + d \longrightarrow (cq + d)^{-1} \longrightarrow (b - ac^{-1}d)(cq + d)^{-1}$$
$$\longrightarrow ac^{-1} + (b - ac^{-1}d)(cq + d)^{-1}.$$

Exercise 4.10.15 Show that, for $\begin{pmatrix} a & b \\ c & d \end{pmatrix} \in SL(2, \mathbb{H})$ with $c \neq 0$, $b - ac^{-1}d$ cannot be zero. **Hint:** Regard $\begin{pmatrix} a & b \\ c & d \end{pmatrix}$ as a linear transformation of $\mathbb{H}^2$ to $\mathbb{H}^2$ and suppose $b - ac^{-1}d = 0$.

Thus, on $\mathbb{H} - \{-c^{-1}d\}$, each of the maps in the above composition is a conformal diffeomorphism and, consequently, so is the composition itself. The conclusion then is that, for $\begin{pmatrix} a & b \\ c & d \end{pmatrix} \in SL(2, \mathbb{H})$, $q \to (aq + b)(cq + d)^{-1}$ is a conformal diffeomorphism on its domain. If $c = 0$, then $a \neq 0$ and it is a linear map of $\mathbb{H}$ onto $\mathbb{H}$. If $c \neq 0$ it maps $\mathbb{H} - \{-c^{-1}d\}$ onto $\mathbb{H} - \{ac^{-1}\}$ and its inverse is $q \to (q(-d) + b)(qc - a)^{-1}$.

Returning to the general development we record a local, smooth version of the familiar Gram-Schmidt orthonormalization process for Riemannian manifolds.

Proposition 4.10.3 *Let X be an n-dimensional smooth manifold with a Riemannian metric $\mathbf{g}$ and an orientation μ. Let U be a connected open subset of X on which are defined smooth vector fields $\mathbf{V}_1, \ldots, \mathbf{V}_n$ such that $\{\mathbf{V}_1(p), \ldots, \mathbf{V}_n(p)\} \in \mu_p$ for each $p \in U$. Then there exist smooth vector fields $\mathbf{E}_1, \ldots, \mathbf{E}_n$ on U such that, for each $p \in U$, $\{\mathbf{E}_1(p), \ldots, \mathbf{E}_n(p)\} \in \mu_p$ and $\mathbf{g}_p(\mathbf{E}_i(p), \mathbf{E}_j(p)) = \delta_{ij}$.*

Proof: For each $i, j = 1, \ldots, n$ and each $p \in U$, let $g_{ij}(p) = \mathbf{g}_p(\mathbf{V}_i(p), \mathbf{V}_j(p))$. Then each g_{ij} is a smooth real-valued function on U. Since $\mathbf{V}_1(p) \neq 0$ for each $p \in U$, g_{11} is a smooth positive function on U so we may define a smooth vector field $\mathbf{E}_1$ on U by

$$\mathbf{E}_1(p) = \left(g_{11}(p)\right)^{-\frac{1}{2}} \mathbf{V}_1(p).$$

Moreover,

$$\mathbf{g}_p\left(\mathbf{E}_1(p), \mathbf{E}_1(p)\right) = 1$$

and

$$\mathbf{g}_p\left(\mathbf{V}_i(p), \mathbf{E}_1(p)\right) = \left(g_{11}(p)\right)^{-\frac{1}{2}} \mathbf{g}_{i1}(p)$$

for each $p \in U$ and each $i = 1, \ldots, n$. The functions $h_{i1}(p) = (g_{11}(p))^{-\frac{1}{2}} \mathbf{g}_{i1}(p)$ are all C^∞ on U.

Now assume inductively that we have defined smooth vector fields $\mathbf{E}_1, \ldots, \mathbf{E}_k$ on U for some $1 \leq k < n$ such that

1. $\mathbf{g}_p(\mathbf{E}_i(p), \mathbf{E}_j(p)) = \delta_{ij}$ for $i, j = 1, \ldots, k$ and $p \in U$.
2. $\text{Span}\{\mathbf{E}_1(p), \ldots, \mathbf{E}_k(p)\} = \text{Span}\{\mathbf{V}_1(p), \ldots, \mathbf{V}_k(p)\}$ for each $p \in U$.
3. The functions $h_{ij}(p) = \mathbf{g}_p(\mathbf{V}_i(p), \mathbf{E}_j(p))$ are smooth on U for $i = 1, \ldots, n$ and $j = 1, \ldots, k$.

We construct $\mathbf{E}_{k+1}$ as follows: Consider the vector field $\mathbf{W}$ on U defined by

$$\mathbf{W}(p) = \mathbf{V}_{k+1}(p) - \sum_{j=1}^{k} h_{k+1\ j}(p)\, \mathbf{E}_j(p).$$

Then W is smooth by the induction hypothesis. Moreover, $\mathbf{W}(p)$ is nonzero for each $p \in U$ since $\sum_{j=1}^{k} h_{k+1\ j}(p)\, \mathbf{E}_j(p) \in \text{Span}\{\mathbf{E}_1(p), \ldots, \mathbf{E}_k(p)\} = \text{Span}\{\mathbf{V}_1(p), \ldots, \mathbf{V}_k(p)\}$, but $\mathbf{V}_{k+1}(p)$ is not in this span. Thus, we may let

$$\mathbf{E}_{k+1}(p) = (\mathbf{g}_p(\mathbf{W}(p), \mathbf{W}(p)))^{-\frac{1}{2}}\, \mathbf{W}(p)$$

for each $p \in U$.

Exercise 4.10.16 Show that $\mathbf{E}_{k+1}$ is a smooth vector field on U satisfying each of the following:

1. $\mathbf{g}_p(\mathbf{E}_i(p), \mathbf{E}_j(p)) = \delta_{ij}$ for $i, j = 1, \ldots, k+1$ and $p \in U$.
2. $\text{Span}\{\mathbf{E}_1(p), \ldots, \mathbf{E}_{k+1}(p)\} = \text{Span}\{\mathbf{V}_1(p), \ldots, \mathbf{V}_{k+1}(p)\}$ for each $p \in U$.
3. The functions $h_{i\ k+1}(p) = \mathbf{g}_p(\mathbf{V}_i(p), \mathbf{E}_{k+1}(p))$ are smooth on U for $i = 1, \ldots, n$.

The induction is therefore complete and we have smooth vector fields $\{\mathbf{E}_1, \ldots, \mathbf{E}_n\}$ on U satisfying $\mathbf{g}_p(\mathbf{E}_i(p), \mathbf{E}_j(p)) = \delta_{ij}$ at each point for $i, j = 1, \ldots, n$. In particular, $\{\mathbf{E}_1(p), \ldots, \mathbf{E}_n(p)\}$ must be a basis for $T_p(X)$ so $p \to [\mathbf{E}_1(p), \ldots, \mathbf{E}_n(p)]$ is an orientation on U. Since U is connected, Lemma 4.9.1 implies that this orientation is the restriction to U of either μ or $-\mu$. If it is the restriction of μ we are done. If it is the restriction of $-\mu$ we replace $\{\mathbf{E}_1, \mathbf{E}_2, \ldots, \mathbf{E}_n\}$ with $\{\mathbf{E}_2, \mathbf{E}_1, \ldots, \mathbf{E}_n\}$. ∎

Turning matters around a bit we have the following.

Exercise 4.10.17

(a) Let $\mathbf{V}_1, \ldots, \mathbf{V}_n$ be smooth vector fields defined on an open set U in the manifold X such that, for each $p \in U$, $\{\mathbf{V}_1(p), \ldots, \mathbf{V}_n(p)\}$ is a basis for $T_p(X)$. Show that there exists a Riemannian metric $\mathbf{g}$ defined on U relative to which $\{\mathbf{V}_1(p), \ldots, \mathbf{V}_n(p)\}$ is an orthonormal basis for each $p \in U$.

(b) Let $\mathbf{V}'_1, \ldots, \mathbf{V}'_n$ be smooth vector fields, also defined on U and also with the property that $\{\mathbf{V}'_1(p), \ldots, \mathbf{V}'_n(p)\}$ is a basis for each $p \in U$. Write $\mathbf{V}'_j(p) = A^i{}_j(p)\,\mathbf{V}_i(p)$ for $j = 1, \ldots, n$. Show that the functions $A^i{}_j(p)$ are C^∞ on U. **Hint:** Consider $\mathbf{g}(\mathbf{V}'_j, \mathbf{V}_i)$, where $\mathbf{g}$ is as in (a).

We saw in Section 4.6 that one may view a 1-form either as a smooth assignment of covectors, or as a $C^\infty(X)$-module homomorphism from $\boldsymbol{\mathcal{X}}(X)$ to $C^\infty(X)$. There is an analogous and equally useful reformulation of the definition of a covariant tensor field of rank 2. Suppose X is a manifold and $\mathbf{A} \in \mathcal{T}^2(X)$. Let $\mathbf{V}$ and $\mathbf{W}$ be two smooth vector fields on X and define a real-valued function $\mathbf{A}(\mathbf{V}, \mathbf{W})$ on X by $\mathbf{A}(\mathbf{V}, \mathbf{W})(p) = \mathbf{A}_p(\mathbf{V}_p, \mathbf{W}_p)$ for all $p \in X$. If (U, φ) is a chart on X and if we write, for each $p \in U$, $\mathbf{A}_p = A_{ij}(p)\,dx^i{}_p \otimes dx^j{}_p$, $\mathbf{V}_p = V^k(p)\frac{\partial}{\partial x^k}|_p$ and $\mathbf{W}_p = W^l(p)\frac{\partial}{\partial x^l}|_p$, then $\mathbf{A}(\mathbf{V}, \mathbf{W})(p) = A_{ij}(p)\,V^i(p)\,W^j(p)$ so $\mathbf{A}(\mathbf{V}, \mathbf{W}) \in C^\infty(X)$. Thus, $\mathbf{A}$ determines a map from $\boldsymbol{\mathcal{X}}(X) \times \boldsymbol{\mathcal{X}}(X)$ to $C^\infty(X)$ which, we claim, is $C^\infty(X)$-bilinear, i.e., satisfies

$$\begin{aligned} \mathbf{A}(\mathbf{V}_1 + \mathbf{V}_2, \mathbf{W}) &= \mathbf{A}(\mathbf{V}_1, \mathbf{W}) + \mathbf{A}(\mathbf{V}_2, \mathbf{W}) \\ \mathbf{A}(\mathbf{V}, \mathbf{W}_1 + \mathbf{W}_2) &= \mathbf{A}(\mathbf{V}, \mathbf{W}_1) + \mathbf{A}(\mathbf{V}, \mathbf{W}_2) \end{aligned} \tag{4.10.1}$$

and

$$\mathbf{A}(f\mathbf{V}, \mathbf{W}) = \mathbf{A}(\mathbf{V}, f\mathbf{W}) = f\,\mathbf{A}(\mathbf{V}, \mathbf{W}) \tag{4.10.2}$$

for all $\mathbf{V}, \mathbf{V}_1, \mathbf{V}_2, \mathbf{W}, \mathbf{W}_1, \mathbf{W}_2 \in \boldsymbol{\mathcal{X}}(X)$ and all $f \in C^\infty(X)$.

Exercise 4.10.18 Prove (4.10.1) and (4.10.2).

Suppose, conversely, that we have a $C^\infty(X)$-bilinear map $A : \boldsymbol{\mathcal{X}}(X) \times \boldsymbol{\mathcal{X}}(X) \to C^\infty(X)$. Fix a $p \in X$. Suppose $\mathbf{V}_1$ and $\mathbf{V}_2$ are two elements of $\boldsymbol{\mathcal{X}}(X)$. We claim that if *either* $\mathbf{V}_1$ or $\mathbf{V}_2$ vanishes at p, then $A(\mathbf{V}_1, \mathbf{V}_2)(p) = 0$. Indeed, suppose $\mathbf{V}_2(p) = 0$. Hold $\mathbf{V}_1$ fixed and define $A_{\mathbf{V}_1} : \boldsymbol{\mathcal{X}}(X) \to C^\infty(X)$ by $A_{\mathbf{V}_1}(\mathbf{W}) = A(\mathbf{V}_1, \mathbf{W})$. Then $A_{\mathbf{V}_1}$ satisfies the hypotheses of Lemma 4.6.1 so $\mathbf{V}_2(p) = 0$ implies $A_{\mathbf{V}_1}(\mathbf{V}_2)(p) = 0$, i.e., $A(\mathbf{V}_1, \mathbf{V}_2)(p) = 0$. Now suppose $\mathbf{V}_1, \mathbf{V}_2, \mathbf{W}_1, \mathbf{W}_2 \in \boldsymbol{\mathcal{X}}(X)$ with $\mathbf{V}_1(p) = \mathbf{W}_1(p)$ and $\mathbf{V}_2(p) = \mathbf{W}_2(p)$. We claim that $A(\mathbf{V}_1, \mathbf{V}_2)(p) = A(\mathbf{W}_1, \mathbf{W}_2)(p)$. To see this note that $A(\mathbf{V}_1, \mathbf{V}_2) - A(\mathbf{W}_1, \mathbf{W}_2) = A(\mathbf{V}_1 - \mathbf{W}_1, \mathbf{V}_2) + A(\mathbf{W}_1, \mathbf{V}_1 - \mathbf{W}_2)$ and, by what we have just shown, both terms on the right-hand side vanish at p. Thus, just as we did for 1-forms, we can define a bilinear map $\mathbf{A}_p : T_p(X) \times T_p(X) \to \mathbb{R}$ as follows: For any $\mathbf{v}_1, \mathbf{v}_2 \in T_p(X)$ select any vector fields $\mathbf{V}_1, \mathbf{V}_2 \in \boldsymbol{\mathcal{X}}(X)$ with $\mathbf{V}_1(p) = \mathbf{v}_1$ and $\mathbf{V}_2(p) = \mathbf{v}_2$ and set $\mathbf{A}_p(\mathbf{v}_1, \mathbf{v}_2) = A(\mathbf{V}_1, \mathbf{V}_2)(p)$ (such $\mathbf{V}_i$'s exist by Exercise 4.6.10).

Exercise 4.10.19 Show that the map $p \to \mathbf{A}_p$ thus defined is a smooth covariant tensor field of rank 2 on X and that the one-to-one correspondence between $\mathcal{T}^2(X)$ and $C^\infty(X)$-bilinear maps of $\boldsymbol{\mathcal{X}}(X) \times \boldsymbol{\mathcal{X}}(X)$ to $C^\infty(X)$ thus established is an isomorphism when the latter set of maps is provided with its natural $C^\infty(X)$-module structure ($(A_1 + A_2)(\mathbf{V}, \mathbf{W}) = A_1(\mathbf{V}, \mathbf{W}) + A_2(\mathbf{V}, \mathbf{W})$ and $(fA)(\mathbf{V}, \mathbf{W}) = fA(\mathbf{V}, \mathbf{W})$).

This alternative view of covariant tensor fields of rank 2 is particularly convenient for introducing our last (and, arguably, most important) means of manufacturing 2-forms from 1-forms. As motivation we first reformulate the results of Exercise 4.6.7. The elements of $C^\infty(X)$ are often referred to as **0-forms** on X (see Section 0.2) so that the operator d of Exercise 4.6.7 carries 0-forms to 1-forms. We would like to define an analogous operator, also denoted d, that carries 1-forms to 2-forms and shares many of the desirable properties of the differential. The properties of most immediate concern are linearity (Exercise 4.6.7(1)), a sort of product rule (Exercise 4.6.7(2)) and nice behavior under pullbacks ((4.6.13)). Furthermore, we will require that the new d's composition with the old d be zero, i.e., that $d(df)$ should be the zero element of $\Lambda^2(X)$.

Remark: We are now taking the first steps toward what is known as *exterior calculus*, which is a generalization to arbitrary smooth manifolds of much of the classical vector calculus of $\mathbb{R}^3$. A lively and elementary discussion of this generalization is available in [**Flan**] to which we refer the reader who may be feeling motivationally challenged at the moment. We add only that, in $\mathbb{R}^3$, 1-forms may be identified with vector fields (e.g., df with grad f) and that our new d operator is intended to assign to any 1-form a 2-form which, in turn, can be identified with the curl of that vector field (see Section 0.2). Thus, our requirement that $d(df) = 0$ is just an exotic version of the familiar vector identity curl (grad f) $= \vec{0}$.

Now we consider an arbitrary 1-form $\boldsymbol{\Theta}$ on X (thought of as an operator that carries $\mathbf{V} \in \boldsymbol{\mathcal{X}}(X)$ to $\boldsymbol{\Theta}(\mathbf{V}) = \boldsymbol{\Theta}\mathbf{V} \in C^\infty(X)$). We define an operator $d\boldsymbol{\Theta} : \boldsymbol{\mathcal{X}}(X) \times \boldsymbol{\mathcal{X}}(X) \to C^\infty(X)$, called the **exterior derivative** of $\boldsymbol{\Theta}$, by

$$d\boldsymbol{\Theta}(\mathbf{V},\mathbf{W}) = \mathbf{V}(\boldsymbol{\Theta}\mathbf{W}) - \mathbf{W}(\boldsymbol{\Theta}\mathbf{V}) - \boldsymbol{\Theta}([\mathbf{V},\mathbf{W}])$$

for all $\mathbf{V},\mathbf{W} \in \boldsymbol{\mathcal{X}}(X)$, where $[\mathbf{V},\mathbf{W}]$ is the Lie bracket of $\mathbf{V}$ and $\mathbf{W}$. Observe that $d\boldsymbol{\Theta}(\mathbf{V}_1+\mathbf{V}_2,\mathbf{W}) = (\mathbf{V}_1+\mathbf{V}_2)(\boldsymbol{\Theta}\mathbf{W}) - \mathbf{W}(\boldsymbol{\Theta}(\mathbf{V}_1+\mathbf{V}_2)) - \boldsymbol{\Theta}([\mathbf{V}_1+\mathbf{V}_2,\mathbf{W}]) = \mathbf{V}_1(\boldsymbol{\Theta}\mathbf{W}) + \mathbf{V}_2(\boldsymbol{\Theta}\mathbf{W}) - \mathbf{W}(\boldsymbol{\Theta}\mathbf{V}_1 + \boldsymbol{\Theta}\mathbf{V}_2) - \boldsymbol{\Theta}([\mathbf{V}_1,\mathbf{W}]+[\mathbf{V}_2,\mathbf{W}]) = \mathbf{V}_1(\boldsymbol{\Theta}\mathbf{W}) + \mathbf{V}_2(\boldsymbol{\Theta}\mathbf{W}) - \mathbf{W}(\boldsymbol{\Theta}\mathbf{V}_1) - \mathbf{W}(\boldsymbol{\Theta}\mathbf{V}_2) - \boldsymbol{\Theta}([\mathbf{V}_1,\mathbf{W}]) - \boldsymbol{\Theta}([\mathbf{V}_2,\mathbf{W}]) = d\boldsymbol{\Theta}(\mathbf{V}_1,\mathbf{W}) + d\boldsymbol{\Theta}(\mathbf{V}_2,\mathbf{W})$. Similarly, $d\boldsymbol{\Theta}(\mathbf{V},\mathbf{W}_1+\mathbf{W}_2) = d\boldsymbol{\Theta}(\mathbf{V},\mathbf{W}_1) + d\boldsymbol{\Theta}(\mathbf{V},\mathbf{W}_2)$. Now, if $f \in C^\infty(X)$,

$$\begin{aligned}
d\boldsymbol{\Theta}(f\mathbf{V},\mathbf{W}) &= (f\mathbf{V})(\boldsymbol{\Theta}\mathbf{W}) - \mathbf{W}(\boldsymbol{\Theta}(f\mathbf{V})) - \boldsymbol{\Theta}([f\mathbf{V},\mathbf{W}]) \\
&= f(\mathbf{V}(\boldsymbol{\Theta}\mathbf{W})) - \mathbf{W}(f(\boldsymbol{\Theta}\mathbf{V})) - \boldsymbol{\Theta}(f[\mathbf{V},\mathbf{W}] - (\mathbf{W}f)\mathbf{V}) \\
&\qquad \text{by } (4.6.5) \\
&= f(\mathbf{V}(\boldsymbol{\Theta}\mathbf{W})) - f(\mathbf{W}(\boldsymbol{\Theta}\mathbf{V})) - (\mathbf{W}f)(\boldsymbol{\Theta}\mathbf{V}) - \boldsymbol{\Theta}(f[\mathbf{V},\mathbf{W}]) \\
&\qquad + \boldsymbol{\Theta}((\mathbf{W}f)\mathbf{V}) \text{ by Exercise 4.6.3 (2)} \\
&= f(\mathbf{V}(\boldsymbol{\Theta}\mathbf{W})) - f(\mathbf{W}(\boldsymbol{\Theta}\mathbf{V})) - (\mathbf{W}f)(\boldsymbol{\Theta}\mathbf{V}) \\
&\qquad - f(\boldsymbol{\Theta}([\mathbf{V},\mathbf{W}])) + (\mathbf{W}f)(\boldsymbol{\Theta}\mathbf{V}) \\
&= f(d\boldsymbol{\Theta}(\mathbf{V},\mathbf{W})).
\end{aligned}$$

Similarly, $d\boldsymbol{\Theta}(\mathbf{V},f\mathbf{W}) = f(d\boldsymbol{\Theta}(\mathbf{V},\mathbf{W}))$. Thus, $d\boldsymbol{\Theta}$ is $C^\infty(X)$-bilinear (note the crucial role of the Lie bracket in establishing this). Furthermore, $d\boldsymbol{\Theta}(\mathbf{W},\mathbf{V}) = \mathbf{W}(\boldsymbol{\Theta}\mathbf{V}) - \mathbf{V}(\boldsymbol{\Theta}\mathbf{W}) - \boldsymbol{\Theta}([\mathbf{W},\mathbf{V}]) = -\mathbf{V}(\boldsymbol{\Theta}\mathbf{W}) + \mathbf{W}(\boldsymbol{\Theta}\mathbf{V}) - \boldsymbol{\Theta}(-[\mathbf{V},\mathbf{W}]) = -\mathbf{V}(\boldsymbol{\Theta}\mathbf{W}) + \mathbf{W}(\boldsymbol{\Theta}\mathbf{V}) + \boldsymbol{\Theta}([\mathbf{V},\mathbf{W}]) = -d\boldsymbol{\Theta}(\mathbf{V},\mathbf{W})$ so $d\boldsymbol{\Theta}$ is skew-symmetric. Thus, $d\boldsymbol{\Theta}$ is a 2-form.

Theorem 4.10.4 *Let X be a smooth manifold and $d : \boldsymbol{\mathcal{X}}^*(X) \to \Lambda^2(X)$ the operator that carries any 1-form $\boldsymbol{\Theta}$ to its exterior derivative $d\boldsymbol{\Theta}$. Then*

1. $d(a\boldsymbol{\Theta}_1 + b\boldsymbol{\Theta}_2) = a\,d\boldsymbol{\Theta}_1 + b\,d\boldsymbol{\Theta}_2$ *for all* $a,b \in \mathbb{R}$ *and* $\boldsymbol{\Theta}_1,\boldsymbol{\Theta}_2 \in \boldsymbol{\mathcal{X}}^*(X)$.
2. $d(f\boldsymbol{\Theta}) = f\,d\boldsymbol{\Theta} + df \wedge \boldsymbol{\Theta}$ *for any* $f \in C^\infty(X)$ *and* $\boldsymbol{\Theta} \in \boldsymbol{\mathcal{X}}^*(X)$.
3. $d(df) = 0$ *for any* $f \in C^\infty(X)$.

4. *If (U, φ) is a chart on X with coordinate functions $x^1, \ldots, x^n$, then*

$$d\left(\Theta_i \, dx^i\right) = d\Theta_i \wedge dx^i = \sum_{i,j=1}^{n} \frac{\partial \Theta_i}{\partial x^j} dx^j \wedge dx^i \, .$$

5. *If $F : X \to Y$ is smooth and $\boldsymbol{\Theta}$ is a 1-form on Y, then $F^*(d\boldsymbol{\Theta}) = d\,(F^*\boldsymbol{\Theta})$.*

Proof:

Exercise 4.10.20 Prove #1.

To prove #2 we just compute

$$\begin{aligned} d\,(f\boldsymbol{\Theta})\,(\mathbf{V},\mathbf{W}) &= \mathbf{V}(\,(f\boldsymbol{\Theta})\,\mathbf{W}\,) - \mathbf{W}(\,(f\boldsymbol{\Theta})\mathbf{V}\,) - (f\boldsymbol{\Theta})\,(\,[\mathbf{V},\mathbf{W}]\,) \\ &= \mathbf{V}(\,f(\boldsymbol{\Theta}\mathbf{W})\,) - \mathbf{W}(\,f(\boldsymbol{\Theta}\mathbf{V})\,) - f(\,\boldsymbol{\Theta}(\,[\mathbf{V},\mathbf{W}]\,)\,) \\ &= f(\mathbf{V}(\,\boldsymbol{\Theta}\mathbf{W})\,) + (\mathbf{V}\,f)\,(\boldsymbol{\Theta}\mathbf{W}) - f(\,\mathbf{W}(\,\boldsymbol{\Theta}\mathbf{V})\,) \\ &\qquad - (\mathbf{W}\,f)\,(\boldsymbol{\Theta}\mathbf{V}) - f(\,\boldsymbol{\Theta}(\,[\mathbf{V},\mathbf{W}]\,)\,) \\ &= f(\,d\boldsymbol{\Theta}(\mathbf{V},\mathbf{W})\,) + (\,(\mathbf{V}\,f)\,(\boldsymbol{\Theta}\mathbf{W}) - (\mathbf{W}\,f)\,(\boldsymbol{\Theta}\mathbf{V})\,) \\ &= (f\,d\boldsymbol{\Theta})\,(\mathbf{V},\mathbf{W}) + (\,(df(\mathbf{V})\,)\,(\boldsymbol{\Theta}\mathbf{W}) - (\,df(\mathbf{W})\,)\,(\boldsymbol{\Theta}\mathbf{V})\,) \\ &= (f\,d\boldsymbol{\Theta})\,(\mathbf{V},\mathbf{W}) + (df \wedge \boldsymbol{\Theta})\,(\mathbf{V},\mathbf{W}) \\ &= (f\,d\boldsymbol{\Theta} + df \wedge \boldsymbol{\Theta})\,(\mathbf{V},\mathbf{W}). \end{aligned}$$

Property #3 is essentially the definition (4.6.1) of the Lie bracket:

$$\begin{aligned} d\,(df)\,(\mathbf{V},\mathbf{W}) &= \mathbf{V}(\,df(\mathbf{W})\,) - \mathbf{W}(\,df(\mathbf{V})\,) - df(\,[\mathbf{V},\mathbf{W}]\,) \\ &= \mathbf{V}(\mathbf{W}\,f) - \mathbf{W}(\mathbf{V}\,f) - [\mathbf{V},\mathbf{W}]\,f = 0\,. \end{aligned}$$

Exercise 4.10.21 Use #2 and #3 to prove #4.

Finally, we prove #5. It is enough to prove this identity at each point $p \in X$ so we may work in coordinates and use #4. Thus, let $\boldsymbol{\Theta} = \Theta_i\,dy^i$ and compute $F^*(d\boldsymbol{\Theta}) = F^*(d\Theta_i \wedge dy^i) = F^*(d\Theta_i) \wedge F^*(dy^i) = d\,(\Theta_i \circ F) \wedge d\,(y^i \circ F) = d\,(\Theta_i \circ F) \wedge dF^i$ and, from (4.6.11), $d\,(F^*\boldsymbol{\Theta}) = d\,(\,F^*(\Theta_i\,dy^i)\,) = d\,(\frac{\partial F^i}{\partial x^j}(\Theta_i \circ F)\,dx^j\,) = d\,(\,(\Theta_i \circ F)\,dF^i\,) = (\Theta_i \circ F)\,d\,(dF^i) + d\,(\Theta_i \circ F) \wedge dF^i = F^*(d\Theta)$ as required. ∎

A particularly important example is obtained as follows: Let G be a matrix Lie group with Lie algebra $\mathcal{G}$. Select a basis $\{e_1, \ldots, e_n\}$ for $\mathcal{G}$ and let $\{\mathbf{e}_1, \ldots, \mathbf{e}_n\}$ denote the corresponding left invariant vector fields. For each $i, j = 1, \ldots, n$, $[e_i, e_j]$ is in $\mathcal{G}$ so there exist constants C^k_{ij}, $k = 1, \ldots, n$, such that $[e_i, e_j] = C^k_{ij}\, e_k$. Since $[\mathbf{e}_i, \mathbf{e}_j]$ is also left invariant, it follows that $[\mathbf{e}_i, \mathbf{e}_j](g) = (L_g)_{*e}([e_i, e_j]) = (L_g)_{*e}(C^k_{ij}\, e_k) = C^k_{ij}\,(L_g)_{*e}(e_k) = C^k_{ij}\, \mathbf{e}_k\,(g)$ so

$$[\,\mathbf{e}_i\,,\,\mathbf{e}_j\,] = C^k_{ij}\,\mathbf{e}_k.$$

The constants C^k_{ij}, $i, j, k = 1, \ldots, n$, are called the **structure constants** of $\mathcal{G}$. Now, let $\{\Theta^1, \ldots, \Theta^n\}$ be the set of left invariant 1-forms on G for which $\{\Theta^1(id), \ldots, \Theta^n(id)\}$ is the dual basis to $\{e_1, \ldots, e_k\}$. We prove next the so-called **Maurer-Cartan equations**: For $k = 1, \ldots, n$,

$$d\Theta^k = -\frac{1}{2}\sum_{i,j=1}^{n} C^k_{ij}\,\Theta^i \wedge \Theta^j = -\sum_{i<j} C^k_{ij}\,\Theta^i \wedge \Theta^j. \qquad (4.10.3)$$

According to Lemma 4.10.2 we need only show that $d\Theta^k\,(\mathbf{e}_i, \mathbf{e}_j) = -C^k_{ij}$. Observe first that if Θ is a left invariant 1-form and $\mathbf{V}$ and $\mathbf{W}$ are left invariant vector fields, then, since $\Theta\mathbf{W}$ and $\Theta\mathbf{V}$ are constant, $d\Theta(\mathbf{V}, \mathbf{W}) = \mathbf{V}(\Theta\,\mathbf{W}) - \mathbf{W}(\Theta\,\mathbf{V}) - \Theta\,(\,[\mathbf{V}, \mathbf{W}]\,) = -\Theta\,(\,[\mathbf{V}, \mathbf{W}]\,)$. Thus, $d\Theta^k\,(\mathbf{e}_i, \mathbf{e}_j) = -\Theta^k\,(\,[\mathbf{e}_i, \mathbf{e}_j]\,) = -\Theta^k\,(C^l_{ij}\,\mathbf{e}_l) = -C^l_{ij}\,\Theta^k\,(\mathbf{e}_l) = -C^l_{ij}\,\delta^k_l = -C^k_{ij}$ as required.

Now let $\mathcal{V}$ be a finite dimensional real vector space and $\mathcal{V}^*$ its dual space. If X is a smooth manifold, then a **$\mathcal{V}$-valued 2-form on** X is a map Ω on X that assigns to every $p \in X$ a map $\Omega(p) = \Omega_p : T_p(X) \times T_p(X) \to \mathcal{V}$ that is bilinear $(\,\Omega_p(a_1\mathbf{v}_1 + a_2\mathbf{v}_2, \mathbf{w}) = a_1\Omega_p(\mathbf{v}_1, \mathbf{w}) + a_2\Omega_p(\mathbf{v}_2, \mathbf{w})$ and $\Omega_p(\mathbf{v}, a_1\mathbf{w}_1 + a_2\mathbf{w}_2) = a_1\Omega_p(\mathbf{v}, \mathbf{w}_1) + a_2\Omega_p(\mathbf{v}, \mathbf{w}_2)\,)$ and skew-symmetric $(\,\Omega_p(\mathbf{w}, \mathbf{v}) = -\Omega_p(\mathbf{v}, \mathbf{w})\,)$. If $\{e_1, \ldots, e_d\}$ is a basis for $\mathcal{V}$, then, for any $\mathbf{v}, \mathbf{w} \in T_p(X)$, we may write $\Omega_p(\mathbf{v}, \mathbf{w}) = \Omega^1_p(\mathbf{v}, \mathbf{w})e_1 + \cdots + \Omega^d_p(\mathbf{v}, \mathbf{w})e_d = \Omega^i_p(\mathbf{v}, \mathbf{w})e_i$. Defining Ω^i on X by $\Omega^i(p) = \Omega^i_p$ we find that each Ω^i is an $\mathbb{R}$-valued 2-form on X. The Ω^i, $i = 1, \ldots, d$, are called the **components** of Ω relative to $\{e_1, \ldots, e_d\}$ and we will say that Ω is **smooth** if each Ω^i is in $\Lambda^2(X)$.

Exercise 4.10.22 Show that this definition does not depend on the choice of basis in $\mathcal{V}$ and, in fact, is equivalent to the requirement that, for every $\lambda \in \mathcal{V}^*$, the $\mathbb{R}$-valued 2-form $\lambda\Omega$ defined on X by $(\lambda\Omega)\,(p) = \lambda \circ \Omega_p$ is smooth.

Conversely, given a family $\Omega^1, \ldots, \Omega^d$ of smooth $\mathbb{R}$-valued 2-forms on X

and a basis $\{e_1, \ldots, e_d\}$ for $\mathcal{V}$ one can build a smooth $\mathcal{V}$-valued 2-form $\boldsymbol{\Omega} = \boldsymbol{\Omega}^i e_i$ on X defined by $\boldsymbol{\Omega}(p)(\mathbf{v}, \mathbf{w}) = \boldsymbol{\Omega}^i(p)(\mathbf{v}, \mathbf{w})e_i$ for each $p \in X$ and all $\mathbf{v}, \mathbf{w} \in T_p(X)$. The collection of all smooth $\mathcal{V}$-valued 2-forms on X is denoted $\Lambda^2(X, \mathcal{V})$. Defining the algebraic operations componentwise, $\Lambda^2(X, \mathcal{V})$ has the structure of a real vector space and, indeed, a $C^\infty(X)$-module.

Exercise 4.10.23 If $f : X \to Y$ is a smooth map and $\boldsymbol{\Omega}$ is a $\mathcal{V}$-valued 2-form on Y, the **pullback of $\boldsymbol{\Omega}$ by f** is the $\mathcal{V}$-valued 2-form $f^*\boldsymbol{\Omega}$ on X defined by $(f^*\boldsymbol{\Omega})_p(\mathbf{v}, \mathbf{w}) = \boldsymbol{\Omega}_{f(p)}(f_{*p}(\mathbf{v}), f_{*p}(\mathbf{w}))$ for all $p \in X$ and $\mathbf{v}, \mathbf{w} \in T_p(X)$. Show that if $\boldsymbol{\Omega} = \boldsymbol{\Omega}^i e_i$, then $f^*\boldsymbol{\Omega} = (f^*\boldsymbol{\Omega}^i)e_i$ so that $f^*\boldsymbol{\Omega}$ is smooth if $\boldsymbol{\Omega}$ is smooth.

Exterior derivatives of $\mathcal{V}$-valued 1-forms are defined componentwise and yield $\mathcal{V}$-valued 2-forms. More precisely, if $\boldsymbol{\omega}$ is a $\mathcal{V}$-valued 1-form on X and $\{e_1, \ldots, e_d\}$ is a basis for $\mathcal{V}$ and if we write $\boldsymbol{\omega} = \omega^i e_i$ for $\omega^i \in \boldsymbol{\mathcal{X}}^*(X)$, $i = 1, \ldots, d$, then the **exterior derivative** $d\boldsymbol{\omega}$ of $\boldsymbol{\omega}$ is defined by

$$d\,\boldsymbol{\omega} = (d\,\omega^i)\,e_i\,.$$

We need only show that this definition does not depend on the choice of basis, i.e., that if $\{\hat{e}_1, \ldots, \hat{e}_d\}$ is another basis for $\mathcal{V}$ and $\boldsymbol{\omega} = \hat{\omega}^j \hat{e}_j$, then $d\hat{\omega}^j(p)(\mathbf{v}, \mathbf{w})\,\hat{e}_j = d\omega^i(p)(\mathbf{v}, \mathbf{w})\,e_i$ for all $\mathbf{v}, \mathbf{w} \in T_p(X)$.

Exercise 4.10.24 Prove this. **Hint:** Write $e_i = A^j{}_i\,\hat{e}_j$ for some *constants* $A^j{}_i$, $i, j = 1, \ldots, d$, and show that $\hat{\omega}^j = A^j{}_i\,\omega^i$.

Wedge products of $\mathcal{V}$-valued 1-forms are a bit troublesome because the notion of a wedge product depends on the existence of some sort of multiplicative structure in the image and this is too much to ask of a vector space in general. However, for the cases of interest to us ($\mathbb{R}$-valued, $\mathbb{C}$-valued, $\mathbb{H}$-valued and Lie algebra-valued forms) one can formulate a meaningful (and useful) notion of the wedge product. We consider, more generally, three vector spaces $\mathcal{U}$, $\mathcal{V}$ and $\mathcal{W}$. Suppose that one is given a bilinear map $\rho : \mathcal{U} \times \mathcal{V} \to \mathcal{W}$. Now, if X is a smooth manifold, $\boldsymbol{\omega}$ is a $\mathcal{U}$-valued 1-form on X and $\boldsymbol{\eta}$ is a $\mathcal{V}$-valued 1-form on X, then one may define their **ρ-wedge product** $\boldsymbol{\omega} \wedge_\rho \boldsymbol{\eta}$ by

$$(\boldsymbol{\omega} \wedge_\rho \boldsymbol{\eta})_p(\mathbf{v}, \mathbf{w}) = \rho(\boldsymbol{\omega}_p(\mathbf{v}), \boldsymbol{\eta}_p(\mathbf{w})) - \rho(\boldsymbol{\omega}_p(\mathbf{w}), \boldsymbol{\eta}_p(\mathbf{v}))$$

for all $p \in X$ and $\mathbf{v}, \mathbf{w} \in T_p(X)$.

Lemma 4.10.5 *Let $\{u_1, \ldots, u_c\}$ and $\{v_1, \ldots, v_d\}$ be bases for $\mathcal{U}$ and*

$\mathcal{V}$, respectively, and $\rho : \mathcal{U} \times \mathcal{V} \to \mathcal{W}$ a bilinear map. Let $\boldsymbol{\omega} = \boldsymbol{\omega}^i u_i$ be a $\mathcal{U}$-valued 1-form on X and $\boldsymbol{\eta} = \boldsymbol{\eta}^j v_j$ a $\mathcal{V}$-valued 1-form on X. Then

$$\boldsymbol{\omega} \wedge_\rho \boldsymbol{\eta} = \sum_{i=1}^{c} \sum_{j=1}^{d} \left(\boldsymbol{\omega}^i \wedge \boldsymbol{\eta}^j \right) \rho(u_i, v_j).$$

Exercise 4.10.25 Prove Lemma 4.10.5. **Remark:** Although this is a simple consequence of the bilinearity of ρ, it should be pointed out that the vectors $\rho(u_i, v_j) \in \mathcal{W}$ will, in general, *not* constitute a basis for $\mathcal{W}$ so that the $\boldsymbol{\omega}^i \wedge \boldsymbol{\eta}^j$ are not components of $\boldsymbol{\omega} \wedge_\rho \boldsymbol{\eta}$.

With the natural differentiable structures on $\mathcal{U}$, $\mathcal{V}$ and $\mathcal{W}$, a bilinear map $\rho : \mathcal{U} \times \mathcal{V} \to \mathcal{W}$ is smooth so, writing out the $\rho(u_i, v_j)$ in terms of some basis for $\mathcal{W}$, it follows that, if $\boldsymbol{\omega}$ and $\boldsymbol{\eta}$ are smooth, so is $\boldsymbol{\omega} \wedge_\rho \boldsymbol{\eta}$. Since bilinearity and skew-symmetry are clear, $\boldsymbol{\omega} \wedge_\rho \boldsymbol{\eta}$ is a $\mathcal{W}$-valued 2-form on X.

In order to solidify these notions we will write out explicitly the examples of interest to us. First we let $\mathcal{U} = \mathcal{V} = \mathcal{W} = \mathrm{Im}\,\mathbb{H}$ and take $\rho : \mathrm{Im}\,\mathbb{H} \times \mathrm{Im}\,\mathbb{H} \to \mathrm{Im}\,\mathbb{H}$ to be the Lie bracket $\rho(x, y) = [x, y] = 2\,\mathrm{Im}(xy)$. Let $\boldsymbol{\omega}$ and $\boldsymbol{\eta}$ be $\mathrm{Im}\,\mathbb{H}$-valued 1-forms on X and take $\{\mathbf{i}, \mathbf{j}, \mathbf{k}\}$ as a basis for $\mathrm{Im}\,\mathbb{H}$. Writing $\boldsymbol{\omega} = \boldsymbol{\omega}^1\mathbf{i} + \boldsymbol{\omega}^2\mathbf{j} + \boldsymbol{\omega}^3\mathbf{k}$ and $\boldsymbol{\eta} = \boldsymbol{\eta}^1\mathbf{i} + \boldsymbol{\eta}^2\mathbf{j} + \boldsymbol{\eta}^3\mathbf{k}$, Lemma 4.10.5 gives

$$\begin{aligned}
\boldsymbol{\omega} \wedge_\rho \boldsymbol{\eta} &= (\boldsymbol{\omega}^1 \wedge \boldsymbol{\eta}^1)[\mathbf{i},\mathbf{i}] + (\boldsymbol{\omega}^1 \wedge \boldsymbol{\eta}^2)[\mathbf{i},\mathbf{j}] + (\boldsymbol{\omega}^1 \wedge \boldsymbol{\eta}^3)[\mathbf{i},\mathbf{k}] + \\
&\quad (\boldsymbol{\omega}^2 \wedge \boldsymbol{\eta}^1)[\mathbf{j},\mathbf{i}] + (\boldsymbol{\omega}^2 \wedge \boldsymbol{\eta}^2)[\mathbf{j},\mathbf{j}] + (\boldsymbol{\omega}^2 \wedge \boldsymbol{\eta}^3)[\mathbf{j},\mathbf{k}] + \\
&\quad (\boldsymbol{\omega}^3 \wedge \boldsymbol{\eta}^1)[\mathbf{k},\mathbf{i}] + (\boldsymbol{\omega}^3 \wedge \boldsymbol{\eta}^2)[\mathbf{k},\mathbf{j}] + (\boldsymbol{\omega}^3 \wedge \boldsymbol{\eta}^3)[\mathbf{k},\mathbf{k}] \\
&= 4\,[\boldsymbol{\omega}^2 \wedge \boldsymbol{\eta}^3\,\mathbf{i} + \boldsymbol{\omega}^3 \wedge \boldsymbol{\eta}^1\,\mathbf{j} + \boldsymbol{\omega}^1 \wedge \boldsymbol{\eta}^2\,\mathbf{k}].
\end{aligned}$$

On the other hand, if $\boldsymbol{\omega}$ and $\boldsymbol{\eta}$ are $\mathbb{H}$-valued 1-forms on X and $\rho : \mathbb{H} \times \mathbb{H} \to \mathbb{H}$ is quaternion multiplication $\rho(x, y) = xy$, then, with $\boldsymbol{\omega} = \boldsymbol{\omega}^0 + \boldsymbol{\omega}^1\,\mathbf{i} + \boldsymbol{\omega}^2\,\mathbf{j} + \boldsymbol{\omega}^3\,\mathbf{k}$ and $\boldsymbol{\eta} = \boldsymbol{\eta}^0 + \boldsymbol{\eta}^1\,\mathbf{i} + \boldsymbol{\eta}^2\,\mathbf{j} + \boldsymbol{\eta}^3\,\mathbf{k}$, one obtains

$$\begin{aligned}
\boldsymbol{\omega} \wedge_\rho \boldsymbol{\eta} = {} & [\boldsymbol{\omega}^0 \wedge \boldsymbol{\eta}^0 - \boldsymbol{\omega}^1 \wedge \boldsymbol{\eta}^1 - \boldsymbol{\omega}^2 \wedge \boldsymbol{\eta}^2 - \boldsymbol{\omega}^3 \wedge \boldsymbol{\eta}^3] + \\
& [\boldsymbol{\omega}^0 \wedge \boldsymbol{\eta}^1 + \boldsymbol{\omega}^1 \wedge \boldsymbol{\eta}^0 + \boldsymbol{\omega}^2 \wedge \boldsymbol{\eta}^3 - \boldsymbol{\omega}^3 \wedge \boldsymbol{\eta}^2]\,\mathbf{i} + \\
& [\boldsymbol{\omega}^0 \wedge \boldsymbol{\eta}^2 + \boldsymbol{\omega}^2 \wedge \boldsymbol{\eta}^0 + \boldsymbol{\omega}^3 \wedge \boldsymbol{\eta}^1 - \boldsymbol{\omega}^1 \wedge \boldsymbol{\eta}^3]\,\mathbf{j} + \\
& [\boldsymbol{\omega}^0 \wedge \boldsymbol{\eta}^3 + \boldsymbol{\omega}^3 \wedge \boldsymbol{\eta}^0 + \boldsymbol{\omega}^1 \wedge \boldsymbol{\eta}^2 - \boldsymbol{\omega}^2 \wedge \boldsymbol{\eta}^1]\,\mathbf{k}.
\end{aligned}$$

Exercise 4.10.26 Let $dq = dq^0 + dq^1\,\mathbf{i} + dq^2\,\mathbf{j} + dq^3\,\mathbf{k}$ and $d\bar{q} = dq^0 - dq^1\,\mathbf{i} - dq^2\,\mathbf{j} - dq^3\,\mathbf{k}$ be the usual $\mathbb{H}$-valued 1-forms on $\mathbb{R}^4 = \mathbb{H}$. Show that, if ρ denotes quaternion multiplication, then

$$dq \wedge_\rho dq = d\bar{q} \wedge_\rho d\bar{q} = 2(dq^2 \wedge dq^3\,\mathbf{i} + dq^3 \wedge dq^1\,\mathbf{j} + dq^1 \wedge dq^2\,\mathbf{k})$$

$$dq \wedge_\rho d\bar{q} = -2\left((dq^0 \wedge dq^1 + dq^2 \wedge dq^3)\,\mathbf{i} + (dq^0 \wedge dq^2 - dq^1 \wedge dq^3)\,\mathbf{j} \right.$$
$$\left. + (dq^0 \wedge dq^3 + dq^1 \wedge dq^2)\,\mathbf{k} \right)$$

$$d\bar{q} \wedge_\rho dq = 2\left((dq^0 \wedge dq^1 - dq^2 \wedge dq^3)\,\mathbf{i} + (dq^0 \wedge dq^2 + dq^1 \wedge dq^3)\,\mathbf{j} \right.$$
$$\left. + (dq^0 \wedge dq^3 - dq^1 \wedge dq^2)\,\mathbf{k} \right).$$

If $\boldsymbol{\omega}$ is an $\mathbb{H}$-valued 1-form and $\boldsymbol{\eta}$ is an $\mathbb{R}$-valued 1-form and $\rho : \mathbb{H} \times \mathbb{R} \to \mathbb{H}$ is given by $\rho(x, r) = xr = x^0 r + x^1 r\,\mathbf{i} + x^2 r\,\mathbf{j} + x^3 r\,\mathbf{k}$, then $\boldsymbol{\omega} \wedge_\rho \boldsymbol{\eta} = \boldsymbol{\omega}^0 \wedge \boldsymbol{\eta} + \boldsymbol{\omega}^1 \wedge \boldsymbol{\eta}\,\mathbf{i} + \boldsymbol{\omega}^2 \wedge \boldsymbol{\eta}\,\mathbf{j} + \boldsymbol{\omega}^3 \wedge \boldsymbol{\eta}\,\mathbf{k}$, where we have used 1 as the basis for $\mathbb{R}$. Dropping the real part, this same formula gives $\boldsymbol{\omega} \wedge_\rho \boldsymbol{\eta}$ when $\boldsymbol{\omega}$ is $\mathrm{Im}\,\mathbb{H}$-valued, $\boldsymbol{\eta}$ is $\mathbb{R}$-valued and $\rho : \mathrm{Im}\,\mathbb{H} \times \mathbb{R} \to \mathrm{Im}\,\mathbb{H}$ is $\rho(x, r) = xr = x^1 r\,\mathbf{i} + x^2 r\,\mathbf{j} + x^3 r\,\mathbf{k}$.

Now we wish to calculate in some detail a number of basic formulas for the cases of most interest to us. Since the bilinear mappings ρ by which we define our wedge products will always be one of those specified above we will henceforth omit references to them and write simply $\wedge$ for all of our wedge products. We consider an $\mathrm{Im}\,\mathbb{H}$-valued 1-form $\boldsymbol{\omega} = \boldsymbol{\omega}^1\,\mathbf{i} + \boldsymbol{\omega}^2\,\mathbf{j} + \boldsymbol{\omega}^3\,\mathbf{k}$ on $\mathbb{R}^4 = \mathbb{H}$ (these will, via stereographic projection, be coordinate expressions for $\mathrm{Im}\,\mathbb{H}$-valued, i.e., $\mathcal{SP}(1)$-valued, 1-forms on S^4). Since each $\boldsymbol{\omega}^i$ is an $\mathbb{R}$-valued 1-form on $\mathbb{H}$ we can write it in standard coordinates as $\boldsymbol{\omega}^i = \omega^i_0 dq^0 + \omega^i_1 dq^1 + \omega^i_2 dq^2 + \omega^i_3 dq^3 = \omega^i_\alpha dq^\alpha$, $i = 1, 2, 3$. Thus, $\boldsymbol{\omega} = (\omega^1_\alpha dq^\alpha)\,\mathbf{i} + (\omega^2_\alpha dq^\alpha)\,\mathbf{j} + (\omega^3_\alpha dq^\alpha)\,\mathbf{k} = (\omega^1_0 dq^0 \mathbf{i} + \omega^2_0 dq^0 \mathbf{j} + \omega^3_0 dq^0 \mathbf{k}) + \cdots + (\omega^1_3 dq^3 \mathbf{i} + \omega^2_3 dq^3 \mathbf{j} + \omega^3_3 dq^3 \mathbf{k}) = (\omega^1_0\,\mathbf{i} + \omega^2_0 \mathbf{j} + \omega^3_0 \mathbf{k})\, dq^0 + \cdots + (\omega^1_3\,\mathbf{i} + \omega^2_3\,\mathbf{j} + \omega^3_3\,\mathbf{k})\, dq^3$. Thus, we may write $\boldsymbol{\omega}$ as

$$\boldsymbol{\omega} = \omega_0 dq^0 + \omega_1 dq^1 + \omega_2 dq^2 + \omega_3 dq^3\,, \tag{4.10.4}$$

where each ω_α is an $\mathrm{Im}\,\mathbb{H}$-valued function on $\mathbb{H}$ given by

$$\omega_\alpha = \omega^1_\alpha\,\mathbf{i} + \omega^2_\alpha\,\mathbf{j} + \omega^3_\alpha\,\mathbf{k}\,, \qquad \alpha = 0, 1, 2, 3. \tag{4.10.5}$$

We compute $\boldsymbol{\omega} \wedge \boldsymbol{\omega} = 4\,(\boldsymbol{\omega}^2 \wedge \boldsymbol{\omega}^3\,\mathbf{i} + \boldsymbol{\omega}^3 \wedge \boldsymbol{\omega}^1\,\mathbf{j} + \boldsymbol{\omega}^1 \wedge \boldsymbol{\omega}^2\,\mathbf{k})$ in these terms by observing that $\boldsymbol{\omega}^2 \wedge \boldsymbol{\omega}^3 = (\omega^2_\alpha\, dq^\alpha) \wedge (\omega^3_\beta\, dq^\beta) = \omega^2_\alpha \omega^3_\beta dq^\alpha \wedge dq^\beta$ and similarly for $\boldsymbol{\omega}^3 \wedge \boldsymbol{\omega}^1$ and $\boldsymbol{\omega}^1 \wedge \boldsymbol{\omega}^2$. Thus,

$$\begin{aligned} \boldsymbol{\omega} \wedge \boldsymbol{\omega} = 4\,(&\omega^2_\alpha\,\omega^3_\beta\, dq^\alpha \wedge dq^\beta\ \mathbf{i} + \omega^3_\alpha\,\omega^1_\beta\, dq^\alpha \wedge dq^\beta\ \mathbf{j} \\ &+ \omega^1_\alpha\,\omega^2_\beta\, dq^\alpha \wedge dq^\beta\ \mathbf{k}). \end{aligned} \tag{4.10.6}$$

To rewrite this last expression we compute the $\mathbb{H}$-product $\omega_\alpha\,\omega_\beta =$

$(\omega_\alpha^1\,\mathbf{i}+\omega_\alpha^2\,\mathbf{j}+\omega_\alpha^3\,\mathbf{k})\,(\omega_\beta^1\,\mathbf{i}+\omega_\beta^2\,\mathbf{j}+\omega_\beta^3\,\mathbf{k}) = -(\omega_\alpha^1\omega_\beta^1+\omega_\alpha^2\omega_\beta^2+\omega_\alpha^3\omega_\beta^3)+(\omega_\alpha^2\omega_\beta^3-\omega_\alpha^3\omega_\beta^2)\,\mathbf{i}+(\omega_\alpha^3\omega_\beta^1-\omega_\alpha^1\omega_\beta^3)\,\mathbf{j}+(\omega_\alpha^1\omega_\beta^2-\omega_\alpha^2\omega_\beta^1)\,\mathbf{k}$. Next observe that

$$\sum_{\alpha,\beta=0}^{3}(\omega_\alpha^1\omega_\beta^1+\omega_\alpha^2\omega_\beta^2+\omega_\alpha^3\omega_\beta^3)\,dq^\alpha\wedge dq^\beta=0\,,$$

$$\sum_{\alpha,\beta=0}^{3}(\omega_\alpha^2\omega_\beta^3-\omega_\alpha^3\omega_\beta^2)\,dq^\alpha\wedge dq^\beta=2\sum_{\alpha,\beta=0}^{3}\omega_\alpha^2\omega_\beta^3\,dq^\alpha\wedge dq^\beta\,,$$

and similarly for $\omega_\alpha^3\omega_\beta^1-\omega_\alpha^1\omega_\beta^3$ and $\omega_\alpha^1\omega_\beta^2-\omega_\alpha^2\omega_\beta^1$. Thus,

$$\begin{aligned}\omega_\alpha\,\omega_\beta\,dq^\alpha\wedge dq^\beta &= 2\,\omega_\alpha^2\omega_\beta^3\,dq^\alpha\wedge dq^\beta\,\mathbf{i}+2\,\omega_\alpha^3\omega_\beta^1\,dq^\alpha\wedge dq^\beta\,\mathbf{j}\\ &\qquad +2\,\omega_\alpha^1\omega_\beta^2\,dq^\alpha\wedge dq^\beta\,\mathbf{k}\\ &= \frac{1}{2}\,\boldsymbol{\omega}\wedge\boldsymbol{\omega}\,.\end{aligned}$$

Finally, observe that $[\omega_\alpha\,\omega_\beta]=\omega_\alpha\,\omega_\beta-\omega_\beta\,\omega_\alpha$ gives

$$\begin{aligned}[\omega_\alpha\,\omega_\beta]\,dq^\alpha\wedge dq^\beta &= \omega_\alpha\,\omega_\beta\,dq^\alpha\wedge dq^\beta-\omega_\beta\,\omega_\alpha\,dq^\alpha\wedge dq^\beta\\ &= 2\,\omega_\alpha\,\omega_\beta\,dq^\alpha\wedge dq^\beta\,.\end{aligned}$$

We conclude then that, for the $\operatorname{Im}\mathbb{H}$-valued 1-form $\boldsymbol{\omega}$ on $\mathbb{R}^4$ given by (4.10.4) and (4.10.5), we have

$$\frac{1}{2}\,\boldsymbol{\omega}\wedge\boldsymbol{\omega}=\omega_\alpha\,\omega_\beta\,dq^\alpha\wedge dq^\beta=\frac{1}{2}[\,\omega_\alpha,\omega_\beta\,]\,dq^\alpha\wedge dq^\beta\,. \tag{4.10.7}$$

Now, for this same $\operatorname{Im}\mathbb{H}$-valued 1-form $\boldsymbol{\omega}$ on $\mathbb{R}^4$ we compute the exterior derivative $d\boldsymbol{\omega}=d\boldsymbol{\omega}^1\,\mathbf{i}+d\boldsymbol{\omega}^2\,\mathbf{j}+d\boldsymbol{\omega}^3\,\mathbf{k}$ as follows:

$$\begin{aligned}d\boldsymbol{\omega} &= d(\omega_\alpha^1\,dq^\alpha)\,\mathbf{i}+d(\omega_\alpha^2\,dq^\alpha)\,\mathbf{j}+d(\omega_\alpha^3\,dq^\alpha)\,\mathbf{k}\\ &= \big(d(\omega_0^1dq^0)\,\mathbf{i}+d(\omega_0^2dq^0)\,\mathbf{j}+d(\omega_0^3dq^0)\,\mathbf{k}\big)+\cdots\\ &\qquad+\big(d(\omega_3^1dq^3)\,\mathbf{i}+d(\omega_3^2dq^3)\,\mathbf{j}+d(\omega_3^3dq^3)\,\mathbf{k}\big)\\ &= \big(d\omega_0^1\wedge dq^0\,\mathbf{i}+d\omega_0^2\wedge dq^0\,\mathbf{j}+d\omega_0^3\wedge dq^0\,\mathbf{k}\big)+\cdots\\ &\qquad+\big(d\omega_3^1\wedge dq^3\,\mathbf{i}+d\omega_3^2\wedge dq^3\,\mathbf{j}+d\omega_3^3\wedge dq^3\,\mathbf{k}\big)\\ &= \big(d\omega_0^1\,\mathbf{i}+d\omega_0^2\,\mathbf{j}+d\omega_0^3\,\mathbf{k}\big)\wedge dq^0+\cdots\\ &\qquad+\big(d\omega_3^1\,\mathbf{i}+d\omega_3^2\,\mathbf{j}+d\omega_3^3\,\mathbf{k}\big)\wedge dq^3\\ &= d\omega_0\wedge dq^0+\cdots+d\omega_3\wedge dq^3\,.\end{aligned}$$

Thus,

$$d\boldsymbol{\omega} = d\,\omega_\beta \wedge dq^\beta \,. \tag{4.10.8}$$

Furthermore, $d\,\omega_\beta = d\,\omega_\beta^1\,\mathbf{i} + d\,\omega_\beta^2\,\mathbf{j} + d\,\omega_\beta^3\,\mathbf{k} = (\partial_\alpha\,\omega_\beta^1\,dq^\alpha)\,\mathbf{i} + (\partial_\alpha\,\omega_\beta^2\,dq^\alpha)\,\mathbf{j} + (\partial_\alpha\,\omega_\beta^3\,dq^\alpha)\,\mathbf{k}$, where, since we have no intention of using any but standard coordinates q^α on $\mathbb{R}^4$, we have taken the liberty of abbreviating $\partial/\partial q^\alpha$ as ∂_α. Thus, $d\,\omega_\beta = (\partial_\alpha\,\omega_\beta^1\,\mathbf{i} + \partial_\alpha\,\omega_\beta^2\,\mathbf{j} + \partial_\alpha\,\omega_\beta^3\,\mathbf{k})dq^\alpha = \partial_\alpha\,\omega_\beta\,dq^\alpha$ (componentwise derivative of ω_β) so we have

$$d\,\boldsymbol{\omega} = \partial_\alpha\,\omega_\beta\,dq^\alpha \wedge dq^\beta = \frac{1}{2}\,(\,\partial_\alpha\,\omega_\beta - \partial_\alpha\,\omega_\alpha\,)\;dq^\alpha \wedge dq^\beta \,. \tag{4.10.9}$$

For reasons that may not become clear until the next chapter (see Theorem 5.2.1) we add $d\,\boldsymbol{\omega}$ and $\frac{1}{2}\boldsymbol{\omega}\wedge\boldsymbol{\omega}$ to obtain an $\mathrm{Im}\,\mathbb{H}$-valued 2-form $d\,\boldsymbol{\omega} + \frac{1}{2}\boldsymbol{\omega}\wedge\boldsymbol{\omega}$ on $\mathbb{R}^4$ and record

$$\begin{aligned} d\,\boldsymbol{\omega} + \frac{1}{2}\boldsymbol{\omega}\wedge\boldsymbol{\omega} &= (\,\partial_\alpha\,\omega_\beta + \omega_\alpha\,\omega_\beta\,)\;dq^\alpha \wedge dq^\beta \\ &= \frac{1}{2}(\,\partial_\alpha\,\omega_\beta - \partial_\beta\,\omega_\alpha + [\,\omega_\alpha\,,\omega_\beta\,]\,)\,dq^\alpha \wedge dq^\beta \,. \end{aligned} \tag{4.10.10}$$

Calculations such as these are admittedly rather dreary, but they are good for the soul and the only means of really making concepts such as these one's own. Here's your chance:

Exercise 4.10.27 Let $f(q) = f^0(q) + f^1(q)\,\mathbf{i} + f^2(q)\,\mathbf{j} + f^3(q)\,\mathbf{k}$ be an arbitrary smooth $\mathbb{H}$-valued function on $\mathbb{H}$. Compute the $\mathbb{H}$-product $f(q)\,dq$ and take the imaginary part to obtain an $\mathrm{Im}\,\mathbb{H}$-valued 1-form $\boldsymbol{\omega} = \mathrm{Im}\,(f(q)\,dq)$ on $\mathbb{R}^4$. Show that $\boldsymbol{\omega} = \mathrm{Im}\,(f(q)\,dq) = \omega_\alpha\,dq^\alpha$, where $\omega_0 = f^1\,\mathbf{i} + f^2\,\mathbf{j} + f^3\,\mathbf{k}$, $\omega_1 = f^0\,\mathbf{i} + f^3\,\mathbf{j} - f^2\,\mathbf{k}$, $\omega_2 = -f^3\,\mathbf{i} + f^0\,\mathbf{j} + f^1\,\mathbf{k}$ and $\omega_3 = f^2\,\mathbf{i} - f^1\,\mathbf{j} + f^0\,\mathbf{k}$. Now show that $d\,\boldsymbol{\omega} = \mathrm{Im}\,(df \wedge dq)$ and $\frac{1}{2}\boldsymbol{\omega}\wedge\boldsymbol{\omega} = \mathrm{Im}\,(\,f(q)\,dq \wedge f(q)\,dq\,)$ so that

$$d\,\boldsymbol{\omega} + \frac{1}{2}\boldsymbol{\omega}\wedge\boldsymbol{\omega} = \mathrm{Im}\;(\,df \wedge dq + f(q)\,dq \wedge f(q)\,dq\,)\,. \tag{4.10.11}$$

Lest the reader feel that the effort expended in Exercise 4.10.27 is not sufficient reward unto itself we will now apply the results of that Exercise to the $\mathrm{Im}\,\mathbb{H}$-valued 1-form

$$\boldsymbol{\mathcal{A}} = \mathrm{Im}\left(\frac{\bar{q}}{1+|\,q\,|^2}\,dq\right)$$

of (4.8.15) (we denote the 1-form $\boldsymbol{\mathcal{A}}$ because it arose as the pullback of another 1-form whose name is $\boldsymbol{\omega}$ and because it will play the role of a potential 1-form in the physical theory). Notice that $\mathcal{A} = \operatorname{Im}(f(q)\,dq)$, where $f(q) = (1+|q|^2)^{-1}\bar{q}$. In particular,

$$f(q)\,dq \wedge f(q)\,dq = \left(1+|q|^2\right)^{-2}(\bar{q}\,dq \wedge \bar{q}\,dq)\,. \tag{4.10.12}$$

Furthermore,

$$\begin{aligned} df &= d\left((1+|q|^2)^{-1}\bar{q}\right) = d\left((1+|q|^2)^{-1}(q^0 - q^1\mathbf{i} - q^2\mathbf{j} - q^3\mathbf{k})\right) \\ &= d\left((1+|q|^2)^{-1}q^0\right) - d\left((1+|q|^2)^{-1}q^1\right)\mathbf{i} \\ &\qquad - d\left((1+|q|^2)^{-1}q^2\right)\mathbf{j} - d\left((1+|q|^2)^{-1}q^3\right)\mathbf{k}. \end{aligned}$$

But $d((1+|q|^2)^{-1}q^\alpha) = (1+|q|^2)^{-1}dq^\alpha + q^\alpha d((1+|q|^2)^{-1})$ so

$$df = \left(1+|q|^2\right)^{-1}d\bar{q} + \bar{q}\,d\left((1+|q|^2)^{-1}\right).$$

Now,

$$\begin{aligned} d\left((1+|q|^2)^{-1}\right) &= \partial_0(1+|q|^2)^{-1}dq^0 + \partial_1(1+|q|^2)^{-1}dq^1 \\ &\qquad + \partial_2(1+|q|^2)^{-1}dq^2 + \partial_3(1+|q|^2)^{-1}dq^3 \\ &= -(1+|q|^2)^{-2}(2q^0\,dq^0 + 2q^1\,dq^1 + 2q^2\,dq^2 + 2q^3\,dq^3) \\ &= -(1+|q|^2)^{-2}(q\,d\bar{q} + \overline{q\,d\bar{q}}) \\ &= -(1+|q|^2)^{-2}(q\,d\bar{q} + dq\,\bar{q}) \end{aligned}$$

so

$$\bar{q}\,d\left((1+|q|^2)^{-1}\right) = -(1+|q|^2)^{-2}(|q|^2 d\bar{q} + \bar{q}\,dq\,\bar{q}).$$

Thus,

$$\begin{aligned} df &= (1+|q|^2)^{-1}d\bar{q} - (1+|q|^2)^{-2}(|q|^2 d\bar{q} + \bar{q}\,dq\,\bar{q}) \\ &= (1+|q|^2)^{-2}(d\bar{q} - \bar{q}\,dq\,\bar{q}). \end{aligned}$$

Next we compute

$$\begin{aligned} df \wedge dq &= (1+|q|^2)^{-2}\left(d\bar{q}\wedge dq - (\bar{q}\,dq\,\bar{q})\wedge dq\right) \\ &= (1+|q|^2)^{-2}\left(d\bar{q}\wedge dq - (\bar{q}\,dq)\wedge(\bar{q}\,dq)\right). \end{aligned}$$

From this and (4.10.12) it follows that $df \wedge dq + f(q)\,dq \wedge f(q)\,dq = (1+|q|^2)^{-2}d\bar{q}\wedge dq$. Since $(1+|q|^2)^{-2}$ is real and, according to Exercise

4.10.26, $d\bar{q} \wedge dq$ is pure imaginary we conclude that

$$\begin{aligned} \boldsymbol{\mathcal{A}} &= \operatorname{Im}\left(\frac{\bar{q}}{1+|\,q\,|^2}\, dq \right) \\ &\Longrightarrow d\boldsymbol{\mathcal{A}} + \frac{1}{2}\boldsymbol{\mathcal{A}} \wedge \boldsymbol{\mathcal{A}} = \frac{1}{(1+|\,q\,|^2)^2}\, d\bar{q} \wedge dq \end{aligned} \tag{4.10.13}$$

(what we have just computed is the field strength of the BPST potential $\boldsymbol{\mathcal{A}}$; see Section 5.2).

Exercise 4.10.28 Let $n \in \mathbb{H}$ and $\lambda > 0$ be fixed. Defined an $\operatorname{Im}\mathbb{H}$-valued 1-form $\boldsymbol{\mathcal{A}}_{\lambda,n}$ on $\mathbb{H}$ by

$$\boldsymbol{\mathcal{A}}_{\lambda,n} = \operatorname{Im}\left(\frac{\bar{q} - \bar{n}}{|\,q - n\,|^2 + \lambda^2}\, dq \right)$$

(we'll tell you where this came from in Section 5.1). Show that

$$d\boldsymbol{\mathcal{A}}_{\lambda,n} + \frac{1}{2}\boldsymbol{\mathcal{A}}_{\lambda,n} \wedge \boldsymbol{\mathcal{A}}_{\lambda,n} = \frac{\lambda^2}{(\,|\,q-n\,|^2 + \lambda^2\,)^2}\, d\bar{q} \wedge dq\,.$$

The potentials $\boldsymbol{\mathcal{A}}_{\lambda,n}$ of Exercise 4.10.28 are called "generic BPST potentials" and each is determined by a point (λ, n) in $(0,\infty) \times \mathbb{H} \subseteq \mathbb{R}^5$. In Section 5.5 we will need to know that, with its usual structure, $(0,\infty) \times \mathbb{H}$ is diffeomorphic (and, in fact, conformally equivalent) to the open 5-dimensional disc. To see this we regard $(0,\infty) \times \mathbb{H}$ as the subset of $\mathbb{R}^6$ with $x^6 = 0$ and $x^1 = \lambda > 0$ (x^2, x^3, ,x^4 and x^5 are the standard coordinates q^0, q^1, q^2 and q^3 in $\mathbb{H}$). We denote by S^4 the equator ($x^6 = 0$) in S^5 and by B^5 the 5-dimensional ball that is the interior of S^4 in $\mathbb{R}^5$, i.e., $B^5 = \{\,(x^1,\ldots,x^5,0) \in \mathbb{R}^6 : (x^1)^2 + \cdots + (x^5)^2 < 1\,\}$. Let $\varphi_S : S^5 - \{N\} \to \mathbb{R} \times \mathbb{H}$ be the stereographic projection from the north pole. Notice that φ_S^{-1} carries $(0,\infty) \times \mathbb{H}$ onto the "front" hemisphere $S^5_F = \{\,(x^1,\ldots,x^6) \in S^5 : x^1 > 0\,\}$ of S^5. Now let R be the rotation of $\mathbb{R}^6$ through $\frac{\pi}{2}$ that leaves $x^2,\ldots,x^5$ fixed and carries $N = (0,0,0,0,0,1)$ onto $(1,0,0,0,0,0)$, i.e., $R(x^1,x^2,x^3,x^4,x^5,x^6) = (x^6,x^2,x^3,x^4,x^5,-x^1)$ for all $(x^1,\ldots,x^6) \in \mathbb{R}^6$. Then R carries S^5_F onto the "lower" hemisphere $S^5_L = \{\,(x^1,\ldots,x^6) \in S^5 : x^6 < 0\,\}$. Finally, note that φ_S carries S^5_L onto B^5.

Exercise 4.10.29 Show that $\varphi_S \circ R \circ \varphi_S^{-1}$ is an orientation preserving, conformal diffeomorphism of $(0,\infty) \times \mathbb{H}$ onto B^5.

We close this section with a general result on Lie groups that will be of use in the next chapter. Thus, we let G be a matrix Lie group with Lie algebra $\mathcal{G}$ and Cartan 1-form Θ. Select a basis $\{e_1, \ldots, e_n\}$ for $\mathcal{G}$ and let $\{\Theta^1, \ldots, \Theta^n\}$ be the left invariant 1-forms on G for which $\{\Theta^1(id), \ldots, \Theta^n(id)\}$ is the dual basis to $\{e_1, \ldots, e_n\}$. According to Lemma 4.8.1, $\Theta = \Theta^k e_k$ so that $d\Theta = d\Theta^k e_k$. The Maurer-Cartan equations (4.10.3) therefore give

$$d\Theta = -\frac{1}{2}\left(\sum_{i,j=1}^{n} C_{ij}^k\, \Theta^i \wedge \Theta^j\right) e_k\,,$$

where C_{ij}^k, $i,j,k = 1,\ldots,n$, are the structure constants of $\mathcal{G}$. On the other hand, defining the wedge product of $\mathcal{G}$-valued 1-forms via the pairing $\rho : \mathcal{G} \times \mathcal{G} \to \mathcal{G}$ given by $\rho(x,y) = [x,y]$, Lemma 4.10.5 gives

$$\begin{aligned}\Theta \wedge \Theta &= \sum_{i,j=1}^{n} (\Theta^i \wedge \Theta^j)\,[\,e_i, e_k\,] = \sum_{i,j=1}^{n} (\Theta^i \wedge \Theta^j)\, C_{ij}^k\, e_k \\ &= \left(\sum_{i,j=1}^{n} C_{ij}^k\, \Theta^i \wedge \Theta^j\right) e_k = -2d\Theta\end{aligned}$$

so we have

$$d\Theta + \frac{1}{2}\,\Theta \wedge \Theta = 0 \quad (\,\Theta = \text{Cartan 1-form for } G\,)\,. \tag{4.10.14}$$

This is often referred to as the **equation of structure** for G.

5

Gauge Fields and Instantons

5.1 Connections and Gauge Equivalence

The $\mathrm{Im}\,\mathbb{H}$-valued 1-form $\omega = \mathrm{Im}\,(\bar{q}^1\,dq^1 + \bar{q}^2\,dq^2)$ will occupy center stage for much of the remainder of our story. We begin by adopting its two most important properties ((4.8.10) and (4.8.11)) as the defining conditions for a connection on a principal bundle.

Let $\mathcal{B} = (P, \mathcal{P}, \sigma)$ be a smooth principal G-bundle over X (we assume G is a matrix Lie group and denote its Lie algebra $\mathcal{G}$). A **connection form** (or **gauge connection**) on $\mathcal{B}$ (or, on P) is a smooth $\mathcal{G}$-valued 1-form ω on P which satisfies the following two conditions:

1. $(\sigma_g)^*\omega = ad_{g^{-1}} \circ \omega$ for all $g \in G$, i.e., for all $g \in G$, $p \in P$ and $\mathbf{v} \in T_{p \cdot g^{-1}}(P)$, $\omega_p(\,(\sigma_g)_{*p \cdot g^{-1}}(\mathbf{v})\,) = g^{-1}\omega_{p \cdot g^{-1}}(\mathbf{v})\,g$.
2. $\omega(A^\#) = A$ for all $A \in \mathcal{G}$, i.e., for all $A \in \mathcal{G}$ and $p \in P$, $\omega_p(\,A^\#(p)\,) = A$.

We have shown that the $\mathrm{Im}\,\mathbb{H}$-valued 1-form $\mathrm{Im}\,(\bar{q}^1\,dq^1 + \bar{q}^2\,dq^2)$ on S^7 is a connection form on the quaternionic Hopf bundle $Sp(1) \to S^7 \to S^4$.

Exercise 5.1.1 Show that the $\mathrm{Im}\,\mathbb{C}$-valued 1-form $\mathbf{i}\,\mathrm{Im}\,(\bar{z}^1\,dz^1 + \bar{z}^2\,dz^2)$ on S^3 is a connection form on the complex Hopf bundle $U(1) \to S^3 \to S^2$.

A local cross-section $s : V \to \mathcal{P}^{-1}(V)$ of the bundle $\mathcal{B}$ is known in the physics literature as a **local gauge**. If ω is a connection form on $\mathcal{B}$, then the pullback $\mathcal{A} = s^*\omega$ of ω to $V \subseteq X$ is called a **local gauge potential** (in the gauge s) on X and these are the objects that physicists arrive at when solving the differential equations of gauge theories (e.g., Maxwell's equations, or the Yang-Mills equations). Choosing a different local gauge, or, what amounts to the same thing (Section 3.3), choosing a different trivialization of the bundle, is known among the physicists as a **local gauge transformation**. The result of such a gauge transformation can be two rather different looking gauge potentials (see, e.g., (4.8.19) and (4.8.20)). Not too different though, since we proved in Lemma 4.8.2 that two such pullbacks must be related by the consistency condition (4.8.23). One can show that this process can be reversed in the sense that a family of local gauge potentials whose domains cover X and that are consistent on the overlaps piece together to give a connection form on the bundle.

Theorem 5.1.1 *Let G be a matrix Lie group with Lie algebra $\mathcal{G}$ and $\mathcal{B} = (P, \mathcal{P}, \sigma)$ a smooth principal G-bundle over X. Let $\{(V_j, \Psi_j)\}_{j\in J}$ be a family of trivializations for $\mathcal{B}$ with $\bigcup_{j\in J} V_j = X$. Suppose that, for each $j \in J$, $\mathcal{A}_j$ is a $\mathcal{G}$-valued 1-form on V_j and that, whenever $V_j \cap V_i \neq \emptyset$,*

$$\mathcal{A}_j = ad_{g_{ij}^{-1}} \circ \mathcal{A}_i + \Theta_{ij} \quad on \;\; V_j \cap V_i, \tag{5.1.1}$$

where $g_{ij} : V_j \cap V_i \to G$ is the transition function and $\Theta_{ij} = g_{ij}^{\;*}\Theta$ is the pullback by g_{ij} of the Cartan 1-form Θ for G. Then there exists a unique connection form ω on P such that, for each $j \in J$, $\mathcal{A}_j = s_j^*\omega$, where $s_j : V_j \to \mathcal{P}^{-1}(V_j)$ is the canonical cross-section associated with the trivialization (V_j, Ψ_j).

We leave the proof of Theorem 5.1.1 to the reader in a sequence of Exercises.

Exercise 5.1.2 Let (V, Ψ) be any trivialization for $\mathcal{B}$ and $s : V \to \mathcal{P}^{-1}(V)$, $s(x) = \Psi^{-1}(x, e)$, the associated cross-section. Show that, for any $(x_0, g_0) \in V \times G$ and any $(\mathbf{v}_1, \mathbf{v}_2) \in T_{(x_0, g_0)}(V \times G)$, $(\Psi^{-1})_{*(x_0,g_0)}(\mathbf{v}_1, \mathbf{v}_2) = (\sigma_{g_0})_{*s(x_0)}(s_{*x_0}(\mathbf{v}_1)) + A^{\#}(s(x_0)\cdot g_0)$, where $A = (L_{g_0^{-1}})_{*g_0}(\mathbf{v}_2)$. In particular, if $g_0 = e$ and $(\mathbf{v}, A) \in T_{(x_0,e)}(V \times G)$, $(\Psi^{-1})_{*(x_0,e)}(\mathbf{v}, A) = s_{*x_0}(\mathbf{v}) + A^{\#}(s(x_0))$ so every element of $T_{s(x_0)}(\mathcal{P}^{-1}(V))$, can be written as $s_{*x_0}(\mathbf{v}) + A^{\#}(s(x_0))$ for some $A \in \mathcal{G}$. **Hint:** The calculations are not unlike those in the proof of Lemma 4.8.2.

Exercise 5.1.3 For each $j \in J$ define ω_j on $\mathcal{P}^{-1}(V_j)$ as follows: If $x_0 \in V_j$, $p = s_j(x_0)$, $\mathbf{v} \in T_{x_0}(X)$ and $A \in \mathcal{G}$, define $\omega_j(p)((s_j)_{*x_0}(\mathbf{v}) + A^{\#}(p)) = \mathcal{A}_j(x_0)(\mathbf{v}) + A$. Now, any point in $\mathcal{P}^{-1}(V_j)$ is $p \cdot g$ for some $p = s_j(x_0)$ and some $g \in G$. For any $\mathbf{w} \in T_{p\cdot g}(\mathcal{P}^{-1}(V_j))$ set

$$\omega_j(p \cdot g)(\mathbf{w}) = ad_{g^{-1}} \circ \omega_j\left((\sigma_{g^{-1}})_{*p\cdot g}(\mathbf{w})\right).$$

Show that ω_j is a connection form on the bundle $\mathcal{P} : \mathcal{P}^{-1}(V_j) \to V_j$.

Exercise 5.1.4 Let (V_i, Ψ_i) and (V_j, Ψ_j) be two of the trivializations in Theorem 5.1.1 with $x_0 \in V_j \cap V_i$ and $\mathbf{v} \in T_{x_0}(X)$. Show that $(s_j)_{*x_0}(\mathbf{v}) = (\sigma_{g_{ij}(x_0)})_{*s_i(x_0)}((s_i)_{*x_0}(\mathbf{v})) + [\Theta_{ij}(\mathbf{v})]^{\#}(s_j(x_0))$. **Hint:** Let $\alpha(t)$ be a smooth curve in X with $\alpha'(0) = \mathbf{v}$ and compute $\frac{d}{dt}s_j(\alpha(t))|_{t=0}$.

Exercise 5.1.5 Show that, on $\mathcal{P}^{-1}(V_j \cap V_i)$, $\omega_j = \omega_i$. **Hint:** It is enough to show that they agree on $s_j(V_j \cap V_i)$. Check vectors of the form

$A^{\#}(p)$ and $(s_j)_{*x_0}(\mathbf{v})$ separately and use Exercise 5.1.4 and (5.1.1) for the latter type.

Exercise 5.1.6 Define ω on P by $\omega \,|\, \mathcal{P}^{-1}(V_j) = \omega_j$ for each $j \in J$. Show that ω is well-defined and has all of the required properties. ∎

A particular consequence of Theorem 5.1.1 is that *any* $\mathcal{G}$-valued 1-form on the manifold X is the pullback to X of a unique connection form on the trivial G-bundle over X. The reason is that one can choose a global trivialization $\{(X, \Psi)\}$ so that the consistency condition (5.1.1) is satisfied vacuously.

Now suppose that one is given a connection form ω on P. For each $p \in P$ we define the **horizontal subspace** $\mathrm{Hor}_p(P)$ **of** $T_p(P)$ determined by ω by

$$\mathrm{Hor}_p(P) = \{\mathbf{v} \in T_p(P) : \omega_p(\mathbf{v}) = 0\}.$$

We claim that

$$T_p(P) = \mathrm{Hor}_p(P) \oplus \mathrm{Vert}_p(P). \tag{5.1.2}$$

To prove this we first observe that, if $\mathbf{v} \in \mathrm{Hor}_p(P) \cap \mathrm{Vert}_p(P)$, then $\mathbf{v} = A^{\#}(p)$ for some $A \in \mathcal{G}$ (Corollary 4.7.9) and $\omega_p(\mathbf{v}) = 0 = \omega_p(A^{\#}(p)) = A$. Thus, $\mathbf{v} = 0$ so $\mathrm{Hor}_p(P) \cap \mathrm{Vert}_p(P) = \{0\}$. It will therefore suffice to show that $\dim \mathrm{Hor}_p(P) + \dim \mathrm{Vert}_p(P) = \dim T_p(P)$. Now, ω_p is a linear transformation of $T_p(P)$ to $\mathcal{G}$ so $\dim T_p(P)$ is the dimension of its kernel plus the dimension of its image (Theorem 3, Chapter $\underline{\mathrm{IV}}$, of [**Lang**]). The kernel of ω_p is just $\mathrm{Hor}_p(P)$. Furthermore, ω_p maps onto $\mathcal{G}$ ($\omega_p(A^{\#}(p)) = A$ for any $A \in \mathcal{G}$) so the dimension of its image is $\dim \mathcal{G} = \dim \mathrm{Vert}_p(P)$ (Corollary 4.7.9 again) and our result follows.

Observe that $\dim T_p(P) = \dim P$ and $\dim \mathrm{Vert}_p(P) = \dim \mathcal{G} = \dim G$. Since $\dim P = \dim X + \dim G$, it follows from (5.1.2) that $\dim \mathrm{Hor}_p(P) = \dim X$. Moreover, $\mathcal{P} : P \to X$ is a submersion (Exercise 4.4.18).

Exercise 5.1.7 Let $\mathcal{P}(p) = x$. Show that $\mathcal{P}_{*p}$ is identically zero on $\mathrm{Vert}_p(P)$ and so must carry $\mathrm{Hor}_p(P)$ isomorphically onto $T_x(X)$.

These horizontal subspaces determined by the connection form ω are also invariant under the action of G on P in the sense that, for any $p \in P$ and any $g \in G$,

$$(\sigma_g)_{*p}(\mathrm{Hor}_p(P)) = \mathrm{Hor}_{p \cdot g}(P). \tag{5.1.3}$$

For the proof we first observe that if $\mathbf{v} \in \mathrm{Hor}_p(P)$, then $\omega_{p \cdot g}((\sigma_g)_{*p}(\mathbf{v})) = \omega_{p \cdot g}((\sigma_g)_{*(p \cdot g) \cdot g^{-1}}(\mathbf{v})) = ad_{g^{-1}}(\omega_p(\mathbf{v})) = ad_{g^{-1}}(0) = 0$ so $(\sigma_g)_{*p}(\mathrm{Hor}_p(P)) \subseteq \mathrm{Hor}_{p \cdot g}(P)$. Next suppose $\mathbf{w} \in \mathrm{Hor}_{p \cdot g}(P)$.

Since $(\sigma_g)_{*p}$ is an isomorphism, there exists a $\mathbf{v} \in T_p(P)$ with $(\sigma_g)_{*p}(\mathbf{v}) = \mathbf{w}$. We need only show that $\mathbf{v} \in \mathrm{Hor}_p(P)$. But $\omega_p(\mathbf{v}) = \omega_p((\sigma_{g^{-1}})_{*p\cdot g}(\mathbf{w})) = ad_g(\omega_{p\cdot g}(\mathbf{w})) = ad_g(0) = 0$ so (5.1.3) is proved.

Thus, a connection form ω on a principal G-bundle $\mathcal{P}: P \to X$ assigns to each $p \in P$ a subspace $\mathrm{Hor}_p(P)$ of $T_p(P)$. We claim that this assignment $p \to \mathrm{Hor}_p(P)$ is smooth in the sense of the following definition. Let P be a smooth manifold of dimension $n+k$. An **n-dimensional distribution $\mathcal{D}$ on P** is an assignment to each $p \in P$ of an n-dimensional subspace $\mathcal{D}(p)$ of $T_p(P)$ that is smooth in the sense that, for each $q \in P$, there exists a nbd U of q and n C^∞ vector fields $\mathbf{V}_1, \ldots, \mathbf{V}_n$ on U such that, for every $p \in U$, $\mathcal{D}(p)$ is spanned by $\{\mathbf{V}_1(p), \ldots, \mathbf{V}_n(p)\}$.

Lemma 5.1.2 *Let $\mathcal{P}: P \to X$ be a smooth principal G-bundle with $\dim X = n$ and $\dim G = k$ and on which is defined a connection form ω. Then the assignment $p \to Hor_p(P) = \{\mathbf{v} \in T_p(P) : \omega_p(\mathbf{v}) = 0\}$ is an n-dimensional distribution on P.*

Proof: Only smoothness remains to be proved. Fix a $q \in P$ and choose C^∞ vector fields $\mathbf{W}_1, \ldots, \mathbf{W}_n, \mathbf{W}_{n+1}, \ldots, \mathbf{W}_{n+k}$ on a nbd U of q which, for each $p \in U$, span $T_p(P)$ (coordinate vector fields will do). Let $A_1, \ldots, A_k$ be a basis for $\mathcal{G}$ so that we may write $\omega = \omega^j A_j$ for C^∞ $\mathbb{R}$-valued 1-forms $\omega^1, \ldots, \omega^k$. Now, for each $i = 1, \ldots, n$, let $\mathbf{V}_i = \mathbf{W}_i - \omega^j(\mathbf{W}_i) A_j^\#$. Each of these is C^∞ on U and each is horizontal at each $p \in U$ because $\omega(\mathbf{V}_i) = \omega(\mathbf{W}_i) - \omega^j(\mathbf{W}_i)\,\omega(A_j^\#) = \omega(\mathbf{W}_i) - \omega^j(\mathbf{W}_i) A_j = \omega(\mathbf{W}_i) - \omega(\mathbf{W}_i) = 0$.

Exercise 5.1.8 Show that, for every $p \in U$, $\{\mathbf{V}_1(p), \ldots, \mathbf{V}_n(p)\}$ spans $\mathrm{Hor}_p(P)$. ∎

The process which led us from the connection form ω to the smooth distribution $p \to \mathrm{Hor}_p(P)$ satisfying (5.1.2) and (5.1.3) can be reversed.

Exercise 5.1.9 Let $\mathcal{P}: P \to X$ be a smooth principal G-bundle with $\dim X = n$ and $\dim G = k$ and on which is defined a smooth n-dimensional distribution $p \to \mathcal{D}(p)$ satisfying $T_p(P) = \mathcal{D}(p) \oplus \mathrm{Vert}_p(P)$ and $(\sigma_g)_{*p}(\mathcal{D}(p)) = \mathcal{D}(p \cdot g)$ for each $p \in P$ and $g \in G$. Define $\omega_p : T_p(P) \to \mathcal{G}$ for each $p \in P$ by $\omega_p(\mathbf{v} + A^\#(p)) = A$, where $\mathbf{v} \in \mathcal{D}(p)$. Show that ω is a connection form on the bundle and that $\mathrm{Hor}_p(P) = \mathcal{D}(p)$ for each $p \in P$.

For this reason a connection on $\mathcal{P}: P \to X$ is often defined to be a smooth distribution $p \to \mathrm{Hor}_p(P)$ on P satisfying (5.1.2) and (5.1.3).

The distribution of horizontal vectors provides a very visual means of relating to a connection form on a bundle. Let's see what it looks like for the BPST instanton connection on the Hopf bundle $Sp(1) \to S^7 \to \mathbb{HP}^1$. Thus, we let $\tilde{\omega} = \mathrm{Im}(\bar{q}^1 dq^1 + \bar{q}^2 dq^2)$ be the $\mathrm{Im}\,\mathbb{H}$-valued 1-form on $\mathbb{H}^2$ whose restriction to S^7 is the connection form ω of interest. Then, for $p = (p^1, p^2) \in \mathbb{H}^2$ and $\mathbf{v} = (v^1, v^2) \in T_p(\mathbb{H}^2)$ we have $\tilde{\omega}_p(\mathbf{v}) = \mathrm{Im}(\bar{p}^1 v^1 + \bar{p}^2 v^2)$ (as usual, we identify tangent vectors to $\mathbb{H}$ with elements of $\mathbb{H}$ via the canonical isomorphism).

Exercise 5.1.10 Show that for $p, v \in \mathbb{H} = \mathbb{R}^4$, the usual real inner product is given by $<p, v> = \mathrm{Re}(\bar{p}v)$ and conclude that, for (p^1, p^2), $(v^1, v^2) \in \mathbb{H}^2 = \mathbb{R}^8$, the usual real inner product is $< (p^1, p^2), (v^1, v^2) >$ $= \mathrm{Re}(\bar{p}^1 v^1 + \bar{p}^2 v^2)$.

Now, suppose $p = (p^1, p^2) \in S^7 \subseteq \mathbb{H}^2$ and consider the vertical part of the tangent space to S^7 at p, i.e., $\mathrm{Vert}_{(p^1,p^2)}(S^7)$. This is the tangent space to the fiber of $\mathcal{P}$ containing (p^1, p^2). Since this fiber is just $\{(p^1 g, p^2 g) : g \in Sp(1)\}$, $\mathrm{Vert}_{(p^1,p^2)}(S^7) = \{(p^1 a, p^2 a) : a \in \mathcal{SP}(1) = \mathrm{Im}\,\mathbb{H}\}$. We compute the real, orthogonal complement of $\mathrm{Vert}_{(p^1,p^2)}(S^7)$ in $\mathbb{H}^2 = \mathbb{R}^8$. Now, $\mathbf{v} = (v^1, v^2)$ is in this orthogonal complement iff, for every $a \in \mathrm{Im}\,\mathbb{H}$, $< (p^1 a, p^2 a), (v^1, v^2) > = 0$. By Exercise 5.1.10 this is the case iff $\mathrm{Re}(\overline{p^1 a}\, v^1 + \overline{p^2 a}\, v^2) = 0$ for every $a \in \mathrm{Im}\,\mathbb{H}$.

Exercise 5.1.11 Show that this is the case iff $\bar{p}^1 v^1 + \bar{p}^2 v^2$ is real, i.e., iff $\mathrm{Im}(\bar{p}^1 v^1 + \bar{p}^2 v^2) = 0$.

Thus, the kernel of $\tilde{\omega}_p$ is precisely the real orthogonal complement of $\mathrm{Vert}_p(S^7)$ in $\mathbb{R}^8$. Since the connection ω on the Hopf bundle is just the restriction of $\tilde{\omega}$ to S^7, its horizontal subspaces are just $\mathrm{Hor}_p(S^7) = T_p(S^7) \cap \ker \tilde{\omega}_p$, i.e., that part of the real orthogonal complement of $\mathrm{Vert}_p(S^7)$ in $\mathbb{R}^8$ that lies in $T_p(S^7)$. More succinctly, $\mathrm{Hor}_p(S^7)$ is the real orthogonal complement of $\mathrm{Vert}_p(S^7)$ in $T_p(S^7)$. Thus, the distribution $p \to \mathrm{Hor}_p(S^7)$ just assigns to each $p \in S^7$ the orthogonal complement of the tangent space to the fiber of $\mathcal{P} : S^7 \to S^4$ containing p. Although it may not have been apparent from its original definition (Exercise 4.8.7), ω really arises quite naturally from the structure of the Hopf bundle (and the way S^7 sets in $\mathbb{R}^8$) and so is often called the **natural connection** on $S^3 \to S^7 \to S^4$. The same arguments yield the same result for the complex Hopf bundle.

Our next result provides not only a machine for the mass production of connection forms, but also the key to deciding when two different connections on the same bundle are sufficiently different that we should distinguish them.

Theorem 5.1.3 *Let $\mathcal{P}_1 : P_1 \to X$ and $\mathcal{P}_2 : P_2 \to X$ be two smooth principal G-bundles over the same base manifold X, $f : P_1 \to P_2$ a smooth bundle map and ω a connection form on P_2. Then $f^*\omega$ is a connection form on P_1.*

Proof: We denote the actions of G on P_1 and P_2 by σ^1 and σ^2, respectively. However, for $A \in \mathcal{G}$ we will use $A^\#$ for the fundamental vector fields on both P_1 and P_2 since it will always be clear from the context which is intended. Thus, we are given that $(\sigma_g^2)^*\omega = ad_{g^{-1}} \circ \omega$ and $\omega(A^\#) = A$ for all $g \in G$ and $A \in \mathcal{G}$ and we must show that

$$\left(\sigma_g^1\right)^* (f^*\omega) = ad_{g^{-1}} \circ (f^*\omega) \tag{5.1.4}$$

and

$$(f^*\omega)\left(A^\#\right) = A \tag{5.1.5}$$

for all $g \in G$ and all $A \in \mathcal{G}$. Since f is a bundle map, $f \circ \sigma_g^1 = \sigma_g^2 \circ f$ so $(\sigma_g^1)^*(f^*\omega) = (\sigma_g^1)^* \circ f^*(\omega) = (f \circ \sigma_g^1)^*\omega = (\sigma_g^2 \circ f)^*\omega$ and therefore (5.1.4) is equivalent to

$$\left(\sigma_g^2 \circ f\right)^* \omega = ad_{g^{-1}} \circ (f^*\omega)\,.$$

To prove this we simply compute, for each $p \in P_1$ and $\mathbf{v} \in T_p(P_1)$,

$$\begin{aligned}
\left((\sigma_g^2 \circ f)^*\omega\right)_p(\mathbf{v}) &= \omega_{\sigma_g^2(f(p))}\left((\sigma_g^2)_{*f(p)}(f_{*p}(\mathbf{v}))\right) \\
&= \omega_{f(p)\cdot g}\left((\sigma_g^2)_{*(f(p)\cdot g)\cdot g^{-1}}(f_{*p}(\mathbf{v}))\right) \\
&= ad_{g^{-1}}\left(\omega_{f(p)}(f_{*p}(\mathbf{v}))\right) \\
&= ad_{g^{-1}}\left((f^*\omega)_p(\mathbf{v})\right) \\
&= \left(ad_{g^{-1}} \circ (f^*\omega)\right)_p(\mathbf{v})
\end{aligned}$$

as required. For (5.1.5) we observe that

$$\begin{aligned}
f_{*p}\left(A^\#(p)\right) &= f_{*p}\left(\frac{d}{dt}\left(\sigma^1_{\exp(tA)}(p)\right)\Big|_{t=0}\right) = f_{*p}\left(\frac{d}{dt}\left(p\cdot\exp(tA)\right)\Big|_{t=0}\right) \\
&= \frac{d}{dt}\left(f(p\cdot\exp(tA))\right)\Big|_{t=0} = \frac{d}{dt}\left(f(p)\cdot\exp(tA)\right)\Big|_{t=0} \\
&= \frac{d}{dt}\left(\sigma^2_{\exp(tA)}(f(p))\right)\Big|_{t=0} = A^\#(f(p))\,.
\end{aligned}$$

Thus,

$$\left((f^* \omega) (A^\#) \right) (p) = \omega_{f(p)} \left(f_{*p} \left(A^\# (p) \right) \right)$$
$$= \omega_{f(p)} \left(A^\# \left(f(p) \right) \right) = A$$

as required. ∎

We first apply Theorem 5.1.3 to the case in which both bundles are the quaternionic Hopf bundle and ω is the natural (BPST) connection form. The bundle maps by which we pull back all arise from a natural left action ρ of $SL(2, \mathbb{H})$ on S^7. Let $g = \begin{pmatrix} a & b \\ c & d \end{pmatrix} \in SL(2, \mathbb{H})$ and $\begin{pmatrix} q^1 \\ q^2 \end{pmatrix} \in S^7 \subseteq \mathbb{H}^2$ (it will be more convenient to write elements of $S^7 \subseteq \mathbb{H}^2$ and tangent vectors to S^7 as column vectors for these calculations). Define $g \cdot \begin{pmatrix} q^1 \\ q^2 \end{pmatrix} \in S^7$ by normalizing

$$\begin{pmatrix} a & b \\ c & d \end{pmatrix} \begin{pmatrix} q^1 \\ q^2 \end{pmatrix} = \begin{pmatrix} aq^1 + bq^2 \\ cq^1 + dq^2 \end{pmatrix},$$

i.e.,

$$g \cdot \begin{pmatrix} q^1 \\ q^2 \end{pmatrix} = \left(|\, aq^1 + bq^2 \,|^2 + |\, cq^1 + dq^2 \,|^2 \right)^{-\frac{1}{2}} \begin{pmatrix} aq^1 + bq^2 \\ cq^1 + dq^2 \end{pmatrix}. \tag{5.1.6}$$

Exercise 5.1.12 Show that (5.1.6) defines a smooth left action of $SL(2, \mathbb{H})$ on S^7.

Thus, for each fixed $g \in SL(2, \mathbb{H})$, the map $\rho_g : S^7 \to S^7$ defined by $\rho_g \begin{pmatrix} q^1 \\ q^2 \end{pmatrix} = g \cdot \begin{pmatrix} q^1 \\ q^2 \end{pmatrix}$ is a diffeomorphism. Notice that ρ_g also respects the Hopf bundle's right $Sp(1)$-action on S^7, i.e., for each $q \in Sp(1)$,

$$\rho_g \left(\begin{pmatrix} q^1 \\ q^2 \end{pmatrix} \cdot q \right) = \rho_g \begin{pmatrix} q^1 q \\ q^2 q \end{pmatrix} = \rho_g \begin{pmatrix} q^1 \\ q^2 \end{pmatrix} \cdot q\,.$$

In particular, ρ_g is a bundle map from the Hopf bundle to itself. Thus, if ω is the natural connection on the Hopf bundle, Theorem 5.1.3 guarantees that ${\rho_g}^* \omega$ is also a connection form on $S^3 \to S^7 \to S^4$. As we did for ω itself in Section 4.8 we wish now to calculate explicitly the gauge potentials $(s_k \circ \varphi_k^{-1})^* \, ({\rho_g}^* \omega)$ (as indicated in Exercise 4.8.12 and the remarks following it, we feel free to identify $s_k^*(\rho_g{}^*\boldsymbol{\omega})$ with $(s_k \circ \varphi_k^{-1})^*(\rho_g{}^*\boldsymbol{\omega}))$. Thus, for example, we let $s = s_2 \circ \varphi_2^{-1} : \mathbb{H} \to \mathcal{P}^{-1}(V_2)$ and fix a $q \in \mathbb{H}$. Exercise 4.8.8 gives $s(q) = (1 + |q|^2)^{-\frac{1}{2}} \left(\begin{smallmatrix} q \\ 1 \end{smallmatrix}\right)$. Each $\mathbf{v}_q \in T_q(\mathbb{H})$ we identify with $\frac{d}{dt}(q + tv)|_{t=0}$, where $dq(\mathbf{v}_q) = v$. Thus,

$$\begin{aligned}\left(s^*(\rho_g{}^*\boldsymbol{\omega})\right)_q(\mathbf{v}_q) &= \left((\rho_g \circ s)^*\boldsymbol{\omega}\right)_q(\mathbf{v}_q) \\ &= \boldsymbol{\omega}_{\rho_g(s(q))}\left((\rho_g \circ s)_{*q}(\mathbf{v}_q)\right),\end{aligned} \tag{5.1.7}$$

where

$$(\rho_g \circ s)_{*q}(\mathbf{v}_q) = \frac{d}{dt}\left((\rho_g \circ s)(q + tv)\right)\Big|_{t=0}. \tag{5.1.8}$$

Now,

$$(\rho_g \circ s)(q + t\mathbf{v}) = g \cdot \begin{pmatrix} (1 + |q + tv|^2)^{-\frac{1}{2}}(q + tv) \\ (1 + |q + tv|^2)^{-\frac{1}{2}} \end{pmatrix}.$$

Let $h(t) = (1 + |q + tv|^2)^{-\frac{1}{2}}$ and compute

$$\begin{pmatrix} a & b \\ c & d \end{pmatrix}\begin{pmatrix} h(t)(q + tv) \\ h(t) \end{pmatrix} = \begin{pmatrix} ah(t)(q + tv) + bh(t) \\ ch(t)(q + tv) + dh(t) \end{pmatrix}.$$

Exercise 5.1.13 Show that

$$\begin{aligned}|ah(t)(q + tv) + bh(t)|^2 &+ |ch(t)(q + tv) + dh(t)|^2 \\ &= (h(t))^2\left(At^2 + Bt + C\right),\end{aligned}$$

where $A = (|a|^2 + |c|^2)|v|^2$, $B = 2\,\mathrm{Re}(b\bar{v}\bar{a} + aq\bar{v}\bar{a} + d\bar{v}\bar{c} + cq\bar{v}\bar{c})$ and $C = |b|^2 + |a|^2|q|^2 + |d|^2 + |c|^2|q|^2 + 2\,\mathrm{Re}(b\bar{q}\bar{a} + d\bar{q}\bar{c})$. Note that these are all real and show also that

$$C = |aq + b|^2 + |cq + d|^2.$$

Notice that $\left(\begin{smallmatrix} a & b \\ c & d \end{smallmatrix}\right)\left(\begin{smallmatrix} q \\ 1 \end{smallmatrix}\right) = \left(\begin{smallmatrix} aq+b \\ cq+d \end{smallmatrix}\right)$. Since $\left(\begin{smallmatrix} a & b \\ c & d \end{smallmatrix}\right) \in SL(2, \mathbb{H})$ is invertible, it follows that C (in Exercise 5.1.13) is strictly positive. Now, we have

$$(\rho_g \circ s)(q + tv) = \left(At^2 + Bt + C\right)^{-\frac{1}{2}}\begin{pmatrix} aq + avt + b \\ cq + cvt + d \end{pmatrix}. \tag{5.1.9}$$

Computing derivatives at $t = 0$ coordinatewise gives, by (5.1.8),

$$(\rho_g \circ s)_{*q}(\mathbf{v}_q) = \begin{pmatrix} C^{-\frac{1}{2}}\, av - \dfrac{1}{2}\, C^{-\frac{3}{2}}\, B\,(aq+b) \\ \\ C^{-\frac{1}{2}}\, cv - \dfrac{1}{2}\, C^{-\frac{3}{2}}\, B\,(cq+d) \end{pmatrix}. \tag{5.1.10}$$

We need to compute $\boldsymbol{\omega}_{\rho_g(s(q))}$ of the tangent vector given by (5.1.10). Now, setting $t = 0$ in (5.1.9) gives

$$\rho_g(s(q)) = \begin{pmatrix} \dfrac{aq+b}{\sqrt{C}} \\ \\ \dfrac{cq+d}{\sqrt{C}} \end{pmatrix}$$

so $\bar{q}^1(\rho_g(s(q))) = \frac{\bar{q}\,\bar{a}+\bar{b}}{\sqrt{C}}$ and $\bar{q}^2(\rho_g(s(q))) = \frac{\bar{q}\,\bar{c}+\bar{d}}{\sqrt{C}}$. Thus,

$$\begin{aligned}
&\left(\bar{q}^1\, dq^1\right)_{\rho_g(s(q))}\left((\rho_g \circ s)_{*q}(\mathbf{v}_q)\right) \\
&\qquad = \frac{\bar{q}\,\bar{a}+\bar{b}}{\sqrt{C}}\left(C^{-\frac{1}{2}}\, av - \frac{1}{2}\, C^{-\frac{3}{2}}\, B\,(aq+b)\right) \\
&\qquad = \frac{(\bar{q}\,\bar{a}+\bar{b})av}{C} - \frac{B(\bar{q}\,\bar{a}+\bar{b})(aq+b)}{2\,C^2} \\
&\qquad = \frac{|a|^2\,\bar{q}v + \bar{b}\,av}{C} - \frac{B\,|aq+b|^2}{2\,C^2}
\end{aligned}$$

and

$$(\bar{q}^2\, dq^2)_{\rho_g(s(q))}\left((\rho_g \circ s)_{*q}(\mathbf{v}_q)\right) = \frac{|c|^2\,\bar{q}v + \bar{d}\,cv}{C} - \frac{B\,|cq+d|^2}{2\,C^2}.$$

Consequently,

$$\begin{aligned}
&\mathrm{Im}\left(\bar{q}^1\, dq^1 + \bar{q}^2\, dq^2\right)_{\rho_g(s(q))}\left((\rho_g \circ s)_{*q}(\mathbf{v}_q)\right) \\
&\qquad = \mathrm{Im}\left(\frac{(|a|^2 + |c|^2)\,\bar{q}v + (\bar{b}\,a + \bar{d}\,c)v}{|aq+b|^2 + |cq+d|^2}\right) \\
&\qquad = \mathrm{Im}\left(\frac{(|a|^2 + |c|^2)\,\bar{q} + (\bar{b}\,a + \bar{d}\,c)}{|aq+b|^2 + |cq+d|^2}\, dq(\mathbf{v}_q)\right).
\end{aligned}$$

We conclude from (5.1.7) that, for $g = \begin{pmatrix} a & b \\ c & d \end{pmatrix} \in SL(2, \mathbb{H})$,

$$\left(s_2 \circ \varphi_2^{-1} \right)^* \left(\rho_g^* \omega \right) = \operatorname{Im} \left(\frac{(|a|^2 + |c|^2)\bar{q} + (\bar{b}a + \bar{d}c)}{|aq+b|^2 + |cq+d|^2} dq \right). \tag{5.1.11}$$

Exercise 5.1.14 Show that

$$\left(s_1 \circ \varphi_1^{-1} \right)^* \left(\rho_g^* \omega \right) = \operatorname{Im} \left(\frac{(|b|^2 + |d|^2)\bar{q} + (\bar{a}b + \bar{c}d)}{|bq+a|^2 + |dq+c|^2} dq \right). \tag{5.1.12}$$

Remarks: We remind the reader once again of the possible confusion inherent in the (traditional) use of the same symbol "q" in both (5.1.11) and (5.1.12) (see the Remark following Exercise 4.8.15). Also note that, according to the uniqueness assertion in Theorem 5.1.1, the connection $\rho_g^* \omega$ is completely determined by the pair $\{s_1^*(\rho_g^* \omega), s_2^*(\rho_g^* \omega)\}$ of pullbacks, i.e., by the 1-forms (5.1.11) and (5.1.12) on $\mathbb{H}$. In this case one can say more, however. Indeed, any connection η on the Hopf bundle is uniquely determined by *either one* of the gauge potentials $s_1^* \eta$ or $s_2^* \eta$ alone. The reason is that if one is given, say, $s_1^* \eta$ on V_1, then the transformation law $s_2^* \eta = ad_{g_{12}^{-1}} \circ s_1^* \eta + \Theta_{12}$ ((4.8.23)) uniquely determines $s_2^* \eta$ on $V_2 \cap V_1$. But $V_2 \cap V_1$ misses only one point of V_2 so continuity then determines $s_2^* \eta$ on all of V_2. For this reason it is common in the literature to find a connection on the Hopf bundle represented by a single $\operatorname{Im} \mathbb{H}$-valued 1-form on $\mathbb{R}^4$. In Section 5.3 we will have something to say about going in the other direction, i.e., about when an $\operatorname{Im} \mathbb{H}$-valued 1-form on $\mathbb{R}^4$ determines a connection on the Hopf bundle.

Needless to say, (5.1.11) and (5.1.12) reduce to $(s_2 \circ \varphi_2^{-1})^* \omega$ and $(s_1 \circ \varphi_1^{-1})^* \omega$, respectively, when $g = \begin{pmatrix} 1 & 0 \\ 0 & 1 \end{pmatrix}$, i.e., $\rho_{id}^* \omega = \omega$. We show now that, in fact, $\rho_g^* \omega = \omega$ for all g in a subgroup of $SL(2, \mathbb{H})$ that is well-known to us (see Exercise 1.6.16). We compute $\rho_g^* \omega$ when $g = \begin{pmatrix} a & b \\ c & d \end{pmatrix} \in Sp(2)$. Then $\bar{g}^T g = id$ implies $|a|^2 + |c|^2 = |b|^2 + |d|^2 = 1$ and $\bar{a}b + \bar{c}d = \bar{b}a + \bar{d}c = 0$. Moreover,

$$\begin{pmatrix} a & b \\ c & d \end{pmatrix} \begin{pmatrix} q \\ 1 \end{pmatrix} = \begin{pmatrix} aq+b \\ cq+d \end{pmatrix} \quad \text{and} \quad \begin{pmatrix} a & b \\ c & d \end{pmatrix} \begin{pmatrix} 1 \\ q \end{pmatrix} = \begin{pmatrix} bq+a \\ dq+c \end{pmatrix}$$

and, since the elements of $Sp(2)$ preserve the bilinear form on $\mathbb{H}^2$, $|aq+b|^2+|cq+d|^2 = |bq+a|^2+|dq+c|^2 = 1+|q|^2$. Substituting all of these into (5.1.11) and (5.1.12) we obtain $(s_2\circ\varphi_2^{-1})^*\omega$ and $(s_1\circ\varphi_1^{-1})^*\omega$, respectively, so that $\rho_g^*\omega = \omega$. Observe that the elements of $Sp(2)$ are, in fact, the *only* elements of $SL(2,\mathbb{H})$ that leave ω invariant, i.e., that $\rho_g^*\omega = \omega$ implies $g \in Sp(2)$. Indeed, if, for example, the right-hand side of (5.1.11) is just $(s_2\circ\varphi_2^{-1})^*\omega$, then $|a|^2+|c|^2 = 1$, and $\bar{b}\,a+\bar{d}\,c = 0$ (so $\bar{a}\,b+\bar{c}\,d = 0$) and $|aq+b|^2+|cq+d|^2 = 1+|q|^2$. With $q = 0$ this last equality implies $|b|^2+|d|^2 = 1$. All of these together simply say

$$\begin{pmatrix} \bar{a} & \bar{c} \\ \bar{b} & \bar{d} \end{pmatrix}\begin{pmatrix} a & b \\ c & d \end{pmatrix} = \begin{pmatrix} 1 & 0 \\ 0 & 1 \end{pmatrix}$$

so $g \in Sp(2)$. We have therefore proved that

$$\rho_g^*\omega = \omega \quad \text{iff} \quad g \in Sp(2)\,. \tag{5.1.13}$$

We wish to examine more closely the connections $\rho_g^*\omega$ for g in two other subgroups of $SL(2,\mathbb{H})$. First consider the set

$$N = \left\{ \begin{pmatrix} 1 & n \\ 0 & 1 \end{pmatrix} : n \in \mathbb{H} \right\}.$$

To see that this is, in fact, a subgroup of $SL(2,\mathbb{H})$ we compute $\phi\left(\begin{smallmatrix} 1 & n \\ 0 & 1 \end{smallmatrix}\right)$ (see (1.1.26)). Let $n = n^1+n^2\mathbf{j}$, where $n^1, n^2 \in \mathbb{C}$. Then

$$\begin{pmatrix} 1 & n \\ 0 & 1 \end{pmatrix} = \begin{pmatrix} 1 & n^1 \\ 0 & 1 \end{pmatrix} + \begin{pmatrix} 1 & n^2 \\ 0 & 0 \end{pmatrix}\mathbf{j}$$

so

$$\phi\begin{pmatrix} 1 & n \\ 0 & 1 \end{pmatrix} = \begin{pmatrix} 1 & n^1 & 0 & n^2 \\ 0 & 1 & 0 & 0 \\ 0 & -\bar{n}^2 & 1 & \bar{n}^1 \\ 0 & 0 & 0 & 1 \end{pmatrix}.$$

Since $\det\phi\left(\begin{smallmatrix} 1 & n \\ 0 & 1 \end{smallmatrix}\right) = 1$, $\left(\begin{smallmatrix} 1 & n \\ 0 & 1 \end{smallmatrix}\right) \in SL(2,\mathbb{H})$.

Exercise 5.1.15 Show that N is closed under matrix multiplication and inversion and so is a subgroup of $SL(2,\mathbb{H})$.

Exercise 5.1.16 Let $A = \{ \begin{pmatrix} \sqrt{\lambda} & 0 \\ 0 & 1/\sqrt{\lambda} \end{pmatrix} : \lambda > 0 \}$. Show that A is a subgroup of $SL(2, \mathbb{H})$.

It follows that the set

$$NA = \left\{ \begin{pmatrix} 1 & n \\ 0 & 1 \end{pmatrix} \begin{pmatrix} \sqrt{\lambda} & 0 \\ 0 & 1/\sqrt{\lambda} \end{pmatrix} : n \in \mathbb{H}, \ \lambda > 0 \right\}$$
$$= \left\{ \begin{pmatrix} \sqrt{\lambda} & n/\sqrt{\lambda} \\ 0 & 1/\sqrt{\lambda} \end{pmatrix} : n \in \mathbb{H}, \ \lambda > 0 \right\}$$

is contained in $SL(2, \mathbb{H})$. For $g = \begin{pmatrix} \sqrt{\lambda} & n/\sqrt{\lambda} \\ 0 & 1/\sqrt{\lambda} \end{pmatrix} \in NA$ we wish to compute $\rho^*_{g^{-1}} \boldsymbol{\omega}$ (the reason for the inverse will become clear shortly). Now,

$$g^{-1} = \begin{pmatrix} 1/\sqrt{\lambda} & -n/\sqrt{\lambda} \\ 0 & \sqrt{\lambda} \end{pmatrix} = \begin{pmatrix} a & b \\ c & d \end{pmatrix},$$

where $a = 1/\sqrt{\lambda}$, $b = -n/\sqrt{\lambda}$, $c = 0$ and $d = \sqrt{\lambda}$. Thus, $|a|^2 + |c|^2 = 1/\lambda$, $\bar{b}\,a + \bar{d}\,c = -\bar{n}/\lambda$, $|aq + b|^2 = \frac{1}{\lambda}|q - n|^2$ and $|cq + d|^2 = \lambda$ so (5.1.11) gives

$$\left(s_2 \circ \varphi_2^{-1}\right)^* \left(\rho^*_{g^{-1}} \boldsymbol{\omega}\right) = \mathrm{Im}\left(\frac{\bar{q} - \bar{n}}{|q - n|^2 + \lambda^2}\, dq\right). \tag{5.1.14}$$

Similarly,

$$\left(s_1 \circ \varphi_1^{-1}\right)^* \left(\rho^*_{g^{-1}} \boldsymbol{\omega}\right) = \mathrm{Im}\left(\frac{(|n|^2 + \lambda^2)\,\bar{q} - n}{|1 - nq|^2 + \lambda^2 |q|^2}\, dq\right). \tag{5.1.15}$$

Because we wish to express each of these in both φ_1- and φ_2-coordinates on $V_1 \cap V_2$, we write them out in more detail. For $p \in V_2$ and $\mathbf{X} \in T_p(S^4)$, (5.1.14) gives

$$\left(s_2^* \left(\rho^*_{g^{-1}} \boldsymbol{\omega}\right)\right)_p \mathbf{X} = \left(\left(s_2 \circ \varphi_2^{-1}\right)^* \left(\rho^*_{g^{-1}} \boldsymbol{\omega}\right)\right)_{\varphi_2(p)} \left((\varphi_2)_{*p} \mathbf{X}\right)$$
$$= \mathrm{Im}\left(\frac{\overline{\varphi_2(p)} - \bar{n}}{|\varphi_2(p) - n|^2 + \lambda^2}\, w\right),$$

where $dq\left((\varphi_2)_{*p}\mathbf{X}\right) = w$. Similarly, for $p \in V_1$ and $\mathbf{X} \in T_p(S^4)$, (5.1.15) gives

$$\left(s_1^* \left(\rho_{g^{-1}}^* \omega \right) \right)_p \mathbf{X} = \left(\left(s_1 \circ \varphi_1^{-1} \right)^* \left(\rho_{g^{-1}}^* \omega \right) \right)_{\varphi_1(p)} \left((\varphi_1)_{*p} \mathbf{X} \right)$$

$$= \operatorname{Im} \left(\frac{(|n|^2 + \lambda^2)\overline{\varphi_1(p)} - n}{|1 - n\varphi_1(p)|^2 + \lambda^2 |\varphi_1(p)|^2} v \right),$$

where $dq((\varphi_1)_{*p}\mathbf{X}) = v$.

Exercise 5.1.17 Show that, for $p \in V_1 \cap V_2$ and $\mathbf{X} \in T_p(S^4)$,

$$\left(s_2^* \left(\rho_{g^{-1}}^* \omega \right) \right)_p \mathbf{X} = \operatorname{Im} \left(\frac{\varphi_1(p)\, \bar{v} + n\, \overline{\varphi_1(p)}\, v\, \overline{\varphi_1(p)}}{|\varphi_1(p)|^2 \left(|1 - n\varphi_1(p)|^2 + \lambda^2 |\varphi_1(p)|^2 \right)} \right)$$

and

$$\left(s_1^* \left(\rho_{g^{-1}}^* \omega \right) \right)_p \mathbf{X} = \operatorname{Im} \left(\frac{(|n|^2 + \lambda^2)\, \varphi_2(p)\, \bar{w} + n\, \overline{\varphi_2(p)}\, w\, \overline{\varphi_2(p)}}{|\varphi_2(p)|^2 \left(|\varphi_2(p) - n|^2 + \lambda^2 \right)} \right),$$

where $dq((\varphi_2)_{*p}\mathbf{X}) = w$ and $dq((\varphi_1)_{*p}\mathbf{X}) = v$.

We have therefore managed to produce a fairly substantial collection of connection forms on $S^3 \to S^7 \to S^4$. In fact, though, we have done more than that. If we denote by Υ the set of all connection forms on $S^3 \to S^7 \to S^4$, then the natural left action of $SL(2, \mathbb{H})$ on S^7 given by (5.1.6) has given rise to a left action of $SL(2, \mathbb{H})$ on Υ defined by

$$(g, \eta) \in SL(2, \mathbb{H}) \times \Upsilon \longrightarrow g \cdot \eta = \rho_{g^{-1}}^* \eta.$$

That this deserves to be called a left action follows from

$$(gh, \eta) \longrightarrow (gh) \cdot \eta = \rho_{(gh)^{-1}}^* \eta = \rho_{h^{-1}g^{-1}}^* \eta$$
$$= \left(\rho_{h^{-1}} \circ \rho_{g^{-1}} \right)^* \eta$$
$$= \rho_{g^{-1}}^* \circ \rho_{h^{-1}}^* (\eta)$$
$$= \rho_{g^{-1}}^* (h \cdot \eta)$$
$$= g \cdot (h \cdot \eta)$$

(the inverse is required to compensate for the pullback). Thus far we have calculated this action only when $g \in Sp(2)$ or $g \in NA$ and only on the natural connection ω for the Hopf bundle. We will eventually see, however, that we have gotten a great deal more than we had any right to expect from such minimal effort. The full story will not emerge for some time, but here's a prologue. Consider the subset $NA\,Sp(2)$ of $SL(2,\mathbb{H})$ consisting of all products $g_1 g_2 g_3$ with $g_1 \in N$, $g_2 \in A$ and $g_3 \in Sp(2)$. If ω is the natural connection on the Hopf bundle, then

$$(g_1 g_2 g_3) \cdot \omega = (g_1 g_2) \cdot (g_3 \cdot \omega) = (g_1 g_2) \cdot \omega$$

because $g_3 \cdot \omega = \rho^*_{g_3^{-1}} \omega = \omega$ by (5.1.13). Since we have already calculated $(g_1 g_2) \cdot \omega$ for $g_1 \in N$ and $g_2 \in A$, it follows that we have, in fact, determined $g \cdot \omega$ for all $g \in NA\,Sp(2)$. The punch line here is that, although it is far from being obvious, $NA\,Sp(2)$ is actually *all* of $SL(2,\mathbb{H})$. This is the so-called **Iwasawa decomposition** of $SL(2,\mathbb{H})$:

$$SL(2,\mathbb{H}) = NA\,Sp(2). \tag{5.1.16}$$

Although it would take us too far afield algebraically to prove this (Iwasawa decompositions are treated in detail in [**Helg**]) we will allow ourselves to conclude from (5.1.16) that we have now identified the entire orbit (and isotropy subgroup) of ω under the action of $SL(2,\mathbb{H})$ on Υ.

Theorem 5.1.3 illuminates another issue of considerable interest to us, i.e., when are two different connections on a bundle sufficiently different that we should distinguish them? Consider, for example, the $\mathrm{Im}\,\mathbb{H}$-valued 1-forms

$$\frac{|q|^2}{1+|q|^2}\,\mathrm{Im}\left(q^{-1}dq\right) \quad \text{and} \quad \frac{1}{1+|q|^2}\,\mathrm{Im}\left(\bar{q}^{-1}d\bar{q}\right)$$

on $\mathbb{H} - \{0\}$. Each of these can (by Theorem 5.1.1) be identified with a unique connection on the trivial $Sp(1)$-bundle over $\mathbb{H} - \{0\}$ and, thought of in this way, they appear rather different. They certainly take different values at tangent vectors to points in $\mathbb{H} - \{0\}$ and even exhibit different asymptotic behaviors as $|q| \to \infty$ (the first approaching $\mathrm{Im}(q^{-1}dq)$ and the second approaching zero). However, we know more about these 1-forms. Indeed, (4.8.19) and (4.8.20) expose them for what they are, namely, pullbacks to $\mathbb{H} - \{0\}$ of the *same* connection form ω on the Hopf bundle via two different cross-sections of that bundle. Consequently, they differ only by what we have called a (local) gauge transformation (see (4.8.26)) and, owing to the one-to-one correspondence between cross-sections and trivializations (Section 3.3), this amounts to differing only by the particular manner in which our trivial bundle is trivialized. As such they should be regarded as two different coordinate expressions for the same underlying geometrical object and so should be deemed "equivalent". In order to formalize (and globalize) the appropriate notion of equivalence here we re-examine local gauge equivalence somewhat more carefully.

A local gauge transformation is a change of cross-section and all such arise in the following way: Let $s : V \to \mathcal{P}^{-1}(V)$ be a local cross-section and $g : V \to G$ a smooth map of V into G. Define $s^g : V \to \mathcal{P}^{-1}(V)$ by $s^g(x) = s(x) \cdot g(x)$ for every $x \in V$. Then s^g is also a local cross-section so

$$\mathcal{P}^{-1}(V) = \bigcup_{x \in V} \{\, s(x) \cdot h : h \in G \,\} = \bigcup_{x \in V} \{\, s^g(x) \cdot h : h \in G \,\} .$$

Thus, we may define a map $f : \mathcal{P}^{-1}(V) \to \mathcal{P}^{-1}(V)$ by $f(s(x) \cdot h) = s^g(x) \cdot h$.

Exercise 5.1.18 Show that f is a (smooth) automorphism of the G-bundle $\mathcal{P} : \mathcal{P}^{-1}(V) \to V$ (Section 3.3).

Conversely, suppose we are given an automorphism f of the bundle $\mathcal{P}^{-1}(V)$ onto itself. If $s : V \to \mathcal{P}^{-1}(V)$ is any cross-section we define another map of V into $\mathcal{P}^{-1}(V)$ by $x \to f^{-1}(s(x))$. Since f^{-1} is also an automorphism of $\mathcal{P}^{-1}(V)$, this is another cross-section on V $(\mathcal{P}(f^{-1}(s(x))) = \mathcal{P}(s(x)) = x)$. Thus, for each $x \in V$ there exists a unique $g(x) \in G$ such that $f^{-1}(s(x)) = s(x) \cdot g(x)$.

Exercise 5.1.19 Show that $g(x)$ is smooth so that the cross-section $x \to f^{-1}(s(x))$ is just $s^g : V \to \mathcal{P}^{-1}(V)$.

Consequently, a local gauge transformation on $V \subseteq X$ can be identified with an automorphism of the bundle $\mathcal{P}^{-1}(V)$. The appropriate global notion is therefore clear. If $\mathcal{P} : P \to X$ is a principal G-bundle over X, then a (**global**) **gauge transformation** is an automorphism of the bundle, i.e., a diffeomorphism $f : P \to P$ of P onto itself that preserves the fibers of $\mathcal{P}$ $(\mathcal{P} \circ f = \mathcal{P})$ and commutes with the action of G on P $(f(p \cdot g) = f(p) \cdot g)$. Since compositions and inverses of automorphisms are also automorphisms, the collection of all gauge transformations of $\mathcal{P} : P \to X$ is a group under composition called the **group of gauge transformations** of $\mathcal{P} : P \to X$ and denoted $\mathcal{G}(P)$ or simply $\mathcal{G}$ (some sources refer to $\mathcal{G}$ as the "gauge group", but we will reserve this terminology for G). Now if ω is a connection form on the bundle and $f \in \mathcal{G}(P)$, then, by Theorem 5.1.3, $f^* \omega$ is also a connection form on the same bundle. Two connection forms ω and η on P are said to be **gauge equivalent** if there exists an $f \in \mathcal{G}(P)$, such that $\eta = f^* \omega$.

Exercise 5.1.20 Show that gauge equivalence does, indeed, define an equivalence relation on the set Υ of all connection forms on P.

The set $\Upsilon/\mathcal{G}$ of gauge equivalence classes of connections on P is called the **moduli space of connections** on the bundle. Such moduli spaces have become, since Donaldson [**Don**], objects of profound significance to topology. To the physicist they represent the configuration spaces of quantum field theories and, as such, are the "manifolds" on which Feynman path integrals are defined (sort of) and evaluated (in a manner of speaking).

Remark: We use the term "space" advisedly here since we have made no attempt to provide Υ or $\Upsilon/\mathcal{G}$ with topologies. Although it is possible to do this in a meaningful way, the study of the resulting moduli space is quite beyond the power of our meager tools. Our goal is much less ambitious. By restricting attention to the quaternionic Hopf bundle and to a particular type of connection on it (called "anti-self-dual") and by appealing to a deep theorem of Atiyah, Hitchin and Singer we will, in Section 5.5, identify topologically a much smaller, much more manageable moduli space.

Notice that if f is an automorphism (so that $\boldsymbol{\omega}$ and $f^*\boldsymbol{\omega}$ are gauge equivalent) and $s : V \to \mathcal{P}^{-1}(V)$ is a cross-section, then $s^*(f^*\boldsymbol{\omega}) = (f \circ s)^*\boldsymbol{\omega}$. Since $f \circ s : V \to \mathcal{P}^{-1}(V)$ is also a cross-section we conclude that $s^*\boldsymbol{\omega}$ and $s^*(f^*\boldsymbol{\omega})$ are, in fact, both gauge potentials for $\boldsymbol{\omega}$) (by different cross-sections). Consequently, they are related by a transformation law of type (4.8.23).

Before turning to the subject of a connection's "curvature" and the precise definition of a "gauge field" we must briefly return to our roots. Our interest in connections was originally motivated (in Chapter 0) by the suggestion that such a structure would provide the unique path lifting procedure whereby one might keep track of the evolution of a particle's internal state (e.g., phase) as it traverses the field established by some other particle (e.g., the electromagnetic field of a magnetic monopole).

Theorem 5.1.4 *Let $\mathcal{P} : P \to X$ be a smooth principal G-bundle over X and $\boldsymbol{\omega}$ a connection form on P. Let $\alpha : [0,1] \to X$ be a smooth curve in X with $\alpha(0) = x_0$ and let $p_0 \in \mathcal{P}^{-1}(x_0)$. Then there exists a unique smooth curve $\tilde{\alpha} : [0,1] \to P$ such that*

1. $\tilde{\alpha}(0) = p_0$,
2. $\mathcal{P} \circ \tilde{\alpha}(t) = \alpha(t)$ *for all* $t \in [0,1]$, *and*
3. $\tilde{\alpha}'(t) \in Hor_{\tilde{\alpha}(t)}(P)$ *for all* $t \in [0,1]$.

Proof: Assume first that $\alpha([0,1]) \subseteq V$ for some trivialization (V, Ψ). Let $s : V \to \mathcal{P}^{-1}(V)$ be the canonical cross-section associated with the trivialization and assume without loss of generality that $s(x_0) = p_0$. Since $\Psi : \mathcal{P}^{-1}(V) \to V \times G$ is a diffeomorphism that preserves the fibers of $\mathcal{P}$, any $\tilde{\alpha} : [0,1] \to \mathcal{P}^{-1}(V)$ satisfying (2) must be of the form $\tilde{\alpha}(t) = \Psi^{-1}(\alpha(t), g(t)) = s(\alpha(t)) \cdot g(t)$ for some smooth curve $g : [0,1] \to G$. In order to satisfy (1) we must have $g(0) = e$ (the identity in G). We need only show that we can find such a g so that (3) is satisfied. Now, by Exercise 5.1.2,

$$\begin{aligned}\tilde{\alpha}'(t) &= \left(\Psi^{-1} \circ (\alpha, g)\right)'(t) = \left(\Psi^{-1}\right)_{*(\alpha(t),g(t))}(\alpha'(t), g'(t)) \\ &= \left(\sigma_{g(t)}\right)_{*s(\alpha(t))}\left(s_{*\alpha(t)}(\alpha'(t))\right) + A^{\#}\left(s(\alpha(t)) \cdot g(t)\right),\end{aligned}$$

where $A = (L_{g(t)^{-1}})_{*g(t)}(g'(t))$. Thus,

$$\begin{aligned}\boldsymbol{\omega}_{\tilde{\alpha}(t)}(\tilde{\alpha}'(t)) &= ad_{g(t)^{-1}}\left(\boldsymbol{\omega}_{s(\alpha(t))}\left(s_{*\alpha(t)}(\alpha'(t))\right)\right) + A \\ &= ad_{g(t)^{-1}}\left((s^*\boldsymbol{\omega})_{\alpha(t)}(\alpha'(t))\right) + \left(L_{g(t)^{-1}}\right)_{*g(t)}(g'(t)).\end{aligned}$$

Now, $\beta(t) = (s^*\boldsymbol{\omega})_{\alpha(t)}(\alpha'(t))$ is a known, smooth curve in the Lie algebra $\mathcal{G}$ so

$$\boldsymbol{\omega}_{\tilde{\alpha}(t)}(\tilde{\alpha}'(t)) = ad_{g(t)^{-1}}(\beta(t)) + \left(L_{g(t)^{-1}}\right)_{*g(t)}(g'(t)).$$

In order to satisfy (3) we must have $\boldsymbol{\omega}_{\tilde{\alpha}(t)}(\tilde{\alpha}'(t)) = 0$ for each t, i.e.,

$$\left(L_{g(t)^{-1}}\right)_{*g(t)}(g'(t)) = -ad_{g(t)^{-1}}(\beta(t)).$$

But $L_{g(t)^{-1}}$ is a diffeomorphism for each t so $(L_{g(t)^{-1}})_{*g(t)}$ is an isomorphism and its inverse is $(L_{g(t)})_{*e}$. Thus, this last equation is equivalent to

$$\begin{aligned}g'(t) &= -\left(L_{g(t)}\right)_{*e}\left(ad_{g(t)^{-1}}(\beta(t))\right) \\ &= -\left(L_{g(t)}\right)_{*e}\Big(\left(L_{g(t)^{-1}}\right)_{*g(t)} \circ \left(R_{g(t)}\right)_{*e}(\beta(t))\Big) \\ &= -\left(R_{g(t)}\right)_{*e}(\beta(t)).\end{aligned}$$

The conclusion then is that (1), (2) and (3) will be satisfied iff $\tilde{\alpha}$ has the form $\tilde{\alpha}(t) = s(\alpha(t)) \cdot g(t)$, where

$$\begin{cases} g'(t) = -\left(R_{g(t)}\right)_{*e}(\beta(t)), \quad 0 \le t \le 1, \\ g(0) = e \end{cases}, \tag{5.1.17}$$

where $\beta(t) = (s^*\boldsymbol{\omega})_{\alpha(t)}(\alpha'(t))$. We obtain such a $g(t)$ by applying Theorem 4.6.2. First we extend α smoothly to an open interval $(0-\delta, 1+\delta)$ for some $\delta > 0$. We use the same symbol for the extension $\alpha : (0-\delta, 1+\delta) \to X$ and assume, as we may, that $\alpha(\,(0-\delta, 1+\delta)\,) \subseteq V$ so that β is defined by the same formula on $(0-\delta, 1+\delta)$. Define a vector field $\mathbf{V}$ on $G \times (0-\delta, 1+\delta)$ by

$$\mathbf{V}\,(g,s) = \left(-\left(R_{g(t)} \right)_{*e} (\beta(s))\,,\ \frac{d}{dt}\bigg|_s \right).$$

The maximal integral curve of $\mathbf{V}$ starting at $(e, 0)$ will be of the form $(\,g(t), t\,)$ with $g(t)$ satisfying (5.1.17) on some interval $[0, t_0]$. Since $\{e\} \times [0,1]$ is compact we may select an $\varepsilon > 0$ such that, for each $r \in [0,1]$, $\mathbf{V}_t\,(e, r)$ is defined for $|\,t\,| < \varepsilon$ (see (4.6.21)). We claim that $g(t)$ can be extended to a solution to (5.1.17) on $[0, t_0 + \varepsilon] \cap [0,1]$ and from this it will follow that it can be extended to a solution on all of $[0,1]$. Let $(h(t), t + t_0)$ be the integral curve of $\mathbf{V}$ starting at (e, t_0). It is defined for $|\,t\,| < \varepsilon$. Let $\tilde{g}(t) = h(t - t_0)\,g(t_0)$ for $|\,t - t_0\,| < \varepsilon$ (the product is in G). Then $\tilde{g}(t_0) = g(t_0)$ and, since $h'\,(t) = -(\,R_{h(t)}\,)_{*e}(\,\beta(t_0 + t)\,)$, we have $\tilde{g}\,'\,(t) = -(\,R_{g(t_0)}\,)_{*h(t-t_0)}(\,(\,R_{h(t-t_0)}\,)_{*e}(\,\beta(t)\,)\,) = -(\,R_{\tilde{g}(t)}\,)_{*e}(\,\beta(t)\,)$ as required.

Exercise 5.1.21 Complete the proof by considering the case in which α does not map entirely into a trivializing nbd. **Hint:** $\alpha(\,[0,1]\,)$ is compact. Use the Theorem 1.4.6. ∎

The existence of the horizontal lifts described in Theorem 5.1.4 provides a means of identifying ("connecting") the fibers above any two points in X that can be joined by a smooth curve. Specifically, let us suppose that $x_0, x_1 \in X$ and $\alpha : [0,1] \to X$ is a smooth curve with $\alpha(0) = x_0$ and $\alpha(1) = x_1$. For any $p_0 \in \mathcal{P}^{-1}\,(x_0)$ there exists a unique smooth curve $\tilde{\alpha}_{\,p_0} : [0,1] \to P$ that lifts α, goes through p_0 at $t = 0$ and has horizontal velocity vector at each point. In particular, $\tilde{\alpha}_{\,p_0}\,(1) \in \mathcal{P}^{-1}\,(x_1)$. Define a map $\tau_\alpha : \mathcal{P}^{-1}\,(x_0) \to \mathcal{P}^{-1}\,(x_1)$, called **parallel translation along** α (determined by the connection $\boldsymbol{\omega}$), by

$$\tau_\alpha\,(\,p_0\,) = \tilde{\alpha}_{\,p_0}\,(\,1\,)\,.$$

Exercise 5.1.22 Show that τ_α commutes with the action σ of G on P, i.e., that $\tau_\alpha \circ \sigma_g = \sigma_g \circ \tau_\alpha$ for each $g \in G$. **Hint:** Use (5.1.3) to show that $\tilde{\alpha}_{\,\sigma_g\,(p_0)} = \sigma_g \circ \tilde{\alpha}_{\,p_0}$.

Exercise 5.1.23 Show that $\tau_{\alpha^\leftarrow} = (\tau_\alpha)^{-1}$ and that if α and β are smooth curves in X with $\alpha(1) = \beta(0)$, then $\tau_{\alpha\beta} = \tau_\beta \circ \tau_\alpha$.

Suppose now that α is a smooth *loop* at x_0 in X, i.e., that $x_0 = \alpha(0) = \alpha(1)$. Then $\tau_\alpha : \mathcal{P}^{-1}(x_0) \to \mathcal{P}^{-1}(x_0)$. Now, G acts transitively on the fibers of $\mathcal{P}$ (Lemma 3.1.1) so, for each $p_0 \in \mathcal{P}^{-1}(x_0)$ there exists a unique $g \in G$ such that $\tau_\alpha(p_0) = p_0 \cdot g$. Holding p_0 fixed, but letting α vary over all smooth loops at $x_0 = \mathcal{P}(p_0)$ in X we obtain a subset $\mathcal{H}(p_0)$ of G consisting of all those g such that p_0 is parallel translated to $p_0 \cdot g$ over some smooth loop at $\mathcal{P}(p_0)$ in X.

Exercise 5.1.24 Show that $\mathcal{H}(p_0)$ is a subgroup of G (called the **holonomy group** of the connection $\boldsymbol{\omega}$ at p_0) and that

$$\mathcal{H}(p_0 \cdot g) = g^{-1} \mathcal{H}(p_0)\, g$$

for any $g \in G$.

5.2 Curvature and Gauge Fields

Throughout this section we consider a smooth principal G-bundle $\mathcal{P} : P \to X$ equipped with a connection form $\boldsymbol{\omega}$. Thus, at each $p \in P$ we have a decomposition $T_p(P) = \mathrm{Hor}_p(P) \oplus \mathrm{Vert}_p(P)$ so any $\mathbf{v} \in T_p(P)$ can be written uniquely as $\mathbf{v} = \mathbf{v}^H + \mathbf{v}^V$, where $\mathbf{v}^H \in \mathrm{Hor}_p(P)$ is the **horizontal part** of $\mathbf{v}$ and $\mathbf{v}^V \in \mathrm{Vert}_p(P)$ is the **vertical part** of $\mathbf{v}$.

The connection form $\boldsymbol{\omega}$ is a Lie algebra-valued 1-form on P and so has an exterior derivative $d\boldsymbol{\omega}$, defined componentwise relative to any basis for $\mathcal{G}$ (Section 4.10). Being a $\mathcal{G}$-valued 2-form on P, $d\boldsymbol{\omega}$ operates on pairs of tangent vectors to P and produces elements of the Lie algebra. One obtains an object of particular interest by having $d\boldsymbol{\omega}$ operate only on horizontal parts. More precisely, we define a $\mathcal{G}$-valued 2-form $\boldsymbol{\Omega}$ on P, called the **curvature** of $\boldsymbol{\omega}$ as follows: For each $p \in P$ and for all $\mathbf{v}, \mathbf{w} \in T_p(P)$ we let

$$\boldsymbol{\Omega}(p)(\mathbf{v}, \mathbf{w}) = \boldsymbol{\Omega}_p(\mathbf{v}, \mathbf{w}) = (d\boldsymbol{\omega})_p(\mathbf{v}^H, \mathbf{w}^H).$$

Exercise 5.2.1 Show that $\boldsymbol{\Omega}_p$ is bilinear and skew-symmetric.

To show that the 2-form $\boldsymbol{\Omega}$ thus defined is smooth one chooses a basis $e_1, \ldots, e_n$ for $\mathcal{G}$. The components of $d\boldsymbol{\omega}$ relative to this basis are smooth so it will suffice to show that for any smooth vector field $\mathbf{V}$ on P, the vector field $\mathbf{V}^H$ defined by $\mathbf{V}^H(p) = (\mathbf{V}(p))^H$ is also smooth. This is an immediate consequence of the following exercise.

Exercise 5.2.2 Show that $\mathbf{V}^H = \mathbf{V} - \boldsymbol{\omega}^i(\mathbf{V})\, e_i^{\#}$. **Hint:** It is enough to show that $\mathbf{V} - \omega^i(\mathbf{V})\, e_i^{\#}$ is horizontal and $\omega^i(\mathbf{V})\, e_i^{\#}$ is vertical at each point.

Remark: The historical evolution of our definition of the curvature form from more familiar notions of curvature (e.g., for curves and surfaces) is not easily related in a few words. Happily, Volume II of [**Sp 2**] is a leisurely and entertaining account of this very story which we heartily recommend to the reader in search of motivation. Our attitude here will be that the proof of the pudding is in the eating and the justification for a definition is in its utility. Our task then is to persuade you that the definition is useful. We point out also that the process of computing the exterior derivative and evaluating only on horizontal parts, by which we arrived at the curvature from the connection, can be applied to any 0-form or any 1-form on P. The result is called the **covariant exterior derivative** of that form. We will encounter such derivatives again in our discussion of matter fields in Section 5.8.

The definition of the curvature form is short and sweet, but not very easy to compute in practice. We remedy this situation with what is called the **Cartan Structure Equation**. The formula we derive for $\boldsymbol{\Omega}$ involves the wedge product $\boldsymbol{\omega} \wedge \boldsymbol{\omega}$ and we remind the reader that, because $\boldsymbol{\omega}$ is $\mathcal{G}$-valued, this wedge product is the one determined by the Lie bracket pairing in $\mathcal{G}$ (Section 4.10). Specifically, $(\boldsymbol{\omega} \wedge \boldsymbol{\omega})_p(\mathbf{v}, \mathbf{w}) = [\boldsymbol{\omega}_p(\mathbf{v}), \boldsymbol{\omega}_p(\mathbf{w})] - [\boldsymbol{\omega}_p(\mathbf{w}), \boldsymbol{\omega}_p(\mathbf{v})] = 2\,[\boldsymbol{\omega}_p(\mathbf{v}), \boldsymbol{\omega}_p(\mathbf{w})]$.

Theorem 5.2.1 (**Cartan Structure Equation**) *Let $\mathcal{P} : P \to X$ be a smooth principal G-bundle with connection form $\boldsymbol{\omega}$ and let $\boldsymbol{\Omega}$ be the curvature of $\boldsymbol{\omega}$. Then*

$$\boldsymbol{\Omega} = d\boldsymbol{\omega} + \frac{1}{2}\boldsymbol{\omega} \wedge \boldsymbol{\omega}\,.$$

Proof: Fix a $p \in P$ and $\mathbf{v}, \mathbf{w} \in T_p(P)$. We must prove that

$$(d\boldsymbol{\omega})_p(\mathbf{v}, \mathbf{w}) = -[\boldsymbol{\omega}_p(\mathbf{v}), \boldsymbol{\omega}_p(\mathbf{w})] + \boldsymbol{\Omega}_p(\mathbf{v}, \mathbf{w})\,. \tag{5.2.1}$$

Exercise 5.2.3 By writing $\mathbf{v} = \mathbf{v}^H + \mathbf{v}^V$ and $\mathbf{w} = \mathbf{w}^H + \mathbf{w}^V$ and using the bilinearity and skew-symmetry of both sides of (5.2.1) show that it will suffice to consider the following three cases:

1. $\mathbf{v}$ and $\mathbf{w}$ both horizontal,
2. $\mathbf{v}$ and $\mathbf{w}$ both vertical, and
3. $\mathbf{v}$ vertical and $\mathbf{w}$ horizontal.

We consider in order the three cases described in Exercise 5.2.3.

1. If $\mathbf{v}$ and $\mathbf{w}$ are both horizontal, then $\omega_p(\mathbf{v}) = \omega_p(\mathbf{w}) = 0$ so $[\omega_p(\mathbf{v}), \omega_p(\mathbf{w})] = 0$. Moreover, $\mathbf{v} = \mathbf{v}^H$ and $\mathbf{w} = \mathbf{w}^H$ so $\boldsymbol{\Omega}_p(\mathbf{v}, \mathbf{w}) = (d\omega)_p(\mathbf{v}^H, \mathbf{w}^H) = (d\omega)_p(\mathbf{v}, \mathbf{w})$ and (5.2.1) is proved in this case.

2. If $\mathbf{v}$ and $\mathbf{w}$ are both vertical, then $\mathbf{v}^H = \mathbf{w}^H = 0$ so $\boldsymbol{\Omega}_p(\mathbf{v}, \mathbf{w}) = 0$. We must show then that $(d\omega)_p(\mathbf{v}, \mathbf{w}) = -[\omega_p(\mathbf{v}), \omega_p(\mathbf{w})]$. By Corollary 4.7.9, there exist $A, B \in \mathcal{G}$ such that $\mathbf{v} = A^\#(p)$ and $\mathbf{w} = B^\#(p)$. Thus, $(d\omega)_p(\mathbf{v}, \mathbf{w}) = (d\omega(A^\#, B^\#))(p)$. But $d\omega(A^\#, B^\#) = A^\#(\omega(B^\#)) - B^\#(\omega(A^\#)) - \omega([A^\#, B^\#]) = -\omega([A^\#, B^\#])$ because $\omega(A^\#)$ and $\omega(B^\#)$ are constant functions. But then, by Theorem 4.7.8, $d\omega(A^\#, B^\#) = -\omega([A, B]^\#) = -[A, B] = -[\omega(A^\#), \omega(B^\#)]$. Thus, $(d\omega)_p(\mathbf{v}, \mathbf{w}) = -[\omega(A^\#), \omega(B^\#)](p) = -[\omega_p(A^\#(p)), \omega_p(B^\#(p))] = -[\omega_p(\mathbf{v}), \omega_p(\mathbf{w})]$ as required.

3. Now we assume $\mathbf{v}$ is vertical and $\mathbf{w}$ is horizontal. Thus, $\boldsymbol{\Omega}_p(\mathbf{v}, \mathbf{w}) = (d\omega)_p(0, \mathbf{w}) = 0$ and $-[\omega_p(\mathbf{v}), \omega_p(\mathbf{w})] = -[\omega_p(\mathbf{v}), 0] = 0$ so it is enough to show that $(d\omega)_p(\mathbf{v}, \mathbf{w}) = 0$ as well. By Corollary 4.7.9, we may write $\mathbf{v} = A^\#(p)$ for some $A \in \mathcal{G}$.

Exercise 5.2.4 Show that there exists a $\mathbf{W} \in \mathcal{X}(P)$ that is horizontal at each point and satisfies $\mathbf{W}(p) = \mathbf{w}$. **Hint:** Exercise 5.2.2.

Thus, $(d\omega)_p(\mathbf{v}, \mathbf{w}) = (d\omega)_p(A^\#(p)), \mathbf{W}(p)) = (d\omega(A^\#, \mathbf{W}))(p)$. But $d\omega(A^\#, \mathbf{W}) = A^\#(\omega(\mathbf{W})) - \mathbf{W}(\omega(A^\#)) - \omega([A^\#, \mathbf{W}]) = -\omega([A^\#, \mathbf{W}])$ because $\omega(\mathbf{W})$ and $\omega(A^\#)$ are both constant (the first is 0 and the second is A). We can therefore conclude the proof by showing that $[A^\#, \mathbf{W}]$ is horizontal. According to (4.6.23),

$$[A^\#, \mathbf{W}]_p = \lim_{t \to 0} \frac{1}{t} \Big((A^\#_{-t})_{*\alpha_p(t)} (\mathbf{W}_{\alpha_p(t)}) - \mathbf{W}_p \Big),$$

where, by Exercise 4.7.19, $\alpha_p(t) = p \cdot \exp(tA)$ and $A^\#_{-t} = \sigma_{\exp(-tA)}$. Thus

$$\left(A^\#_{-t}\right)_{*\alpha_p(t)} = \left(\sigma_{\exp(-tA)}\right)_{*\alpha_p(t)} = \left(\sigma_{\exp(-tA)}\right)_{*p \cdot \exp(-tA)}$$

and this, by (5.1.3), carries $\mathrm{Hor}_{\alpha_p(t)}(P)$ onto $\mathrm{Hor}_p(P)$. Consequently, $(A^\#_{-t})_{*\alpha_p(t)}(\mathbf{W}_{\alpha_p(t)})$ is horizontal. Since $\mathrm{Hor}_p(P)$ is a linear subspace of

$T_p(P)$, $\frac{1}{t}\left(\left(A^{\#}_{-t}\right)\left(\mathbf{W}_{\alpha_p(t)}\right) - \mathbf{W}_p\right) \in \mathrm{Hor}_p(P)$ for every t.

Exercise 5.2.5 Conclude that $[A^{\#}, \mathbf{W}]_p$ is horizontal and thereby complete the proof. ∎

In the terminology of the physics literature, the curvature $\boldsymbol{\Omega}$ of a connection 1-form $\boldsymbol{\omega}$ on a principal G-bundle $\mathcal{P} : P \to X$ is called a **gauge field** on that bundle, whereas the pullback $s^* \boldsymbol{\Omega}$ of $\boldsymbol{\Omega}$ by some local cross-section s is called the **local field strength** (in gauge s) and is often denoted $\mathcal{F}_s$, or simply $\mathcal{F}$ if there is no ambiguity as to which gauge is intended. In order to calculate some specific examples we will need a local version of the Cartan Structure Equation which relates a gauge potential $\mathcal{A} = s^* \boldsymbol{\omega}$ to the corresponding local field strength $\mathcal{F} = s^* \boldsymbol{\Omega}$.

Lemma 5.2.2 *Let $\boldsymbol{\omega}$ be a connection 1-form on the principal G-bundle $\mathcal{P} : P \to X$ with curvature $\boldsymbol{\Omega}$. Then, for every $g \in G$, $\sigma_g^* \boldsymbol{\Omega} = ad_{g^{-1}} \circ \boldsymbol{\Omega}$.*

Proof: From the Structure Equation we have $\sigma_g^* \boldsymbol{\Omega} = \sigma_g^*\left(d\boldsymbol{\omega} + \frac{1}{2}\boldsymbol{\omega} \wedge \boldsymbol{\omega}\right) = \sigma_g^*(d\boldsymbol{\omega}) + \frac{1}{2}\sigma_g^*(\boldsymbol{\omega} \wedge \boldsymbol{\omega}) = d(\sigma_g^* \boldsymbol{\omega}) + \frac{1}{2}\sigma_g^*(\boldsymbol{\omega} \wedge \boldsymbol{\omega}) = d(ad_{g^{-1}} \circ \boldsymbol{\omega}) + \frac{1}{2}\sigma_g^*(\boldsymbol{\omega} \wedge \boldsymbol{\omega})$.

Exercise 5.2.6 Show that $\sigma_g^*(\boldsymbol{\omega} \wedge \boldsymbol{\omega}) = (\sigma_g^* \boldsymbol{\omega}) \wedge (\sigma_g^* \boldsymbol{\omega})$.

Thus, $\sigma_g^* \boldsymbol{\Omega} = d(ad_{g^{-1}} \circ \boldsymbol{\omega}) + \frac{1}{2}(\sigma_g^* \boldsymbol{\omega}) \wedge (\sigma_g^* \boldsymbol{\omega}) = d(ad_{g^{-1}} \circ \boldsymbol{\omega}) + \frac{1}{2}(ad_{g^{-1}} \circ \boldsymbol{\omega}) \wedge (ad_{g^{-1}} \circ \boldsymbol{\omega})$.

Exercise 5.2.7 Complete the proof by showing that, for any fixed $g \in G$,

$$\sigma_g^* \boldsymbol{\Omega} = ad_{g^{-1}} \circ \left(d\boldsymbol{\omega} + \frac{1}{2}\boldsymbol{\omega} \wedge \boldsymbol{\omega}\right) = ad_{g^{-1}} \circ \boldsymbol{\Omega}.$$ ∎

Exercise 5.2.8 Show that if $s : V \to P$ is a local cross-section of $\mathcal{P} : P \to X$, then

$$s^* \boldsymbol{\Omega} = d(s^* \boldsymbol{\omega}) + \frac{1}{2}(s^* \boldsymbol{\omega}) \wedge (s^* \boldsymbol{\omega}).$$

Writing the gauge potential as $\mathcal{A} = s^* \boldsymbol{\omega}$ and the field strength as $\mathcal{F} = s^* \boldsymbol{\Omega}$, the result of Exercise 5.2.8 assumes the form

$$\mathcal{F} = d\mathcal{A} + \frac{1}{2}\mathcal{A} \wedge \mathcal{A}. \tag{5.2.2}$$

By shrinking V if necessary we may assume that it is the coordinate nbd for a chart $\varphi : V \to \mathbb{R}^n$. Observe that (5.2.2) also expresses the relationship between $\boldsymbol{\mathcal{A}}$ and $\boldsymbol{\mathcal{F}}$ when these are identified with their coordinate expressions $(s \circ \varphi^{-1})^* \boldsymbol{\omega}$ and $(s \circ \varphi^{-1})^* \boldsymbol{\Omega}$ because

$$\begin{aligned}(s \circ \varphi^{-1})^* \boldsymbol{\Omega} &= (\varphi^{-1})^* (s^* \boldsymbol{\Omega}) \\ &= (\varphi^{-1})^* \left(d(s^* \boldsymbol{\omega}) + \frac{1}{2} (s^* \boldsymbol{\omega}) \wedge (s^* \boldsymbol{\omega}) \right) \\ &= d((s \circ \varphi^{-1})^* \boldsymbol{\omega}) + \frac{1}{2} ((s \circ \varphi^{-1})^* \boldsymbol{\omega}) \wedge ((s \circ \varphi^{-1})^* \boldsymbol{\omega}).\end{aligned}$$

As a result, Exercise 4.10.28 presents us with a nice collection of examples of gauge potentials and field strengths. For any $\lambda > 0$ and any $n \in \mathbb{H}$,

$$\boldsymbol{\mathcal{A}}_{\lambda,n} = \operatorname{Im} \left(\frac{\bar{q} - \bar{n}}{|q-n|^2 + \lambda^2} \, dq \right)$$

is the gauge potential for a connection on the Hopf bundle $S^3 \to S^7 \to S^4$ (see (5.1.14)) whose field strength $\boldsymbol{\mathcal{F}}_{\lambda,n}$ is given by

$$\boldsymbol{\mathcal{F}}_{\lambda,n} = \frac{\lambda^2}{(|q-n|^2 + \lambda^2)^2} \, d\bar{q} \wedge dq.$$

Now suppose that the chart (V, φ) has coordinate functions $x^1, \ldots, x^n$ and, on V, write $\boldsymbol{\mathcal{A}} = \mathcal{A}_\alpha \, dx^\alpha$ and $\boldsymbol{\mathcal{F}} = \frac{1}{2} \mathcal{F}_{\alpha\beta} \, dx^\alpha \wedge dx^\beta$, where the $\mathcal{A}_\alpha$ and $\mathcal{F}_{\alpha\beta}$ are $\mathcal{G}$-valued functions on V (cf., (4.10.4) for $\operatorname{Im} \mathbb{H}$-valued 1-forms on $\mathbb{R}^4$).

Exercise 5.2.9 Show that $\mathcal{F}_{\alpha\beta} = \partial_\alpha \mathcal{A}_\beta - \partial_\beta \mathcal{A}_\alpha + [\mathcal{A}_\alpha, \mathcal{A}_\beta]$, where we have written ∂_α for $\frac{\partial}{\partial x^\alpha}$ and these derivatives are computed componentwise in $\mathcal{G}$.

Theorem 5.2.3 *Let $\boldsymbol{\omega}$ be a connection form on the principal G-bundle $\mathcal{P} : P \to X$ with curvature $\boldsymbol{\Omega}$. Let $s_1 : \mathbf{V}_1 \to \mathcal{P}^{-1}(V_1)$ and $s_2 : \mathbf{V}_2 \to \mathcal{P}^{-1}(V_2)$ be two local cross-sections with $V_2 \cap V_1 \neq \emptyset$ and let $g_{12} : V_2 \cap V_1 \to G$ be the corresponding transition function ($s_2(x) = s_1(x) \cdot g_{12}(x)$). Then, on $V_2 \cap V_1$,*

$$s_2^* \boldsymbol{\Omega} = ad_{g_{12}^{-1}} \circ s_1^* \boldsymbol{\Omega}. \tag{5.2.3}$$

Proof: Fix an $x_0 \in V_2 \cap V_1$ and $\mathbf{v}, \mathbf{w} \in T_{x_0}(X)$. Then

$$(s_2^* \boldsymbol{\Omega})_{x_0} (\mathbf{v}, \mathbf{w}) = \boldsymbol{\Omega}_{s_2(x_0)} ((s_2)_{*x_0}(\mathbf{v}), (s_2)_{*x_0}(\mathbf{w})).$$

Now, appealing to Exercise 5.1.4 and dropping the vertical parts involving $[\boldsymbol{\Theta}_{12}(\mathbf{v})]^{\#}$ and $[\boldsymbol{\Theta}_{12}(\mathbf{w})]^{\#}$ we obtain

$$\begin{aligned}(s_2^*\boldsymbol{\Omega})_{x_0}(\mathbf{v},\mathbf{w}) &= \boldsymbol{\Omega}_{s_2(x_0)}\left((\sigma_{g_{12}(x_0)})_{*s_1(x_0)}(\mathbf{v}),(\sigma_{g_{12}(x_0)})_{*s_1(x_0)}(\mathbf{w})\right)\\ &= \left(\sigma_{g_{12}(x_0)}^{\;*}\boldsymbol{\Omega}\right)_{s_1(x_0)}(\mathbf{v},\mathbf{w})\\ &= ad_{g_{12}(x_0)^{-1}}\circ\boldsymbol{\Omega}(\mathbf{v},\mathbf{w})\end{aligned}$$

by Lemma 5.2.2. ■

Exercise 5.2.10 The transformation law (5.2.3) for the gauge field strength can be written $s_2^*\boldsymbol{\Omega} = ad_{g_{21}}\circ s_1^*\boldsymbol{\Omega}$. Show that the same transformation law relates the coordinate expressions $(s_k\circ\varphi^{-1})^*\boldsymbol{\Omega}$, i.e., prove that

$$(s_2\circ\varphi^{-1})^*\boldsymbol{\Omega} = ad_{g_{21}\circ\varphi^{-1}}\circ(s_1\circ\varphi^{-1})^*\boldsymbol{\Omega}$$

for any chart (U,φ) with $U\subseteq V_2\cap V_1$.

Writing $s_1^*\boldsymbol{\Omega} = \boldsymbol{\mathcal{F}}_1$ and $s_2^*\boldsymbol{\Omega} = \boldsymbol{\mathcal{F}}_2$ and using the fact that, for matrix Lie groups, $ad_g(A) = g\,A\,g^{-1}$, the conclusion of Theorem 5.2.3 is simply

$$\boldsymbol{\mathcal{F}}_2 = g_{12}^{-1}\boldsymbol{\mathcal{F}}_1 g_{12}\,. \tag{5.2.4}$$

It follows from Exercise 5.2.10 that (5.2.4) remains valid when field strengths $\boldsymbol{\mathcal{F}}$ are identified with their coordinate expressions $(s\circ\varphi^{-1})^*\boldsymbol{\Omega}$ (of course, g_{12} is now $g_{12}\circ\varphi^{-1}$). Thus, by comparison with the gauge potential, whose transformation under change of gauge is given by (4.8.23), the gauge field strength obeys a relatively simple transformation law under local gauge transformation. We will see that this transformation law has important consequences for the physics of gauge theory.

Exercise 5.2.11 Let $\mathcal{P}_1 : P_1 \to X$ and $\mathcal{P}_2 : P_2 \to X$ be two smooth principal G-bundles over the same base manifold X, $f : P_1 \to P_2$ a smooth bundle map and $\boldsymbol{\omega}$ a connection form on P_2 with curvature $\boldsymbol{\Omega}$. Show that the curvature of $f^*\boldsymbol{\omega}$ is $f^*\boldsymbol{\Omega}$.

The result of Exercise 5.2.11 is, in particular, true for a bundle automorphism of $\mathcal{P} : P \to X$, in which case $\boldsymbol{\omega}$ and $f^*\boldsymbol{\omega}$ are gauge equivalent. In this case, if $s : V \to \mathcal{P}^{-1}(V)$ is a cross-section, then the corresponding field strengths are $s^*\boldsymbol{\Omega}$ and $s^*(f^*\boldsymbol{\Omega})$. But $s^*(f^*\boldsymbol{\Omega}) = (f\circ s)^*\boldsymbol{\Omega}$ and $f\circ s : V \to \mathcal{P}^{-1}(V)$ is also a cross-section of $\mathcal{P} : P \to X$, so these two field strengths are both pullbacks of $\boldsymbol{\Omega}$ (by different sections). Consequently, field strengths for gauge equivalent connections are also related by a transformation law of the form (5.2.4), where g_{12} is the transition function relating the sections s and $f \circ s$.

We discussed at some length in Chapter 0 the fact that, even in classical electromagnetic theory (where the gauge group is $U(1)$), the potential 1-form is not uniquely determined. A gauge transformation produces a new local potential (see (0.4.6)). The field itself, however, *is* uniquely determined and the reason for this is now clear. $U(1)$ is *Abelian* so that the factors in (5.2.4) commute, the g_{12}^{-1} and g_{12} cancel and one is left with $\mathcal{F}_2 = \mathcal{F}_1$. On the other hand, when the gauge group is non-Abelian (e.g., when $G = SU(2) \cong Sp(1)$ as in the original Yang-Mills theory of [**YM**]) there is no permuting the factors in (5.2.4) so that, in general, $\mathcal{F}_2 \neq \mathcal{F}_1$. In order to have available a single, well-defined object with which to represent the physical field under consideration one must opt for the curvature itself (defined on the bundle space) rather than a field strength (defined on the base manifold). Since the base manifold is often physical space, or spacetime, whereas the bundle space is a rather less concrete space of internal states, this represents a significant departure from the classical point of view in mathematical physics.

Note, however, that, in at least one special case, the field strength *is* gauge invariant. Should there happen to exist a local gauge s_1 in which the field strength is zero ($\mathcal{F}_1 = 0$), then (5.2.4) ensures that the field strength in any other gauge will be zero as well. Naturally, this will occur if some local potential $\mathcal{A}_1$ happens to be zero, but, just as in the case of electromagnetism, one cannot infer from this that the potential $\mathcal{A}_2$ in another gauge will also be zero. Indeed, gauge potentials transform according to (4.8.23) so $\mathcal{A}_2 = g_{12}^{-1} \mathcal{A}_1 g_{12} + g_{12}^* \boldsymbol{\Theta} = g_{12}^{-1} \mathbf{0}\, g_{12} + g_{12}^* \boldsymbol{\Theta} = g_{12}^* \boldsymbol{\Theta}$, where $\boldsymbol{\Theta}$ is the Cartan 1-form for the Lie algebra $\mathcal{G}$ of G. A potential of the form $\mathcal{A} = g^* \boldsymbol{\Theta}$, where $g : V \to G$ is some smooth function defined on the open subset V of X is said to be **pure gauge** because it is the gauge transform of zero (by g) and so has field strength zero. According to Exercise 4.8.17, the $\mathrm{Im}\,\mathbb{H}$-valued 1-form $\mathrm{Im}\,(q^{-1}\,dq)$ is just such a pure gauge potential for the trivial $Sp(1)$-bundle over $\mathbb{H} - \{0\}$. An interesting observation, to which we will return in the next section, is that, by Exercise 4.8.16,

$$\mathcal{A} = \mathrm{Im}\left(\frac{\bar{q}}{1 + |q|^2}\, dq \right) = \frac{|q|^2}{1 + |q|^2}\, \mathrm{Im}\,\left(q^{-1}\,dq\right)$$

on $\mathbb{H} - \{0\}$ and that this approaches $\mathrm{Im}\,(q^{-1}\,dq)$ as $|q| \to \infty$. In the physics literature it is common to say that $\mathcal{A}$ is "asymptotically pure gauge".

A connection ω on a principal bundle $\mathcal{P} : P \to X$ is said to be **flat** if its curvature 2-form $\boldsymbol{\Omega}$ is identically zero. These are easy to produce on trivial bundles. Indeed, if we let $P = X \times G$ be the trivial G-bundle over X, $\pi : P \times G \to G$ the projection onto the second factor and $\boldsymbol{\Theta}$ the Cartan 1-form on G, then the pullback $\omega = \pi^* \boldsymbol{\Theta}$ is a $\mathcal{G}$-valued 1-form on P.

Exercise 5.2.12 Show that $\omega = \pi^* \boldsymbol{\Theta}$ is a connection form on $P = X \times G$ whose horizontal subspace $\text{Hor}_{(x,g)}(P)$ at any $(x, g) \in P$ is the tangent space to the submanifold $X \times \{g\}$ of P.

To show that $\omega = \pi^* \boldsymbol{\Theta}$ is flat we use the Structure Equation for G ((4.10.14)) to compute $d\omega$,

$$\begin{aligned} d\omega &= d\,(\pi^* \boldsymbol{\Theta}) = \pi^*\,(d\boldsymbol{\Theta}) \\ &= \pi^* \left(-\frac{1}{2} \boldsymbol{\Theta} \wedge \boldsymbol{\Theta} \right) = -\frac{1}{2}\,(\pi^* \boldsymbol{\Theta}) \wedge (\pi^* \boldsymbol{\Theta}) \\ &= -\frac{1}{2} \omega \wedge \omega\,. \end{aligned}$$

Thus,

$$\boldsymbol{\Omega} = d\omega + \frac{1}{2} \omega \wedge \omega = 0\,.$$

We mention in passing, although we will not require the result, that flat connections cannot exist on bundles whose base manifold is simply connected (see Corollary 9.2 of [**KN 1**]). In particular, the Hopf bundles $S^1 \to S^3 \to S^2$ and $S^3 \to S^7 \to S^4$ admit *no* flat connections.

5.3 The Yang-Mills Functional

In 1932, Werner Heisenberg suggested the possibility that the known nucleons (the proton and the neutron) were, in fact, just two different "states" of the same particle and proposed a mathematical device for modeling this so-called **isotopic spin** state of a nucleon. Just as the phase of a charged particle is represented by a complex number of modulus 1 and phase changes are accomplished by the action of $U(1)$ on S^1 (rotation) so the isotopic spin of a nucleon is represented by a *pair* of complex numbers whose squared moduli sum to 1 and changes in the isotopic spin state are accomplished by an action of $SU(2)$ on S^3. In 1954, C. N. Yang and R. L. Mills set about constructing a theory of isotopic spin that was strictly analogous to classical electromagnetic theory. They were led to consider matrix-valued potential functions (denoted B_μ in [**YM**]) and corresponding fields ($F_{\mu\nu}$ in [**YM**]) constructed from the derivatives of the potential functions. The underlying physical assumption of the theory (**gauge invariance**) was that, when electromagnetic effects can be neglected, interactions between nucleons should be invariant under arbitrary and independent "rotation" of the isotopic spin state at each spacetime point. This is entirely analogous to the invariance of classical electromagnetic interactions under arbitrary phase changes (see Chapter 0) and has the effect of dictating the transformation properties of the potential functions B_μ under a change of gauge and suggesting the appropriate combination of the B_μ and their derivatives to act as the field $F_{\mu\nu}$. We quote briefly from [**YM**]:

> " Let Ψ be a two-component wave function describing a field with isotopic spin $\frac{1}{2}$. Under an isotopic gauge transformation it transforms by
>
> $$\Psi = S\Psi', \tag{1}$$
>
> where S is a 2×2 unitary matrix with determinant unity. ... we obtain the isotopic gauge transformation on B_μ:
>
> $$B'_\mu = S^{-1} B_\mu S + \tfrac{i}{\epsilon} S^{-1} \frac{\partial S}{\partial x_\mu}. \tag{3}$$
>
> ... In analogy to the procedure of obtaining gauge invariant field stengths in the electromagnetic case, we define now
>
> $$F_{\mu\nu} = \frac{\partial B_\mu}{\partial x_\nu} - \frac{\partial B_\nu}{\partial x_\mu} + i\epsilon\,(B_\mu B_\nu - B_\nu B_\mu). \tag{4}$$
>
> One easily shows from (3) that
>
> $$F'_{\mu\nu} = S^{-1} F_{\mu\nu} S \tag{5}$$
>
> ... "

Although we have not yet encountered anything in our study corresponding to a "two-component wave function" (see Section 5.8), one cannot help but be struck by the similarity between (3), (4) and (5) and our results on the gauge transformation of a gauge potential ($\boldsymbol{\mathcal{A}}^g = g^{-1}\boldsymbol{\mathcal{A}}\,g + g^{-1}\,dg$), the component expression for the gauge field ($\mathcal{F}_{\alpha\beta} = \partial_\alpha \mathcal{A}_\beta - \partial_\beta \mathcal{A}_\alpha + [\mathcal{A}_\alpha, \mathcal{A}_\beta]$) and the transformation equation for the gauge field ($\mathcal{F}^g = g^{-1}\mathcal{F}\, g$), respectively.

The physics of isotopic spin led Yang and Mills to propose certain differential equations (about which we will have a bit more to say shortly) that the potential functions B_μ should satisfy. In 1975, Belavin, Polyakov, Schwartz and Tyupkin [**BPST**] found a number of remarkable solutions to these equations that they christened "pseudoparticles". More remarkable still is the fact that these solutions formally coincide with the pullbacks (5.1.14) to $\mathbb{R}^4$ of connections on the Hopf bundle (only the $n = 0$ case appears explicitly in [**BPST**]). This observation was made explicit and generalized by Trautman [**Trau**] and further generalized by Nowakowski and Trautman [**NT**]. The subsequent deluge of research on the relationship between Yang-Mills theory and the geometry and topology of connections has produced not only some of the deepest and most beautiful mathematics of this era, but also profound insights into the structure of fundamental physical theories. While most of this material lies in greater depths than we are equipped to explore, a survey of its logical underpinnings is possible if we temporarily revert to the more casual attitude we adopted in Chapter 0. For the remainder of this section then we will feel free to use the occasional term that we have not rigorously defined, appeal to a theorem now and then that we have not proved and do the odd calculation by the seat of our pants. The reader who wishes to see all of this done honestly should consult [**FU**] and [**Law**].

We begin by temporarily suppressing the Hopf bundle altogether and considering again the $\mathrm{Im}\,\mathbb{H}$-valued (i.e., $\mathcal{SU}(2)$-valued) 1-form $\boldsymbol{\mathcal{A}}$ on $\mathbb{H}$ given by

$$\boldsymbol{\mathcal{A}} = \mathrm{Im}\left(\frac{\bar{q}}{1+|q|^2}\,dq\right). \tag{5.3.1}$$

By Theorem 5.1.1 we may (but need not) identify $\boldsymbol{\mathcal{A}}$ with the gauge potential for a connection 1-form on the trivial $Sp(1)$-bundle (i.e., $SU(2)$-bundle) over $\mathbb{H}$. We have already seen that, on $\mathbb{H} - \{0\}$, $\boldsymbol{\mathcal{A}}$ can be written

$$\boldsymbol{\mathcal{A}} = \frac{|q|^2}{1+|q|^2}\,\mathrm{Im}\,(q^{-1}dq) = \frac{|q|^2}{1+|q|^2}\,g^*\boldsymbol{\Theta}, \tag{5.3.2}$$

where $g : \mathbb{H} - \{0\} \to S^3 \cong Sp(1) \cong SU(2)$ is given by $g(q) = q/|q|$ and $\boldsymbol{\Theta}$ is the Cartan 1-form for $Sp(1) \cong SU(2)$. From this it is clear that $\boldsymbol{\mathcal{A}}$ is "asymptotically pure gauge", i.e., that as $|q| \to \infty$, $\boldsymbol{\mathcal{A}} \to g^*\boldsymbol{\Theta}$, where $g^*\boldsymbol{\Theta}$ is now thought of as the gauge potential for a flat connection on the trivial $Sp(1)$-bundle over $\mathbb{H} - \{0\}$ (Section 5.2). The field strength of $\boldsymbol{\mathcal{A}}$ has been computed ((4.10.12)) and is given by

$$\begin{aligned}\mathcal{F} &= \frac{1}{(1+|q|^2)^2}\, d\bar{q} \wedge dq \\ &= \frac{2}{(1+|q|^2)^2}\,\big((dq^0 \wedge dq^1 - dq^2 \wedge dq^3)\,\mathbf{i} \\ &\qquad + (dq^0 \wedge dq^2 + dq^1 \wedge dq^3)\,\mathbf{j} \\ &\qquad + (dq^0 \wedge dq^3 - dq^1 \wedge dq^2)\,\mathbf{k}\big) \\ &= \frac{2\,\mathbf{i}}{(1+|q|^2)^2}\, dq^0 \wedge dq^1 + \frac{-2\,\mathbf{i}}{(1+|q|^2)^2}\, dq^2 \wedge dq^3 + \cdots \end{aligned} \tag{5.3.3}$$

There is a standard construction, using the inner product determined by the Killing form (Exercise 4.7.18), for assigning a numerical measure of the total field strength at each point q. Specifically, we define, at each $q \in \mathbb{H}$, the squared norm $\|\mathcal{F}(q)\|^2$ of $\mathcal{F}(q)$ to be the sum of the squared norms (in $\mathcal{SU}(2) = \operatorname{Im}\mathbb{H}$, relative to the Killing form) of the components of $\mathcal{F}(q)$ relative to $dq^\alpha \wedge dq^\beta$. By Exercise 4.7.18 (b) and the component expression for $\mathcal{F}(q)$ in (5.3.3) we therefore have

$$\|\mathcal{F}(q)\|^2 = 6\left(2\left(\frac{4}{(1+|q|^2)^4}\right)\right) = \frac{48}{(1+|q|^2)^4}. \tag{5.3.4}$$

A global measure of the total field strength is then obtained by integrating $\|\mathcal{F}(q)\|^2$ over $\mathbb{R}^4 = \mathbb{H}$. Thus, we define

$$\|\mathcal{F}\|^2 = \int_{\mathbb{R}^4} \|\mathcal{F}(q)\|^2 = 48\int_{\mathbb{R}^4} \frac{1}{(1+|q|^2)^4}. \tag{5.3.5}$$

Remark: Theorem 3-12 of [**Sp1**] defines $\int_A f$, where $A \subseteq \mathbb{R}^n$ is open and f is bounded on some nbd of each point in A and continuous almost everywhere. Calculations are performed using the change of variables formula (Theorem 3-10 of [**Sp1**]) and Fubini's Theorem (Theorem 3-13 of [**Sp1**]).

As it happens, the integral in (5.3.5) is quite elementary. One introduces standard spherical coordinates on $\mathbb{R}^4$ defined by

$$\begin{aligned} q^0 &= \rho \sin\chi \sin\phi \cos\theta \\ q^1 &= \rho \sin\chi \sin\phi \sin\theta \\ q^2 &= \rho \sin\chi \cos\phi \\ q^3 &= \rho \cos\chi \end{aligned}$$

where $\rho = |\,q\,| \geq 0,\ 0 \leq \chi \leq \pi,\ 0 \leq \phi \leq \pi$ and $0 \leq \theta \leq 2\pi$. Then

$$
\begin{aligned}
\|\,\mathcal{F}\,\|^2 &= 48 \int_{\mathbb{R}^4} \frac{1}{(1+|\,q\,|^2)^4} \\
&= 48 \int_0^{2\pi}\int_0^{\pi}\int_0^{\pi}\int_0^{\infty} \frac{1}{(1+\rho^2)^4}\, \rho^3 \sin^2\chi\, \sin\phi\, d\rho\, d\chi\, d\phi\, d\theta \\
&= 48 \left(\int_0^{\infty} \frac{\rho^3}{(1+\rho^2)^4}\, d\rho \right) \left(\int_0^{2\pi}\int_0^{\pi}\int_0^{\pi} \sin^2\chi\, \sin\phi\, d\chi\, d\phi\, d\theta \right) \\
&= 48 \left(\frac{1}{12} \right) (2\pi^2) \\
&= 8\pi^2 .
\end{aligned}
$$

Exercise 5.3.1 Let $n \in \mathbb{H}$, $\lambda > 0$ and $\mathcal{A}_{\lambda,n} = \mathrm{Im}\left(\frac{\bar{q}-\bar{n}}{|\,q-n\,|^2+\lambda^2}\, dq \right)$. By Exercise 4.10.28,

$$
\mathcal{F}_{\lambda,n} = \frac{\lambda^2}{(|\,q-n\,|^2+\lambda^2)^2}\, d\bar{q} \wedge dq .
$$

Show that

$$
\|\,\mathcal{F}_{\lambda,n}(q)\,\|^2 = \frac{48\,\lambda^2}{(|\,q-n\,|^2+\lambda^2)^4}
$$

and

$$
\|\,\mathcal{F}_{\lambda,n}\,\|^2 = \int_{\mathbb{R}^4} \|\,\mathcal{F}_{\lambda,n}(q)\,\|^2 = 8\pi^2 .
$$

It follows from Exercise 5.3.1 that all of the gauge potentials described there (including the $n=0$, $\lambda=1$ case considered earlier) have the same total field strength $8\pi^2$. Observe that, for a fixed n, $\|\,\mathcal{F}_{\lambda,n}(q)\,\|^2 = 48\,\lambda^2/(|\,q-n\,|^2+\lambda^2)^4$ has a maximum value of $48/\lambda^2$ at $q=n$. As $\lambda \to 0$ this maximum value approaches infinity in such a way that the integrals over $\mathbb{R}^4$ of the $\|\,\mathcal{F}_{\lambda,n}(q)\,\|^2$ for various λ remain constant (see Figure 5.3.1). Thus, as $\lambda \to 0$ the field strength concentrates more and more at $q=n$. We shall refer to n as the **center** and λ as the **scale** (or **spread**) of the potential $\mathcal{A}_{\lambda,n}$.

Now let us consider more generally an arbitrary gauge potential $\mathcal{A}$ on the trivial $Sp(1)$-bundle over $\mathbb{R}^4$ and let $\mathcal{F}$ be its field strength. At each $q \in \mathbb{R}^4$ we define $\|\,\mathcal{F}(q)\,\|^2$ to be the sum of the squared norms (relative

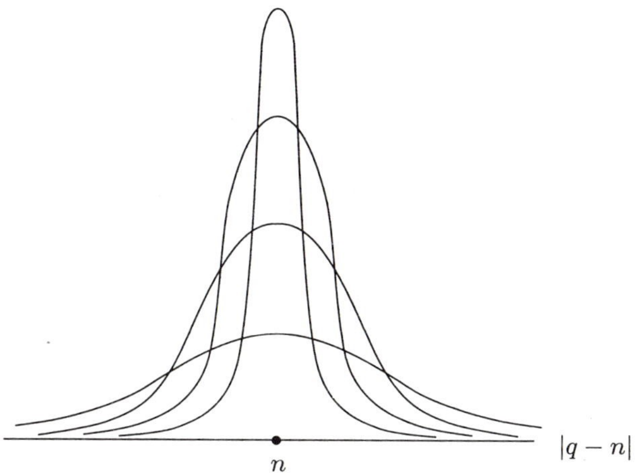

Figure 5.3.1

to the Killing form on $\mathcal{SP}(1)$) of the components of $\mathcal{F}(q)$ relative to $dq^\alpha \wedge dq^\beta$. From the definition of the Killing form on $\mathcal{SP}(1)$ (Exercise 4.7.17 (b)) and the transformation law (5.2.4) for the field strength it follows at once that $\|\mathcal{F}(q)\|^2$ is **gauge invariant**, i.e., that if g is a gauge transformation on some open set in $\mathbb{R}^4$ and $\mathcal{F}^g = g^{-1}\mathcal{F}g$ is the corresponding field strength, then $\|\mathcal{F}^g(q)\|^2 = \|\mathcal{F}(q)\|^2$ for each q. Now define the total field strength of $\boldsymbol{\mathcal{A}}$ by

$$\|\mathcal{F}\|^2 = \int_{\mathbb{R}^4} \|\mathcal{F}(q)\|^2 .$$

$\|\mathcal{F}\|^2$ is also called the **Yang-Mills action** of $\boldsymbol{\mathcal{A}}$ and denoted $\mathcal{YM}(\boldsymbol{\mathcal{A}})$. The functional $\mathcal{YM}$ that assigns to each such potential $\boldsymbol{\mathcal{A}}$ its Yang-Mills action $\mathcal{YM}(\boldsymbol{\mathcal{A}})$ is called the **Yang-Mills functional** on $\mathbb{R}^4$.

Physical considerations impose certain restrictions on the class of potentials $\boldsymbol{\mathcal{A}}$ that are of interest. Since $\mathcal{YM}(\boldsymbol{\mathcal{A}})$ represents a total field strength one is led to consider only **finite action** potentials, i.e., those $\boldsymbol{\mathcal{A}}$ for which

$$\mathcal{YM}(\boldsymbol{\mathcal{A}}) = \int_{\mathbb{R}^4} \|\mathcal{F}(q)\|^2 < \infty .$$

This requires that $\|\mathcal{F}(q)\|^2$ decay "sufficiently fast" as $|q| \to \infty$. In a fixed gauge/section/trivialization this simply means that the squared (Killing) norms of the components of $\mathcal{F}$ decay "sufficiently fast". The component expressions in Exercise 5.2.9 would then seem to require a similar rate of decay for the components of $\mathcal{A}$ and their first derivatives. Remarkably, this is not the case. Indeed, for the potential $\mathcal{A}$ given by (5.3.1) we have already computed $\mathcal{YM}(\mathcal{A}) = 8\pi^2 < \infty$. The components of $\mathcal{F}$ are given by (5.3.3) and decay quite rapidly (like $|q|^{-4}$) as $|q| \to \infty$, but the components of $\mathcal{A}$ itself decay much less rapidly. The explanation for this rather unusual phenomenon is to be found in the gauge invariance of $\| \mathcal{F}(q) \|^2$. In order to ensure that $\mathcal{YM}(\mathcal{A})$ is finite one need only be able to find some local gauge transformation g, defined for sufficiently large $|q|$, such that the potentials *in this gauge* decay "sufficiently fast" (g need only be defined for large $|q|$ because the integral over any compact set in $\mathbb{R}^4$ is necessarily finite). For (5.3.1) we have already seen (in (5.3.2)) that the appropriate g is defined on $\mathbb{R}^4 - \{0\}$ by $g(q) = q/|q|$ since $\mathcal{A}$ is asymptotically the gauge transform of zero by this g (equivalently, applying the gauge transformation g^{-1}, defined on $\mathbb{R}^4 - \{0\}$ by $g^{-1}(q) = (g(q))^{-1} = \bar{q}/|q|$, to $\mathcal{A}$ gives zero asymptotically). The essential point here is that these gauge transformations g need not and, indeed, cannot, in general, be defined on all of $\mathbb{R}^4$. To see this let S_r^3 be a 3-sphere about the origin in $\mathbb{R}^4$ of sufficiently large radius r that it is contained in the domain of g. Consider the map

$$g \mid S_r^3 : S_r^3 \longrightarrow Sp(1).$$

Since S_r^3 and $Sp(1)$ are both topologically 3-spheres, $g|S_r^3$ can be regarded as a map of S^3 to S^3. By Exercise 2.3.18, $g|S_r^3$ can be continuously extended to $|q| \leq r$ iff it is *nullhomotopic*.

Exercise 5.3.2 Show that, if $0 < r_1 \leq r_2$ are sufficiently large that S_r, is contained in the domain of g whenever $r_1 \leq r \leq r_2$, then $g|S_{r_1}^3$ and $g|S_{r_2}^3$ are homotopic.

For a given g, $g|S_r^3$ may or may not be nullhomotopic, but, in any case, it determines an element of $\pi_3(S^3) \cong \mathbb{Z}$. For the gauge potential $\mathcal{A}$ defined by (5.3.1), g is given by $g(q) = q/|q|$ and this, when restricted to the unit 3-sphere, is the identity map. Thus, $g|S_r^3$ is *not* nullhomotopic since $\deg(id_{S^3}) = 1$ (see Section 4.9). We will see shortly that the integer k corresponding to a given g is directly related to the "rate of decay" of the field strength as $|q| \to \infty$.

The potentials $\mathcal{A}$ of most interest in physics are those which (locally) minimize the Yang-Mills functional. One can apply standard techniques from the calculus of variations to write down differential equations (the

Euler-Lagrange equations) that must be satisfied by the stationary points of $\mathcal{YM}$. The resulting equations for $\boldsymbol{\mathcal{A}}$ are called the **Yang-Mills equations**. In standard coordinates on $\mathbb{H} = \mathbb{R}^4$ (i.e., q^0, q^1, q^2, q^3) they are given by

$$\sum_{\alpha=0}^{3} (\partial_\alpha \mathcal{F}_{\alpha\beta} + [\mathcal{A}_\alpha, \mathcal{F}_{\alpha\beta}]) = 0, \quad \beta = 0, 1, 2, 3,$$

where $\mathcal{A}_\alpha$ and $\mathcal{F}_{\alpha\beta}$ are as in Exercise 5.2.9. This is a system of second order, nonlinear partial differential equations for the components $\mathcal{A}_\alpha$ of the potential $\boldsymbol{\mathcal{A}}$. The nonlinearity of the equations is viewed as representing a "self-interaction" of the Yang-Mills field with itself, something that is not present in classical electromagnetic theory (because the gauge group is $U(1)$, which is Abelian, so all of the Lie brackets are zero). The BPST pseudoparticle potentials $\boldsymbol{\mathcal{A}}_{\lambda,n}$ are all solutions to the Yang-Mills equations.

For reasons that lie rather deep in the Feynman path integral approach to quantum field theory and the peculiar quantum mechanical phenomenon of tunneling (see [**Guid**]), the absolute minima of $\mathcal{YM}$ are of particular significance. These absolute minima (called **instantons**) are also the objects whose study has led to the Donaldson-inspired revolution in low dimensional topology. They are, of course, solutions to the Yang-Mills equations, but can also be characterized as the solutions to another, much simpler, set of equations that we wish now to briefly describe. In order to do so it will be necessary to anticipate a few results from the next section.

In Section 5.4 we will show how to associate with every 2-form $\boldsymbol{\Omega}$ on a 4-dimensional, oriented, Riemannian manifold X another 2-form $^*\boldsymbol{\Omega}$ called its "Hodge dual" (the definition depends on a choice of orientation and Riemannian metric which, for $\mathbb{R}^4$ and S^4 we take here to be the standard ones). Then $\boldsymbol{\Omega}$ is said to be "self-dual" (respectively, "anti-self-dual") if $\boldsymbol{\Omega} = {}^*\boldsymbol{\Omega}$ (respectively, $\boldsymbol{\Omega} = -{}^*\boldsymbol{\Omega}$). Furthermore, any $\boldsymbol{\Omega}$ can be uniquely written as $\boldsymbol{\Omega} = \boldsymbol{\Omega}_+ + \boldsymbol{\Omega}_-$, where $\boldsymbol{\Omega}_+$ is self-dual and $\boldsymbol{\Omega}_-$ is anti-self-dual so that $\boldsymbol{\Omega}$ is self-dual iff $\boldsymbol{\Omega} = \boldsymbol{\Omega}_+$ and anti-self-dual iff $\boldsymbol{\Omega} = \boldsymbol{\Omega}_-$. If $\boldsymbol{\mathcal{A}}$ is a gauge potential and $\boldsymbol{\mathcal{F}}$ is its field strength, then it is customary to refer to $\boldsymbol{\mathcal{A}}$ itself as self-dual (respectively, anti-self-dual) if $\boldsymbol{\mathcal{F}} = \boldsymbol{\mathcal{F}}_+$ (respectively, $\boldsymbol{\mathcal{F}} = \boldsymbol{\mathcal{F}}_-$). In this case we will show also that

$$\mathcal{YM}(\boldsymbol{\mathcal{A}}) = \|\boldsymbol{\mathcal{F}}\|^2 = \|\boldsymbol{\mathcal{F}}_+\|^2 + \|\boldsymbol{\mathcal{F}}_-\|^2 . \tag{5.3.6}$$

(Anti-) self-duality is a symmetry condition that we will find is rather easy to check for any given gauge potential. In particular, all of the potentials on $\mathbb{R}^4$ described in Exercise 5.3.1 (including (5.3.1) which corresponds to

$\lambda = 1$ and $n = 0$) are easily seen to be anti-self-dual. We will also find that reversing the orientation of X interchanges the notions of self-dual and anti-self-dual so that the distinction is simply a matter of convention and of no real significance.

The relevance of all this is that the absolute minima of $\mathcal{YM}$ on $\mathbb{R}^4$ (i.e., the instantons on $\mathbb{R}^4$) correspond precisely to the (anti-) self-dual connections. In order to glimpse the reason this is true we must wade briefly in some rather deep waters, but we trust the reader will find the dip invigorating. Those inclined to take the plunge are encouraged to consult [**FU**] and [**Law**].

Let us begin by recalling that the finite action BPST gauge potential $\mathcal{A}$ defined by (5.3.1) is not "just" a gauge potential on the trivial $Sp(1)$-bundle over $\mathbb{R}^4$. It is, in fact, the pullback to $\mathbb{R}^4$ of a connection on a nontrivial $Sp(1)$-bundle over S^4. Turning matters about, one might say that the connection on the trivial $Sp(1)$-bundle over $\mathbb{R}^4$ corresponding to $\mathcal{A}$ "extends to S^4" in the sense that S^4 is the one-point compactification of $\mathbb{R}^4$ and, due to the (gauge) asymptotic behavior of $\mathcal{A}$ as $|q| \to \infty$, the connection extends to the point at infinity. Notice, however, that this extension process involves not only the connection, but also the bundle itself. The connection on the trivial $Sp(1)$-bundle over $\mathbb{R}^4$ corresponding to $\mathcal{A}$ extends to the natural connection on the (nontrivial) Hopf bundle over S^4. Such an extension could certainly not exist for a potential on $\mathbb{R}^4$ whose Yang-Mills action was not finite (S^4 is compact so integrals over it are necessarity finite). A remarkable theorem of Karen Uhlenbeck [**Uhl**] asserts that this is, in fact, the *only* obstruction to extending a Yang-Mills connection on $\mathbb{R}^4$ to some $Sp(1)$-bundle over S^4. This **Removable Singularities Theorem** of Uhlenbeck is very general, but the special case of interest to us is easy to state: Let $\mathcal{A}$ be an $\operatorname{Im}\mathbb{H}$-valued gauge potential on $\mathbb{R}^4$ that satisfies the Yang-Mills equations and whose action $\mathcal{YM}(\mathcal{A}) = \int_{\mathbb{R}^4} \| \mathcal{F}(q) \|^2$ is finite. Then there exists a unique $Sp(1)$-bundle $\mathcal{P} : P \to S^4$ over S^4, a connection 1-form $\boldsymbol{\omega}$ on P and a cross-section $s : S^4 - \{N\} \to \mathcal{P}^{-1}(S^4 - \{N\})$ such that $\mathcal{A} = (s \circ \varphi_S^{-1})^* \boldsymbol{\omega}$ where $\varphi_S : S^4 - \{N\} \to \mathbb{R}^4$ is stereographic projection from the north pole N.

Now recall (Theorem 3.4.3) that the principal $Sp(1)$-bundles over S^4 are characterized topologically by an integer, i.e., by an element of $\pi_3(Sp(1)) \cong \pi_3(S^3) \cong \mathbb{Z}$. We have seen that such an integer invariant can be obtained as the degree of the characteristic map $T = g|S^3$ of the bundle, where g is a transition function and S^3 is the equatorial 3-sphere in S^4. There is, however, another way of calculating an integer k that uniquely determines the equivalence class of the bundle, provided by a deep and beautiful branch of topology known as the theory of **characteristic**

classes (see Chapter XII of [**KN2**]). Although this subject is beyond our level here, much insight is to be gained by simply recording, without proof, the relevant formula for computing our topological invariant k. The so-called "Chern-Weil formula" gives

$$\| \mathcal{F}_+ \|^2 - \| \mathcal{F}_- \|^2 = 8\pi^2 k, \tag{5.3.7}$$

where $\mathcal{F}$ is the field strength of $\mathcal{A}$. If $\mathcal{F}$ happens to be anti-self-dual, $\mathcal{F} = \mathcal{F}_-$ and $\mathcal{F}_+ = 0$ so this gives

$$k = -\frac{1}{8\pi^2} \int_{\mathbb{R}^4} \| \mathcal{F}(q) \|^2 = -\frac{1}{8\pi^2} \mathcal{YM}(\mathcal{A}) \tag{5.3.8}$$

($\mathcal{F}$ anti-self-dual). In particular, for the BPST connection $\mathcal{A}$ given by (5.3.1) we have computed $\mathcal{YM}(\mathcal{A}) = 8\pi^2$ so $k = -1$ for the Hopf bundle $S^3 \to S^7 \to S^4$ (this is often called the **instanton number**, or **topological charge** of the Hopf bundle). Notice that (5.3.7) implies that self-dual connections cannot exist on a bundle with $k < 0$, while anti-self-dual connections cannot exist if $k > 0$. Observe also that a simple algebraic combination of (5.3.6) and (5.3.7) yields

$$\mathcal{YM}(\mathcal{A}) \leq 8\pi^2 |k|, \tag{5.3.9}$$

and

$$\mathcal{YM}(\mathcal{A}) = 8\pi^2 |k| \quad \text{iff} \quad \mathcal{F} = (\operatorname{sign} k)({}^*\mathcal{F}). \tag{5.3.10}$$

An immediate consequence of (5.3.9) and (5.3.10) is that a gauge potential $\mathcal{A}$ on $\mathbb{R}^4$ is an absolute minimum for the Yang-Mills functional iff $\mathcal{F}$ is either flat ($k = 0$), self-dual ($k > 0$), or anti-self-dual ($k < 0$). Flat connections have field strength zero and extend only to the ($k = 0$) trivial bundle over S^4 and we will not consider them any further. Since self-dual and anti-self-dual can be interchanged by switching orientation, we may restrict our attention to one or the other. Because the Hopf bundle has $k = -1$ we prefer to henceforth consider only the anti-self-dual case (some sources refer to these as **anti-instantons**). Thus, (5.3.8) gives the topological invariant k (instanton number) of the $Sp(1)$-bundle over S^4 to which a gauge potential $\mathcal{A}$ on $\mathbb{R}^4$ extends as a multiple of the total field strength. But the total field strength of a finite action potential is determined entirely by the "rate of decay" of $\| \mathcal{F}(q) \|^2$ as $|q| \to \infty$. It is really quite remarkable that the asymptotic behavior of the field strength can be directly encoded in this way into the topology of the bundle over S^4 to which the gauge potential extends. Equally remarkable is the fact that these minimum field strengths emerge "quantized", i.e., parametrized by the integers, so that one is naturally presented with something akin to a generalized Dirac quantization condition.

With these attempts at motivation behind us we turn once more to the business of doing mathematics. In the next two sections of this chapter we define the Hodge dual of a 2-form, focus our attention on the anti-self-dual connections on the Hopf bundle and, with the assistance of Atiyah, Hitchin and Singer [**AHS**], describe the moduli space of such connections or, what amounts to the same thing, the equivalence classes of potentials $\boldsymbol{\mathcal{A}}$ on $\mathbb{R}^4$ with $\mathcal{YM}(\boldsymbol{\mathcal{A}}) = 8\pi^2$.

5.4 The Hodge Dual for 2-Forms in Dimension Four

We begin with some linear algebra. Throughout this section $\mathcal{V}$ will denote an oriented, 4-dimensional, real vector space on which is defined an inner product $< , >$, i.e., a nondegenerate, symmetric, positive definite bilinear form. We let $\{e_1, e_2, e_3, e_4\}$ be an oriented, orthonormal basis for $\mathcal{V}$ and $\{e^1, e^2, e^3, e^4\}$ its dual basis for $\mathcal{V}^*$. We wish to extend the orientation and inner product on $\mathcal{V}$ to $\mathcal{V}^*$ by taking $\{e^1, e^2, e^3, e^4\}$ to be oriented, defining $< e^i, e^j > = < e_i, e_j >$ and extending $< , >$ on $\mathcal{V}^*$ by bilinearity ($< v_i e^i, w_j e^j > = v_i w_j < e^i, e^j > = v_i w_j \delta^{ij} = \sum_{i=1}^4 v_i w_i$).

Exercise 5.4.1 Let $\{\hat{e}_1, \hat{e}_2, \hat{e}_3, \hat{e}_4\}$ be another oriented, orthonormal basis for $\mathcal{V}$ with dual basis $\{\hat{e}^1, \hat{e}^2, \hat{e}^3, \hat{e}^4\}$. Show that there exists a 4×4 real matrix $A = (A^i{}_j)_{i,j=1,2,3,4}$ with $A\,A^T = A^T A = id$ and $\det A = 1$ such that $\hat{e}_j = A^i{}_j\, e_i$ for $j = 1,2,3,4$ and $\hat{e}^i = A^i{}_j\, e^j$ for $i = 1,2,3,4$. Conclude that the definitions of the orientation and inner product on $\mathcal{V}^*$ are independent of the choice of $\{e_1, e_2, e_3, e_4\}$.

Now let $\Lambda^2(\mathcal{V})$ denote the space of skew-symmetric bilinear forms on $\mathcal{V}$ (Section 4.10). According to Lemma 4.10.2, $\{e^1 \wedge e^2, e^1 \wedge e^3, e^1 \wedge e^4, e^2 \wedge e^3, e^2 \wedge e^4, e^3 \wedge e^4\}$ is a basis for $\Lambda^2(\mathcal{V})$ and any $\Omega \in \Lambda^2(\mathcal{V})$ can be written as $\Omega = \sum_{i<j} \Omega_{ij}\, e^i \wedge e^j = \frac{1}{2}\Omega_{ij}\, e^i \wedge e^j$, where $\Omega_{ij} = \Omega(e_i, e_j)$. We extend the inner product to $\Lambda^2(\mathcal{V})$ by defining, for all $\varphi^1, \varphi^2, \xi^1, \xi^2 \in \mathcal{V}^*$,

$$< \varphi^1 \wedge \varphi^2 ,\, \xi^1 \wedge \xi^2 > = \begin{vmatrix} < \varphi^1, \xi^1 > & < \varphi^1, \xi^2 > \\ < \varphi^2, \xi^1 > & < \varphi^2, \xi^2 > \end{vmatrix}$$

and extend by bilinearity.

Exercise 5.4.2 Show that this does, indeed, define a inner product on $\Lambda^2(\mathcal{V})$ and that, relative to it, $\{e^1\wedge e^2, e^1\wedge e^3, e^1\wedge e^4, e^2\wedge e^3, e^2\wedge e^4, e^3\wedge e^4\}$ is an orthonormal basis for any oriented, orthonormal basis $\{e_1, e_2, e_3, e_4\}$ for $\mathcal{V}$.

Now we define a mapping $^*\colon \Lambda^2(\mathcal{V}) \to \Lambda^2(\mathcal{V})$, called the **Hodge star operator**, by $^*(e^i \wedge e^j) = e^k \wedge e^l$, where $ijkl$ is an even permutation of 1234, and extending by linearity. In more detail,

$$\begin{array}{ll} ^*(e^1\wedge e^2) = e^3\wedge e^4 & ^*(e^1\wedge e^3) = -e^2\wedge e^4 \\ ^*(e^1\wedge e^4) = e^2\wedge e^3 & ^*(e^2\wedge e^3) = e^1\wedge e^4 \\ ^*(e^2\wedge e^4) = -e^1\wedge e^3 & ^*(e^3\wedge e^4) = e^1\wedge e^2 \end{array}$$

so that, if $\Omega = \sum_{i<j} \Omega_{ij}\, e^i\wedge e^j = \Omega_{12}\, e^1\wedge e^2 + \Omega_{13}\, e^1\wedge e^3 + \Omega_{14}\, e^1\wedge e^4 + \Omega_{23}\, e^2\wedge e^3 + \Omega_{24}\, e^2\wedge e^4 + \Omega_{34}\, e^3\wedge e^4$, then

$$\begin{aligned} ^*\Omega = \Omega_{34}\, e^1\wedge e^2 - \Omega_{24}\, e^1\wedge e^3 + \Omega_{23}\, e^1\wedge e^4 \\ +\Omega_{14}\, e^2\wedge e^3 - \Omega_{13}\, e^2\wedge e^4 + \Omega_{12}\, e^3\wedge e^4\,. \end{aligned} \tag{5.4.1}$$

$^*\Omega$ is called the **Hodge dual** of $\Omega \in \Lambda^2(V)$. To verify that this definition does not depend on the choice of $\{e_1, e_2, e_3, e_4\}$ we first derive an alternative formula for computing it. For this we introduce the **Levi-Civita symbol** ϵ_{abcd} defined, for $a, b, c, d = 1, 2, 3, 4$, by

$$\epsilon_{abcd} = \begin{cases} 1 & \text{if } abcd \text{ is an even permutation of } 1234 \\ -1 & \text{if } abcd \text{ is an odd permutation of } 1234 \\ 0 & \text{otherwise} \end{cases}\,.$$

This symbol first arose in the theory of determinants from which we borrow the following easily verified fact: For any 4×4 matrix $A = (A^\alpha{}_\beta)_{\alpha,\beta=1,2,3,4}$,

$$A^\alpha{}_a\, A^\beta{}_b\, A^\gamma{}_c\, A^\delta{}_d\, \epsilon_{\alpha\beta\gamma\delta} = \epsilon_{abcd}\,(\det A)\,. \tag{5.4.2}$$

For any $i, j = 1, 2, 3, 4$, we now have

$$^*(e^i\wedge e^j) = \frac{1}{2}\,\epsilon_{ijkl}\; e^k\wedge e^l\,. \tag{5.4.3}$$

For example, $\frac{1}{2}\,\epsilon_{24\,kl}\, e^k\wedge e^l = \frac{1}{2}(-e^1\wedge e^3 + e^3\wedge e^1) = \frac{1}{2}(-e^1\wedge e^3 - e^1\wedge e^3) = -e^1\wedge e^3 = {}^*(e^2\wedge e^4)$ and the rest are proved similarly. Thus,

$$
\begin{aligned}
{}^*\Omega &= {}^*\left(\frac{1}{2}\sum_{i,j=1}^{4}\Omega\,(\,e_i,e_j\,)\,e^i\wedge e^j\right) = \frac{1}{2}\sum_{i,j=1}^{4}\Omega\,(\,e_i,e_j\,)^*\,(\,e^i\wedge e^j\,)\\
&= \frac{1}{2}\sum_{i,j=1}^{4}\Omega\,(\,e_i,e_j\,)\,\left(\frac{1}{2}\,\epsilon_{\,ijkl}\,e^k\wedge e^l\right)
\end{aligned}
$$

$$
{}^*\Omega = \frac{1}{4}\sum_{i,j,k,l=1}^{4}\Omega\,(\,e_i,e_j\,)\,\epsilon_{\,ijkl}\,e^k\wedge e^l\,. \tag{5.4.4}
$$

Now suppose $\{\hat{e}_1,\hat{e}_2,\hat{e}_3,\hat{e}_4\}$ is another oriented, orthonormal basis for $\mathcal{V}$. We must show that

$$
\frac{1}{4}\sum_{i,j,k,l=1}^{4}\Omega(\,\hat{e}_i,\hat{e}_j\,)\,\epsilon_{\,ijkl}\,\hat{e}^k\wedge\hat{e}^l = \frac{1}{4}\sum_{i,j,k,l=1}^{4}\Omega(\,e_i,e_j\,)\,\epsilon_{\,ijkl}\,e^k\wedge e^l. \tag{5.4.5}
$$

Letting $A = (A^i{}_j)_{i,j=1,2,3,4}$ be as in Exercise 5.4.1 we have

$$
\begin{aligned}
&\frac{1}{4}\sum_{i,j,k,l=1}^{4}\Omega\,(\,\hat{e}_i,\hat{e}_j\,)\,\epsilon_{\,ijkl}\,\hat{e}^k\wedge\hat{e}^l\\
&\quad= \frac{1}{4}\sum_{i,j,k,l=1}^{4}\Omega\,\left(A^\alpha{}_i\,e_\alpha,A^\beta{}_j\,e_\beta\right)\,\epsilon_{\,ijkl}\,\left(\,A^k{}_\gamma\,e^\gamma\,\right)\wedge\left(\,A^l{}_\delta\,e^\delta\,\right)\\
&\quad= \frac{1}{4}\sum_{i,j,k,l=1}^{4}\;\sum_{\alpha,\beta,\gamma,\delta=1}^{4}A^\alpha{}_i\,A^\beta{}_j\,A^k{}_\gamma\,A^l{}_\delta\,\epsilon_{\,ijkl}\,\Omega\,(\,e_\alpha,e_\beta\,)\,e^\gamma\wedge e^\delta\\
&\quad= \frac{1}{4}\sum_{\alpha,\beta,\gamma,\delta=1}^{4}\left(\sum_{i,j,k,l=1}^{4}A^\alpha{}_i\,A^\beta{}_j\,A^k{}_\gamma\,A^l{}_\delta\,\epsilon_{\,ijkl}\right)\Omega\,(\,e_\alpha,e_\beta\,)\,e^\gamma\wedge e^\delta\,.
\end{aligned}
$$

Thus, it will suffice to show that

$$
\sum_{i,j,k,l=1}^{4}A^\alpha{}_i\,A^\beta{}_j\,A^k{}_\gamma\,A^l{}_\delta\,\epsilon_{\,ijkl} = \epsilon_{\alpha\beta\gamma\delta}\,. \tag{5.4.6}
$$

For this we recall that $A^T A = A\,A^T = id$ and $\det A = 1$ so (5.4.2) gives

$$
A^\alpha{}_i\,A^\beta{}_j\,A^\gamma{}_k\,A^\delta{}_l\,\epsilon_{\alpha\beta\gamma\delta} = \epsilon_{\,ijkl}
$$

and

$$\begin{aligned}
&\sum_{i,j,k,l=1}^{4} A^{\alpha}{}_{i}\, A^{\beta}{}_{j}\, A^{k}{}_{\gamma}\, A^{l}{}_{\delta}\, \epsilon_{ijkl} \\
&= \sum_{i,j,k,l=1}^{4} A^{\alpha}{}_{i}\, A^{\beta}{}_{j}\, A^{k}{}_{\gamma}\, A^{l}{}_{\delta} \left(A^{\alpha}{}_{i}\, A^{\beta}{}_{j}\, A^{\gamma}{}_{k}\, A^{\delta}{}_{l}\, \epsilon_{\alpha\beta\gamma\delta} \right) \\
&= \sum_{i,j,k,l=1}^{4} \left(A^{\alpha}{}_{i}\, A^{\alpha}{}_{i}\right) \left(A^{\beta}{}_{j}\, A^{\beta}{}_{j}\right) \left(A^{k}{}_{\gamma}\, A^{\gamma}{}_{k}\right) \left(A^{l}{}_{\delta}\, A^{\delta}{}_{l}\right) \epsilon_{\alpha\beta\gamma\delta} \\
&= \left(\sum_{i=1}^{4} (A^{\alpha}{}_{i})^2\right) \left(\sum_{j=1}^{4} (A^{\beta}{}_{j})^2\right) \left(\sum_{k=1}^{4} A^{k}{}_{\gamma}\, A^{\gamma}{}_{k}\right) \left(\sum_{l=1}^{4} A^{l}{}_{\delta}\, A^{\delta}{}_{l}\right) \epsilon_{\alpha\beta\gamma\delta} \\
&= \epsilon_{\alpha\beta\gamma\delta}
\end{aligned}$$

as required.

Thus, $^* : \Lambda^2(\mathcal{V}) \to \Lambda^2(\mathcal{V})$ is a well-defined linear transformation. It has a number of properties of interest to us. For example, it is its own inverse, i.e.,

$$* \circ * = id_{\Lambda^2(\mathcal{V})}\,. \tag{5.4.7}$$

To prove (5.4.7) it is enough to show that $* \circ *$ leaves basis elements fixed and these can be checked one at a time, e.g., $^*(^*(e^2 \wedge e^4)) = {}^*(-e^1 \wedge e^3) = -{}^*(e^1 \wedge e^3) = -(-e^2 \wedge e^4) = e^2 \wedge e^4$. Furthermore, * has eigenvalues ± 1 since

$$\begin{aligned}
{}^*\Omega = \lambda\,\Omega &\implies {}^*({}^*\Omega) = {}^*(\lambda\,\Omega) = \lambda\,{}^*\Omega = \lambda\,(\lambda\,\Omega) = \lambda^2\,\Omega \\
&\implies \Omega = \lambda^2\,\Omega \\
&\implies \lambda = \pm 1\,.
\end{aligned}$$

Exercise 5.4.3 Show that $^* : \Lambda^2(\mathcal{V}) \to \Lambda^2(\mathcal{V})$ is an isometry with respect to the inner product $< \, , \, >$ on $\Lambda^2(\mathcal{V})$, i.e., that $< {}^*\Omega, {}^*\Psi > \, = \, < \Omega, \Psi >$ for all $\Omega, \Psi \in \Lambda^2(\mathcal{V})$.

From Exercise 5.4.3 and (5.4.7) it follows that * is a symmetric linear transformation with respect to $< \, , \, >$, i.e., that

$$< {}^*\Omega\,, \Psi > \, = \, < \Omega\,, {}^*\Psi > \tag{5.4.8}$$

for all $\Omega, \Psi \in \Lambda^2(\mathcal{V})$. Indeed, $< {}^*\Omega, \Psi > \, = \, < {}^*\Omega, {}^*({}^*\Psi) > \, = \, < \Omega, {}^*\Psi >$. Consequently, since * has eigenvalues ± 1, $\Lambda^2(\mathcal{V})$ has an orthogonal decomposition (Exercise 18, Section 2, Chapter $\overline{\text{XI}}$, of [**Lang**])

$$\Lambda^2(\mathcal{V}) = \Lambda^2_+(\mathcal{V}) \oplus \Lambda^2_-(\mathcal{V})\,, \tag{5.4.9}$$

where $\Lambda^2_{\pm}(\mathcal{V})$ is the eigenspace of ± 1. In fact,

$$\Lambda^2_+(\mathcal{V}) = \text{Span}\,\{\, e^1 \wedge e^2 + e^3 \wedge e^4 \,,\; e^1 \wedge e^3 + e^4 \wedge e^2 \,,\; e^1 \wedge e^4 + e^2 \wedge e^3 \,\} \tag{5.4.10}$$

$$\Lambda^2_-(\mathcal{V}) = \text{Span}\,\{\, e^1 \wedge e^2 - e^3 \wedge e^4 \,,\; e^1 \wedge e^3 - e^4 \wedge e^2 \,,\; e^1 \wedge e^4 - e^2 \wedge e^3 \,\} \tag{5.4.11}$$

Exercise 5.4.4 Prove (5.4.10) and (5.4.11).

Thus, every $\Omega \in \Lambda^2(\mathcal{V})$ has a unique decomposition $\Omega = \Omega_+ + \Omega_-$, where $\Omega_\pm \in \Lambda^2_\pm(\mathcal{V})$. Furthermore, since the decomposition is orthogonal,

$$\|\,\Omega\,\|^2 \;=\; <\Omega\,,\Omega> \;=\; \|\,\Omega_+\,\|^2 \;+\; \|\,\Omega_-\,\|^2 \;. \tag{5.4.12}$$

Notice also that ${}^*\Omega = {}^*(\Omega_+ + \Omega_-) = {}^*\Omega_+ + {}^*\Omega_- = \Omega_+ - \Omega_-$ so that

$$\Omega_+ = \frac{1}{2}\,(\Omega + {}^*\Omega) \;\;\text{and}\;\; \Omega_- = \frac{1}{2}\,(\Omega - {}^*\Omega)\;. \tag{5.4.13}$$

In particular, ${}^*\Omega = \Omega$ iff $\Omega_- = 0$, in which case we say that Ω is **self-dual (SD)** and ${}^*\Omega = -\Omega$ iff $\Omega_+ = 0$, in which case we say that Ω is **anti-self-dual (ASD)**. Thus, for any Ω, Ω_+ is SD and Ω_- is ASD.

Next we observe that the star operator ${}^* : \Lambda^2(\mathcal{V}) \to \Lambda^2(\mathcal{V})$, which depends on the choice of orientation and inner product $<\,,\,>$ for $\mathcal{V}$, is actually a conformal invariant in the sense that if $<\,,\,>$ is replaced by $<\,,\,>' = \lambda <\,,\,>$, where $\lambda > 0$, but the orientation is not changed, then, for every $\Omega \in \Lambda^2(\mathcal{V})$, one obtains the same element of $\Lambda^2(\mathcal{V})$ by computing ${}^*\Omega$ relative to either inner product. The reason is simple. If $\{e_1, e_2, e_3, e_4\}$ is an oriented, $<\,,\,>$-orthonormal basis for $\mathcal{V}$, then $\{\tilde{e}_1, \tilde{e}_2, \tilde{e}_3, \tilde{e}_4\}$, where $\tilde{e}_i = \frac{1}{\sqrt{\lambda}}\, e_i$, $i = 1,2,3,4$, is an oriented, $<\,,\,>'$-orthonormal basis for $\mathcal{V}$. Furthermore, $\tilde{e}^{\,i} = \sqrt{\lambda}\, e^i$ for $i = 1,2,3,4$ so

$$\begin{aligned}
&\frac{1}{4}\sum_{i,j=1}^{4} \Omega\,(\tilde{e}_i, \tilde{e}_j)\;\epsilon_{ijkl}\;\tilde{e}^{\,k} \wedge \tilde{e}^{\,l} \\
&\quad = \frac{1}{4}\sum_{i,j=1}^{4} \Omega\left(\frac{1}{\sqrt{\lambda}}\,e_i, \frac{1}{\sqrt{\lambda}}\,e_j\right)\epsilon_{ijkl}\left(\sqrt{\lambda}\,e^k\right) \wedge \left(\sqrt{\lambda}\,e^l\right) \\
&\quad = \frac{1}{4}\sum_{i,j=1}^{4} \frac{1}{\lambda}\Omega\,(e_i, e_j)\;\epsilon_{ijkl}\;\lambda\, e^k \wedge e^l \\
&\quad = \frac{1}{4}\sum_{i,j=1}^{4} \Omega\,(e_i, e_j)\;\epsilon_{ijkl}\; e^k \wedge e^l
\end{aligned}$$

as required.

We use this last result to prove the crucial fact that, under orientation preserving conformal isomorphism, the Hodge star commutes with pullback. More precisely, we let $\mathcal{V}_1$ and $\mathcal{V}_2$ be oriented, 4-dimensional, real vector spaces with inner products $< \, , \, >_1$ and $< \, , \, >_2$, respectively. Let $L : \mathcal{V}_1 \to \mathcal{V}_2$ be an orientation preserving isomorphism that is conformal in the sense that $L^* < \, , \, >_2 = \lambda < \, , \, >_1$ for some $\lambda > 0$, where the pullback $L^* < \, , \, >_2$ is defined by $(L^* < \, , \, >_2)(v, w) = < L(v), L(w) >_2$ for all $v, w \in \mathcal{V}_1$. For any $\Omega \in \Lambda^2(\mathcal{V}_2)$ define the pullback $L^*\Omega$ by $(L^*\Omega)(v, w) = \Omega(L(v), L(w))$ for all $v, w \in \mathcal{V}_1$. At the risk of trying the reader's patience we bow to tradition and use $*$ also to denote the Hodge star (in both $\mathcal{V}_1$ and $\mathcal{V}_2$). Our claim then is that

$$^*(L^*\Omega) = L^*(^*\Omega) \tag{5.4.14}$$

(the notation really is unambiguous). According to what we have just proved, we may calculate $^*(L^*\Omega)$ relative to a basis $\{e_1, e_2, e_3, e_4\}$ that is oriented and orthonormal with respect to $L^* < \, , \, >_2 = \lambda < \, , \, >_1$. Then $\{L(e_1), L(e_2), L(e_3), L(e_4)\} = \{\tilde{e}_1, \tilde{e}_2, \tilde{e}_3, \tilde{e}_4\}$ is an oriented basis for $\mathcal{V}_2$ that is orthonormal with respect to $< \, , \, >_2$ so we may use it to calculate $^*\Omega$. Notice that if $\{e^1, e^2, e^3, e^4\}$ and $\{\tilde{e}^1, \tilde{e}^2, \tilde{e}^3, \tilde{e}^4\}$ are the corresponding dual bases, then $L^*\tilde{e}^k$ is given by $(L^*\tilde{e}^k)(v) = \tilde{e}^k(L(v))$ so, in particular, $(L^*\tilde{e}^k)(e_l) = \tilde{e}^k(L(e_l)) = \tilde{e}^k(\tilde{e}_l) = \delta^k_l$. Thus, $L^*\tilde{e}^k = e^k$. Now we compute

$$\begin{aligned} ^*(L^*\Omega) &= \frac{1}{4} \sum_{i,j,k,l=1}^{4} (L^*\Omega)(e_i, e_j)\, \epsilon_{ijkl}\, e^k \wedge e^l \\ &= \frac{1}{4} \sum_{i,j,k,l=1}^{4} \Omega(\tilde{e}_i, \tilde{e}_j)\, \epsilon_{ijkl}\, e^k \wedge e^l \end{aligned}$$

and

$$\begin{aligned} L^*(^*\Omega) &= L^* \left(\frac{1}{4} \sum_{i,j,k,l=1}^{4} \Omega(\tilde{e}_i, \tilde{e}_j)\, \epsilon_{ijkl}\, \tilde{e}^k \wedge \tilde{e}^l \right) \\ &= \frac{1}{4} \sum_{i,j,k,l=1}^{4} \Omega(\tilde{e}_i, \tilde{e}_j)\, \epsilon_{ijkl}\, L^*(\tilde{e}^k \wedge \tilde{e}^l) \\ &= \frac{1}{4} \sum_{i,j,k,l=1}^{4} \Omega(\tilde{e}_i, \tilde{e}_j)\, \epsilon_{ijkl}\, (L^*\tilde{e}^k) \wedge (L^*\tilde{e}^l) = {}^*(L^*\Omega) \end{aligned}$$

as required.

Next observe that if $\Omega = \Omega_+ + \Omega_-$, then $L^*\Omega = L^*\Omega_+ + L^*\Omega_-$, so $^*(L^*\Omega) = ^*(L^*\Omega_+) + ^*(L^*\Omega_-) = L^*(^*\Omega_+) + L^*(^*\Omega_-) = L^*\Omega_+ - L^*\Omega_-$. Now, if Ω is SD, then $\Omega = \Omega_+$ and $\Omega_- = 0$ so this gives $^*(L^*\Omega) = L^*\Omega_+ = L^*\Omega$ and $L^*\Omega$ is also SD. Similarly, if Ω is ASD, then so is $L^*\Omega$. In particular, $L^*\Omega_+$ is SD and $L^*\Omega_-$ is ASD so

$$(L^*\Omega)_+ = L^*\Omega_+ \quad \text{and} \quad (L^*\Omega)_- = L^*\Omega_- \tag{5.4.15}$$

for any orientation preserving conformal isomorphism $L : \mathcal{V}_1 \to \mathcal{V}_2$ and any $\Omega = \Omega_+ + \Omega_-$ in $\Lambda^2(\mathcal{V}_2)$.

Now we extend all of this to 2-forms on oriented, Riemannian 4-manifolds. For such a manifold X, each $T_p(X)$ has an orientation determined by that of X. Furthermore, the Riemannian metric $\mathbf{g}$ on X assigns to each $T_p(X)$ an inner product $\mathbf{g}_p = <\,,\,>_p$. Thus, we may define a Hodge star operator on each $\Lambda^2(T_p(X))$, all of which will be denoted

$$^* : \Lambda^2(T_p(X)) \longrightarrow \Lambda^2(T_p(X)),$$

although they do depend on p. Since a 2-form $\boldsymbol{\Omega}$ on X assigns a skew-symmetric bilinear form $\boldsymbol{\Omega}_p \in \Lambda^2(T_p(X))$ to each $p \in X$, one has a $^*\boldsymbol{\Omega}_p \in \Lambda^2(T_p(X))$ for each $p \in X$. We show that $p \to {^*\boldsymbol{\Omega}_p}$ is a 2-form, denoted $^*\boldsymbol{\Omega}$, by proving that it is smooth. This we may do locally so let (U, φ) be any chart on X consistent with the orientation μ of X (Exercise 4.9.9) and with U connected. If the coordinate functions of (U, φ) are x^1, x^2, x^3, x^4, then the vector fields $\mathbf{V}_i = \frac{\partial}{\partial x^i}$, $i = 1, 2, 3, 4$, satisfy the hypotheses of Proposition 4.10.3. Thus, there exist smooth vector fields $\mathbf{E}_1$, $\mathbf{E}_2$, $\mathbf{E}_3$ and $\mathbf{E}_4$ on U such that, for each $p \in U$, $\{\mathbf{E}_1(p), \mathbf{E}_2(p), \mathbf{E}_3(p), \mathbf{E}_4(p)\} \in \mu_p$, and $<\mathbf{E}_i(p), \mathbf{E}_j(p)>_p = \delta_{ij}$. Since any oriented, orthonormal basis for $T_p(X)$ may be used to compute $^*\boldsymbol{\Omega}_p$, we have

$$\begin{aligned} ^*\boldsymbol{\Omega}_p &= \frac{1}{4} \sum_{i,j,k,l=1}^{4} \boldsymbol{\Omega}_p(\mathbf{E}_i(p), \mathbf{E}_j(p))\, \epsilon_{ijkl}\, \mathbf{E}^k(p) \wedge \mathbf{E}^l(p) \\ &= \frac{1}{4} \sum_{i,j,k,l=1}^{4} \left(\boldsymbol{\Omega}(\mathbf{E}_i, \mathbf{E}_j)\, \epsilon_{ijkl}\, \mathbf{E}^k \wedge \mathbf{E}^l \right)(p). \end{aligned}$$

Exercise 5.4.5 Conclude that $p \to {^*\boldsymbol{\Omega}_p}$ is smooth so that $^*\boldsymbol{\Omega} \in \Lambda^2(X)$.

Thus, for any 2-form $\boldsymbol{\Omega}$ on the 4-dimensional, oriented, Riemannian manifold X we have another 2-form $^*\boldsymbol{\Omega}$ called its **Hodge dual.** $\boldsymbol{\Omega}$ is said to be **self-dual (SD)** if $^*\boldsymbol{\Omega} = \boldsymbol{\Omega}$ and **anti-self-dual (ASD)** if $^*\boldsymbol{\Omega} = -\boldsymbol{\Omega}$. Defining $\boldsymbol{\Omega}_+ = \frac{1}{2}(\boldsymbol{\Omega} + ^*\boldsymbol{\Omega})$ and $\boldsymbol{\Omega}_- = \frac{1}{2}(\boldsymbol{\Omega} - ^*\boldsymbol{\Omega})$ we have

$$\boldsymbol{\Omega} = \boldsymbol{\Omega}_+ + \boldsymbol{\Omega}_-,$$

where $\boldsymbol{\Omega}_+$ is SD (the **self-dual part** of $\boldsymbol{\Omega}$) and $\boldsymbol{\Omega}_-$ is ASD (the **anti-self-dual part** of $\boldsymbol{\Omega}$). Thus, $\boldsymbol{\Omega}$ is SD iff $\boldsymbol{\Omega} = \boldsymbol{\Omega}_+$ and $\boldsymbol{\Omega}$ is ASD iff $\boldsymbol{\Omega} = \boldsymbol{\Omega}_-$ and, as $C^\infty(X)$-modules,

$$\Lambda^2(X) = \Lambda^2_+(X) \oplus \Lambda^2_-(X),$$

where $\Lambda^2_+(X)$ consists of all the SD 2-forms on X and $\Lambda^2_-(X)$ consists of all the ASD elements of $\Lambda^2(X)$. Defining the Hodge dual of a vector-valued 2-form on X componentwise, all of these notions extend immediately to this context as well.

Suppose (U, φ) is a chart for X, consistent with the orientation of X and such that, at each point of U, the coordinate vector fields $\{\frac{\partial}{\partial x^i}\}_{i=1,2,3,4}$ are orthonormal relative to the Riemannian metric on X. This situation occurs, for example, for the standard orientation, metric and coordinates on $\mathbb{R}^4 = \mathbb{H}$. Then, by (5.4.10) and (5.4.11),

$$\begin{aligned} &dx^1 \wedge dx^2 + dx^3 \wedge dx^4, \\ &dx^1 \wedge dx^3 + dx^4 \wedge dx^2, \\ &dx^1 \wedge dx^4 + dx^2 \wedge dx^3 \end{aligned} \tag{5.4.16}$$

span the SD 2-forms on U and

$$\begin{aligned} &dx^1 \wedge dx^2 - dx^3 \wedge dx^4, \\ &dx^1 \wedge dx^3 - dx^4 \wedge dx^2, \\ &dx^1 \wedge dx^4 - dx^2 \wedge dx^3 \end{aligned} \tag{5.4.17}$$

span the ASD 2-forms on U. Thus, it follows from Exercises 4.10.28 and 4.10.26 that all of the $\operatorname{Im}\mathbb{H}$-valued 2-forms on $\mathbb{R}^4 = \mathbb{H}$ given by

$$\boldsymbol{\mathcal{F}}_{\lambda,n} = \frac{\lambda^2}{(|q-n|^2 + \lambda^2)^2}\, d\bar{q} \wedge dq$$

for some $\lambda > 0$ and $n \in \mathbb{H}$ are ASD. These are, of course, just the gauge field strengths for the generic BPST gauge potentials

$$\boldsymbol{\mathcal{A}}_{\lambda,n} = \operatorname{Im}\left(\frac{\bar{q} - \bar{n}}{|q-n|^2 + \lambda^2}\, dq\right)$$

on $\mathbb{R}^4$.

Remark: The anti-self-dual equation $^*\mathcal{F} = -\mathcal{F}$ for a gauge field on $\mathbb{R}^4$ is a system of differential equations for the components of the corresponding potential $\boldsymbol{\mathcal{A}}$ and the BPST potentials were originally found as solutions to this system. As cultural information we offer the following indication of how this might be done (for $n = 0$). We seek an $\mathrm{Im}\,\mathbb{H}$-valued 1-form $\boldsymbol{\mathcal{A}}$ on $\mathbb{H}$ whose field strength $\mathcal{F}$ is ASD and which is asymptotically pure gauge in the sense that $\boldsymbol{\mathcal{A}}(q) = A(\,|\,q\,|\,)\, g^*\boldsymbol{\Theta}$, where $A(\,|\,q\,|\,)$ decreases to 0 as $|\,q\,| \to \infty$, $g : \mathbb{H} - \{0\} \to Sp\,(1)$ is given by $g(q) = q/\,|\,q\,|$ and $\boldsymbol{\Theta}$ is the Cartan 1-form for $Sp\,(1)$ (choosing this specific g amounts to selecting a desired homotopy class of asymptotic behavior for the pure gauge; see Section 5.3). Applying Exercise 4.8.17 we may write $\boldsymbol{\mathcal{A}}\,(q) = A\,(\,|\,q\,|\,)\,\mathrm{Im}\,(\,q^{-1}\,dq\,) = \mathrm{Im}\,(\,(\,A(\,|\,q\,|\,)\,q^{-1}\,)\,dq\,) = \mathrm{Im}\,(\,f\,(q)\,dq\,)$, where $f\,(q) = A\,(\,|\,q\,|\,)\,q^{-1} = |\,q\,|^{-2}\,A\,(\,|\,q\,|\,)\,\bar{q}$.

Exercise 5.4.6 Use Exercise 4.10.27 and argue as in the derivation of (4.10.12) to show that

$$\mathcal{F} = \frac{1}{2\,|\,q\,|}\,A'\,(\,|\,q\,|\,)\,d\,\bar{q}\wedge dq$$
$$+\left(\frac{|\,q\,|}{2}\,A'(\,|\,q\,|\,) + A(\,|\,q\,|\,)\,(\,A(\,|\,q\,|\,) - 1\,)\right)\,d\,(\,g^*\,\boldsymbol{\Theta}\,)\,.$$

Hint: A' denotes the ordinary derivative of $A(\rho)$, where $\rho = |\,q\,|$. You will need (5.2.2).

Exercise 5.4.7 Conclude from Exercise 5.4.6 that $\mathcal{F}$ will be ASD if $A'(\,|\,q\,|\,) = -\frac{2}{|\,q\,|}\,A(\,|\,q\,|\,)\,(\,A(\,|\,q\,|\,) - 1\,)$ and solve this equation to obtain

$$A\,(\,|\,q\,|\,) = \frac{|\,q\,|^2}{|\,q\,|^2 + \lambda^2}$$

so that $\boldsymbol{\mathcal{A}}(q) = \dfrac{|\,q\,|^2}{|\,q\,|^2 + \lambda^2}\,\mathrm{Im}\,(q^{-1}\,dq)$. Then use Exercise 5.4.6 to obtain

$$\mathcal{F} = \frac{\lambda^2}{(\,|\,q - n\,|^2 + \lambda^2\,)^2}\,d\,\bar{q}\wedge dq\,.$$

Next suppose that $(X_1, \mathbf{g}_1)$ and $(X_2, \mathbf{g}_2)$ are two Riemannian 4-manifolds with orientations μ_1 and μ_2. Suppose also that $f : X_1 \to X_2$ is an orientation preserving, conformal diffeomorphism of X_1 onto X_2. Thus, at each $p \in X_1$, $f_{*p} : T_p(X_1) \to T_{f(p)}(X_2)$ is an orientation preserving, conformal isomorphism. Since the pullback and the Hodge dual are both defined pointwise, it follows from (5.4.14) that

$$^*(f^*\Omega) = f^*(^*\Omega) \tag{5.4.18}$$

for all $\Omega \in \Lambda^2(X_2)$. In particular, the pullback by an orientation preserving, conformal diffeomorphism of an ASD 2-form is ASD. Consider, for example, $\mathbb{H} = \mathbb{R}^4$ and S^4, both with their standard Riemannian metrics and orientations. Let $U_S = S^4 - \{N\}$ and let $\varphi_S : U_S \to \mathbb{H}$ be the chart on S^4 that stereographically projects from the north pole N. Then φ_S and $\varphi_S^{-1} : \mathbb{H} \to U_S$ are both orientation preserving, conformal diffeomorphisms. Consequently, a 2-form on U_S is ASD iff its pullback to $\mathbb{H}$ by $(\varphi_S^{-1})^*$ is ASD and we know the ASD 2-forms on $\mathbb{H}$ (they are generated by the 2-forms in (5.4.17)). Now, by continuity, a 2-form Ω on S^4 is ASD (i.e., satisfies $\Omega_+ = 0$) iff its restriction to U_S is ASD on U_S. The conclusion then is that anti-self-duality on S^4 is quite easy to check in practice: Restrict to $S^4 - \{N\}$, pull back to $\mathbb{H}$ by φ_S^{-1} and see if you are in the span of (5.4.17).

We carry this analysis one step further by now considering a connection form ω on the Hopf bundle $S^3 \to S^7 \to S^4$. Now, the curvature Ω of ω is a 2-form on S^7 so the notion of anti-self-duality is not (yet) defined for Ω. However, $U_S \subseteq S^4$ is also the domain of a canonical cross-section $s_S : U_S \to \mathcal{P}^{-1}(U_S)$ for the Hopf bundle (it corresponds to $V_2 \subseteq \mathbb{HP}^1$ under the diffeomorphism in Exercise 4.3.10). The corresponding gauge field strength $\mathcal{F}_S = s_S^*\Omega$ is a 2-form on U_S and we will say that Ω (and the connection ω itself) is **anti-self-dual (ASD)** if $\mathcal{F}_S$ is ASD. As we have just seen this will be the case iff the coordinate expression $(s_S \circ \varphi_S^{-1})^*\Omega = (\varphi_S^{-1})^*\mathcal{F}_S$ is ASD.

Exercise 5.4.8 Conclude that the natural connection ω on the Hopf bundle as well as all of the connections $g \cdot \omega = \rho_{g^{-1}}^*\omega$ for $g \in SL(2, \mathbb{H})$ are ASD.

5.5 The Moduli Space

We now have at hand a fairly substantial collection of ASD connections on the quaternionic Hopf bundle (Exercise 5.4.8). Each arises (as Section 5.3 assures us it must) from a gauge potential $\mathcal{A}_{\lambda,n}$ on $\mathbb{R}^4$ with ASD field strength $\mathcal{F}_{\lambda,n}$ and Yang-Mills action $\mathcal{YM}(\mathcal{A}_{\lambda,n}) = 8\pi^2$. It is not, however, the connections themselves that interest us, but rather their gauge equivalence classes (Section 5.1). We define the **moduli space of ASD connections on the Hopf bundle** $S^3 \to S^7 \to S^4$ (or, the **moduli space of instantons on** S^4 **with instanton number** -1) to be the set $\mathcal{M}$ of gauge equivalence classes of ASD connection forms on $S^3 \to S^7 \to S^4$. Thus, the natural connection $\boldsymbol{\omega}$, as well as each $g \cdot \boldsymbol{\omega}$ for $g \in SL(2, \mathbb{H})$, determines a point in the moduli space (consisting of all ASD connections on the Hopf bundle that are gauge equivalent to it). They do not all determine different points, of course, since we know, for example, that $g \cdot \boldsymbol{\omega} = \boldsymbol{\omega}$ if $g \in Sp(2)$ (so, being equal, these are certainly gauge equivalent).

Exercise 5.5.1 Let $\boldsymbol{\omega}$ be the natural connection on $S^3 \to S^7 \to S^4$ and let $g, g' \in SL(2, \mathbb{H})$. Show that $g' \cdot \boldsymbol{\omega} = g \cdot \boldsymbol{\omega}$ iff $g^{-1} g' \in Sp(2)$, i.e., iff g' is in the coset $g\, Sp(2)$.

In particular, the map of $SL(2, \mathbb{H})$ to $\mathcal{M}$ that sends g to the gauge equivalence class of $g \cdot \boldsymbol{\omega}$ is constant on each coset $g\, Sp(2)$ of $Sp(2)$ in $SL(2, \mathbb{H})$. This map therefore descends to a map of the quotient $SL(2, \mathbb{H})/Sp(2)$ to $\mathcal{M}$ that sends $[g] \in SL(2, \mathbb{H})/Sp(2)$, to the gauge equivalence class of $g \cdot \boldsymbol{\omega}$ in $\mathcal{M}$ (denoted $[g \cdot \boldsymbol{\omega}]$).

The Atiyah-Hitchin-Singer Theorem ([**AHS**]): *The map* $[g] \to [g \cdot \boldsymbol{\omega}]$ *of* $SL(2, \mathbb{H})/Sp(2)$ *to the moduli space* $\mathcal{M}$ *of* ASD *connections on the Hopf bundle* $S^3 \to S^7 \to S^4$ *is a bijection.*

Remarks: That the map $[g] \to [g \cdot \boldsymbol{\omega}]$ is one-to-one follows from observations we made in Section 5.3. One need only show that $g' \cdot \boldsymbol{\omega}$ is gauge equivalent to $g \cdot \boldsymbol{\omega}$ only if g' is in the coset $g\, Sp(2)$. But $g' \cdot \boldsymbol{\omega}$ and $g \cdot \boldsymbol{\omega}$ are both uniquely determined by gauge potentials $\mathcal{A}_{\lambda',n'}$ and $\mathcal{A}_{\lambda,n}$ with corresponding field strengths $\mathcal{F}_{\lambda',n'}$ and $\mathcal{F}_{\lambda,n}$, respectively. Now, if $g' \cdot \boldsymbol{\omega}$ and $g \cdot \boldsymbol{\omega}$ are gauge equivalent, then the field strengths are related by a transformation law of the form (5.2.4) (see the remarks following Exercise 5.2.11). But from this it follows that $\| \mathcal{F}_{\lambda',n'}(q) \|^2 = \| \mathcal{F}_{\lambda,n}(q) \|^2$ for each $q \in \mathbb{R}^4$ and this is possible only if $\lambda' = \lambda$ and $n' = n$ (Exercise 5.3.1). Thus, $\mathcal{A}_{\lambda',n'} = \mathcal{A}_{\lambda,n}$ so $g' \cdot \boldsymbol{\omega} = g \cdot \boldsymbol{\omega}$ and Exercise 5.5.1 gives $g' \in g\, Sp(2)$, as required. The *surjectivity* of $[g] \to [g \cdot \boldsymbol{\omega}]$, i.e., the fact that *every* gauge equivalence class of ASD connections on the Hopf bundle is represented by some $g \cdot \boldsymbol{\omega}$, where $g \in SL(2, \mathbb{H})$ and $\boldsymbol{\omega}$ is the natural connection, is quite another matter. The proof of this is quite beyond our powers here, involving, as it does, deep results from sheaf cohomology and algebraic geometry. Atiyah has written a very nice set of lecture notes

[**Atiy**] to which we refer those eager to see how it is done. Having done so we unabashedly appropriate the result for our own purposes.

Concrete descriptions of the moduli space $\mathcal{M}$ can be obtained by parametrizing the elements of $SL(2,\mathbb{H})/Sp(2)$ in some convenient way. We already have one such parametrization at our disposal. As we remarked in Section 5.1, the Iwasawa decomposition (5.1.16) of $SL(2,\mathbb{H})$ shows that every $g\cdot\boldsymbol{\omega}$ is uniquely determined by a BPST potential $\boldsymbol{\mathcal{A}}_{\lambda,n}$ and therefore by a pair (λ,n), where $\lambda>0$ and $n\in\mathbb{H}$. The Atiyah-Hitchin-Singer Theorem and the argument in the Remarks that follow it imply that the points in the moduli space $\mathcal{M}$ are in one-to-one correspondence with the set of all such pairs (λ,n), with $\lambda>0$ and $n\in\mathbb{H}$, i.e., with the points in the open half-space $(0,\infty)\times\mathbb{H}$ in $\mathbb{R}^5$. Intuitively, "nearby" points (λ,n) and (λ',n') in $(0,\infty)\times\mathbb{H}$ give rise to connections whose gauge potentials are "close" in the sense that their field strengths are centered at nearby points and have approximately the same scale (see Figure 5.3.1). It would seem appropriate then to identify $\mathcal{M}$ topologically with $(0,\infty)\times\mathbb{H}$ as well. As we pointed out in Section 4.3, any two differentiable structures on $\mathbb{R}^5$ (which is homeomorphic to $(0,\infty)\times\mathbb{H}$) are necessarily diffeomorphic so there is no ambiguity as to the appropriate differentiable structure for $\mathcal{M}$. Thus, we identify the moduli space $\mathcal{M}$ as a topological space and as a differentiable manifold with $(0,\infty)\times\mathbb{H}$.

Remark: It is not such a simple matter to justify supplying $\mathcal{M}$ with the Riemannian metric and orientation of $(0,\infty)\times\mathbb{H}$ and, indeed, there are other natural choices (see, e.g., [**Mat**]). Since the physicists wish to define path integrals on such moduli spaces, their geometry is crucial, but much work remains to be done (see Section 9.5 of [**MM**]).

This picture we have of $\mathcal{M}$ as $(0,\infty)\times\mathbb{H}$ is remarkably simple, but it is not the most instructive. To arrive at another view of $\mathcal{M}$, more in the spirit of Donaldson's general theory, we begin by recalling that $(0,\infty)\times\mathbb{H}$, with its usual structure, is diffeomorphic to the 5-dimensional ball B^5 (Exercise 4.10.29). What we would like to do is find a parametrization of the points of $\mathcal{M}$ which naturally identifies it with B^5 (i.e., we want to introduce "spherical coordinates" on $\mathcal{M}$). We begin, as we did in (5.1.16), by borrowing a rather nontrivial matrix decomposition theorem, this one called the **Cartan decomposition** of $SL(2,\mathbb{H})$ (see [**Helg**]):

$$SL(2,\mathbb{H}) = Sp(2)\,A\,Sp(2). \tag{5.5.1}$$

The assertion here is that every element of $SL(2,\mathbb{H})$ can be written as a product $g_1 a g_2$, where $g_1, g_2 \in Sp(2)$, and $a \in A$ (see Exercise 5.1.16). In order to put this into a usable form we must digress briefly.

We begin with the left action of $SL(2, \mathbb{H})$ on S^7 given by (5.1.6). Since this respects the Hopf bundle's right $Sp(1)$-action on S^7 it descends to a left action of $SL(2, \mathbb{H})$ on the quotient $S^7/Sp(1) = \mathbb{HP}^1 \cong S^4$ which can be described as follows: If $g = \begin{pmatrix} a & b \\ c & d \end{pmatrix} \in SL(2, \mathbb{H})$ then

$$g \cdot \begin{bmatrix} q \\ 1 \end{bmatrix} = \begin{bmatrix} aq+b \\ cq+d \end{bmatrix} \quad \text{and} \quad g \cdot \begin{bmatrix} 1 \\ 0 \end{bmatrix} = \begin{bmatrix} a \\ c \end{bmatrix}$$

for all $q \in \mathbb{H}$. In particular, $Sp(2) \subseteq SL(2, \mathbb{H})$ has a natural left action on $\mathbb{HP}^1$. We show that this left action of $Sp(2)$ on $\mathbb{HP}^1$ is transitive as follows: For each $m \in \mathbb{H} \cup \{\infty\}$ $(\cong S^4)$ define $g_m \in Sp(2)$ by

$$g_m = \begin{cases} \dfrac{1}{\sqrt{1+|m|^2}} \begin{pmatrix} 1 & m \\ -\bar{m} & 1 \end{pmatrix} & \text{if } m \in \mathbb{H} \\ \begin{pmatrix} 0 & 1 \\ -1 & 0 \end{pmatrix} & \text{if } m = \infty \end{cases} .$$

Then

$$g_m \cdot \begin{bmatrix} q \\ 1 \end{bmatrix} = \begin{cases} \begin{bmatrix} q+m \\ -\bar{m}\,q+1 \end{bmatrix} & \text{if } m \in \mathbb{H} \\ \begin{bmatrix} 1 \\ -q \end{bmatrix} & \text{if } m = \infty \end{cases}$$

and

$$g_m \cdot \begin{bmatrix} 1 \\ 0 \end{bmatrix} = \begin{cases} \begin{bmatrix} 1 \\ -\bar{m} \end{bmatrix} & \text{if } m \in \mathbb{H} \\ \begin{bmatrix} 0 \\ -1 \end{bmatrix} & \text{if } m = \infty \end{cases} .$$

In particular,

$$g_m \cdot \begin{bmatrix} 0 \\ 1 \end{bmatrix} = \begin{cases} \begin{bmatrix} m \\ 1 \end{bmatrix} & \text{if } m \in \mathbb{H} \\ \begin{bmatrix} 1 \\ 0 \end{bmatrix} & \text{if } m = \infty \end{cases}$$

so every element of $\mathbb{HP}^1$ is in the orbit of $\begin{bmatrix} 0 \\ 1 \end{bmatrix}$ and the action is transitive.

Next we compute the isotropy subgroup of $\left[\begin{smallmatrix} 0 \\ 1 \end{smallmatrix}\right]$ under this left action of $Sp(2)$ on $\mathbb{HP}^1$. If $g = \left(\begin{smallmatrix} a & b \\ c & d \end{smallmatrix}\right)$, then $g \cdot \left[\begin{smallmatrix} 0 \\ 1 \end{smallmatrix}\right] = \left[\begin{smallmatrix} b \\ d \end{smallmatrix}\right]$ so $g \cdot \left[\begin{smallmatrix} 0 \\ 1 \end{smallmatrix}\right] = \left[\begin{smallmatrix} 0 \\ 1 \end{smallmatrix}\right]$ implies $b = 0$ so $g \cdot \left[\begin{smallmatrix} 0 \\ 1 \end{smallmatrix}\right] = \left[\begin{smallmatrix} 0 \\ d \end{smallmatrix}\right]$, where $d \neq 0$. Now, $\left(\begin{smallmatrix} a & 0 \\ c & d \end{smallmatrix}\right) \in Sp(2)$ implies $\left(\begin{smallmatrix} a & 0 \\ c & d \end{smallmatrix}\right)\left(\begin{smallmatrix} \bar{a} & \bar{c} \\ 0 & \bar{d} \end{smallmatrix}\right) = \left(\begin{smallmatrix} 1 & 0 \\ 0 & 1 \end{smallmatrix}\right)$ so $|a|^2 = 1$, $a\bar{c} = 0$ (so $c = 0$) and $|d|^2 = 1$. Thus, the isotropy subgroup of $\left[\begin{smallmatrix} 0 \\ 1 \end{smallmatrix}\right]$ is

$$\left\{ \begin{pmatrix} a & 0 \\ 0 & d \end{pmatrix} : |a|^2 = |d|^2 = 1 \right\}$$

and this is isomorphic to $Sp(1) \times Sp(1)$. Since $Sp(2)$ is compact, Theorem 1.6.6 gives

$$\mathbb{HP}^1 \cong S^4 \cong Sp(2)/Sp(1) \times Sp(1). \tag{5.5.2}$$

Now, $Sp(2)$ is the union of the cosets $g\,(Sp(1) \times Sp(1))$ for $g \in Sp(2)$. But these cosets are in one-to-one correspondence with the points of $\mathbb{HP}^1$ and we have already seen that every element of $\mathbb{HP}^1$ is $g_m \cdot \left[\begin{smallmatrix} 0 \\ 1 \end{smallmatrix}\right]$ for some $m \in \mathbb{H} \cup \{\infty\}$. Thus,

$$Sp(2) = \bigcup_{m \in \mathbb{H} \cup \{\infty\}} g_m\,(Sp(1) \times Sp(1))\,. \tag{5.5.3}$$

Exercise 5.5.2 Use (5.5.1) and (5.5.3) and the fact that the elements of $Sp(1) \times Sp(1)$ commute with those of A to show that

$$SL(2, \mathbb{H}) = \bigcup_{m \in \mathbb{H} \cup \{\infty\}} g_m\, A\, Sp(2)\,. \tag{5.5.4}$$

Since the elements of $Sp(2)$ leave the natural connection ω on $S^3 \to S^7 \to S^4$ fixed it follows from (5.5.4) that each point in the moduli space $\mathcal{M}$ is $g \cdot \omega$ for some g of the form $g = g_m\, a$, where $m \in \mathbb{H} \cup \{\infty\}$ and $a \in A$.

Exercise 5.5.3 Show that, if $m \in \mathbb{H}$ and $\lambda > 0$, then

$$\left(\frac{1}{\sqrt{1 + |m|^2}} \begin{pmatrix} 1 & m \\ -\bar{m} & 1 \end{pmatrix} \begin{pmatrix} \sqrt{\lambda} & 0 \\ 0 & 1/\sqrt{\lambda} \end{pmatrix} \right)^{-1}$$

$$= (1 + |m|^2)^{-3/2} \begin{pmatrix} 1/\sqrt{\lambda} & -m/\sqrt{\lambda} \\ \sqrt{\lambda}\,\bar{m} & \sqrt{\lambda} \end{pmatrix}$$

and

$$\left(\begin{pmatrix} 0 & 1 \\ -1 & 0 \end{pmatrix} \begin{pmatrix} \sqrt{\lambda} & 0 \\ 0 & 1/\sqrt{\lambda} \end{pmatrix} \right)^{-1} = \begin{pmatrix} 0 & -1/\sqrt{\lambda} \\ \sqrt{\lambda} & 0 \end{pmatrix} .$$

Exercise 5.5.4 Let $g = g_m \, a_\lambda$, where $a_\lambda = \begin{pmatrix} \sqrt{\lambda} & 0 \\ 0 & 1/\sqrt{\lambda} \end{pmatrix} \in A$. Use Exercise 5.5.3 and (5.1.11) to show that

$$(s_2 \circ \varphi_2^{-1})^* (g \cdot \boldsymbol{\omega}) = \mathrm{Im} \left(\frac{(1 + \lambda^2 \,|\, m \,|^2) \, \bar{q} \, dq + (\lambda^2 - 1) \, \bar{m} \, dq}{|\, q - m \,|^2 + \lambda^2 \,|\, \bar{m} \, q + 1 \,|^2} \right) \tag{5.5.5}$$

if $m \in \mathbb{H}$ and

$$(s_2 \circ \varphi_2^{-1})^* (g \cdot \boldsymbol{\omega}) = \mathrm{Im} \left(\frac{\lambda^2 \, \bar{q} \, dq}{1 + \lambda^2 \,|\, q \,|^2} \right) \tag{5.5.6}$$

if $m = \infty$.

Exercise 5.5.5 Suppose $m \in \mathbb{H}$, $\lambda > 0$ and $g = g_m \, a_\lambda$, as in Exercise 5.5.4. Let $n = -\bar{m}^{-1}$, $\mu = 1/\lambda$ and $g' = g_n \, a_\mu$, where $a_\mu = \begin{pmatrix} \sqrt{\mu} & 0 \\ 0 & 1/\sqrt{\mu} \end{pmatrix}$. Show that $(s_2 \circ \varphi_2^{-1})^* (g' \cdot \boldsymbol{\omega}) = (s_2 \circ \varphi_2^{-1})^* (g \cdot \boldsymbol{\omega})$ and conclude that $g' \cdot \boldsymbol{\omega} = g \cdot \boldsymbol{\omega}$.

Exercise 5.5.6 Suppose $m = \infty \in \mathbb{H} \cup \{ \infty \}$ and $\lambda > 0$. Let $\mu = 1/\lambda$, $g = g_0 \, a_\lambda$, and $g' = g_\infty \, a_\mu$. Show that $g' \cdot \boldsymbol{\omega} = g \cdot \boldsymbol{\omega}$.

The bottom line here is that every point in the moduli space $\mathcal{M}$ is represented by a gauge potential of the form (5.5.5) or (5.5.6) and that, to obtain a complete set of representatives, it is sufficient to consider only λ's in the interval $0 < \lambda \leq 1$. Some redundancy still remains, however, since, as we now show, $\lambda = 1$ specifies the natural connection $\boldsymbol{\omega}$ for *any* $m \in \mathbb{H} \cup \{ \infty \}$. For $m = \infty$ this is clear from (5.5.6) and Exercise 4.8.15. For $m \in \mathbb{H}$ and $\lambda = 1$, (5.5.5) reduces to

$$\mathrm{Im} \left(\frac{(1 + \,|\, m \,|^2) \, \bar{q} \, dq}{|\, q - m \,|^2 + \,|\, \bar{m} \, q + 1 \,|^2} \right) .$$

But $|\, q - m \,|^2 + \,|\, \bar{m} \, q + 1 \,|^2 = (\bar{q} - \bar{m})(q - m) + (\bar{m} \, q + 1)(\bar{q} \, m + 1) = |\, q \,|^2 - \bar{q} \, m - \bar{m} \, q + \,|\, m \,|^2 + \,|\, m \,|^2 \,|\, q \,|^2 + \bar{m} \, q + \bar{q} \, m + 1 = 1 + \,|\, m \,|^2 + \,|\, q \,|^2 + \,|\, m \,|^2 \,|\, q \,|^2 = (1 + \,|\, m \,|^2)(1 + \,|\, q \,|^2)$ so this too is just the basic BPST instanton.

A curvature calculation such as that in Exercises 4.10.28 and 5.3.1 shows that distinct points (λ, m) in $(0,1)\times(\mathbb{H}\cup\{\infty\})$ give rise to gauge inequivalent connections $(g_m\, a_\lambda)\cdot\boldsymbol{\omega}$. Thus, every point in the moduli space $\mathcal{M}$ except the class $[\boldsymbol{\omega}]$ of the natural connection is represented uniquely by one of these, i.e., by a point $(\lambda, m)\in(0,1)\times(\mathbb{H}\cup\{\infty\})$. Now, $\mathbb{H}\cup\{\infty\}$ is homeomorphic to S^4 and $(0,1)\times S^4$ is homeomorphic to the open 5-ball B^5 with the origin removed. Thus, we may identify $\mathcal{M}-\{\,[\boldsymbol{\omega}]\,\}$ with $B^5-\{0\}$ and view $r=1-\lambda$ as a "radial" coordinate in $\mathcal{M}$. Then $\lambda=1$ fills in the "center" of $\mathcal{M}$ with $[\boldsymbol{\omega}]$ (see Figure 5.5.1).

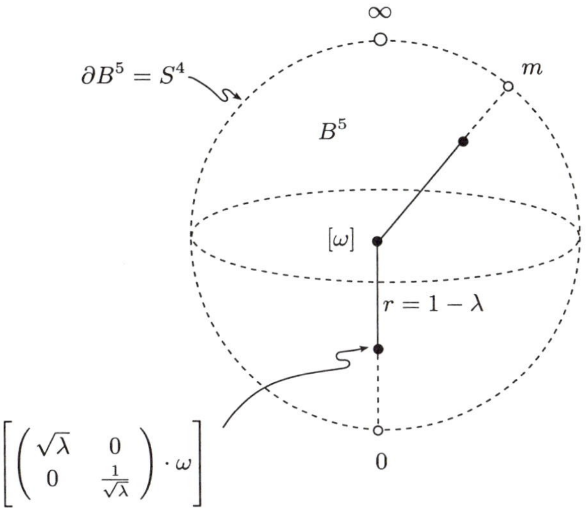

Figure 5.5.1

A particularly pleasing aspect of this picture is that the base manifold S^4 of the underlying bundle appears naturally as a boundary of the moduli space $\mathcal{M}$ in some "compactification" of $\mathcal{M}$ (the closed 5-dimensional disc D^5). Furthermore, these boundary points have a simple physical interpretation. We have already seen that the gauge field strengths corresponding to any of the connections in the class $\left[\begin{pmatrix} \sqrt{\lambda} & 0 \\ 0 & 1/\sqrt{\lambda} \end{pmatrix} \cdot \omega\right]$ become increasingly concentrated at 0 as $\lambda \to 0$, i.e., as $r \to 1$ and one approaches the boundary of the moduli space along the radius of D^5 containing 0 (the south pole). This boundary point can then, at least intuitively, be identified with a connection/gauge field concentrated entirely at the point 0. Any point m on the boundary admits a similar interpretation. Indeed, the point a radial distance $r = 1 - \lambda$ along the radius from $[\omega]$ to m is represented by the gauge potential which, in coordinates obtained by stereographically projecting from the antipodal point $-m$, is the BPST potential whose curvature concentrates at m.

Exercise 5.5.7 Carry out the curvature calculation mentioned at the beginning of the last paragraph and the stereographic projection described at the end.

And so, where do we stand? At this point we have (at least, modulo the Atiyah-Hitchin-Singer Theorem [**AHS**]) a complete geometrical description of the gauge equivalence classes of ASD connections on the Hopf bundle (instanton number -1). Where does one go from here? Of course, there are other instanton numbers corresponding, in the manner described in Section 5.3, to other possible gauge asymptotic behaviors for Yang-Mills fields on $\mathbb{R}^4$. All of these instantons are known and can, in fact, be described entirely in terms of linear algebra, although the motivation lies rather deep in algebraic geometry. (Atiyah's lecture notes [**Atiy**] provide the most leisurely exposition available; see, in particular, the Theorem in Section 2 of Chapter II.) One might then attempt generalizations to other (4-dimensional, compact, oriented, Riemannian) base manifolds X and/or other (compact, semisimple) Lie structure groups G (see, for example, [**Buch**]).

By far the most influential generalizations, however, have been those spawned by the work of Donaldson in [**Don**]. Although we are not equipped to deal with this material in any detail, we offer a brief synopsis. Thus, we let X denote a compact, simply connected, oriented, smooth 4-manifold (one can actually show that a compact, simply connected 4-manifold is necessarily orientable). Every smoothly embedded, oriented surface (2-manifold) $\sum$ in X determines an element $[\sum]$ of what is called the *second (integer) homology group* $H_2(X;\mathbb{Z})$ of X ($[\sum]$ is actually a certain equivalence class of surfaces in X and $H_2(X;\mathbb{Z})$ turns out to be a finitely generated free Abelian group). The *second cohomology group* $H^2(X;\mathbb{Z})$ is the "dual" of $H_2(X;\mathbb{Z})$, i.e., the group of integer-valued homomorphisms on $H_2(X;\mathbb{Z})$, and is also a free Abelian group (in fact, isomorphic to $H_2(X;\mathbb{Z})$). Defined on $H_2(X;\mathbb{Z})$ is a certain symmetric, bilinear form

$$Q_X : H_2(X;\mathbb{Z}) \times H_2(X;\mathbb{Z}) \longrightarrow \mathbb{Z}$$

called the *intersection form* of X. Very roughly, the idea is as follows. Given α_1 and α_2 in $H_2(X;\mathbb{Z})$ one may select oriented surfaces $\sum_1$ and $\sum_2$ in X representing α_1 and α_2, respectively, and which intersect transversally in a finite number of isolated points ("transversally" means that, at each intersection point, the tangent spaces to $\sum_1$ and $\sum_2$ span the tangent space to X). An intersection point p is assigned the value $+1$ if an oriented basis for $T_p(\sum_1)$ together with an oriented basis for $T_p(\sum_2)$ gives an oriented basis for $T_p(X)$; otherwise, it is assigned the value -1. Then $Q_X(\alpha_1, \alpha_2)$ is the sum of these values over all the intersection points. One can show that Q_X is bilinear and symmetric and that, moreover, it is *unimodular* (i.e., if $\alpha_1, \ldots, \alpha_t$ is a basis for $H_2(X;\mathbb{Z})$ over $\mathbb{Z}$, then the matrix $(Q_X(\alpha_i, \alpha_j))$ has determinant ± 1).

The intersection form Q_X is said to be *even* if $Q_X(\alpha, \alpha)$ is an even integer for all $\alpha \in H_2(X;\mathbb{Z})$; otherwise, Q_X is *odd*. Q_X is *positive definite* if $Q_X(\alpha, \alpha) \geq 0$ for all α and *negative definite* if $Q_X(\alpha, \alpha) \leq 0$ for all α. If Q_X is either positive definite or negative definite, it is said to be *definite*; otherwise, it is *indefinite*. The *rank* of Q_X, denoted $b_2(X)$ or simply b_2, is the rank of the finitely generated free Abelian group $H_2(X;\mathbb{Z})$ on which it is defined. Then b_2 can be written $b_2 = b^+ + b^-$, where b^+ (respectively, b^-) is the maximal dimension of a subspace of $H_2(X;\mathbb{Z})$ on which Q_X is positive (respectively, negative) definite. Finally, the *signature* σ of Q_X (or of X) is defined by $\sigma = b^+ - b^- = b_2 - 2b^- = 2b^+ - b_2$.

It has been known for some time that the intersection form is a basic invariant for compact 4-manifolds. Indeed, in 1949, Whitehead [**Wh**] proved that two compact, simply connected 4-manifolds X_1 and X_2 have the same homotopy type iff their intersection forms are *equivalent* in the sense that one can find bases for $H_2(X_1;\mathbb{Z})$ and $H_2(X_2;\mathbb{Z})$ relative to which the matrices of Q_{X_1} and Q_{X_2} are the same. It is of some interest then to understand both the algebraic structure of unimodular, symmetric, bilinear forms and which such forms can actually occur as intersection forms of a 4-manifold. The algebraic classification of indefinite forms is relatively straightforward: They are completely characterized (up to equivalence) by their rank, signature and type (even/odd). The classification of definite forms is much more difficult and the number of equivalence classes grows very rapidly with the rank (if the rank is 40 there are at least 10^{51} such classes). Donaldson's remarkable result relieves the differential topologist of the need to penetrate this great morass of definite forms.

Donaldson's Theorem ([**Don**]): *Let X be a compact, simply connected, smooth 4-manifold with positive definite intersection form Q_X. Then Q_X is equivalent to the standard diagonal form, i.e., there exists a basis for $H_2(X;\mathbb{Z})$ relative to which the matrix of Q_X is the $b_2 \times b_2$ identity matrix.*

Remark: One cannot appreciate just how extraordinary this result is without contrasting it with a theorem of Freedman [**Fr**], proved just a year earlier. Although we will not enter into the details here, one can actually define the intersection form for *topological* as well as smooth 4-manifolds. In a remarkable *tour de force*, Freedman showed that compact, simply connected topological 4-manifolds are completely determined up to homeomorphism by their intersection form and one additional bit of information (an element of $\mathbb{Z}_2$ called the manifold's *Kirby-Siebenmann invariant*) and that, moreover, *every* integral, unimodular, symmetric, bilinear form is the intersection form of at least one such manifold (see [**FL**] for an exposition of Freedman's work). In concert with Donaldson's Theorem this immediately serves up a huge supply of compact, simply connected topological 4-manifolds that can admit no compatible differentiable structure. A more subtle combination of the two theorems implies the existence of fake $\mathbb{R}^4$'s, i.e., smooth 4-manifolds that are homeomorphic, but not diffeomorphic to $\mathbb{R}^4$ with its standard structure (see [**Ster**], [**FU**], or [**Law**]). As we have already mentioned (Section 4.3), this phenomenon occurs *only* in dimension four.

The most remarkable thing about Donaldson's Theorem, however, is its proof. Whereas Freedman established his classification theorem through a subtle and brilliant manipulation and extension of traditional topological techniques, Donaldson startled the mathematical community by extracting his purely topological conclusion from an analysis of moduli spaces of Yang-Mills instantons. Briefly, the story goes something like this: Begin with an arbitrary compact, simply connected, oriented 4-manifold X with positive definite intersection form Q_X. Now consider the $k = -1$ principal $SU(2)$-bundle $\mathcal{P} : P \to X$ over X (unlike in our earlier discussions of particles and fields, this bundle is now to be regarded as simply an auxiliary structure the purpose of which is to facilitate the study of X). Next introduce a Riemannian metric $\mathbf{g}$ on X. (This can always be done, generally in many different ways; $\mathbf{g}$ is also to be regarded as an auxiliary structure to aid in the study of X.) From $\mathbf{g}$ (and the given orientation of X) one obtains a Hodge star operator and thereby a notion of anti-self-dual connection for the bundle P. (There are technical difficulties here stemming from the fact that, when $X \neq S^4$, a connection is generally not determined by a single gauge potential. This necessitates regarding the curvature as a "bundle-valued 2-form.") We need more than the "notion", however. It

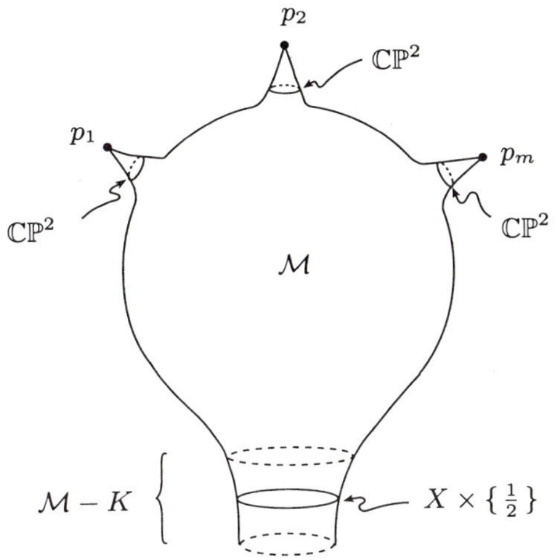

Figure 5.5.2

is a highly nontrivial result of Taubes [**Tau1**] that, for manifolds X with positive definite intersection form, the bundle $\mathcal{P} : P \to X$ actually admits ASD connections. Thus, one may, without fear of discoursing at length on the empty set, introduce and study the moduli space $\mathcal{M} = \mathcal{M}(\mathbf{g})$ of such connections (the moduli space *does* depend on the choice of $\mathbf{g}$).

For a randomly chosen Riemannian metric $\mathbf{g}$, the moduli space $\mathcal{M}(g)$ may not be very nice. However, one can show that, for some choice of $\mathbf{g}$ (indeed, for "almost every" choice of $\mathbf{g}$), all of the following are true (see Figure 5.5.2)

1. If m denotes half the number of homology classes $\alpha \in H_2(X; \mathbb{Z})$ for which $Q_X(\alpha, \alpha) = 1$, then there exist points $p_1, \ldots, p_m$ in $\mathcal{M}$ such that $\mathcal{M} - \{p_1, \ldots, p_m\}$ is a smooth, orientable manifold of dimension 5.

2. Each p_i, $i = 1, \ldots, m$, has a nbd in $\mathcal{M}$ that is homeomorphic to the cone over $\mathbb{CP}^2$ with p_i at the vertex. (The cone over $\mathbb{CP}^2$ is the quotient of $\mathbb{CP}^2 \times [0,1]$ obtained by identifying all points of the form $(p, 1)$.)

3. There is a compact set $K \subseteq \mathcal{M}$ such that $\mathcal{M} - K$ is a submanifold

of $\mathcal{M} - \{p_1, \ldots, p_m\}$ diffeomorphic to $X \times (0,1)$.

Our real interest lies in a subspace $\mathcal{M}_0$ of $\mathcal{M}$ obtained by deleting from $\mathcal{M}$ the following open sets: (i) The subset of $\mathcal{M} - K$ corresponding to $X \times (0, \frac{1}{2})$. (ii) For each $i = 1, \ldots, m$, the open "top half" of the cone containing p_i (i.e., the points in the quotient corresponding to $\mathbb{CP}^2 \times (\frac{1}{2}, 1]$). $\mathcal{M}_0$ is compact (because K is compact), but it cannot be a manifold because deleting an open set from a manifold will generally leave behind an "edge". In fact, $\mathcal{M}_0$ is what is known as a "manifold with boundary." The "boundary" is just the union of the edges left behind. Specifically, it consists of a copy of X (i.e., $X \times \{\frac{1}{2}\}$) at the bottom and a disjoint union of copies of $\mathbb{CP}^2$ at the top (see Figure 5.5.3). The subset $X \times [\frac{1}{2}, 1)$ is called a *collar* of X in $\mathcal{M}_0$.

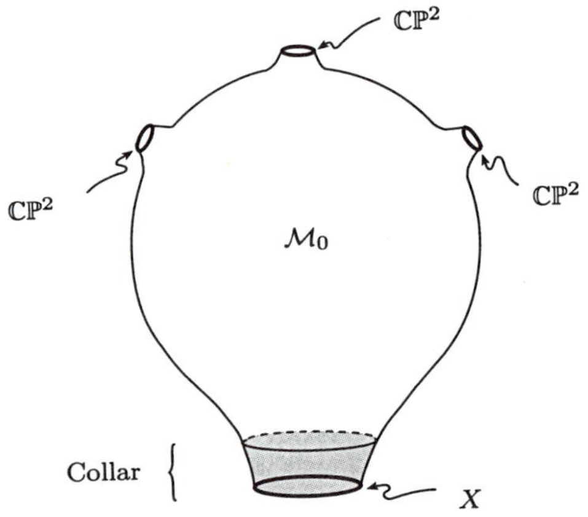

Figure 5.5.3

In general, if X_1 and X_2 are two n-dimensional manifolds and if there exists an $(n+1)$-dimensional manifold with boundary M whose boundary is the disjoint union of X_1 and X_2, then X_1 and X_2 are said to be *cobordant* and M is called a *cobordism* between X_1 and X_2. As it happens, the signature of a 4-manifold is a *cobordism invariant.* The conclusion then is that the signature of X is the same as the signature of the disjoint union of the $\mathbb{CP}^2$'s in $\mathcal{M}_0$. This fact, together with the positive definiteness of X and some elementary counting arguments gives $m = b_2(X)$. The rest is essentially just (integer) linear algebra. Start with any $\alpha_1 \in H_2(X;\mathbb{Z})$ for which $Q_X(\alpha_1,\alpha_1) = 1$. Then there is an orthogonal (with respect to Q_X) decomposition $H_2(X;\mathbb{Z}) = \mathbb{Z}\alpha_1 \oplus G_1$ given by writing any $\beta \in H_2(X;\mathbb{Z})$ as $\beta = Q_X(\beta,\alpha_1)\,\alpha_1 + (\beta - Q_X(\beta,\alpha_1)\,\alpha_1)$. Now, for any $\alpha_2 \in H_2(X;\mathbb{Z})$ with $Q_X(\alpha_2,\alpha_2) = 1$ and $\alpha_2 \neq \pm\alpha_1$, the Schwartz inequality gives $(Q_X(\alpha_1,\alpha_2))^2 < 1$. But $Q_X(\alpha_1,\alpha_2)$ is an integer so $Q_X(\alpha_1,\alpha_2) = 0$ and $\alpha_2 \in G_1$. Now repeat the process in G_1 and continue inductively to obtain a basis $\alpha_1,\ldots,\alpha_{b_2(X)}$, with $Q_X(\alpha_i,\alpha_i) = 1$ for each $i = 1,\ldots,b_2(X)$. Relative to this basis the matrix of Q_X is the $b_2 \times b_2$ identity matrix as required. This completes our sketch of the ideas behind Donaldson's Theorem (those who wish to see how all of this is actually proved are referred to [**FU**] and [**Law**]).

It goes without saying that the appearence of Donaldson's Theorem in 1983 caused quite a stir and unleashed a furious storm of activity centered on the application of techniques from gauge theory to topology. Much of this material is described in [**DK**]. More recently, yet another revolution has taken place that may well supersede Donaldson theory altogether. Remarkably, this too has sprung from the soil of theoretical physics. For an accessible introduction to this so-called **Seiberg-Witten theory** (and a pleasant afternoon's entertainment) we recommend [**Tau2**].

5.6 Matter Fields: Motivation

Let us return once again to our roots in Chapter 0. There we began with the classical description of the field of a magnetic monopole. We found that there was no globally defined vector potential for such a field and observed that, from the point-of-view of quantum mechanics, this is rather inconvenient. Indeed, the motion of a charged particle in an electromagnetic field is decribed in quantum mechanics by solving the Schroedinger equation for the wavefunction ψ of the particle and this equation involves the vector potential of the field in an essential way. This was our motivation for seeking an alternative mathematical device that does the job of the vector potential, but is in some sense globally defined. This "job" is keeping track of the phase of the charged particle traversing the field. We were therefore led to the notion of a principal bundle of phases over the space in which the particle moves and a unique path lifting procedure for keeping track of the phase of the particle as it traverses its path. Due to a peculiarity in the monopole's local vector potentials (when expressed as 1-forms in spherical coordinates) it sufficed to build the bundle of phases over S^2. This (and Theorem 5.1.4) is our motivation for studying connections on principal $U(1)$-bundles over S^2. We also mentioned the analogous Yang-Mills problem of keeping track of a nucleon's isotopic spin state, which leads to a consideration of connections on $SU(2)$-bundles (or, equivalently, $Sp(1)$-bundles).

Remark: In real physics the base manifold for such bundles is "spacetime". Neglecting gravitational effects this means "Minkowski spacetime", where each event in the nucleon's history (worldline) is represented by four numbers (x, y, z, t) specifying its spatial location (x, y, z) and the time t at which it occupies this location. As a differentiable manifold this is simply $\mathbb{R}^4$, but Minkowski spacetime cannot be fully identified with $\mathbb{R}^4$, because physically meaningful relationships between events are expressed in terms of a certain indefinite inner product $(x^1 x^2 + y^1 y^2 + z^1 z^2 - t^1 t^2)$ rather than the usual Euclidean inner product (see [**N3**]). This distinction is significant since our discussion of Yang-Mills fields in Section 5.3 relied heavily on the Hodge dual and this, in turn, was defined in terms of the Riemannian (i.e., inner product) structure of $\mathbb{R}^4$. Unfortunately, relatively little is known about Yang-Mills fields on Minkowski spacetime and, worse yet, the basic objects of interest in quantum field theory (Feynman path integrals) are extraordinarily difficult to make any sense of in this indefinite context. The minus sign in the Minkowski inner product is rather troublesome. Not to be deterred by such a minor inconvenience, the physicists do the only reasonable thing under the circumstances—they change the sign! To lend an air of respectability to this subterfuge, however, they give it a name. Introducing an imaginary time coordinate $\tau = \mathbf{i}\,t$ is designated a **Wick rotation** and has the laudable effect of transforming Minkowski spacetime into $\mathbb{R}^4$ $(x^1 x^2 + y^1 y^2 + z^1 z^2 - t^1 t^2 = x^1 x^2 + y^1 y^2 + z^1 z^2 + \tau^1 \tau^2)$. What more could you ask? Well, of course, a pedant might ask whether or not any physics survives this transformation. This is a delicate issue and not one that we are prepared to address. The answer would seem to be in the affirmative, but the reader will have to consult the physics literature to learn why (see Section 13.7 of [**Guid**]). Whether or not there is any physics in this positive definite context is quite beside the point for mathematics, of course. It is *only* in the positive definite case that (anti-) self-dual connections exist and it is an understanding of the moduli space of these that pays such handsome topological dividends. Granting then that Yang-Mills fields over $\mathbb{R}^4$ are worth looking for and that those of finite action (i.e., finite total field strength) are particularly desirable, the discussion in Section 5.3 leads us to our interest in connections on $Sp(1)$-bundles over S^4.

A gauge field is therefore something akin to a modern version of the

Newtonian notion of a "force" in that it is something to which particles (electrons, nucleons, etc.) "respond" by experiencing changes in their internal states (phase, isotopic spin, etc.). It is the "atmosphere" mediating changes in the internal spin state of the ping-pong ball in Section 0.5. The mathematical models for such fields (connections on principal bundles) we now have well in hand, but the reader may be wondering where the charged particles and nucleons (the particles responding to, i.e., "coupled to", the gauge field) appear in our picture. The answer is that they do not (yet) and our task now is to remedy this deficiency.

As motivation we consider again the classical quantum mechanical picture of a charged particle moving through a region of space permeated by an electromagnetic field. The charge is described by its wavefunction ψ (a complex-valued function of x, y, z and t) obtained by solving the Schroedinger equation. The particular field through which the charge moves enters the Schroedinger equation by way of its scalar and vector potential functions. If these exist on the entire region X of interest, then all is well. One solves the differential equation and begins the process of sorting out what the solution $\psi : X \to \mathbb{C}$ means physically.

As the example of the magnetic monopole has made apparent, however, nature is not always so magnanimous as to provide us with a vector potential defined on the entire region of interest. In this case, at least two vector potentials, defined on overlapping regions, are required. On each of these regions one can solve the corresponding Schroedinger equation to obtain a local wavefunction. On the intersection of their domains the potentials differ by a gradient and so the wavefunctions differ by a phase factor. This same sort of phenomenon occurs for the field of the solenoid in the Aharonov-Bohm experiment (Section 0.2) and it is precisely the interference of the different phases when the two partial beams (which experience different local potentials) recombine that accounts for the rather startling outcome.

All of this has a rather familiar ring to it. One has at hand a principal $U(1)$-bundle and a connection on it representing an electromagnetic field. Each fiber is a copy of S^1 representing the possible values of the phase for a charged particle whose trajectory passes through the projection of the fiber into the base. Choosing a local cross-section of the bundle amounts to selecting a "reference phase" at each point in some open subset of the base. Relative to this choice one obtains a local representation for the connection, i.e., a gauge potential, and, from this, a local wavefunction which assigns to each point in the trivializing nbd a complex number. Another local cross-section (i.e., another choice of "zero phase angle" at all points in some open set) yields another local wavefunction which, on the intersection of the two trivializing nbds, will not agree with the first, but will differ from it only by a $U(1)$-action (see (0.2.8)). The wavefunctions appear to be playing the role of cross-sections of some locally trivial bundle whose fibers are copies of the complex plane $\mathbb{C}$. For the monopole one might view this new bundle as having been manufactured from the complex Hopf bundle by replacing each S^1-fiber with a copy of $\mathbb{C}$. Of course, one must also build into this picture the transformation law for the local wavefunctions and this involves both the right action of $U(1)$ on the Hopf bundle, which describes the relationship between the local potentials, as well as the action of $U(1)$ on $\mathbb{C}$ mentioned above that describes the relationship between the corresponding local wavefunctions. In the analogous situation of a particle with isotopic spin coupled to a Yang-Mills field, the wavefunctions have two complex components $\begin{pmatrix} \psi^1 \\ \psi^2 \end{pmatrix}$ so the fibers of the new bundle are $\mathbb{C}^2$ and the effect of a gauge transformation is represented by an action of $SU(2)$ on $\mathbb{C}^2$ (equivalently, one may regard the wavefunction as $\mathbb{H}$-valued and the action as being by $Sp(1)$).

In its most general form the situation we seek to model is the following: A particle has an internal structure, the states of which are represented by the elements of some Lie group G (e.g., $U(1)$, or $SU(2) \cong Sp(1)$). The wavefunction of the particle takes values in some vector space $\mathcal{V}$ (for our purposes, $\mathcal{V}$ will be some $\mathbb{C}^k$). The particle is coupled to (i.e., experiences the effects of) a gauge field which is represented by a connection on a principal G-bundle. The connection describes (via Theorem 5.1.4) the evolution of the particle's internal state. The response of the wavefunction at each point to a gauge transformation will be specified by a left action (representation) of G on $\mathcal{V}$. $\mathcal{V}$ and this left action of G on $\mathcal{V}$ determine an "associated vector bundle" obtained by replacing the G-fibers of the principal bundle with copies of $\mathcal{V}$. The local cross-sections of this bundle then represent local wavefunctions of the particle coupled to the gauge field. Because of the manner in which the local wavefunctions respond to a gauge transformation the corresponding local cross-sections piece together to give a global cross-section of the associated vector bundle and this, we will find, can be identified with a certain type of $\mathcal{V}$-valued function on the original principal bundle space. Finally, the connection on the principal bundle representing the gauge field gives rise to a natural gauge invariant differentiation process for such wavefunctions. In terms of this derivative one can then postulate differential equations (field equations) that describe the quantitative response of the particle to the gauge field (selecting these equations is, of course, the business of the physicists).

5.7 Associated Fiber Bundles

We now describe in detail the procedure, hinted at in the last section, for surgically replacing the G-fibers of a principal bundle with copies of some vector space on which G acts on the left. Indeed, since it is just as easy to do so, we will carry out a more general procedure that replaces the G-fibers with copies of any manifold on which G acts on the left. The result of the construction will be a locally trivial bundle (called a "fiber bundle") which looks much like a principal bundle except that its fibers are no longer diffeomorphic to the underlying Lie group G.

Let $\mathcal{P} : P \to X$ be a smooth principal G-bundle over X with right action $\sigma : P \times G \to P$, $\sigma(p,g) = p \cdot g$. Now let F be a manifold on which G acts smoothly on the left (the image of $(g,\xi) \in G \times F$ under this action will be written $g \cdot \xi$).

Remark: Whether one chooses to work with group actions on the right or on the left is very often a matter of convenience or personal taste. If, for example, $(g,\xi) \to g \cdot \xi$ is a left action, then $(\xi, g) \to \xi \odot g = g^{-1} \cdot \xi$ defines a right action because $(\xi \odot g_1) \odot g_2 = g_2^{-1} \cdot (\xi \odot g_1) = g_2^{-1} \cdot (g_1^{-1} \cdot \xi) = (g_2^{-1} g_1^{-1}) \cdot \xi = (g_1 g_2)^{-1} \cdot \xi = \xi \odot (g_1 g_2)$. Similarly, any right action $(\xi, g) \to \xi \cdot g$ gives rise to a left action $(g,\xi) \to g \odot \xi = \xi \cdot g^{-1}$ and these operations are inverses of each other. Some actions appear more natural in one guise than another, however (e.g., the definition of matrix multiplication dictates viewing a representation of a group on a vector space as acting on the left).

Exercise 5.7.1 Show that the map from $(P \times F) \times G$ to $P \times F$ defined by

$$((p,\xi), g) \longrightarrow (p \cdot g, g^{-1} \cdot \xi)$$

defines a smooth right action of G on $P \times G$.

At the risk of abusing the privilege we will use the same dot $\cdot$ to indicate the smooth right action of G on $P \times F$ described in Exercise 5.7.1:

$$(p,\xi) \cdot g = (p \cdot g, g^{-1} \cdot \xi)$$

We denote by $P \times_G F$ the orbit space of $P \times F$ modulo this action, i.e., the quotient space of $P \times F$ obtained by identifying to points all of the orbits. More precisely, we define an equivalence relation $\sim$ on $P \times F$ as follows: $(p_1, \xi_1) \sim (p_2, \xi_2)$ iff there exists a $g \in G$ such that $(p_2, \xi_2) = (p_1, \xi_1) \cdot g$, i.e., $p_2 = p_1 \cdot g$ and $\xi_2 = g^{-1} \cdot \xi_1$. The equivalence class of (p,ξ) is denoted $[p,\xi] = \{ (p \cdot g, g^{-1} \cdot \xi) : g \in G \}$ and, as a set, $P \times_G F = \{ [p,\xi] : (p,\xi) \in P \times F \}$. If $\mathcal{Q} : P \times F \to P \times_G F$ is the quotient map ($\mathcal{Q}(p,\xi) = [p,\xi]$), then $P \times_G F$ is given the quotient topology determined by $\mathcal{Q}$.

Next we define a mapping $\mathcal{P}_G : P \times_G F \to X$ by $\mathcal{P}_G([p,\xi]) = \mathcal{P}(p)$.

Observe that $\mathcal{P}_G$ is well-defined because $\mathcal{P}(p \cdot g) = \mathcal{P}(p)$ for each $g \in G$. Moreover, since $\mathcal{P}_G \circ \mathcal{Q} = \mathcal{P}$, Lemma 1.2.1 implies that $\mathcal{P}_G$ is continuous.

Exercise 5.7.2 Let x be any point in X. Select any fixed point $p \in \mathcal{P}^{-1}(x)$. Show that $\mathcal{P}_G^{-1}(x) = \{\,[p, \xi] : \xi \in F\,\}$.

Now fix an $x_0 \in X$ and select an open nbd V of x_0 in X and a diffeomorphism $\Psi : \mathcal{P}^{-1}(V) \to V \times G$ with $\Psi(p) = (\mathcal{P}(p), \psi(p))$, $\psi(p \cdot g) = \psi(p)\, g$, for all $p \in \mathcal{P}^{-1}(V)$ and $g \in G$. Let $s : V \to \mathcal{P}^{-1}(V)$ be the associated canonical cross-section $(s(x) = \Psi^{-1}(x, e))$. Define a map $\tilde{\Phi} : V \times F \to \mathcal{P}_G^{-1}(V)$ by

$$\tilde{\Phi}\,(x\,,\,\xi\,) = [\,s\,(x)\,,\,\xi\,] \tag{5.7.1}$$

(note that $\mathcal{P}_G(\tilde{\Phi}\,(x, \xi)) = \mathcal{P}_G([s(x), \xi]) = \mathcal{P}(s(x)) = x$ so $\tilde{\Phi}$ does, indeed, map into $\mathcal{P}_G^{-1}(V)$). We claim that $\tilde{\Phi}$ is a homeomorphism. First we show that it is a bijection. From Exercise 5.7.2 we obtain

$$\begin{aligned} \mathcal{P}_G^{-1}(V) &= \bigcup_{x \in V} \mathcal{P}_G^{-1}(x) \\ &= \bigcup_{x \in V} \{\,[p, \xi] : \mathcal{P}(p) = x,\ \xi \in F\,\} = \mathcal{Q}\left(\mathcal{P}^{-1}(V) \times F\right). \end{aligned} \tag{5.7.2}$$

To prove that $\tilde{\Phi}$ is onto, let $[p, \xi]$ be in $\mathcal{P}_G^{-1}(V)$. Set $\mathcal{P}(p) = x$. Then $\mathcal{P}(p) = \mathcal{P}(s(x))$ so there exists a $g \in G$ for which $p = s(x) \cdot g$. Thus, $(s(x) \cdot g, g^{-1} \cdot \xi') \sim (s(x), \xi')$ for all $\xi' \in F$, i.e., $(p, g^{-1} \cdot \xi') \sim (s(x), \xi')$ for all $\xi' \in F$. Consequently, $[p, g^{-1} \cdot \xi'] \sim [s(x), \xi']$ for all $\xi' \in F$. Taking $\xi' = g \cdot \xi$ gives $\tilde{\Phi}\,(x, g \cdot \xi) = [s(x), g \cdot \xi] = [p, \xi]$ as required. To see that $\tilde{\Phi}$ is one-to-one suppose $\tilde{\Phi}\,(x', \xi') = \tilde{\Phi}\,(x, \xi)$ for some $(x', \xi'), (x, \xi) \in V \times F$. Then $(s(x'), \xi') \sim (s(x), \xi)$ so, in particular, $\mathcal{P}(s(x')) = \mathcal{P}(s(x))$, i.e., $x' = x$. As a result, $(s(x), \xi') \sim (s(x), \xi)$ so there is a $g \in G$ for which $s(x) = s(x) \cdot g$ and $\xi' = g^{-1} \cdot \xi$. The first of these implies $g = e$ by Exercise 3.1.1 so the second gives $\xi' = \xi$. Thus, $(x', \xi') = (x, \xi)$ as required and we have shown that $\tilde{\Phi}$ is bijective. It is continuous since it is the composition $(x, \xi) \to (s(x), \xi) \to \mathcal{Q}(s(x), \xi) = [s(x), \xi]$ of continuous maps. Finally, the inverse $\tilde{\Psi} : \mathcal{P}_G^{-1}(V) \to V \times F$ is given by

$$\tilde{\Psi}\,([\,s(x), \xi\,]) = (\,x, \xi\,) \tag{5.7.3}$$

for all $x \in V$ (or $\tilde{\Psi}\,([p, \xi]) = (\mathcal{P}(p), g \cdot \xi)$, where $p = s(\mathcal{P}(p)) \cdot g$) and this is continuous by Lemma 1.2.1 since $\tilde{\Psi} \circ \mathcal{Q}$ is given on $\mathcal{Q}^{-1}(\mathcal{P}_G^{-1}(V)) =$

$\mathcal{P}^{-1}(V) \times F$ by $(p, \xi) \to [p, \xi] = (\mathcal{P}(p), g \cdot \xi)$.

Since $\mathcal{P}_G \circ \tilde{\Phi}(x, \xi) = \mathcal{P}_G([s(x), \xi]) = \mathcal{P}(s(x)) = x$, we have shown thus far that $(P \times_G F, X, \mathcal{P}_G, F)$ is a locally trivial bundle. In particular, $P \times_G F$ is a Hausdorff space by Exercise 1.3.23. We wish now to supply $P \times_G F$ with a differentiable structure relative to which $\mathcal{P}_G$ is smooth and all of the maps $\tilde{\Psi} : \mathcal{P}_G^{-1}(V) \to V \times F$ are diffeomorphisms (here $V \times F$ is an open submanifold of the product manifold $X \times F$). Toward this end we suppose that (V_i, Ψ_i) and (V_j, Ψ_j) are two local trivializations of the principal G-bundle $\mathcal{P} : P \to X$ with $V_i \cap V_j \neq \emptyset$. We let $s_i : V_i \to \mathcal{P}^{-1}(V_i)$ and $s_j : V_j \to \mathcal{P}^{-1}(V_j)$ be the corresponding cross-sections and $g_{ji} : V_i \cap V_j \to G$ the transition function. Each of these trivializations gives rise, as above, to a homeomorphism $\tilde{\Psi}_i : \mathcal{P}_G^{-1}(V_i) \to V_i \times F$ and $\tilde{\Psi}_j : \mathcal{P}_G^{-1}(V_j) \to V_j \times F$ and we now compute

$$\tilde{\Psi}_j \circ \tilde{\Psi}_i^{-1} : (V_i \cap V_j) \times F \longrightarrow (V_i \cap V_j) \times F.$$

Let $(x, \xi) \in (V_i \cap V_j) \times F$. Then

$$\begin{aligned}
\tilde{\Psi}_j \circ \tilde{\Psi}_i^{-1}(x, \xi) &= \tilde{\Psi}_j\left(\tilde{\Psi}_i^{-1}(x, \xi)\right) = \tilde{\Psi}_j\left(\tilde{\Phi}_i(x, \xi)\right) \\
&= \tilde{\Psi}_j([s_i(x), \xi]) = \tilde{\Psi}_j([s_j(x) \cdot g_{ji}(x), \xi]) \\
&= \tilde{\Psi}_j\left(\left[s_j(x) \cdot g_{ji}(x), (g_{ji}(x))^{-1} \cdot (g_{ji}(x) \cdot \xi)\right]\right) \\
&= \tilde{\Psi}_j\left(\left[s_j(x), g_{ji}(x) \cdot \xi\right]\right) \\
&= (x, g_{ji}(x) \cdot \xi).
\end{aligned}$$

Exercise 5.7.3 Conclude that $\tilde{\Psi}_j \circ \tilde{\Psi}_i^{-1}$ is a diffeomorphism and therefore that there is a unique differentiable structure on $P \times_G F$ relative to which each $\tilde{\Psi} : \mathcal{P}_G^{-1}(V) \to V \times F$ is a diffeomorphism. Show also that, relative to this differentiable structure, $\mathcal{P}_G : P \times_G F \to X$ is smooth.

The conclusion then is that $(P \times_G F, X, \mathcal{P}_G, F)$ is a smooth locally trivial bundle whose trivializing nbds in X are the same as those of the principal G-bundle $\mathcal{P} : P \to X$ from which it was constructed. The fibers, however, are now copies of F. We shall call $(P \times_G F, X, \mathcal{P}_G, F)$ the **fiber bundle associated** with $\mathcal{P} : P \to X$ by the given left action of G on F.

Some cases of particular interest to us are the following: Let $\mathcal{P} : P \to X$ be a principal $U(1)$-bundle over X. There is a natural left action of $U(1)$ on $\mathbb{C}$ defined by $(g, \xi) \in U(1) \times \mathbb{C} \to g\xi \in \mathbb{C}$ and thus an associated fiber bundle $P \times_{U(1)} \mathbb{C}$ whose fibers are copies of $\mathbb{C}$. Similarly, if $\mathcal{P} : P \to X$ is an $SU(2)$-bundle over X, the natural left action of $SU(2)$ on $\mathbb{C}^2$ given by

$$\left(\begin{pmatrix} \alpha & \beta \\ \gamma & \delta \end{pmatrix}, \begin{pmatrix} \xi^1 \\ \xi^2 \end{pmatrix} \right) \in SU \times \mathbb{C}^2 \longrightarrow \begin{pmatrix} \alpha & \beta \\ \gamma & \delta \end{pmatrix} \begin{pmatrix} \xi^1 \\ \xi^2 \end{pmatrix}$$

$$= \begin{pmatrix} \alpha\,\xi^1 + \beta\,\xi^2 \\ \gamma\,\xi^1 + \delta\,\xi^2 \end{pmatrix}$$

determines an associated bundle $P \times_{SU(2)} \mathbb{C}^2$ whose fibers are $\mathbb{C}^2$. Left multiplication of $Sp(1)$ on $\mathbb{H}$ associates with every $Sp(1)$-bundle $\mathcal{P} : P \to X$ a fiber bundle $P \times_{Sp(1)} \mathbb{H}$ whose fibers are copies of the quaternions. Finally, we let G be an arbitrary matrix Lie group and $\mathcal{P} : P \to X$ a principal G-bundle over X. Let $g \to ad_g$ be the adjoint representation of G on $\mathcal{G}$ (Section 4.7). For each $A \in \mathcal{G}$, $ad_g(A) = g\,Ag^{-1}$ by Lemma 4.7.7. We define a left action of G on $\mathcal{G}$ by

$$(g, A) \in G \times \mathcal{G} \longrightarrow g \cdot A = ad_g(A) = g\,Ag^{-1}. \qquad (5.7.4)$$

Exercise 5.7.4 Show that if $\mathcal{G}$ is given its natural differentiable structure as a vector space (Section 4.2), then (5.7.4) defines a smooth left action of G on $\mathcal{G}$.

The fiber bundle $P \times_{ad} \mathcal{G}$ associated with $\mathcal{P} : P \to X$ by the adjoint action (5.7.4) of G on $\mathcal{G}$ is often denoted $ad\,P$ and called the **adjoint bundle** of $\mathcal{P} : P \to X$.

There is an important general procedure for constructing examples of this sort which we now describe. Thus, we let $\mathcal{P} : P \to X$ be a principal G-bundle, $\mathcal{V}$ a finite dimensional vector space and ρ a smooth representation on G on $\mathcal{V}$ (Section 4.7). We define a left action of G on $\mathcal{V}$ by

$$(g, \xi) \in G \times \mathcal{V} \longrightarrow g \cdot \xi = (\rho(g))(\xi),$$

where $(\rho(g))(\xi)$ is the value of the nonsingular linear transformation $\rho(g)$ on the vector ξ in $\mathcal{V}$. The associated bundle is denoted

$$\mathcal{P}_\rho : P \times_\rho \mathcal{V} \longrightarrow X$$

and called the **vector bundle** associated with $\mathcal{P} : P \to X$ by the representation ρ. From our point-of-view the most important special case arises as follows: We let $\mathbb{F}$ denote one of $\mathbb{R}$, $\mathbb{C}$ or $\mathbb{H}$ and $\mathcal{V} = \mathbb{F}^k$ for some positive integer k (in the quaternionic case, $\mathbb{H}^k$ is the "right vector space" described in Section 1.1). Then any smooth map $\rho : G \to GL(k, \mathbb{F})$ determines a representation of G on $\mathbb{F}^k$ (also denoted ρ) by matrix multiplication, i.e., $(\rho(g))(\xi) = \rho(g)\xi$, where $\rho(g)\xi$ is the matrix product of $\rho(g)$ and the (column vector) ξ. From this representation one obtains an associated vector bundle $P \times_\rho \mathbb{F}^k$. When $k = 1$ the vector bundle $P \times_\rho \mathbb{F}$ is called a **real, complex**, or **quaternionic line bundle** according as $\mathbb{F} = \mathbb{R}$, $\mathbb{C}$, or $\mathbb{H}$.

Remark: One often sees the notion of a "vector bundle" defined abstractly without any reference to a principal bundle associated with it. That there is no loss of generality in the definition we have given is then proved by showing that any such "vector bundle" can, in fact, be regarded as the vector bundle associated to some principal bundle by some representation of its structure group.

Being, in particular, a smooth, locally trivial bundle, any vector bundle associated with a principal bundle by some representation of the structure group admits smooth (local) cross-sections. In the physics literature, cross-sections of complex line bundles are known as "complex scalar fields" and play the role of classical wavefunctions. Cross-sections of bundles such as $P \times_\rho \mathbb{C}^2$ are the "2-component wavefunctions" appropriate, for example, to the particles of isotopic spin $\frac{1}{2}$ referred to in [**YM**]. As a rule, the underlying principal bundle P is over space, or spacetime, and is equipped with a connection representing some gauge field with which the above "matter fields" are interacting. The precise nature of such an interaction is specified by postulating field equations which, in order to ensure the gauge invariance of the theory, are expressed in terms of a natural "covariant derivative" determined by the connection. We formalize all of this in the next section.

5.8 Matter Fields and Their Covariant Derivatives

Although the program of defining covariant derivatives and formulating field equations can be carried out entirely in terms of cross-sections of associated vector bundles, it is computationally simpler and closer to the spirit of the physics literature to employ instead an equivalent notion which we now introduce. Let $\mathcal{P} : P \to X$ be an arbitrary principal G-bundle and $\mathcal{P}_G : P \times_G F \longrightarrow X$ an associated fiber bundle corresponding to some left action of G on the manifold F. Let V be an open subset of X. Then a smooth map $\phi : \mathcal{P}^{-1}(V) \to F$ is said to be **equivariant** (with respect to the given actions of G on P and F) if

$$\phi(p \cdot g) = g^{-1} \cdot \phi(p)$$

for all $p \in \mathcal{P}^{-1}(V)$ and all $g \in G$. Given such a map one can define a local cross-section $s_\phi : V \to P \times_G F$ of $P \times_G F$ as follows: Let x be a point in V. Select some $p \in \mathcal{P}^{-1}(x)$. The pair $(p, \phi(p))$ determines a point $[p, \phi(p)]$ in $P \times_G F$. This point is independent of the choice of $p \in \mathcal{P}^{-1}(x)$ since, if $p' \in \mathcal{P}^{-1}(x)$, then $p' = p \cdot g$ for some $g \in G$ so $[p', \phi(p')] = [p \cdot g, \phi(p \cdot g)] = [p \cdot g, g^{-1} \cdot \phi(p)] = [p, \phi(p)]$. Thus, we may define

$$s_\phi(x) = [p, \phi(p)],$$

where p is *any* point in $\mathcal{P}^{-1}(x)$.

Exercise 5.8.1 Show that $s_\phi : V \to \mathcal{P}_G^{-1}(V)$ is a smooth cross-section of $P \times_G F$.

Thus, every equivariant map on P gives rise to a cross-section of the associated bundle $P \times_G F$. Now, we reverse the process. Suppose we are given a smooth local cross-section $s : V \to P \times_G F$ of the associated bundle. We define a map $\phi_s : \mathcal{P}^{-1}(V) \to F$ as follows: Let $p \in \mathcal{P}^{-1}(V)$. Then $x = \mathcal{P}(p)$ is in V so $s(x)$ is in $\mathcal{P}_G^{-1}(x)$.

Exercise 5.8.2 Show that there is a unique element $\phi_s(p)$ of F such that $s(x) = [p, \phi_s(p)]$. **Hint:** Use Exercises 5.7.2 and 3.1.1.

Thus, we have defined a map $\phi_s : \mathcal{P}^{-1}(V) \to F$. Observe that, for any $p \in \mathcal{P}^{-1}(V)$ and any $g \in G$, $\phi_s(p \cdot g)$ is defined as follows: Since $\mathcal{P}(p \cdot g) = \mathcal{P}(p) = x$, $\phi_s(p \cdot g)$ is the unique element of F for which $s(x) = [p \cdot g, \phi_s(p \cdot g)]$. But $s(x) = [p, \phi_s(p)] = [p \cdot g, g^{-1} \cdot \phi_s(p)]$ so $\phi_s(p \cdot g) = g^{-1} \cdot \phi_s(p)$.

Exercise 5.8.3 Show that $\phi_s : \mathcal{P}^{-1}(V) \to F$ is smooth.

Exercise 5.8.4 Show that $\phi \to s_\phi$ and $s \to \phi_s$ are inverses of each other, i.e., that, for any equivariant map ϕ and any cross-section s of $P \times_G F$, $s_{\phi_s} = s$ and $\phi_{s_\phi} = \phi$.

We have therefore established a one-to-one correspondence between the local cross-sections $s : V \to \mathcal{P}_G^{-1}(V)$ of the fiber bundle $P \times_G F$ associated with some principal bundle $\mathcal{P} : P \to X$ and the equivariant F-valued maps $\phi : \mathcal{P}_G^{-1}(V) \to F$ on P. Of course, if either of these objects is globally defined ($s : X \to P$ or $\phi : P \to F$), then so is the other.

The case of most interest in physics arises in the following way: The principal bundle $G \to P \to X$ has defined on it a connection ω representing a gauge potential. The particles coupled to this gauge field have wavefunctions taking values in some $\mathbb{C}^k$. A local section s gives rise to a local gauge potential $\mathcal{A}$, a local field strength $\mathcal{F}$ and thereby a local wavefunction ψ. Under a change of gauge $(s \to s \cdot g)$, one obtains a new potential $\mathcal{A}^g = ad_{g^{-1}} \circ \mathcal{A} + g^* \Theta$, a new field strength $\mathcal{F}^g = ad_{g^{-1}} \circ \mathcal{F}$ and a new local wavefunction $\psi^g = (\rho(g^{-1}))\psi = g^{-1} \cdot \psi$, where ρ is a representation of G on $\mathbb{C}^k$ and is characteristic of the particular class of particles under consideration. Regarding the local wavefunctions as cross-sections of $P \times_\rho \mathbb{C}^k$ $(x \to [s(x), \psi(x)]$ and $x \to [s(x) \cdot g(x), g(x)^{-1} \cdot \psi(x)] = [s(x), \psi(x)])$ one finds that they agree on the intersections of their domains and so piece together into a single global cross-section of $P \times_\rho \mathbb{C}^k$. This, in turn, determines an equivariant $\mathbb{C}^k$-valued map ϕ on all of P.

If $G \to P \xrightarrow{\mathcal{P}} X$ is a principal bundle, k is a positive integer and ρ is a representation of G on $\mathbb{C}^k$ (with corresponding left action $g \cdot \xi = (\rho(g))\xi$), then an equivariant $\mathbb{C}^k$-valued map ϕ on P is called a **matter field**. If $k = 1$, ϕ is referred to as a **complex scalar field**, while, if $k = 2$, it will be called a **2-component wavefunction**.

Remark: There are circumstances (which we will not encounter) in which a more general definition of "matter field" is appropriate. For example, a cross-section of the adjoint bundle $ad\,P$ (or, equivalently, a $\mathcal{G}$-valued map on P that is equivariant under the adjoint representation) is referred to in the physics literature as a **Higgs field** (see Section 7.3 and Chapter 10 of [**MM**]).

A matter field is, in particular, a $\mathbb{C}^k$-valued 0-form on P and so has a (componentwise) exterior derivative $d\phi$ (see Exercise 4.8.21 and identify $\mathbb{C}^k$ with $\mathbb{R}^{2k}$ as in Section 1.1). Assuming now that the principal bundle $\mathcal{P} : P \to X$ has defined on it a connection 1-form ω, we define the **covariant exterior derivative** of ϕ, denoted $d^\omega \phi$, by having $d\phi$ operate only on horizontal parts (cf., the definition of the curvature Ω, of ω, in Section 5.2). More precisely, we define the $\mathbb{C}^k$-valued 1-form $d^\omega \phi$ as follows: For each $p \in P$ and all $\mathbf{v} \in T_p(P)$ we let

$$(d^\omega \phi)_p(\mathbf{v}) = (d\phi)_p(\mathbf{v}^H).$$

Exercise 5.8.5 Show that $d^\omega \phi$, is a smooth $\mathbb{C}^k$-valued 1-form on P.

As usual, we denote by $\sigma : P \times G \to P$, $\sigma(p,g) = p \cdot g$, the right action of G on P and, for each fixed $g \in G$, define $\sigma_g : P \to P$ by $\sigma_g(p) = p \cdot g$ for each p in P. We claim that

$$\sigma_g^*\left(d^{\omega}\phi\right) = g^{-1} \cdot d\phi \tag{5.8.1}$$

(cf., Lemma 5.2.2). In more detail, what is being asserted here is that, for each $p \in P$ and each $\mathbf{v} \in T_{p\cdot g^{-1}}(P)$,

$$\left(d^{\omega}\phi\right)_p\left((\sigma_g)_{*p\cdot g^{-1}}(\mathbf{v})\right) = g^{-1} \cdot (d\phi)_{p\cdot g^{-1}}(\mathbf{v}),$$

i.e.,

$$(d\phi)_p\left(\left((\sigma_g)_{*p\cdot g^{-1}}(\mathbf{v})\right)^H\right) = g^{-1} \cdot (d\phi)_{p\cdot g^{-1}}(\mathbf{v}). \tag{5.8.2}$$

To prove this we first observe that

$$\left((\sigma_g)_{*p\cdot g^{-1}}(\mathbf{v})\right)^H = (\sigma_g)_{*p\cdot g^{-1}}(\mathbf{v}^H). \tag{5.8.3}$$

Indeed,

$$\begin{aligned}(\sigma_g)_{*p\cdot g^{-1}}(\mathbf{v}) &= (\sigma_g)_{*p\cdot g^{-1}}(\mathbf{v}^H + \mathbf{v}^V)\\ &= (\sigma_g)_{*p\cdot g^{-1}}(\mathbf{v}^H) + (\sigma_g)_{*p\cdot g^{-1}}(\mathbf{v}^V)\end{aligned}$$

and, by (5.1.3), $(\sigma_g)_{*p\cdot g^{-1}}(\mathrm{Hor}_{p\cdot g^{-1}}(P)) = \mathrm{Hor}_p(P)$ so $(\sigma_g)_{*p\cdot g^{-1}}(\mathbf{v}^H)$ is horizontal. Furthermore, since $\mathbf{v}^V$ is vertical, Corollary 4.7.9 implies that $\mathbf{v}^V = A^{\#}(p \cdot g^{-1})$ for some $A \in \mathcal{G}$ and it follows from (4.7.8) that $(\sigma_g)_{*p\cdot g^{-1}}(\mathbf{v}^V)$ is vertical. Thus, (5.8.3) follows from (5.1.2). Now we let α be a smooth curve in P with $\alpha'(0) = \mathbf{v}^H$ and compute

$$\begin{aligned}(d\phi)_p\left(\left((\sigma_g)_{*p\cdot g^{-1}}(\mathbf{v})\right)^H\right) &= (d\phi)_p\left((\sigma_g)_{*p\cdot g^{-1}}(\mathbf{v}^H)\right)\\ &= \left((\sigma_g)_{*p\cdot g^{-1}}(\mathbf{v}^H)\right)(\phi)\\ &= \left((\sigma_g)_{*p\cdot g^{-1}}(\alpha'(0))\right)(\phi)\\ &= (\sigma_g \circ \alpha)'(0)(\phi)\\ &= D_1(\phi \circ \sigma_g \circ \alpha)(0).\end{aligned}$$

But

$$\begin{aligned}(\phi \circ \sigma_g \circ \alpha)(t) &= \phi(\sigma_g(\alpha(t))) = \phi(\alpha(t) \cdot g)\\ &= g^{-1} \cdot \phi(\alpha(t))\\ &= g^{-1} \cdot (\phi \circ \alpha)(t)\end{aligned}$$

so

$$(d\phi)_p\left(\left(\left(\sigma_g\right)_{*p\cdot g^{-1}}(\mathbf{v})\right)^H\right) = g^{-1}\cdot(\phi\circ\alpha)'(0)$$
$$= g^{-1}\cdot(\alpha'(0)(\phi))$$
$$= g^{-1}\cdot\left(\mathbf{v}^H(\phi)\right)$$
$$= g^{-1}\cdot(d\phi)_{p\cdot g^{-1}}\left(\mathbf{v}^H\right).$$

This completes the proof of (5.8.2) which, in turn, proves (5.8.1).

In order to derive a computationally efficient formula for $d^{\boldsymbol{\omega}}\phi$ (analogous to the Cartan Structure Equation for the curvature $\boldsymbol{\Omega}$ of $\boldsymbol{\omega}$) we must first introduce an auxiliary notion. We let $\rho: G \to GL(k, \mathbb{C})$ be a smooth homomorphism with corresponding left action of G on $\mathbb{C}^k$ defined by $(g,\xi)\in G\times\mathbb{C}^k \to g\cdot\xi = \rho(g)\,\xi\in\mathbb{C}^k$. Now, for each A in the Lie algebra $\mathcal{G}$ of G and each $\xi\in\mathbb{C}^k$ we define $A\cdot\xi$ in $\mathbb{C}^k$ by

$$A\cdot\xi = \frac{d}{dt}\left(\exp(tA)\cdot\xi\right)\Big|_{t=0} = \frac{d}{dt}\left(\rho(\exp(tA))\,\xi\right)\Big|_{t=0}.$$

Thus, if $\beta(t) = \exp(tA)\cdot\xi = \rho(\exp(tA))\,\xi$ for sufficiently small $|t|$, then $A\cdot\xi = \beta'(0)$ (identified with an element of $\mathbb{C}^k$).

Exercise 5.8.6 Show that, if G is a matrix Lie group in $GL(k,\mathbb{C})$ and ρ is the inclusion map, then $A\cdot\xi = A\,\xi$ (matrix product) for each $A\in\mathcal{G}$.

Now, if ϕ is a $\mathbb{C}^k$-valued 0-form and $\boldsymbol{\omega}$ is a $\mathcal{G}$-valued 1-form we may define a $\mathbb{C}^k$-valued 1-form $\boldsymbol{\omega}\cdot\phi$ by

$$(\boldsymbol{\omega}\cdot\phi)_p(\mathbf{v}) = \boldsymbol{\omega}_p(\mathbf{v})\cdot\phi(p).$$

We claim now that

$$d^{\boldsymbol{\omega}}\phi = d\phi + \boldsymbol{\omega}\cdot\phi. \tag{5.8.4}$$

To prove this we must show that, for all $\mathbf{v}\in T_p(P)$,

$$(d\phi)_p\left(\mathbf{v}^H\right) = (d\phi)_p(\mathbf{v}) + \boldsymbol{\omega}_p(\mathbf{v})\cdot\phi(p). \tag{5.8.5}$$

Since both sides are linear in $\mathbf{v}$ and since $T_p(P) = \mathrm{Hor}_p(P)\oplus\mathrm{Vert}_p(P)$ it will suffice to consider separately the cases in which $\mathbf{v}$ is horizonal and $\mathbf{v}$ is vertical. Now, if $\mathbf{v}$ is horizontal, then $\mathbf{v}=\mathbf{v}^H$ and $\boldsymbol{\omega}_p(\mathbf{v}) = 0$ so (5.8.5) is clearly satisfied. Suppose then that $\mathbf{v}$ is vertical. Then $\mathbf{v}^H = 0$ so the left-hand side of (5.8.5) is zero. Furthermore, by Corollary 4.7.9, there is a unique element A of $\mathcal{G}$ with $\mathbf{v} = A^{\#}(p)$. Thus, $(d\phi)_p(\mathbf{v}) = (d\phi)_p(A^{\#}(p)) = A^{\#}(\phi)(p)$ and $\boldsymbol{\omega}_p(\mathbf{v}) = \boldsymbol{\omega}_p(A^{\#}(p)) = A$ so we must prove that

$$A^{\#}(\phi)(p) = -A \cdot \phi(p).$$

To compute $A^{\#}(\phi)(p)$ we let $\beta(t) = p \cdot \exp(tA)$. Then

$$A^{\#}(\phi)(p) = A^{\#}(p)(\phi) = \beta'(0)(\phi) = (\phi \circ \beta)'(0).$$

But $(\phi \circ \beta)(t) = \phi(\beta(t)) = \phi(p \cdot \exp(tA)) = \exp(-tA) \cdot \phi(p)$ so $(\phi \circ \beta)'(0) = -A \cdot \phi(p)$ and this completes the proof.

In order to write out the covariant exterior derivative $d^{\boldsymbol{\omega}}\phi$ of a matter field in coordinates one must first choose a gauge, i.e., local section/trivialization of the bundle $\mathcal{P} : P \to X$. We will assume (by shrinking, if necessary) that the domain V of our section $s : V \to \mathcal{P}^{-1}(V)$ is also a coordinate nbd for X. Observe that

$$\begin{aligned} s^*\left(d^{\boldsymbol{\omega}}\phi\right) &= s^*(d\phi + \boldsymbol{\omega} \cdot \phi) = s^*(d\phi) + s^*(\boldsymbol{\omega} \cdot \phi) \\ &= d(\phi \circ s) + (s^*\boldsymbol{\omega}) \cdot (\phi \circ s). \end{aligned}$$

Writing $\boldsymbol{\mathcal{A}}$ for the gauge potential $s^*\boldsymbol{\omega}$ on V we therefore have

$$s^*\left(d^{\boldsymbol{\omega}}\phi\right) = d(\phi \circ s) + \boldsymbol{\mathcal{A}} \cdot (\phi \circ s). \tag{5.8.6}$$

Exercise 5.8.7 Let $\varphi : V \to \mathbb{R}^n$ be a chart on V with coordinate functions $x^i = \mathcal{P}^i \circ \varphi$, $i = 1, \ldots, n$. Show that

$$(s \circ \varphi^{-1})^*(d^{\boldsymbol{\omega}}\phi) = d(\phi \circ (s \circ \varphi^{-1})) + ((s \circ \varphi^{-1})^*\boldsymbol{\omega}) \cdot (\phi \circ (s \circ \varphi^{-1})).$$

Let us write out in detail the two special cases of most interest to us. First we suppose that $\mathcal{P} : P \to X$ is a principal $U(1)$-bundle over X and $\rho : U(1) \hookrightarrow GL(1, \mathbb{C})$ is the inclusion map. Identifying 1×1 matrices with their sole entries the corresponding left action of $U(1)$ on $\mathbb{C}$ is just complex multiplication: $g \cdot \xi = g\xi$ for each $g \in U(1)$ and each $\xi \in \mathbb{C}$. The Lie algebra $\mathcal{U}(1)$ of $U(1)$ is identified with the set $\mathbf{i}\mathbb{R}$ of pure imaginary numbers (Exercise 4.7.12). Thus, for every $\mathbf{i}A \in \mathcal{U}(1)$ and every $\xi \in \mathbb{C}$, Exercise 5.8.6 gives $(\mathbf{i}A) \cdot \xi = \mathbf{i}A\xi$. Now let ϕ be a complex scalar field, i.e., an equivariant $\mathbb{C}$-valued map on P. As above, we let $s : V \to \mathcal{P}^{-1}(V)$ be a local cross-section and assume that V is also a coordinate nbd for X with $\varphi : V \to \mathbb{R}^n$ a chart and $x^1, \ldots, x^n$ its coordinate functions. In order to avoid notational excesses we will write $\phi(x^1, \ldots, x^n)$ for the coordinate expression

$\phi \circ (s \circ \varphi^{-1})(x^1, \ldots, x^n)$. The gauge potential $(s \circ \varphi^{-1})^* \boldsymbol{\omega}$ will be written $\boldsymbol{\mathcal{A}}(x^1, \ldots, x^n) = \mathcal{A}_\alpha(x^1, \ldots, x^n)\, dx^\alpha$ in local coordinates. Here each $\mathcal{A}_\alpha(x^1, \ldots, x^n)$ is $\mathcal{U}(1)$-valued and so may be written $\mathcal{A}_\alpha(x^1, \ldots, x^n) = -\mathbf{i}\, A_\alpha(x^1, \ldots, x^n)$ for real-valued functions A_α (the minus sign is traditional). Thus, $\boldsymbol{\mathcal{A}}(x^1, \ldots, x^n) = -\mathbf{i}\, A_\alpha(x^1, \ldots, x^n)\, dx^\alpha$ and so the corresponding coordinate expression for $d^{\boldsymbol{\omega}} \phi$ is

$$\left(\frac{\partial \phi}{\partial x^{\,\alpha}} - \mathbf{i}\, A_\alpha \phi \right) dx^\alpha . \tag{5.8.7}$$

Exercise 5.8.8 Let q be an integer and define $\rho_q : U(1) \to GL(1, \mathbb{C})$ by $\rho_q(g) = g^q$. Show that ρ_q is a smooth homomorphism and carry out a discussion analogous to that above (for $q = 1$) to obtain the following local coordinate expression for the covariant exterior derivative of a matter field ϕ (of course, "equivariance" now means with respect to the action of $U(1)$ on $\mathbb{C}$ determined by ρ_q):

$$\left(\frac{\partial \phi}{\partial x^{\,\alpha}} - \mathbf{i}\, q\, A_\alpha \phi \right) dx^\alpha .$$

In the physics literature one often sees the basis 1-forms dx^α omitted, the components written as

$$\left(\frac{\partial}{\partial x^{\,\alpha}} - \mathbf{i}\, q\, A_\alpha \right) \phi \tag{5.8.8}$$

and the "operator" $\partial / \partial x^\alpha - \mathbf{i}\, q\, A_\alpha$ referred to as the (α^{th} component of the) "covariant derivative". In this context, X would generally denote spacetime (so that $n = 4$), A_α would be identified with the components of the potential function for the electromagnetic field described by the connection $\boldsymbol{\omega}$ (three components for the vector potential and one for the scalar potential) and ϕ would be the wavefunction for a particle of charge q (more precisely, q times the basic charge of the electron). Notice that this covariant derivative automatically "couples" the gauge field and the matter field since both are involved in its definition. The so-called **coupling constant** q in (5.8.8) is, however, often absorbed into the A_α and this bit of cosmetic surgery gives all covariant derivatives the appearence of those in (5.8.7).

As a final example we consider a principal $SU(2)$-bundle $\mathcal{P} : P \to X$ and let $\rho : SU(2) \hookrightarrow GL(2, \mathbb{C})$ be the inclusion map. The corresponding left action of $SU(2)$ on $\mathbb{C}^2$ is then just the matrix product of $g \in SU(2)$ with a column vector in $\mathbb{C}^2$. The Lie algebra $\mathcal{SU}(2)$ of $SU(2)$ is the set of all 2×2 complex matrices that are skew-Hermitian and tracefree (Exercise

4.7.12). For every such matrix $A \in \mathcal{SU}(2)$ and every (column vector) $\xi \in \mathbb{C}^2$, Exercise 5.8.6 implies that $A \cdot \xi$ is, again, just the matrix product $A\,\xi$. Now let $\phi = \begin{pmatrix} \phi^1 \\ \phi^2 \end{pmatrix}$ be an equivariant $\mathbb{C}^2$-valued map on P. In local coordinates we again write $\phi\,(x^1, \ldots, x^n)$ for $\phi \circ (s \circ \varphi^{-1})\,(x^1, \ldots, x^n)$. The exterior derivative $d\phi$ is componentwise (i.e., entrywise) so we may write

$$d\,\phi = \begin{pmatrix} d\phi^1 \\ d\phi^2 \end{pmatrix} = \begin{pmatrix} \dfrac{\partial \phi^1}{\partial x^{\,\alpha}}\, dx^\alpha \\ \dfrac{\partial \phi^2}{\partial x^{\,\alpha}}\, dx^\alpha \end{pmatrix}$$

$$= \begin{pmatrix} \dfrac{\partial \phi^1}{\partial x^{\,\alpha}} \\ \dfrac{\partial \phi^2}{\partial x^{\,\alpha}} \end{pmatrix} dx^\alpha = \frac{\partial}{\partial x^{\,\alpha}} \begin{pmatrix} \phi^1 \\ \phi^2 \end{pmatrix} dx^\alpha .$$

As usual, the gauge potential $(s \circ \varphi^{-1})^* \boldsymbol{\omega}$ is written as $\boldsymbol{\mathcal{A}}(x^1, \ldots, x^n) = \mathcal{A}_\alpha\,(x^1, \ldots, x^n)\,dx^\alpha$, where each $\mathcal{A}_\alpha$ is $\mathcal{SU}(2)$-valued. Since $\mathcal{A}_\alpha \cdot \phi = \mathcal{A}_\alpha\,\phi$ the local expression for $d^{\boldsymbol{\omega}}\,\phi$ is

$$\left(\frac{\partial}{\partial x^{\,\alpha}} \begin{pmatrix} \phi^1 \\ \phi^2 \end{pmatrix} + \mathcal{A}_\alpha \begin{pmatrix} \phi^1 \\ \phi^2 \end{pmatrix} \right) dx^\alpha .$$

As before, the physics community would tend to omit the dx^α, write the components as

$$\left(\frac{\partial}{\partial x^{\,\alpha}} + \mathcal{A}_\alpha \right) \phi$$

and call the operator $\partial / \partial x^\alpha + \mathcal{A}_\alpha$ the (α^{th} component of the) "covariant derivative".

Exercise 5.8.9 Show that a complex 2×2 matrix $\mathcal{A}$ is skew-Hermitian and tracefree iff $\mathcal{A} = -\mathbf{i}\,\mathcal{B}$ for some unique Hermitian and tracefree matrix $\mathcal{B}$.

Using Exercise 5.8.9 to write each $\mathcal{A}_\alpha$ as $-\mathbf{i}\,\mathcal{B}_\alpha$ for some Hermitian, tracefree matrix $\mathcal{B}_\alpha$, the components of $d^{\boldsymbol{\omega}}\,\phi$ become

$$\left(\frac{\partial}{\partial x^{\,\alpha}} - \mathbf{i}\,\mathcal{B}_\alpha \right) \phi$$

and it is this form that is most frequently encounted in physics (see, e.g.,

[YM]).

For detailed calculations the matrix expressions for $d^{\omega}\phi$ derived above are rather cumbersome and one would prefer expressions that involve the components of ϕ and the entries in the matrices $\mathcal{A}_\alpha$. This is accomplished by choosing a basis for the Lie algebra $\mathcal{SU}(2)$. We describe next that particular choice most favored by the physicists. Begin with the so-called **Pauli spin matrices:**

$$\sigma_1 = \begin{pmatrix} 0 & 1 \\ 1 & 0 \end{pmatrix}, \ \sigma_2 = \begin{pmatrix} 0 & -\mathbf{i} \\ \mathbf{i} & 0 \end{pmatrix}, \ \sigma_3 = \begin{pmatrix} 1 & 0 \\ 0 & -1 \end{pmatrix}.$$

Observe that each of these is Hermitian and tracefree. It follows from Exercise 5.8.9 that the matrices

$$T_1 = -\frac{1}{2}\mathbf{i}\,\sigma_1 \ , \ T_2 = -\frac{1}{2}\mathbf{i}\,\sigma_2 \ , \ T_3 = -\frac{1}{2}\mathbf{i}\,\sigma_3 \ , \tag{5.8.9}$$

are skew-Hermitian and tracefree (the reason for the peculiar "$\frac{1}{2}$" will emerge shortly). In fact, T_1, T_2 and T_3 are just $-\frac{1}{2}$ times the basic quaternions **k**, **j** and **i**, respectively, in the matrix model for $\mathbb{H}$ described in Section 1.1. In particular, they are linearly independent and so form a basis for $\mathcal{SU}(2)$.

Exercise 5.8.10 Show that, relative to the Killing inner product on $\mathcal{SU}(2)$ (Exercise 4.7.17), $\{T_1, T_2, T_3\}$ is an *orthonormal* basis. (This is the reason for the "$\frac{1}{2}$" in (5.8.9).)

Now, each $\mathcal{A}_\alpha$ in the covariant derivative $\partial/\partial x^\alpha + \mathcal{A}_\alpha$ takes values in $\mathcal{SU}(2)$ and so can be written as a linear combination of T_1, T_2 and T_3, i.e.,

$$\mathcal{A}_\alpha = A^k{}_\alpha T_k \, ,$$

where the $A^k{}_\alpha$, $k = 1,2,3$, $\alpha = 1,\ldots,n$, are real-valued functions of $x^1,\ldots,x^n$. Next let us denote the (constant) entries in T_k by $T_k{}^i{}_j$, $i,j = 1,2$. Then the ith component of

$$\mathcal{A}_\alpha \phi = A^k{}_\alpha T_k \begin{pmatrix} \phi^1 \\ \phi^2 \end{pmatrix}$$

is

$$(\mathcal{A}_\alpha \phi)^i = A^k{}_\alpha T_k{}^i{}_j \phi^j$$

and so the ith component of $d^{\omega}\phi$ is

$$\frac{\partial \phi^i}{\partial x^\alpha} + A^k{}_\alpha T_k{}^i{}_j \phi^j \ , \quad i = 1,2\,.$$

Covariant derivatives such as those we have been discussing arose first in physics, not as special cases of our general, geometrically rooted notion of a covariant exterior derivative, but rather as *ad hoc* devices for ensuring the gauge invariance of various field equations. The rationale was to proceed in strict analogy with the case of electromagnetism. Here a change in gauge (section/trivialization) was identified with the resulting change in the potential functions ($A \longrightarrow A + d\Omega$, or, in coordinates, $A_\alpha \longrightarrow A_\alpha + \frac{\partial}{\partial x^\alpha}(\Omega)$). The wavefunction ϕ thereby experiences a phase shift ($\phi \longrightarrow e^{\mathbf{i}\,q\,\Omega}\phi$) and one observes that

$$\left(\frac{\partial}{\partial x^\alpha} - \mathbf{i}\,q\,A_\alpha\right)\phi \longrightarrow \left(\frac{\partial}{\partial x^\alpha} - \mathbf{i}\,q\left(A_\alpha + \frac{\partial}{\partial x^\alpha}(\Omega)\right)\right)\left(e^{\mathbf{i}\,q\,\Omega}\phi\right)$$

$$= \frac{\partial}{\partial x^\alpha}\left(e^{\mathbf{i}\,q\,\Omega}\phi\right) - \mathbf{i}\,q\,A_\alpha\left(e^{\mathbf{i}\,q\,\Omega}\phi\right) - \mathbf{i}\,q\,\frac{\partial}{\partial x^\alpha}(\Omega)\left(e^{\mathbf{i}\,q\,\Omega}\phi\right)$$

$$= e^{\mathbf{i}\,q\,\Omega}\frac{\partial}{\partial x^\alpha}(\phi) + \mathbf{i}\,q\,\frac{\partial}{\partial x^\alpha}(\Omega)\,e^{\mathbf{i}\,q\,\Omega}\phi - \mathbf{i}\,q\,A_\alpha\left(e^{\mathbf{i}\,q\,\Omega}\phi\right) - \mathbf{i}\,q\,\frac{\partial}{\partial x^\alpha}(\Omega)\left(e^{\mathbf{i}\,q\,\Omega}\phi\right)$$

$$= e^{\mathbf{i}\,q\,\Omega}\left(\frac{\partial}{\partial x^\alpha} - \mathbf{i}\,q\,A_\alpha\right)\phi.$$

Thus, the wavefunction ϕ and the covariant derivative $\left(\frac{\partial}{\partial x^\alpha} - \mathbf{i}\,q\,A_\alpha\right)\phi$ (but *not* the ordinary derivative $\frac{\partial}{\partial x^\alpha}(\phi)$) transform *in the same way* under a gauge transformation. As a result, field equations involving covariant rather than ordinary derivatives have some chance of being gauge invariant, i.e., having the property that, if the pair A_α, ϕ is a solution, then so is the pair $A_\alpha + \frac{\partial}{\partial x^\alpha}(\Omega), e^{\mathbf{i}\,q\,\Omega}\phi$. In the same way, one reads in [**YM**] that

> " ... we require, in analogy with the electromagnetic case, that all derivatives of ψ appear in the following combination:
>
> $$(\partial_\mu - i\,\epsilon\,B_\mu)\,\psi. \quad "$$

The view that the gauge field and the matter field responding to it should be coupled *only* through the use of covariant derivatives in the field equations is called the principle of **minimal coupling** (or **minimal replacement**).

Having thus decided that only covariant derivatives of the wavefunction should appear in the field equations that will describe the interaction, one is still faced with the problem of selecting the "right" equations (i.e., those which yield predictions in accord with experiment). This, of course, is a problem in physics, not mathematics, but the search for such equations is almost always conducted in the same way. On the basis of one's physical intuition and experience one conjures up a likely candidate for what is called the "Lagrangian" of the system under consideration. An appeal to the so-called "Principle of Least Action" and a basic result from the Calculus of Variations (the "Euler-Lagrange Equations") serves up a set of differential equations which, if the gods are favorably disposed, have solutions whose behavior mimics what actually goes on in the world. All of this is quite another story, however, and not one of topology and geometry, but rather of physics and analysis. To those interested in pursuing these matters further we heartily recommend [**Bl**].

Appendix

$SU(2)$ and $SO(3)$

The special unitary group $SU(2)$ and its *alter ego* $Sp(1)$ have figured prominently in our story and we have, at several points (e.g., Sections 0.5 and 5.3), intimated that the reasons lie in physics. To understand something of this one must be made aware of a remarkable relationship between $SU(2)$ and the rotation group $SO(3)$. It is this service that we hope to perform for our reader here.

$SO(3)$ is, of course, the group of 3×3 real matrices A satisfying $A^{-1} = A^T$ and $\det A = 1$. Its elements are just the matrices, relative to orthonormal bases for $\mathbb{R}^3$, of the linear transformations that rotate $\mathbb{R}^3$. We will require somewhat more precise information, however. Thus, we let ϕ denote a real number and $\vec{n} = < n^1, n^2, n^3 >$ a vector of length 1 in $\mathbb{R}^3$ (so that $\| \vec{n} \|^2 = (n^1)^2 + (n^2)^2 + (n^3)^2 = 1$). Define a 3×3 matrix $R(\phi, \vec{n})$ by

$$R(\phi, \vec{n}) = id + (\sin\phi)\, N + (1 - \cos\phi)\, N^2 , \tag{A.1}$$

where id is the 3×3 identity matrix and

$$N = \begin{pmatrix} 0 & -n^3 & n^2 \\ n^3 & 0 & -n^1 \\ -n^2 & n^1 & 0 \end{pmatrix} . \tag{A.2}$$

This matrix $R(\phi, \vec{n})$ arises from geometrical considerations. It is, in fact, the matrix, relative to the standard basis for $\mathbb{R}^3$, of the rotation through angle ϕ about an axis along $\vec{n}$ in the sense determined by the right-hand rule from the direction of $\vec{n}$. There is another useful way of looking at $R(\phi, \vec{n})$, however. Note that N is skew-symmetric and therefore it, as well as any ϕN, lies in the Lie algebra $\mathcal{SO}(3)$ of $SO(3)$.

Exercise A.1 Show that

$$N^2 = \begin{pmatrix} -((n^2)^2 + (n^3)^2) & n^1 n^2 & n^1 n^3 \\ n^1 n^2 & -((n^1)^2 + (n^3)^2) & n^2 n^3 \\ n^1 n^3 & n^2 n^3 & -((n^1)^2 + (n^2)^2) \end{pmatrix} \tag{A.3}$$

and

$$N^3 = -N \ , \quad N^4 = -N^2 \ , \quad N^5 = N \ , \ \ldots \ . \tag{A.4}$$

Exercise A.2 Use Exercise A.1 to manipulate the series expansion for $\exp(\phi N)$ and show that

$$\exp(\phi N) = e^{\phi N} = R(\phi, \vec{n}) .$$

Use this to show that $R(\phi, \vec{n}) \in O(3)$.

It follows from $R(\phi, \vec{n}) \in O(3)$ that $\det R(\phi, \vec{n}) = \pm 1$ (Section 1.1). But, for any fixed $\vec{n}$, $\det R(\phi, \vec{n})$ is clearly a continuous function of ϕ. Since $\mathbb{R}$ is connected, the image of this function is either $\{1\}$ or $\{-1\}$. But $\det R(0, \vec{n}) = \det(id) = 1$ so we must have $\det R(\phi, \vec{n}) = 1$ for all ϕ and $\vec{n}$. Thus, $R(\phi, \vec{n}) \in SO(3)$. We propose to show that *every* element of $SO(3)$ is an $R(\phi, \vec{n})$ for some ϕ and some $\vec{n}$.

Lemma A.1 *For any* $\phi \in \mathbb{R}$ *and any* $\vec{n} = \langle n^1, n^2, n^3 \rangle$ *with* $\| \vec{n} \|^2 = (n^1)^2 + (n^2)^2 + (n^3)^2 = 1$, $R(\phi, \vec{n})$ *is in* $SO(3)$. *Conversely, given an* $R = (R_{ij})_{i,j=1,2,3}$ *in* $SO(3)$ *there exists a unique real number* ϕ *satisfying* $0 \le \phi \le \pi$ *and a unit vector* $\vec{n} = \langle n^1, n^2, n^3 \rangle$ *in* $\mathbb{R}^3$ *such that* $R = R(\phi, \vec{n})$. *Moreover,*

(a) *If* $0 < \phi < \pi$, *then* $\vec{n}$ *is unique.*

(b) *If* $\phi = 0$, *then* $R = R(0, \vec{n})$ *for any unit vector* $\vec{n}$ *in* $\mathbb{R}^3$.

(c) *If* $\phi = \pi$, *then* $R = R(\pi, \vec{m})$ *iff* $\vec{m} = \pm \vec{n}$.

Proof: The first assertion has already been proved. For the converse, suppose $R = (R_{ij})_{i,j=1,2,3}$ is in $SO(3)$. We intend to just set $R = R(\phi, \vec{n})$ and attempt to solve for ϕ, n^1, n^2 and n^3 with $0 \le \phi \le \pi$ and $(n^1)^2 + (n^2)^2 + (n^3)^2 = 1$. Taking the trace of both sides of (A.1) and noting that $\operatorname{trace}(id) = 3$, $\operatorname{trace}(N) = 0$, $\operatorname{trace}(N^2) = -2((n^1)^2 + (n^2)^2 + (n^3)^2) = -2$

and $-1 \leq \text{trace}(R) \leq 3$ since $R \in SO(3)$ gives

$$\cos\phi = \tfrac{1}{2}\left(\text{trace}(R) - 1\right). \tag{A.5}$$

This determines a unique ϕ in $0 \leq \phi \leq \pi$, namely, $\cos^{-1}\left(\frac{1}{2}(\text{trace}(R) - 1)\right)$. Next observe that, since id and N^2 are both symmetric, we must have, for any $j, k = 1, 2, 3$,

$$R_{jk} - R_{kj} = 0 + (\sin\phi)(N_{jk} - N_{kj}) + (1 - \cos\phi)\cdot 0 = (\sin\phi)(N_{jk} - N_{kj}).$$

This provides no information if $j = k$, but, for example, when $j = 1$ and $k = 2$, we obtain $R_{12} - R_{21} = (\sin\phi)(N_{12} - N_{21}) = (\sin\phi)(-2n^3)$ so that

$$(\sin\phi)\, n^3 = -\tfrac{1}{2}\left(R_{12} - R_{21}\right). \tag{A.6}$$

Similarly,

$$(\sin\phi)\, n^2 = -\tfrac{1}{2}\left(R_{13} - R_{31}\right), \tag{A.7}$$

and

$$(\sin\phi)\, n^1 = -\tfrac{1}{2}\left(R_{23} - R_{32}\right). \tag{A.8}$$

Exercise A.3 Show that, if $\sin\phi \neq 0$, (A.6), (A.7) and (A.8) can be solved uniquely for n^1, n^2 and n^3 satisfying $(n^1)^2 + (n^2)^2 + (n^3)^2 = 1$.

Since $0 \leq \phi \leq \pi$, $\sin\phi$ is zero only when $\phi = 0$ or $\phi = \pi$ and so Exercise A.3 completes the proof of part (a) of Lemma A.1. If $\phi = 0$, then (A.5) gives $\text{trace}(R) = 3$ so $R = id$. Consequently, any unit vector $\vec{n}$ gives $R = id = id + (\sin 0)\, N + (1 - \cos 0)\, N^2$ and we have proved part (b). Finally, suppose $\phi = \pi$. Then we are attempting to solve $R = id + 2N^2$ for n^1, n^2 and n^3 with $(n^1)^2 + (n^2)^2 + (n^3)^2 = 1$. Note that

$$\begin{aligned}\frac{1}{2}\left(1 + (R(\pi, \vec{n}))_{11}\right) &= \frac{1}{2}\left(2 - 2\left((n^2)^2 + (n^3)^2\right)\right)\\ &= 1 - (n^2)^2 + (n^3)^2 = (n^1)^2\end{aligned}$$

so $R_{11} = (R(\pi, \vec{n}))_{11}$ requires that $(n^1)^2 = \frac{1}{2}(1 + R_{11})$. Similarly,

$$(n^i)^2 = \tfrac{1}{2}\left(1 + R_{ii}\right), \quad i = 1, 2, 3. \tag{A.9}$$

Now, at least one of the $\frac{1}{2}(1+R_{11})$ must be nonzero for otherwise trace $(R) = -3$ and this contradicts (A.5) with $\phi = \pi$. Assume, without loss of generality, that $(n^1)^2 = \frac{1}{2}(1+R_{11}) \neq 0$. Then there are two possibilities for n^1:

$$n^1 = \pm\sqrt{\frac{1}{2}(1+R_{11})} \ .$$

For each choice the equations $R_{12} = 2n^1 n^2$ and $R_{13} = 2n^1 n^3$ determine n^2 and n^3. Consequently, if $\phi = \pi$ there are two solutions for $\vec{n}$ which differ by a sign and this is part (c) of the Lemma. ■

Although we will have another proof shortly the reader may wish to use Lemma A.1 to show directly that $SO(3)$ is homeomorphic to real projective 3-space $\mathbb{RP}^3$.

Exercise A.4 Let $\overline{U_\pi(0)}$ be the closed ball of radius π about the origin in $\mathbb{R}^3$. Any $p \in \overline{U_\pi(0)}$ other than the origin uniquely determines a unit vector $\vec{n}$ in $\mathbb{R}^3$ and a ϕ in $0 < \phi \leq \pi$ such that $p = \phi\vec{n}$. If p is the origin, then $p = 0 \cdot \vec{n}$ for any unit vector $\vec{n}$. Define $f : \overline{U_\pi(0)} \to SO(3)$ by $f(p) = R(\phi, \vec{n})$ if p is not the origin and $f(0) = id$. Identify $\mathbb{RP}^3$ with the quotient of $\overline{U_\pi(0)}$ obtained by identifying antipodal points on the boundary 2-sphere and show that f determines a map $\bar{f} : \mathbb{RP}^3 \to SO(3)$ that is a homeomorphism. **Hint:** Lemma 1.2.2 and Theorem 1.4.4.

Now, $\mathbb{RP}^3$ is also the quotient of S^3 obtained by identifying antipodal points (Section 1.2). Moreover, S^3 is homeomorphic to $SU(2)$ (Theorem 1.4). Our next objective is to construct a model $SU(2) \to SO(3)$ of the quotient map $S^3 \to \mathbb{RP}^3$ that is, in addition, a group homomorphism. This construction is most efficiently carried out by first building a "matrix model" $\mathcal{R}^3$ of $\mathbb{R}^3$ analogous to the model $\mathcal{R}^4$ of $\mathbb{R}^4$ constructed in Section 1.1. Thus, we consider the collection $\mathcal{R}^3$ of all 2×2 complex matrices $X = \begin{pmatrix} \alpha & \beta \\ \gamma & \delta \end{pmatrix}$ that are Hermitian $(\bar{X}^T = X)$ and tracefree $(\text{trace}(X) = 0)$.

Exercise A.5 Show that every X in $\mathcal{R}^3$ can be written in the form

$$X = \begin{pmatrix} x^3 & x^1 - x^2\mathbf{i} \\ x^1 + x^2\mathbf{i} & -x^3 \end{pmatrix} = x^1\sigma_1 + x^2\sigma_2 + x^3\sigma_3 \ , \qquad \text{(A.10)}$$

where x^1, x^2 and x^3 are real numbers and $\sigma_1 = \begin{pmatrix} 0 & 1 \\ 1 & 0 \end{pmatrix}$, $\sigma_2 = \begin{pmatrix} 0 & -\mathbf{i} \\ \mathbf{i} & 0 \end{pmatrix}$

and $\sigma_3 = \begin{pmatrix} 1 & 0 \\ 0 & -1 \end{pmatrix}$ are the Pauli spin matrices.

Exercise A.6 Show that the Pauli spin matrices satisfy the following **commutation relations**:

$$\sigma_j^2 = \sigma_j \sigma_j = \begin{pmatrix} 1 & 0 \\ 0 & 1 \end{pmatrix}, \quad j = 1, 2, 3,$$

$$\sigma_j \sigma_k = -\sigma_k \sigma_j = \mathbf{i} \sigma_i \quad , \text{ where } ijk \text{ is an even permutation of } 123 .$$

$\mathcal{R}^3$ is clearly closed under sums and real scalar multiples and so can be regarded as a real vector space. From Exercise A.5 it follows that $\{\sigma_1, \sigma_2, \sigma_3\}$ is a basis for this vector space so $\dim \mathcal{R}^3 = 3$ and $\mathcal{R}^3$ is linearly isomorphic to $\mathbb{R}^3$. Now define an inner product on $\mathcal{R}^3$ by declaring that $\{\sigma_1, \sigma_2, \sigma_3\}$ is an orthonormal basis so that $< X, Y > = < x^1 \sigma_1 + x^2 \sigma_2 + x^3 \sigma_3, y^1 \sigma_1 + y^2 \sigma_2 + y^3 \sigma_3 > = x^1 y^1 + x^2 y^2 + x^3 y^3$. Thus, the **natural isomorphism** $x^1 \sigma_1 + x^2 \sigma_2 + x^3 \sigma_3 \to (x^1, x^2, x^3)$ of $\mathcal{R}^3$ onto $\mathbb{R}^3$ is an isometry. The corresponding norm on $\mathcal{R}^3$ is then given by

$$\| X \|^2 = < X, X > = (x^1)^2 + (x^2)^2 + (x^3)^2 = -\det X . \quad \text{(A.11)}$$

Exercise A.7 Show that $\mathcal{R}^3$ is *not* closed under matrix multiplication so that, unlike the situation in $\mathbb{R}^4$ (Section 1.1), our matrix model of $\mathbb{R}^3$ does not yield a "bonus" multiplicative structure.

Now notice that if $U \in SU(2)$ and $X \in \mathcal{R}^3$, then $U X \bar{U}^T = U X U^{-1}$ is also in $\mathcal{R}^3$. Indeed, $\text{trace}(U X U^{-1}) = \text{trace}(U(X U^{-1})) = \text{trace}((X U^{-1}) U) = \text{trace}(X (U U^{-1})) = \text{trace}(X) = 0$ so $U X \bar{U}^T$ is tracefree.

Exercise A.8 Show that the conjugate transpose of $U X \bar{U}^T$ is $U X \bar{U}^T$.

Thus, for each $U \in SU(2)$ we may define a map

$$R_U : \mathcal{R}^3 \longrightarrow \mathcal{R}^3$$

by $R_U(X) = U X \bar{U}^T = U X U^{-1}$. R_U is clearly a linear transformation and, moreover,

$$\| R_U(X) \|^2 = -\det(R_U(X)) = -\det(U X U^{-1}) = -\det X = \| X \|^2$$

so R_U is, in fact, an orthogonal transformation. We claim that R_U is actually a rotation, i.e., has determinant 1 (as opposed to -1). To see this (and for other purposes as well) we ask the reader to write out explicitly the matrix of R_U relative to the orthonormal basis $\{\sigma_1, \sigma_2, \sigma_3\}$ for $\mathcal{R}^3$.

Exercise A.9 Let $U = \begin{pmatrix} a+b\,\mathbf{i} & c+d\,\mathbf{i} \\ -c+d\,\mathbf{i} & a-b\,\mathbf{i} \end{pmatrix}$ and show that the matrix of $R_U : \mathcal{R}^3 \to \mathcal{R}^3$ relative to $\{\sigma_1, \sigma_2, \sigma_3\}$ is

$$\begin{pmatrix} a^2 - b^2 - c^2 + d^2 & 2ab + 2cd & -2ac + 2bd \\ -2ab + 2cd & a^2 - b^2 + c^2 - d^2 & 2ad + 2bc \\ 2ac - 2bd & 2bc - 2ad & a^2 + b^2 - c^2 - d^2 \end{pmatrix}. \qquad (A.12)$$

Now, one could compute the determinant of this matrix directly and it would, indeed, be 1 (because $a^2 + b^2 + c^2 + d^2 = 1$). Alternatively, one could argue as follows: Being the determinant of an orthogonal matrix, the value must be either 1 or -1 for every $U \in SU(2)$. But this determinant is clearly a continuous function of a, b, c and d which takes the value 1 at $U = \begin{pmatrix} 1 & 0 \\ 0 & 1 \end{pmatrix} \in SU(2)$. Since $SU(2)$ is homeomorphic to S^3, it is connected so the value must be 1 for all $U \in SU(2)$.

With this we may define the **spinor map**

$$\text{Spin} : SU(2) \longrightarrow SO(3)$$

whose value at any $U \in SU(2)$ is the matrix (A.12) of $R_U : \mathcal{R}^3 \to \mathcal{R}^3$ relative to $\{\sigma_1, \sigma_2, \sigma_3\}$.

Theorem A.2 *The spinor map* Spin : $SU(2) \to SO(3)$ *is a smooth, surjective group homomorphism with kernel* $\pm \begin{pmatrix} 1 & 0 \\ 0 & 1 \end{pmatrix}$.

Proof: Smoothness is clear since the coordinate (i.e., entry) functions in (A.12) are polynomials in the coordinates in $SU(2)$. To prove that Spin is a homomorphism, i.e., that $\text{Spin}(U_1 U_2) = (\text{Spin}(U_1))(\text{Spin}(U_2))$, it will suffice to show that $R_{U_1 U_2} = R_{U_1} \circ R_{U_2}$. But, for any $X \in \mathcal{R}^3$, $R_{U_1 U_2}(X) = (U_1 U_2) X (U_1 U_2)^{-1} = (U_1 U_2) X (U_2^{-1} U_1^{-1}) = U_1 (U_2 X U_2^{-1}) U_1^{-1} = U_1 (R_{U_2}(X)) U_1^{-1} = R_{U_1}(R_{U_2}(X)) = (R_{U_1} \circ R_{U_2})(X)$, as required.

Exercise A.10 Show that $\text{Spin}(U) = \begin{pmatrix} 1 & 0 & 0 \\ 0 & 1 & 0 \\ 0 & 0 & 1 \end{pmatrix}$ iff $U = \pm \begin{pmatrix} 1 & 0 \\ 0 & 1 \end{pmatrix}$.

Thus, the kernel of Spin is $\pm\begin{pmatrix} 1 & 0 \\ 0 & 1 \end{pmatrix}$ so Spin is precisely two-to-one, carrying $\pm U$ onto the same element of $SO(3)$.

Finally, we show that Spin is surjective. By Lemma A.1 we need only show that, for every ϕ in $[0,\pi]$ and every unit vector $\vec{n}$ in $\mathbb{R}^3$ there is an element $U(\phi,\vec{n})$ of $SU(2)$ for which $\text{Spin}(U(\phi,\vec{n})) = R(\phi,\vec{n})$. We claim that the following matrix does the job.

$$U(\phi,\vec{n}) = \begin{pmatrix} \cos\frac{\phi}{2} - (n^3\sin\frac{\phi}{2})\,\mathbf{i} & -n^2\sin\frac{\phi}{2} - (n^1\sin\frac{\phi}{2})\,\mathbf{i} \\ n^2\sin\frac{\phi}{2} - (n^1\sin\frac{\phi}{2})\,\mathbf{i} & \cos\frac{\phi}{2} + (n^3\sin\frac{\phi}{2})\,\mathbf{i} \end{pmatrix} \tag{A.13}$$

$$= \left(\cos\frac{\phi}{2}\right) id - \mathbf{i}\sin\frac{\phi}{2}\left(n^1\sigma_1 + n^2\sigma_2 + n^3\sigma_3\right)$$

Exercise A.11 Show that $U(\phi,\vec{n})$ is in $SU(2)$.

To show that $\text{Spin}(U(\phi,\vec{n})) = R(\phi,\vec{n})$ one need only substitute $a = \cos\frac{\phi}{2}$, $b = -n^3\sin\frac{\phi}{2}$, $c = -n^2\sin\frac{\phi}{2}$ and $d = -n^1\sin\frac{\phi}{2}$ into (A.12) and compare the result with (A.1). For example,

$$\begin{aligned} a^2 - b^2 - c^2 + d^2 &= \cos^2\tfrac{\phi}{2} + \sin^2\tfrac{\phi}{2}\left(-(n^3)^2 - (n^2)^2 + (n^1)^2\right) \\ &= \cos^2\tfrac{\phi}{2} + \sin^2\tfrac{\phi}{2}\left(-(n^3)^2 - (n^2)^2 + 1 - (n^2)^2 - (n^3)^2\right) \\ &= \cos^2\tfrac{\phi}{2} + \sin^2\tfrac{\phi}{2}\left(1 - 2\left((n^2)^2 + (n^3)^2\right)\right) \\ &= 1 - 2\sin^2\tfrac{\phi}{2}\left((n^2)^2 + (n^3)^2\right) \\ &= 1 - (1 - \cos\phi)\left((n^2)^2 + (n^3)^2\right) \end{aligned}$$

which is, indeed, the (1,1)-entry of $R(\phi,\vec{n})$.

Exercise A.12 Complete the proof. ∎

Now let us identify $\mathbb{Z}_2$ with the subgroup of $SU(2)$ generated by $\left\{\begin{pmatrix} 1 & 0 \\ 0 & 1 \end{pmatrix}, -\begin{pmatrix} 1 & 0 \\ 0 & 1 \end{pmatrix}\right\} = \ker(\text{Spin})$. Then, algebraically, the quotient group $SU(2)/\mathbb{Z}_2$ is isomorphic to $SO(3)$. More explicitly, if $\mathcal{Q}: SU(2) \to SU(2)/\mathbb{Z}_2$ is the usual quotient homomorphism, then there exists a unique isomorphism $h: SU(2)/\mathbb{Z}_2 \to SO(3)$ for which the following diagram commutes:

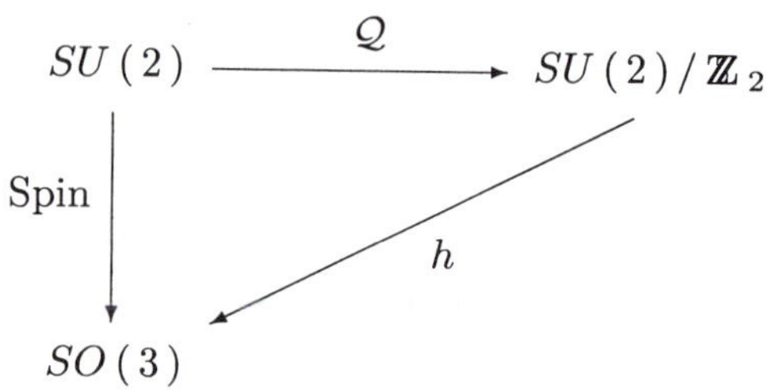

Now, $SU(2)$ is homeomorphic to S^3. Since the cosets in $SU(2)/\mathbb{Z}_2$ are pairs of "antipodal" points in $SU(2)$, if we provide $SU(2)/\mathbb{Z}_2$ with the quotient topology determined by $\mathcal{Q}$, it is just $\mathbb{RP}^3$. Since Spin is continuous, Lemma 1.2.1 implies that h is continuous. But h is also bijective and $SU(2)/\mathbb{Z}_2$ is compact so, by Theorem 1.4.4, h is a homeomorphism. Thus, we have another proof of the result established by the reader in Exercise A.4:

$$SO(3) \cong \mathbb{RP}^3$$

This time, however, $\mathbb{RP}^3$ has a natural group structure and the homeomorphism is also an algebraic isomorphism.

Exercise A.13 Show that Spin : $SU(2) \to SO(3)$ is a covering space. **Hint:** We have already shown that the projection of S^3 onto $\mathbb{RP}^3$, i.e., $\mathcal{Q} : SU(2) \to SU(2)/\mathbb{Z}_2$, is a covering space.

If follows, in particular, from Exercise A.13 that $SU(2)$ and $SO(3)$ are *locally homeomorphic* and it is here that we uncover the real significance of $SU(2)$ for physics. $SO(3)$, being the group of rotations of physical 3-space, plays a role in the formulation of any physical theory that one would have be "invariant under rotation". Specifically, the underlying physical quantities of such a theory should be modeled by the elements of a vector space on which some representation of $SO(3)$ acts (these are the vectors and tensors of classical physics). $SU(2)$ is locally indistinguishable from $SO(3)$, but its global topological structure and the existence of the spinor map present opportunities that are not available within $SO(3)$. To understand what is meant by this, first consider a representation $h : SO(3) \to GL(n, \mathbb{F})$ of $SO(3)$. Composing with Spin : $SU(2) \to SO(3)$ then gives a representation $\tilde{h} = h \circ \text{Spin} : SU(2) \to GL(n, \mathbb{F})$ of $SU(2)$. Every representation of $SO(3)$ "comes from" a representation of $SU(2)$. The converse is not true, however. That is, a given representation $\tilde{h} : SU(2) \to GL(n, \mathbb{F})$ of

$SU(2)$ will not induce a representation of $SO(3)$ unless $\tilde{h}$ is constant on the fibers of Spin, i.e., unless $\tilde{h}(-U) = \tilde{h}(U)$ for every $U \in SU(2)$. Representations of $SU(2)$ that *do not* satisfy this condition are sometimes referred to in the physics literature as "2-valued representations" of $SO(3)$, although they are not representations of $SO(3)$ at all, of course.

Thus, there are strictly more representations of $SU(2)$ than there are of $SO(3)$ and this suggests the possibility of enlarging the collection of "underlying physical quantities" with which to build a rotationally invariant physical theory. That there might be some good reason for wanting to do this, however, was not apparent before the advent of quantum mechanics.

An elementary particle such as a neutron (or any Fermion) has a quantum mechanical property known as "spin $\frac{1}{2}$" which can take precisely two values that we will call $+$ and $-$ and that changes sign when the particle is rotated through $360°$ about any axis, but returns to its original value after a $720°$ rotation. This seems rather peculiar, of course, since rotations through $360°$ and $720°$ both return an object to whatever configuration in space it occupied initially (Paul Dirac devised an ingenious demonstration of such "peculiar" behavior in the more familiar macroscopic world; see Appendix B of **[N3]**). Furthermore, this spin state, being quantized, does not change continuously from $+$ to $-$ or $-$ to $+$ during these rotations; there is nothing "in between" $+$ and $-$. All of this sounds like nonsense, of course, but that's quantum mechanics for you. From our point of view the only revelant issue is that a physical quantity that behaves in this fashion cannot be modeled by any representation of $SO(3)$ for the simple reason that $SO(3)$ cannot tell the difference between $360°$ and $720°$: $R(4\pi, \vec{n}) = R(2\pi, \vec{n})$ for any $\vec{n}$.

We conclude by briefly describing how the spinor map and the global topology of $SU(2)$ remedy this difficulty. The basic topological difference between $SU(2)$ and $SO(3)$ is that $SU(2)$, being homeomorphic to S^3, is simply connected, whereas $\pi_1(SO(3)) \cong \pi_1(\mathbb{RP}^3) \cong \mathbb{Z}_2$. Now, consider the path $\alpha : [0,1] \to SO(3)$ given by

$$\alpha(s) = \begin{pmatrix} \cos 2\pi s & -\sin 2\pi s & 0 \\ \sin 2\pi s & \cos 2\pi s & 0 \\ 0 & 0 & 1 \end{pmatrix}.$$

This is a loop at $\alpha(0) = \alpha(1) = id \in SO(3)$. For each fixed s in $[0,1]$, $\alpha(s)$ represents a rotation through $2\pi s$ about the z-axis. Furthermore, $\alpha^2 = \alpha\,\alpha : [0,1] \to SO(3)$ is given by

$$\alpha^2(s) = \begin{pmatrix} \cos 4\pi s & -\sin 4\pi s & 0 \\ \sin 4\pi s & \cos 4\pi s & 0 \\ 0 & 0 & 1 \end{pmatrix}.$$

This is also a loop at $id \in SO(3)$ and each $\alpha^2(s)$ represents a rotation through $4\pi s$ about the z-axis. A simple calculation with (A.12) shows that, if $\tilde{\alpha} : [0,1] \to SU(2)$ is given by

$$\tilde{\alpha}(s) = \begin{pmatrix} e^{-\pi s \mathbf{i}} & 0 \\ 0 & e^{\pi s \mathbf{i}} \end{pmatrix},$$

then $\alpha(s) = \mathrm{Spin} \circ \tilde{\alpha}(s)$. Moreover,

$$\tilde{\alpha}^2(s) = \begin{pmatrix} e^{-2\pi s \mathbf{i}} & 0 \\ 0 & e^{2\pi s \mathbf{i}} \end{pmatrix},$$

and $\alpha^2(s) = \mathrm{Spin} \circ \tilde{\alpha}^2(s)$. Exercise 2.4.14 (and the discussion immediately preceding it) show that α^2 is homotopically trivial, but α itself is not.

Now suppose we are given a physical object (e.g., a Fermion) and select some configuration of the object in space as its "initial configuration". Then any element of $SO(3)$ can be identified with another configuration of the object (that obtained by applying the corresponding rotation). Thus, a path in $SO(3)$ may be thought of as a continuous sequence of configurations of the object. In particular, the loop α defined above represents a continuous rotation of the object through 360° about the z-axis, whereas α^2 represents a rotation of the object through 720° about the same axis. Each begins and ends at the initial configuration (i.e., $id \in SO(3)$) and so any representation of $SO(3)$ takes the same value at the beginning and end of both rotations. However, α lifts to a *path* $\tilde{\alpha}$ in $SU(2)$ from $\begin{pmatrix} 1 & 0 \\ 0 & 1 \end{pmatrix}$ to $-\begin{pmatrix} 1 & 0 \\ 0 & 1 \end{pmatrix}$ and so a representation of $SU(2)$ need not take the same values at $\tilde{\alpha}(0)$ and $\tilde{\alpha}(1)$ (indeed, the identity representation $SU(2) \hookrightarrow GL(2, \mathbb{C})$ certainly does not since it *changes sign* between $\tilde{\alpha}(0)$ and $\tilde{\alpha}(1)$). On the other hand, α^2 lifts to a *loop* $\tilde{\alpha}^2$ at $\begin{pmatrix} 1 & 0 \\ 0 & 1 \end{pmatrix}$ in $SU(2)$ so a representation of $SU(2)$ necessarily takes the same value at the beginning and end of a 720° rotation. Just what the doctor ordered!

Exercise A.14 Rotate $\mathbb{R}^3$ through $90°$ counterclockwise about the positive z-axis and then rotate through $90°$ counterclockwise about the (new) positive x-axis. What single rotation will have the same effect? **Hint:** Let $R_i = R(\phi_i, \vec{n}_i)$, where $\phi_1 = \frac{\pi}{2}$, $\vec{n}_1 = <0,0,1>$, $\phi_2 = \frac{\pi}{2}$, $\vec{n}_2 = <1,0,0>$. Show that $U_1 = U(\phi_i, \vec{n}_i) = \frac{\sqrt{2}}{2}(id - \mathbf{i}\,\sigma_3)$ and $U_2 = U(\phi_2, \vec{n}_2) = \frac{\sqrt{2}}{2}(id - \mathbf{i}\,\sigma_1)$ and compute $U_2 U_1$ using the commutation relations in Exercise A.6 to obtain

$$U_2 U_1 = \left(\cos\frac{\pi}{3}\right) id - \mathbf{i}\,\sin\frac{\pi}{3}\left(\frac{1}{\sqrt{3}}\sigma_1 - \frac{1}{\sqrt{3}}\sigma_2 + \frac{1}{\sqrt{3}}\sigma_3\right).$$

Now use the fact that Spin is a homomorphism.

References

[AB] Aharonov, Y. and D. Bohm, "Significance of electromagnetic potentials in the quantum theory", Phys. Rev., 115(1959), 485-491.

[Apos] Apostol, Tom M., *Mathematical Analysis*, 2nd Ed., Addison-Wesley, Reading, MA, 1974.

[Atiy] Atiyah, M.F., *Geometry of Yang-Mills Fields*, Lezioni Fermiane, Accademia Nazionale dei Lincei Scuola Normale Superione, Pisa, 1979.

[AHS] Atiyah, M.F., N.J. Hitchin and I.M. Singer, "Self duality in four dimensional Riemannian geometry", Proc. Roy. Soc. Lond., A362(1978), 425-461.

[BPST] Belavin, A., A. Polyakov, A. Schwartz and Y. Tyupkin, "Pseudoparticle solutions of the Yang-Mills equations", Phys. Lett., 59B(1975), 85-87.

[Bl] Bleecker, D., *Gauge Theory and Variational Principles*, Addison-Wesley, Reading, MA, 1981.

[Buch] Buchdahl, N.P., "Instantons on CP_2", J. Diff. Geo., 24(1986), 19-52.

[DV] Daniel, M. and C.M. Viallet, "The geometrical setting of gauge theories of the Yang-Mills type", Rev. Mod. Phys., 52(1980), 175-197.

[Dir1] Dirac, P.A.M., "Quantised singularities in the electromagnetic field", Proc. Roy. Soc., A133(1931), 60-72.

[Dir2] Dirac, P.A.M., "The theory of magnetic poles", Phys. Rev., 74(1948), 817-830.

[Don] Donaldson, S.K., "An application of gauge theory to four dimensional topology", J. Diff. Geo., 18(1983), 279-315.

[DK] Donaldson, S.K., and P.B. Kronheimer, *The Geometry of Four-Manifolds*, Clarendon Press, Oxford, 1991.

[Dug] Dugundji, J., *Topology*, Allyn and Bacon, Boston, 1966.

[Ehr] Ehresmann, C., "Les connexions infinitésimales dans un espace fibré différentiable", in *Colloque de Topologie, Bruxelles* (1950), 29-55, Masson, Paris, 1951.

[Flan] Flanders, H., *Differential Forms with Applications to the Physical Sciences*, Academic Press, New York, 1963.

[FU] Freed, D.S., and K.K. Uhlenbeck, *Instantons and Four-Manifolds*, MSRI Publications, Springer-Verlag, New York, Berlin, 1984.

[Fr] Freedman, M., "The topology of four-dimensional manifolds", J. Diff. Geo., 17(1982), 357-454.

[FL] Freedman, M. and F. Luo, *Selected Applications of Geometry to Low-Dimensional Topology*, University Lecture Series, AMS, Providence, RI, 1989.

[Gra] Gray, B., *Homotopy Theory: An Introduction to Algebraic Topology*, Academic Press, New York, 1975.

[Gre] Greenberg, M., *Lectures on Algebraic Topology*, W.A. Benjamin, New York, 1966.

[Guid] Guidry, M., *Gauge Field Theories*, John Wiley & Sons, Inc., New York, 1991.

[Helg] Helgason, S., *Differential Geometry and Symmetric Spaces*, Academic Press, New York, 1962.

[Hir] Hirsch, M.W., *Differential Topology*, Springer-Verlag, GTM #33, New York, Berlin, 1976.

[Hopf] Hopf, H., "Über die abbildungen der 3-sphere auf die kugelfläache", Math. Annalen, 104(1931), 637-665.

[Howe] Howe, R., "Very basic Lie theory", Amer. Math. Monthly, November, 1983, 600-623.

[Hu] Hu, S.T., *Homotopy Theory*, Academic Press, New York, 1959.

[Hure] Hurewicz, W., *Lectures on Ordinary Differential Equations*, John Wiley and Sons, Inc., and MIT Press, New York and Cambridge, MA, 1958.

[KN1] Kobayashi, S. and K. Nomizu, *Foundations of Differential Geometry*, Vol. 1, Wiley-Interscience, New York, 1963.

[KN2] Kobayashi, S. and K. Nomizu, *Foundations of Differential Geometry*, Vol. 2, Wiley-Interscience, New York, 1969.

[Lang] Lang, S., *Linear Algebra*, 2nd Edition, Addison-Wesley, Reading, MA, 1971.

[Law] Lawson, H.B., *The Theory of Gauge Fields in Four Dimensions*, Regional Conference Series in Mathematics #58, Amer. Math. Soc., Providence, RI, 1985.

[MM] Marathe, K.B. and G. Martucci, *The Mathematical Foundations of Gauge Theories*, North-Holland, Amsterdam, 1992.

[Mat] Matumoto, T., "Three Riemannian metrics on the moduli space of BPST instantons over S^4", Hiroshima Math. J., 19(1989), 221-224.

[Miln] Milnor, J., *Topology from the Differentiable Viewpoint*, University of Virginia Press, Charlottesville, VA, 1966.

[MS] Milnor, J.W. and J.D. Stasheff, *Characteristic Classes*, Princeton University Press, Princeton, NJ, 1974.

[N1] Naber, G.L., *Topological Methods in Euclidean Spaces*, Cambridge University Press, Cambridge, England, 1980.

[N2] Naber, G.L., *Spacetime and Singularities*, Cambridge University Press, Cambridge, England, 1988.

[N3] Naber, G.L., *The Geometry of Minkowski Spacetime*, Springer-Verlag, Applied Mathematical Sciences Series #92, New York, Berlin, 1992.

[NT] Nowakowski, J. and A. Trautman, "Natural connections on Stiefel bundles are sourceless gauge fields", J. Math. Phys., Vol. 19, No. 5 (1978), 1100-1103.

[Rav] Ravenel, D.C., *Complex Cobordism and Stable Homotopy Groups of Spheres*, Academic Press, Orlando, 1986.

[Sp1] Spivak, M., *Calculus on Manifolds*, W. A. Benjamin, Inc., New York, 1965.

[Sp2] Spivak, M., *A Comprehensive Introduction to Differential Geometry*, Volumes I-V, Publish or Perish, Inc., Boston, 1979.

[St] Steenrod, N., *The Topology of Fibre Bundles*, Princeton Univ. Press, Princeton, N.J., 1951.

[Ster] Stern, R.J., "Instantons and the topology of 4-manifolds", Mathematical Intelligencer, Vol. 5, No. 3 (1983), 39-44.

[Tau1] Taubes, C.H., "Self-dual connections on non-self-dual 4-manifolds", J. Diff. Geo., 17(1982), 139-170.

[Tau2] Taubes, C.H., *The Seiberg-Witten Invariants* (Videotape), American Mathematical Society, 1995.

[t´H] t´Hooft, G., "Gauge theories of the forces between elementary particles", Sci. American, 242(1980), No. 6, 104-138.

[Trau] Trautman, A., "Solutions of the Maxwell and Yang-Mills equations associated with Hopf fibrings", Inter. J. Theo. Phys., Vol. 16, No. 8 (1977), 561-565.

[Uhl] Uhlenbeck, K., "Removable singularities in Yang-Mills fields", Comm. Math. Phys., 83(1982), 11-30.

[Warn] Warner, F.W., *Foundations of Differentiable Manifolds and Lie Groups*, Springer-Verlag, GTM #94, New York, Berlin, 1983.

[Wh] Whitehead, J.H.C., "On simply connected 4-dimensional polyhedra", Comment. Math. Helv., 22(1949), 48-92.

[YM] Yang, C.N. and R.L. Mills, "Conservation of isotopic spin and isotopic gauge invariance", Phys. Rev., 96(1954), 191-195.

Symbols

What follows is a list of the those symbols that are used consistently throughout the text, a brief description of their meaning and/or a reference to the page on which such a description can be found.

$\mathbb{R}$	real numbers
$\mathbb{R}^n$	real n-space, 27
$\mathbf{i}$	$\sqrt{-1}$
$\mathbb{C}$	complex numbers
$\mathbb{C}^n$	complex n-space, 40
$\mathbb{H}$	quaternions, 36
$\mathbf{1}, \mathbf{i}, \mathbf{j}, \mathbf{k}$	basis quaternions, 35, 36
$\mathbb{H}^n$	quaternionic n-space, 40
$\mathbb{F}$	$\mathbb{R}$, $\mathbb{C}$, or $\mathbb{H}$, 40
$\mathbb{F}^n$	$\mathbb{R}^n$, $\mathbb{C}^n$, or $\mathbb{H}^n$, 40
S^n	n-sphere, 34
Re	real part of a complex number or quaternion, 37
Im	imaginary part of a complex number or quaternion, 37
Im $\mathbb{C}$	set of pure imaginary complex numbers
Im $\mathbb{H}$	set of pure imaginary complex quaternions, 37
$\mathbb{Z}$	integers
$\mathbb{Q}$	rational numbers
$\mathbb{Z}_2$	integers mod 2
$\mathbb{C}^*$	extended complex numbers, 12
$\bar{x}$	conjugate of x, 37
$\vert x \vert$	modulus of x, 37
$\Vert x \Vert$	Euclidean norm of $x \in \mathbb{R}^n$, 27
x^{-1}	inverse of x, 37
$\mathcal{R}^4$	matrix model of $\mathbb{R}^4$, 35
I^n	n-dimensional cube, 56, 143
∂I^n	boundary of I^n, 56, 143
D^n	n-dimensional disc (ball), 56
∂D^n	boundary of D^n, 56
iff	if and only if

$\ker T$	kernel (null space) of the linear transformation T
$\det$	determinant function
$\delta_{ij} = \delta^{ij} = \delta^i_j$	Kronecker delta, 44
$< \, , \, >$	bilinear form on $\mathbb{F}^n$, 41
id	identity map or identity matrix, 30, 44
$\iota : X \hookrightarrow X'$	inclusion map of X into X', 29
$\hookrightarrow$	inclusion map, 29
$f \vert X$	restriction of f to X
$\sim$	equivalence relation, 49
$[\, x\,]$	equivalence class containing x, 49
$X/\sim$	set of equivalence classes of $\sim$ in X, 49
A^T	transpose of the matrix A
$\bar{A}^T$	conjugate transpose of A, 38
$\operatorname{trace}(A)$	sum of the diagonal entries in A
$\operatorname{Span}\{v_1, \ldots, v_k\}$	linear subspace spanned by $v_1, \ldots, v_k$
$GL(n, \mathbb{F})$	$\mathbb{F}$-general linear group, 42
$U(n, \mathbb{F})$	$\mathbb{F}$-unitary group, 44
$O(n)$	real orthogonal group, 45
$U(n)$	unitary group, 45
$Sp(n)$	symplectic group, 45, 47
$SO(n)$	special orthogonal group, 45
$SU(n)$	special unitary group, 46
$SL(n, \mathbb{H})$	quaternionic special linear group, 47
(U, φ)	chart, 32
(U_S, φ_S)	stereographic projection chart on S^n, 34
(U_N, φ_N)	stereographic projection chart on S^n, 34
$X \cong Y$	X is homeomorphic to Y, 29
$\mathbb{F}\mathbb{P}^{n-1}$	$\mathbb{F}$-projective space of dimension $n-1$, 50
$\mathcal{P}^i$	projection onto the ith factor, 62
$f^i = \mathcal{P}^i \circ f$	i^{th} coordinate function, 63
$f_1 \times f_2$	product map, 63
$(P, X, \mathcal{P}, Y)$	locally trivial bundle, 66
$Y \to P \xrightarrow{\mathcal{P}} X$	locally trivial bundle, 66
$Y \to P \to X$	locally trivial bundle, 66
$(V, \boldsymbol{\Phi})$	local trivialization, 66
$S^1 \to S^3 \to S^2$	complex Hopf bundle, 69

$\alpha + \beta$	sum of n-loops for $n \geq 2$, 143
deg	degree of a map, 134, 163, 269
$(P, \mathcal{P}, \sigma)$	principal bundle, 165, 197
$G \to P \to X$	principal bundle, 165
$\mathcal{P} : P \to X$	principal bundle, 165
g_{ji}	transition function, 168
$D_i f$	partial derivative, 185
C^∞	smooth, 185
$f'(a)$	Jacobian of f at a, 186
$\dim X$	dimension of X, 191
$C^\infty(X)$	smooth, real-valued functions on X, 198
$C^\infty(p)$	smooth, real-valued functions on a nbd of p, 198
$\mathbf{v}, \mathbf{w}, \ldots$	tangent vectors, 198
$T_p(X)$	tangent space, 199
$\frac{\partial}{\partial x^i}\Big\vert_p$	coordinate vector at p, 200
$\frac{\partial f}{\partial x^i}\Big\vert_p$	$\frac{\partial}{\partial x^i}\Big\vert_p (f)$, 200
f_{*p}	derivative of f at p, 203
$\frac{d}{dt}\Big\vert_{t_0}$	coordinate vector on $\mathbb{R}$, 205
$df_p = df(p)$	differential of f at p, 205
$T_p^*(X)$	cotangent space, 205
$\mathbf{V}, \mathbf{W}, \ldots$	vector fields, 215
$\mathbf{V}_p = \mathbf{V}(p)$	value of $\mathbf{V}$ at p, 215
$\boldsymbol{\mathcal{X}}(X)$	module of smooth vector fields on X, 215
$\frac{\partial}{\partial x^i}$	coordinate vector field, 216
$\mathbf{V} f = \mathbf{V}(f)$	$\mathbf{V}$ acting on $f \in C^\infty(X)$, 216
$[\mathbf{V}, \mathbf{W}]$	Lie bracket $\mathbf{V}$ and $\mathbf{W}$, 216
$\boldsymbol{\Theta}, \boldsymbol{\omega}, \ldots$	1-forms, 218
$\boldsymbol{\Theta}_p = \boldsymbol{\Theta}(p)$	value of $\boldsymbol{\Theta}$ at p, 218
df	differential of f, 218
$\boldsymbol{\mathcal{X}}^*(X)$	module of 1-forms on X, 218
$\boldsymbol{\Theta}(\mathbf{V}) = \boldsymbol{\Theta}\mathbf{V}$	$\boldsymbol{\Theta}$ acting on $\mathbf{V}$, 219
$f_*\mathbf{V}$	$\mathbf{V}$ pushed forward by the diffeomorphism f, 220
$f^*\boldsymbol{\Theta}$	pullback of $\boldsymbol{\Theta}$ by the smooth map f, 221, 246, 272

$\mathcal{F}$	local field strength, 312
$\mathcal{YM}$	Yang-Mills functional, 321
$^*\boldsymbol{\Omega}$	Hodge dual of $\boldsymbol{\Omega}$, 327, 333
ϵ_{abcd}	Levi-Civita symbol, 327
$\Lambda^2_\pm(\mathcal{V})$	(anti-) self-dual elements of $\Lambda^2(\mathcal{V})$, 329
$\boldsymbol{\Omega}_\pm$	(anti-) self-dual parts of $\boldsymbol{\Omega}$, 330, 333
SD	self-dual, 330
ASD	anti-self-dual, 330
$\Lambda^2_\pm(X)$	(anti-) self-dual elements of $\Lambda^2(X)$, 333
$P \times_G F$	associated fiber bundle, 353
$ad\,P$	adjoint bundle of P, 354
$d^{\boldsymbol{\omega}}\phi$	covariant exterior derivative of ϕ, 357
$\sigma_1, \sigma_2, \sigma_3$	Pauli spin matrices, 363
Spin	spinor map, 372

Index

D

E